Biocatalysis in Organic Synthesis

The Retrosynthesis Approach

Biocatalysis in Organic Synthesis

The Retrosynthesis Approach

Nicholas J. Turner

University of Manchester, UK
Email: Nicholas.Turner@manchester.ac.uk

and

Luke Humphreys

Gilead Sciences, Edmonton, Canada
Email: Luke.Humphreys@gilead.com

THE QUEEN'S AWARDS
FOR ENTERPRISE:
INTERNATIONAL TRADE
2013

Print ISBN: 978-1-78262-530-8

EPUB ISBN: 978-1-78801-342-0

A catalogue record for this book is available from the British Library

The Royal Society of Chemistry is a charity, registered in England and Wales, Number 207890, and a company incorporated in England by Royal Charter (Registered No. RC000524), registered office: Burlington House, Piccadilly, London W1J 0BA, UK, Telephone: +44 (0) 207 4378 6556.

Visit our website at www.rsc.org/books

Printed in the United Kingdom by CPI Group (UK) Ltd, Croydon, CR0 4YY, UK

Preface

This book, entitled **Biocatalysis in Organic Synthesis: The Retrosynthesis Approach**, was conceived by the authors out of a need to provide guidance and education in the application of biocatalysts in organic synthesis. During the past 20 years, biocatalysis has increasingly assumed greater prominence as a tool for the synthesis of target molecules, particularly in industry where the desire to develop sustainable and economic processes is now a high priority. The much greater availability of a broader range of biocatalysts has resulted in many more options for application in organic synthesis. However, with this greater number of options comes a need to think more strategically about where biocatalysis best fits in a synthetic sequence. The application of retrosynthesis in the planning of target molecule synthesis is now a well-established approach, but one that has not previously been extended to biocatalysis. In this book, we have attempted to place both 'forward' and 'reverse' biocatalytic transformations side-by-side in order to illustrate both what is possible using biocatalysis and where these reactions may be best applied in a strategic sense. It is hoped that this book will encourage a greater awareness of what is possible using biocatalysis and thereby lead to a greater uptake of this emerging technology.

Nicholas J. Turner, University of Manchester, UK
Luke Humphreys, Gilead Sciences, Canada

Biocatalysis in Organic Synthesis: The Retrosynthesis Approach
By Nicholas J. Turner and Luke Humphreys
© Nicholas J. Turner and Luke Humphreys 2018
Published by the Royal Society of Chemistry, www.rsc.org

Contents

Biocatalysis in Organic Synthesis: The Retrosynthesis Approach
By Nicholas J. Turner and Luke Humphreys
© Nicholas J. Turner and Luke Humphreys 2018
Published by the Royal Society of Chemistry, www.rsc.org

8   C–C Bond Formation   217

9   Miscellaneous Biocatalysts   254

10   Biocatalytic Disconnections and Functional Group Interconversions   268

1 Introduction and Aims of the Book

1.1 Introduction

Biocatalysis, by which we mean the use of enzymes and micro-organisms as catalysts for chemical transformations, has had a long and distinguished history, particularly in the brewing, baking and animal feed industries. In addition, the development of fermentation-based processes for antibiotic production relies squarely on the fundamental properties of enzymes for the production of *natural products*. In the latter half of the 20th century, biocatalysis started to be viewed as a potential route to *unnatural* or *synthetic* compounds. For example, scientists in Schering in Berlin in Germany discovered that semi-synthetic steroids could be manipulated with exquisite selectivity using enzymes. In the 1970s, the use of hydrolytic enzymes (*e.g.* lipases and proteases) and oxidoreductases (*e.g.* alcohol dehydrogenases) started to gain popularity for the synthesis of chiral building blocks, particularly within the pharmaceutical industry where stereochemical purity in the final product was a high priority. The discovery that some enzymes (*e.g.* lipases and proteases) could be used under low water conditions in organic solvents extended the range of these biocatalysts to non-polar organic substrates. However, by the early 1990s, the range of biocatalysts being employed by organic chemists was still relatively narrow and confined to hydrolases and oxidoreductases. A typical international conference on biocatalysis at that time would focus *ca.* 80–90% of the presentations on these two classes of enzymes.

Biocatalysis in Organic Synthesis: The Retrosynthesis Approach
By Nicholas J. Turner and Luke Humphreys
© Nicholas J. Turner and Luke Humphreys 2018
Published by the Royal Society of Chemistry, www.rsc.org

In the 1990s, several major developments occurred within the field of biocatalysis that together changed the face of the discipline and resulted in a rapid diversification of the enzymes available to synthetic chemists.[1] Firstly the use of molecular biology protocols as tools to clone, express and manipulate novel genes, and hence produce new enzymes, became much more widespread and meant that there was increasingly less reliance on the use of wild-type crude cell extracts for biocatalysis. Secondly, the use of protein engineering and directed evolution emerged as powerful strategies for optimisation of the properties of enzymes, particularly in the context of improving stability, stereoselectivity, substrate tolerance and catalytic activity. Thirdly, the availability of sequenced genomes meant that there was an explosion in the number of gene/protein sequences that could be mined and used as the basis for discovering new enzymes. At the beginning of 2000, the cost of DNA sequencing and gene synthesis began to drop considerably, as a result of technological advances, which meant that ordering synthetic genes became cost effective and a rapid way of generating novel biocatalysts for screening. Together, these new technologies resulted in a significantly broader range of enzymes becoming available to organic chemists. Nowadays, we have a plethora of different methods and approaches available to us for the discovery and development of new biocatalysts (Figure 1.1). The challenge has shifted from simply discovering new enzymes to trying to curate the existing databases in order to try to understand which

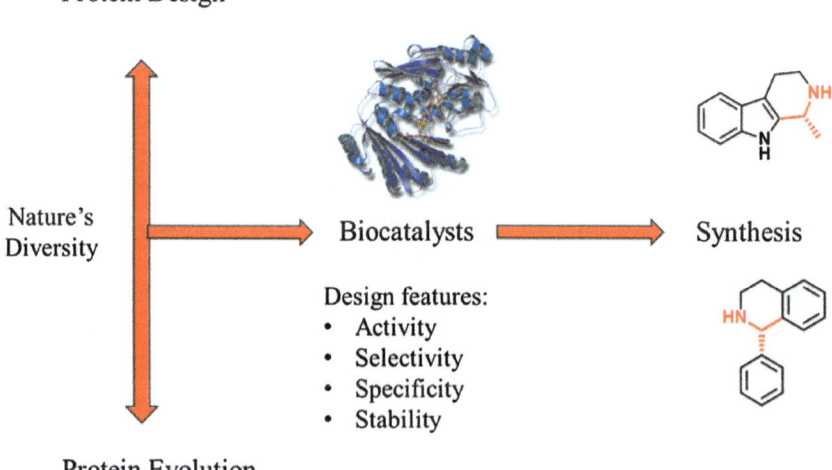

Figure 1.1   Modern technologies available for enzyme discovery, development and application in organic synthesis.

sequences might code for which enzyme activity. *De novo* protein design has made major strides forward, and allows us to generate new enzyme activities from alternative scaffolds. Enzyme evolution has developed to the point where it is now routinely applied to optimise enzyme performance and represents one of the most powerful algorithms available to those interested in the development of new biocatalysts. Other scientific advances in the scale-up of biocatalytic processes provided greater levels of confidence that once a biocatalyst had been identified for a specific chemical transformation, there was a good chance of developing a practical large-scale process. Biocatalysis is now a maturing technology for the manufacture of a range of chemical products across a wide range of industries from pharmaceuticals and biofuels to polymers and personal healthcare products. Compared to traditional synthesis methodologies, biocatalysis offers a number of advantages that can contribute to more sustainable manufacturing processes. In addition, the emerging field of synthetic biology offers the potential for increasingly efficient synthesis by combining multiple biocatalytic reactions all within a single host organism.

Today, biocatalysts are increasingly being considered as an option when planning synthetic strategies for the construction of target organic molecules, particularly in cases where it is important to try to achieve some type of selectivity (*e.g.* stereo-, regio-, and chemoselectivity). The use of biocatalysts as an alternative to chemical reagents and catalysts can also provide benefits in terms of a reduced number of synthetic steps, lower costs of goods, reduced use of harmful solvents and improved safety profiles. Together, these benefits can lead to a more sustainable overall process with a lower overall environmental impact. However, a major barrier remains preventing the more widespread application of biocatalysis, namely a general lack of awareness and understanding amongst the synthetic organic community regarding the types of biocatalysts that are available and the various ways in which they can be applied in target molecule synthesis.

During the past ten years, a number of excellent books and reviews have been published highlighting the various biocatalysts that are now available and detailing the types of transformations that can be accomplished.[2] Typically, these books are organised according to different classes of enzyme (*e.g.* hydrolases, dehydrogenases, transferases and lyases) and the various reactions are presented based upon the type of synthetic transformation that can be achieved (*e.g.* hydrolysis, reverse hydrolysis, C–C bond formation, C–X bond formation, oxidation, *etc.*). This approach provides the reader with an excellent overview of where specific biocatalysts can now be applied in organic synthesis, together with an insight into the tolerance of en-zymes for unnatural substrates, which allows consequently for the synthesis of non-natural target molecules. However, as more new bio-catalysts become available for application in synthesis, inevitably the question arises as to what is the best way of preparing a particular target molecule, or a building block with specific arrangements of function-ality and chirality, given the fact that there may be several available options. This situation is particularly true for the synthesis of chiral alcohols, amines, carboxylic acids, esters, amides, *etc.*, for which there are now several available biocatalytic systems that could in principle be used, and certainly many more than were available ten years ago. As more distinct biocatalytic platforms are developed, this expansion of options will continue and indeed accelerate. Indeed, our ability to discover new biocatalysts, and subsequently engineer them for broader substrate tolerance or new reaction chemistry, is rapidly increasing due to advances in technologies such metagenomics, genome sequencing, protein engineering, directed evolution and high-throughput screening.

Despite the now widespread application of retrosynthesis in everyday organic synthesis, there has been no systematic attempt to include biocatalytic reactions in the whole process of synthetic planning and design. This omission is more remarkable in view of the increasing interest in using biocatalysts for synthesis in both industry and academia. Students are typically taught retrosynthetic analysis at around the same time as more traditional C–C bond forming methods. As a result, the connection between biocatalysis and syn-thesis tends to be made towards the end of their studies, generating the perception that enzymes are specialised and not suitable for mainstream organic synthesis. If biocatalysis is to be fully incorpor-ated into the synthetic chemist's toolbox, then the teaching of bio-catalytic methodology and retrosynthesis must be concurrent, as it is with more traditional organic synthesis.

1.2   Biocatalytic Retrosynthesis

In the mid-1960s, synthetic organic chemists began to apply the principles of retrosynthetic analysis during the planning of new routes to target molecules. Retrosynthetic analysis, which was ori-ginally proposed by E. J. Corey, fundamentally changed the way in which synthetic chemists conceived and developed approaches to the total synthesis of natural and unnatural products.[3] An important principle of retrosynthetic design is that each disconnection in the reverse sense must represent a feasible transformation in the forward synthesis direction. Students are generally taught to disconnect molecules by a formalised process of bond cleavage of either C–C or C–X bonds in the reverse direction. In addition, functional group interconversions (FGI's) are used to prepare the molecule for the desired bond cleavage. An important by-product of this method of analysis has been that synthetic chemists have discovered disconnections for which the corresponding forward synthesis methodologies were unknown and new additions to the synthesis toolbox have been developed as a result.

The invention of new reagents and increasingly catalysts for selective organic synthesis has resulted in the situation where today there is (almost) no molecule that is beyond the reach of the synthetic organic chemist. Organic synthesis using synthetic reagents and catalysts is the predominant technology for the manufacture of a broad range of chemical products, ranging from pharmaceuticals and agrochemicals to polymers and plastics, and including a diverse range of specialty chemicals in between (Figure 1.2). Organic synthesis by and large uses feedstocks derived from the petrochemical industry and indeed to a large extent the two disciplines have 'co-evolved' – one provides the starting materials for the other – and new chemical

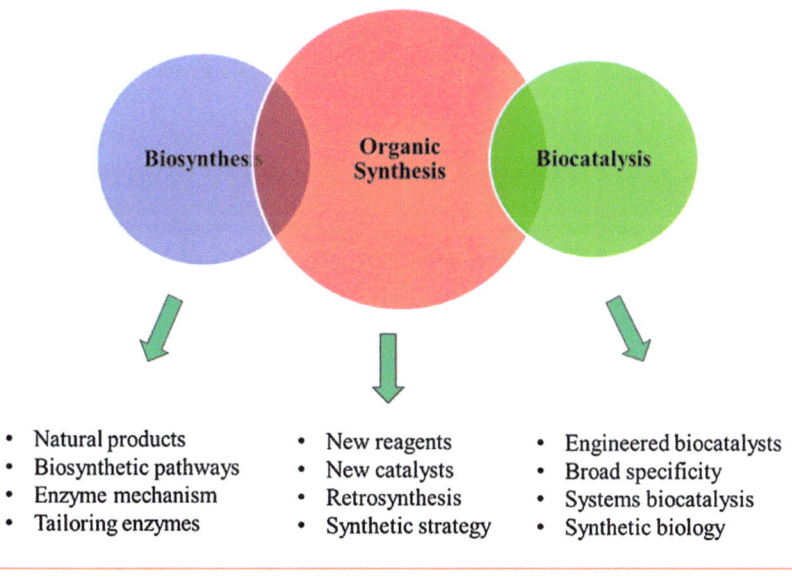

Figure 1.2 The 'three' approaches to organic synthesis.

reagents and catalysts are designed to be compatible with oil-derived starting materials.

Figure 1.2 presents two alternative scenarios for undertaking organic synthesis. As noted earlier, many semi-synthetic pharma-ceuticals are derived from natural products derived from fermen-tation processes. Over millions of years, Nature has evolved complex biosynthetic pathways for the production of a truly diverse set of natural products, including alkaloids, polyketides, terpenes, carbo-hydrates, amino acids, *etc.* Even some of the simpler natural products, such as the alkaloid (−)-sparteine, are fantastic examples of the way in which biosynthetic pathways are able to convert simple building blocks, in this case cadaverine, to structurally more complex target molecules using simple chemical transformations (Figure 1.3). In-spection of this pathway reveals a typical scenario in the biosynthesis of natural products, *i.e.* the enzymes involved are catalysing simple chemical transformations, *e.g.* oxidation, reduction, and hydrolysis.

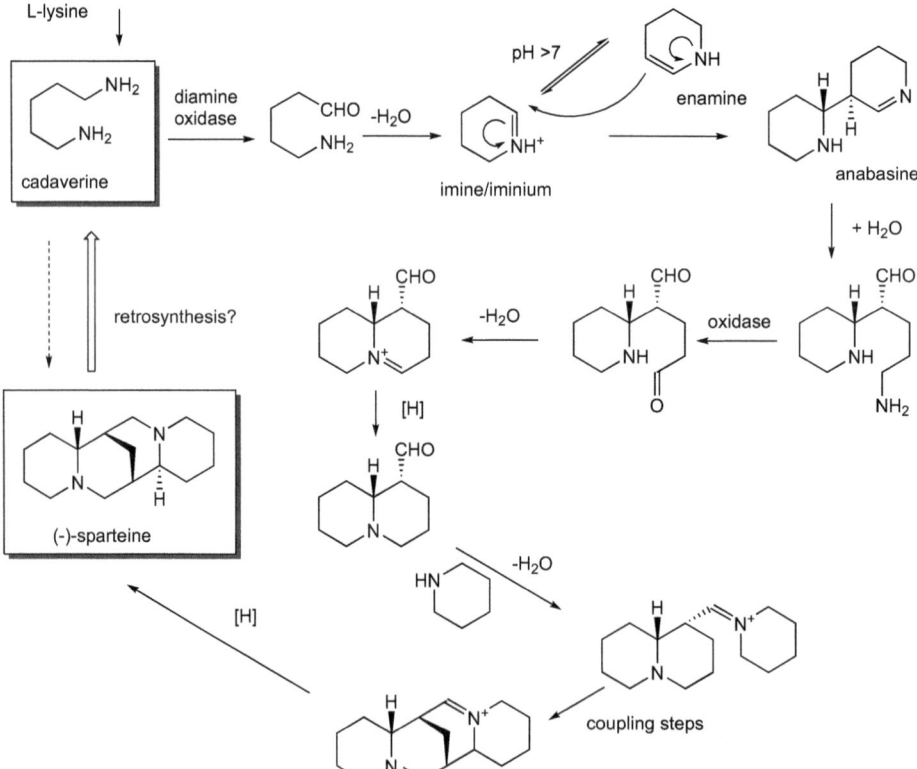

Figure 1.3   Biosynthetic pathway for the conversion of cadaverine to (−)-sparteine.

The bond forming and bond breaking processes involved are not sophisticated, but it is the way in which the reactions are orchestrated that results in the rapid construction of molecular complexity from simple achiral precursors. And of course, all of the reactions are subjected to catalysis by an enzyme – that is Nature's way. Imagine being asked to make a molecule such as (−)-sparteine in the laboratory using only catalysts for the synthesis! If you are truly to emulate Nature, then you would not be allowed to use protecting groups. Moreover, imagine being given an exam question in which you are asked to generate a synthesis of (−)-sparteine using cadaverine as the starting material!

However, despite the abundance of natural products and biosynthetic pathways in Nature, the enzymes involved in secondary metabolism have not found their way in a general sense into the organic chemist's toolbox. There are a number of reasons for this situation. Firstly, enzymes of secondary metabolism often have relatively low activity, probably as a consequence of the fact that natural products are often produced in low concentrations and hence there is no need for enzymes with high catalytic turnover. Secondly, these enzymes are often quite specific for particular intermediates along the biosynthetic pathway and hence do not lend themselves to being applied in a general sense to a broader range of substrates. Biosynthetic enzymes have, however, provided an enormously rich playground of molecules and associated enzymes in order to gain fundamental understanding of the enzyme mechanism and the ways in which complex molecules can be assembled.

This book is about exploring a 'third' approach to organic synthesis, namely the use of engineered biocatalysts. Unlike natural product biosynthesis, these enzymes will predominantly be derived from primary metabolism, and hence should have inherently higher catalytic rates since their natural function is to process large quantities of building blocks and intermediates that are essential to cellular metabolism (*e.g.* amino acids, carbohydrates, and nucleotides). The challenge here is to develop an increasingly broader range of distinct biocatalysts such that a wider range of organic molecules come into play and become potential targets for biocatalytic synthesis. Clearly at the same time as the biocatalytic toolbox expands, it is important to gain a greater understanding of when and how to apply these new engineered enzymes, hence the need for 'biocatalytic retrosynthesis'.

The biocatalytic toolbox has expanded significantly during the past ten years to the point where today more than 50 different classes of enzymes are commercially available, offering chemists an attractive alternative to chemical reagents or chemocatalysts for chemical transformations. Moreover, given the current rate of development of new engineered biocatalysts, it is likely that this number will double in the next few years. At present, the use of biocatalysts is focused primarily on functional group interconversions (FGI's) for the preparation of (chiral) building blocks and intermediates. Thus, there are now myriads of applications, including in industry, of the use of ketoreductases (ketone to alcohol), lipases and esterases (esters to acids and the reverse), and increasingly transaminases (ketone to amine). However, other areas are less well developed, including C–C and C–X (X = N or O) bond formation, as well as selective oxidation reactions for remote C–H activation leading to hydroxylation and other types of functionalisation. The current focus on a relatively small sub-set of chemical transformations is partly due to the lack of available biocatalysts, although this situation is rapidly changing, but also due to a lack of awareness.

Based on the reasons outlined above, we believe that it is now timely to consider the development of guidelines or rules for "biocatalytic retrosynthesis",[4] in which molecules are disconnected with consideration of applying biocatalysts, as well as chemical reagents & chemocatalysts, in the forward synthesis direction. We believe that such a development will stimulate both new ways of teaching the art and science of biocatalysis and also research into areas for which new methodology is required. In its most elementary form, biocatalytic retrosynthesis might simply involve replacing a chemical reagent used for a particular transformation with a biocatalyst. This type of scenario is illustrated in Figure 1.4 for the manufacture of the drug montelukast. Montelukast is now a generic active pharmaceutical ingredient (API) and has been the subject of several attempts to reduce the cost of its production. One of the original manufacturing routes to montelukast employed the reagent (*S*)-(DIP)-chloride to carry out the key step of asymmetric reduction of the ketone to give the enantiomerically enriched chiral secondary alcohol. Recently, the use of ketoreductases with cofactor recycling has become a more favoured route and indeed variations of this process are now operated in several countries for montelukast production.[5] The table in Figure 1.4 provides a comparison of the chemical and biocatalytic processes with respect to various different parameters, such as substrate loading, enantioselectivity, solvent usage and waste. This comparison highlights both the significant differences that can be encountered when comparing chemical and biocatalytic processes, together with

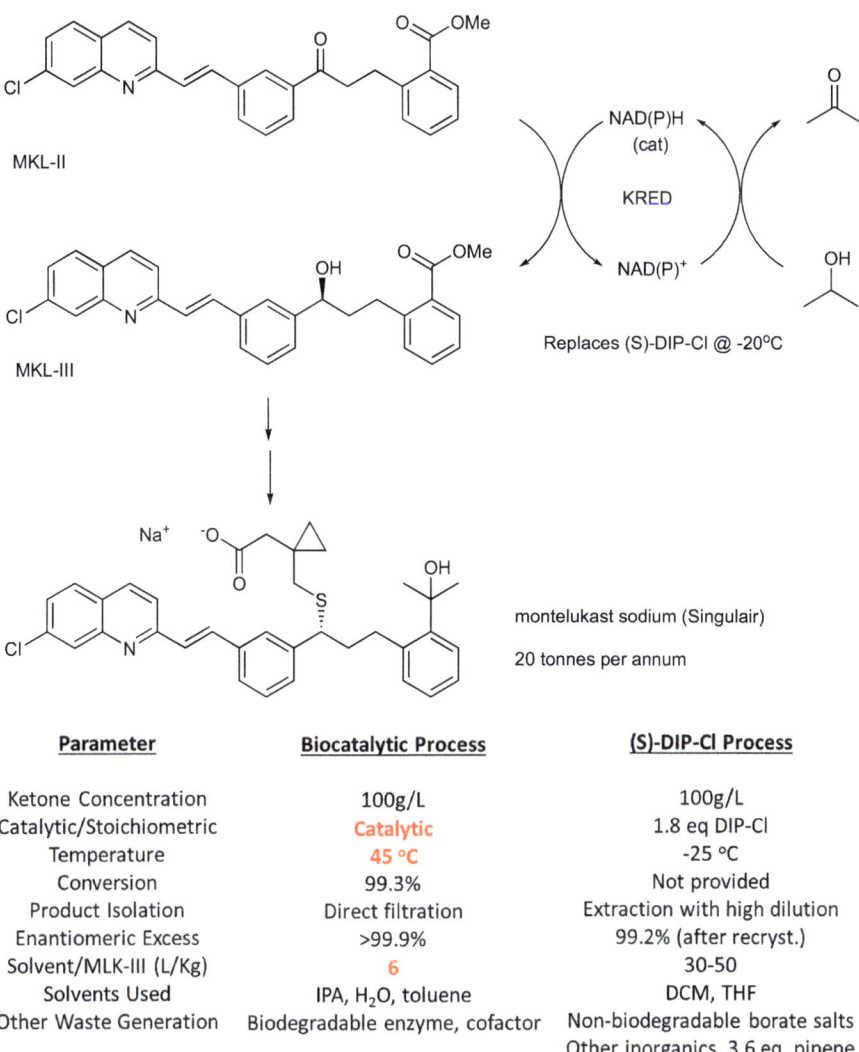

Figure 1.4 Comparison of the chemical and biocatalytic routes to montelukast.

Parameter	Biocatalytic Process	(S)-DIP-Cl Process
Ketone Concentration	100g/L	100g/L
Catalytic/Stoichiometric	Catalytic	1.8 eq DIP-Cl
Temperature	45 °C	-25 °C
Conversion	99.3%	Not provided
Product Isolation	Direct filtration	Extraction with high dilution
Enantiomeric Excess	>99.9%	99.2% (after recryst.)
Solvent/MLK-III (L/Kg)	6	30-50
Solvents Used	IPA, H₂O, toluene	DCM, THF
Other Waste Generation	Biodegradable enzyme, cofactor	Non-biodegradable borate salts Other inorganics, 3.6 eq. pinene

the reasons why the biocatalytic process may be preferred in terms of improved solvent regime and a reduction in overall waste.

However the full impact of biocatalytic retrosynthesis can only be realised when the introduction of one of more biocatalytic steps into a synthesis enables a major redesign of the synthesis of the target molecule. The use of a biocatalyst might result in a shorter route, with a different and less expensive starting material, and might also lead to greater selectivity and reduced waste. The aim of this book is to

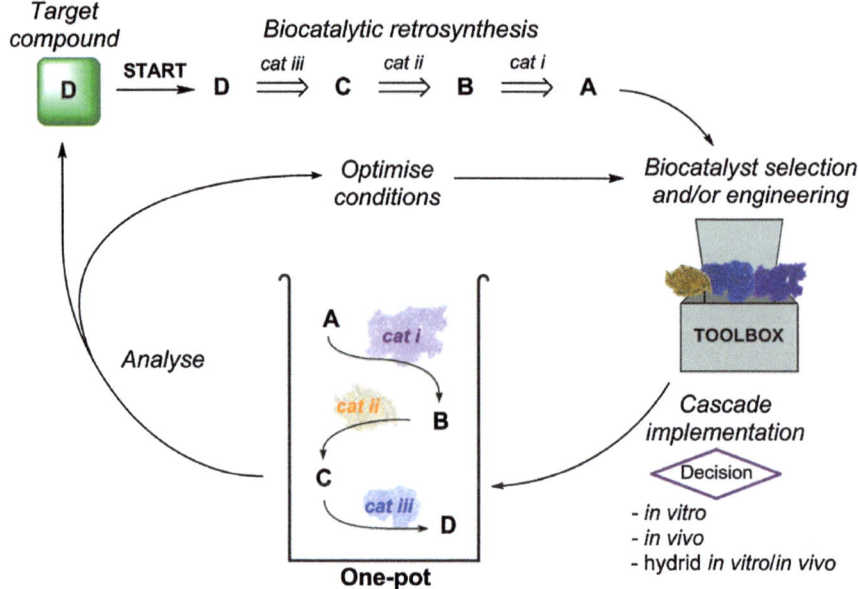

Figure 1.5 Combination of cascade biocatalysis and biocatalytic retrosynthesis.

explore such ideas and develop a better understanding of how the use of biocatalysis might effect such a step change in target molecule synthesis.

An increasing opportunity for biocatalysis is the development of cascade processes in which two or more engineered enzymes are combined *in vitro* or *in vivo* to enable multi-step conversion of simple starting materials to more complex targets.[6] Figure 1.5 illustrates how biocatalytic retrosynthesis can be integrated with cascade reactions to assist with the planning and development of multi-enzyme transformations. In this approach, the target compound D is subjected to a detailed analysis to identify the required biocatalysts, which are then engineered individually followed by construction of the cascade, either *in vitro* or *in vivo*. This cascade is then subjected to rounds of optimisation in order to achieve the target yield of the product. Such design–build–test systems will increasingly feature in the future, particular as the full power of synthetic biology is leveraged.

In summary, the aim of **Biocatalysis in Organic Synthesis: The Retrosynthesis Approach** is to (i) illustrate the current applications of biocatalysis using worked examples and case studies and (ii) develop new guidelines for identifying where biocatalysts can be applied in organic synthesis. The book contains a complete description of the current biocatalyst classes that are available for use and also suggests areas where new enzymes are likely to be developed in the next few years. The book also contains worked examples that enable the reader to practice disconnecting target molecules to find the 'hidden' biocatalytic reactions that can be potentially applied in the synthetic direction.

1.3   Structure of the Book

Chapter 1 outlines the aims of the book and briefly reviews the current landscape that has resulted in a significant increase in the application of biocatalysis, particularly for the synthesis of pharmaceuticals, fine chemicals, agrochemicals, polymers, flavours & fragrances (including for nutraceuticals and personal healthcare). In this chapter, the concept of 'biocatalytic retrosynthesis' is introduced as a means of identifying opportunities for introducing enzymes into syntheses in order to redesign synthetic routes to target molecules.

Chapter 2 provides a basic introduction to enzymes, including their structure, production, properties and considerations for their use, and their potential advantages over chemocatalysts. This chapter also briefly covers biocatalytic technologies and some practical aspects, *e.g.* isolated enzymes, immobilised enzymes, whole cells, cofactor recycling, protein engineering and directed evolution of enzymes, and high-throughput screening.

Chapters 3–9 provide a detailed account of the principal classes of different biocatalysts and how they can be used to synthesise particular motifs or building blocks that are of interest in target molecule synthesis. Information can also be found here regarding the mechanism of particular enzymes, especially as it relates to their application.

- Chapter 3 – Hydrolysis of esters (lipases, esterases), amides (acylases, proteases), nitriles (nitrilases), epoxides (epoxide hydrolases), sulfate esters (sulfatases) and carbon–halogen bonds (dehalogenases).
- Chapter 4 – Reverse hydrolysis, including the application of hydrolytic enzymes for the synthesis of esters, amides, lactones, lactams, *etc.*
- Chapter 5 – Reduction of C=O (KREDs, ADHs), C=C (ene reductases), C=N (imine reductases, amine dehydrogenases, amino acid dehydrogenases), CO_2H (carboxylic acid reductases), nitroaromatics, azides, N=N, sulfoxides and N-oxides.

- Chapter 6 – Oxidation of C–H bonds (P450 monooxygenases, peroxygenases), Baeyer–Villiger monooxygenases, epoxidation (peroxidases, monooxygenases), heteroatom oxidation, allylic oxidation (lipoxygenases), oxidation of amines and alcohols using amines/alcohol/amino acid oxidases, C–Hal bond synthesis (halogenases, fluorinases), oxidation of imines and aldehydes (xanthine DHs and aldehyde oxidases) and dealkylation of N-, O-, and S-ethers.
- Chapter 7 – C–X bond formation, including C–N bond synthesis (transaminases, ammonia lyases) and C–O bond synthesis (fumarase, hydratase).
- Chapter 8 – C–C bond formation including aldolases, TPP-dependent lyases, cyanohydrin lyases, alkyltransferases, Pictet-Spenglerases, carboxylation, cyclopropanation (engineered P450s) and terpene cyclases.
- Chapter 9 – Miscellaneous biocatalysts, including racemases (amino acid, mandelate).

Chapter 10 introduces the idea of developing a structured approach to the disconnection of target molecules based upon biocatalytic retrosynthesis. The reader will be guided through the various disconnections that are possible, both for acyclic and cyclic systems, in order to gain an understanding of where biocatalysts can be applied in organic synthesis. The various disconnections possible are organised into one of five different groups:

- Chapter 10.1 – Acyclic systems: substituted alcohols, amines, carboxylic acids, ketones, *etc.* (1 functional group).
- Chapter 10.2 – Acyclic systems: 1,2-, 1,3- and 1,4-diols, hydroxy-carbonyls, dicarbonyls, *etc.* (2 functional groups).
- Chapter 10.3 – 4-, 5-, 6-, and 7-membered carbocyclic rings.
- Chapter 10.4 – 4-, 5-, 6-, and 7-membered rings containing one or more heteroatoms.
- Chapter 10.5 – Substituted aromatic and heteroaromatic rings.

Chapter 11 provides a comparison of different biocatalytic routes to target molecules. The reader is taken through a series of well-known pharmaceutical targets where multiple different biocatalytic routes have been developed and in some cases scaled up for commercial applications. The aim of this chapter is to begin to gain some insight into the way in which target molecules can be disconnected back to simpler precursors, which can then be transformed using biocatalysis.

Chapter 12 concludes the book by giving readers the opportunity to test their understanding of biocatalysis and gain experience in disconnecting target molecules based on the principles of biocatalytic retrosynthesis. 25 worked examples (with answers), of increasing difficulty, are provided to enable students to develop their skills and apply their knowledge.

References

1. U. T. Bornscheuer, G. W. Huisman, R. J. Kazlauskas, S. Lutz, J. C. Moore and K. Robins, *Nature*, 2012, **485**, 185.
2. K. Faber, W.-D. Fessner and N. J. Turner, *Science of Synthesis: Biocatalysis in Organic Synthesis*, Thieme, Stuttgart, 2015, vol. 1–3; K. Faber, *Biotransformations in Organic Chemistry*, Springer, 6th edn, 2011; *Organic Synthesis Using Biocatalysis*, ed. A. Goswami and J. D. Stewart, Elsevier, Amsterdam, 2016.
3. X. Cheng and E. J. Corey, *The Logic of Chemical Synthesis*, Wiley, New York, 1989; E. J. Corey and L. Kürti, *Enantioselective Chemical Synthesis Methods, Logic and Practice*, Direct Book, Dallas, 2010.
4. N. J. Turner and E. O'Reilly, *Nat. Chem. Biol.*, 2013, **9**, 285; A. P. Green and N. J. Turner, *Perspect. Sci.*, 2016, **9**, 42; H. M. Hönig, P. Sondermann, N. J. Turner and E. M. Carreira, *Angew. Chem. Int. Ed.*, 2017, DOI: 10.1002/ange.201612462; U. T. Bornscheuer, R. O. M. A. de Souza and L. S. M. Miranda, *Chem. – Eur. J.*, 2017, DOI: 10.1002/chem.201702235.
5. J. D. Rozzell and J. Liang, *Spec. Chem. Mag.*, 2008, 36.
6. S. L. Flitsch, S. P. France, L. Hepworth and N. J. Turner, *ACS Catal.*, 2017, 710.

2 Biocatalysis Basics and Principles

2.1 Enzyme Basics

As we saw in Chapter 1, enzymes have been used for thousands of years by humans to catalyse chemical reactions. Enzymes are described as a group of proteins that are capable of catalysing chemical reactions. In this chapter, we will very briefly give you some detail on the basics of enzymes, such as their structure and function, and how as a chemist you can access enzymes for synthesis and some practical considerations of their use. We will also touch upon strategies to improve the properties of an enzyme, if it is unsuitable for our purposes.

Like all catalysts, enzymes increase the rate of a reaction by lowering the activation energy required for the reaction to occur. In addition, they are not affected by the reaction itself (meaning only a small amount of enzyme is required) and they are very selective, typically catalysing very specific reactions. So what is it about an enzyme's structure and function that enables it to act as a catalyst? Like all proteins, enzymes are made up of a long chain of amino acids that is assembled and then folded into a three-dimensional structure, as shown in Figure 2.1. This is known as the enzyme's primary structure, and although unique, we will see that enzymes share many common three-dimensional structural features. The amino acids making up the enzyme's primary structure are linked together *via* amide bonds forming a polypeptide. Lengths or regions of a poly-peptide typically take on certain structural motifs due to the hydrogen

Biocatalysis in Organic Synthesis: The Retrosynthesis Approach
By Nicholas J. Turner and Luke Humphreys
© Nicholas J. Turner and Luke Humphreys 2018
Published by the Royal Society of Chemistry, www.rsc.org

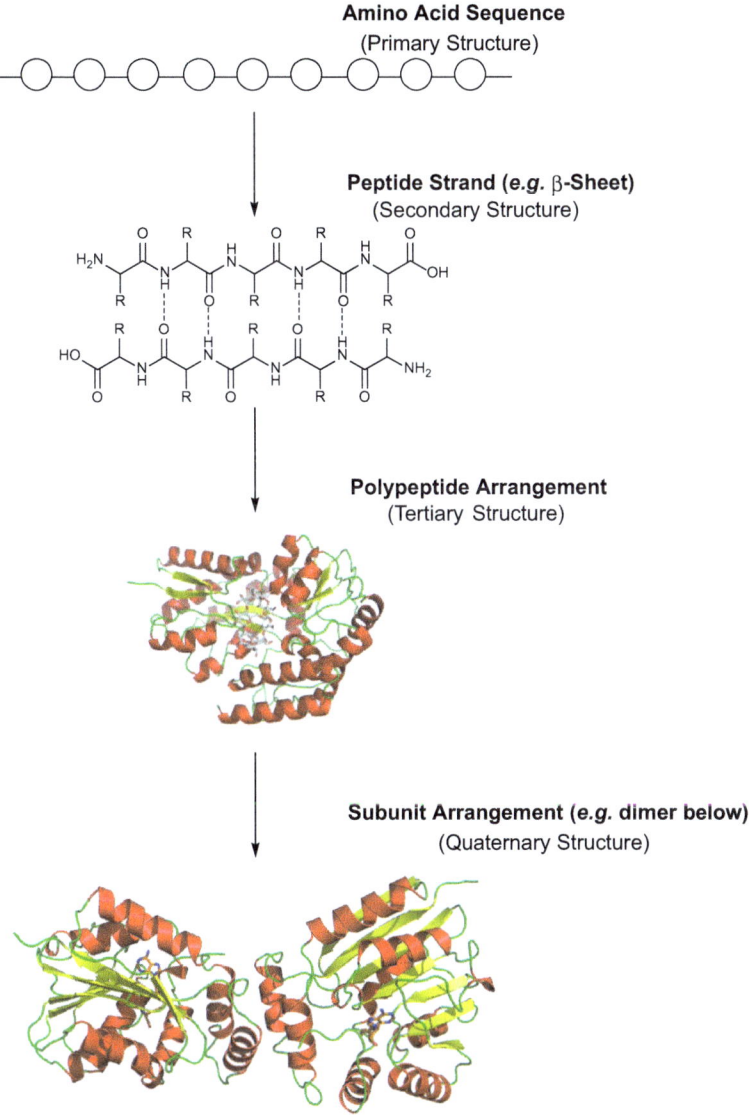

Figure 2.1 The structure of enzymes.

bonds formed between the amide hydrogens of the polypeptide backbone and the carbonyls that are adjacent in space. You may recall some of the elements from previous studies, such as α-helices or β-sheets. These motifs are referred to as an enzyme's secondary structure. The enzyme's secondary structural elements are then folded, twisted and arranged into the lowest possible energy conformation to form the three-dimensional form of the enzyme. This is known as the tertiary structure of the enzyme, and families of enzymes that catalyse the same reaction (albeit on different substrates) typically have very similar or the same tertiary structures. Different peptide chains may be stabilised to hold them in place through interactions between amino acid side-chains, such as disulfide bridges (between two cysteines) or salt bridges (between positively and negatively charged side-chains). Finally, the enzyme may be divided into subunits, such as a dimer, that may be the same or different. The way in which these subunits interact is known as the quaternary structure.

The structure of an enzyme is very important as it plays a vital role in its biological function. As a result of the way in which the polypeptide chains fold to form a three-dimensional structure, the reactive part of the enzyme, the active site, is formed. It is in this part of the enzyme that the reaction catalysis takes place. The active site may be at the surface of an enzyme or buried within the enzyme structure. In either case, when the reaction starting material approaches (the substrate), the enzyme typically undergoes a change in shape that allows the substrate to dock or bind in the active site. The amino acid side-chains then present in the active site help to catalyse the reaction and once complete the product is released. This mechanism is known as the "induced fit hypothesis" as the approach of the substrate induces the change in the shape of the enzyme, which makes it catalytically active.

As we saw in Figure 2.1, the interactions holding the proteins in the correct structure are quite weak in general chemistry terms. This means that if we are not careful, then under the wrong conditions (*e.g.* pH or temperature) the enzyme structure will be disrupted. This in turn may lead the enzyme to lose its structure, which is termed denaturation. As we have already seen, an enzyme's biological function is typically linked very strongly to its structure and as a result, denatured proteins are often inactive. This is something that we must be aware of or overcome when using enzymes in organic synthesis.

2.2 Production of Enzymes

We have seen that enzymes are used in a variety of industries, enabling the production of a diverse range of materials. We will now consider how, as chemists, we can obtain or produce enzymes to

Table 2.1 The advantages and disadvantages of obtaining and using different enzyme preparations.

Type	Pros	Cons
Commercial enzyme screening panels or kits.	Easy to use (no microbiology knowledge required) Easy to obtain and many useful enzyme classes available Good chance of success as likely to have been tested on a range of substrates	Can be expensive Not all enzymes available May require a license for use on a large scale
Microorganisms from a culture collection	Relatively easy to use Inexpensive and easy to obtain	May require some microbiology knowledge to culture Using the whole cell organism may result in unwanted by-products May require a license for use on a large scale
Produce enzyme "in-house"	Freedom to operate (no need to pay for a license) More knowledge of the enzyme sequence (in case of evolution) Freedom to produce desired enzymes	Requires microbiology knowledge More difficult to pick appropriate enzyme(s) from the many sequences known

enable us to carry out chemical syntheses. Each of these approaches has its own advantages and disadvantages, as shown in Table 2.1. In terms of obtaining ready to use enzymes, we have a number of options available to us. The first option is to obtain a microorganism from a culture collection. Examples in the UK include the United Kingdom National Culture Collection (UKNCC – http://www.ukncc.co.uk/) and the National Collection of Type Cultures (NCTC – https://www.phe-culturecollections.org.uk/). These are reference collections of microorganisms (almost like the equivalent of a library), from which chemists can order and use different strains of bacteria and other microorganisms. Assuming that a collection holds a microorganism containing the enzyme we are interested in using, then it may be possible to order a sample. We may be able to use the sample in the format that it arrives in, or we might need to culture it on a solid medium such as an agar plate prior to use. The alternative to accessing an enzyme through a culture collection is to order a suitable enzyme from a commercial supplier. The number of suppliers and the number of enzymes being offered commercially has grown remarkably in the past 5–10 years with large numbers and many different enzyme classes now available. In order to make this process even easier, many suppliers now offer "screening kits" for enzymes (often in 96-well plate format), to enable rapid screening and identification of a suitable enzyme for a reaction. Whilst this is very convenient, the cost of such kits may be prohibitive to some users.

An alternative approach is to produce the enzyme of interest ourselves. The cost and ease of producing enzymes has decreased rapidly over the past decade (which is why many enzymes are commercially available) enabling laboratory researchers to access a huge number of potential catalysts. This has become possible due to the cost of deoxyribonucleic acid (DNA) synthesis, which is a fraction of the price it was a decade ago, in conjunction with an explosion in bioinformatics, which has resulted in millions of enzyme sequences being deposited in databases. Scientists now have the ability to make countless enzymes; the problem is knowing which one might catalyse the reaction on the substrate of interest. We will now look in very brief detail at the production of enzymes in a standard laboratory micro-organism. This is by no means a full description, but we have included it here to give you an understanding of how this process works. For a more detailed understanding, readers are encouraged to read further in the area. Before going further, we first need to understand a theory known as the "central dogma of biology". This was first proposed by Francis Crick in 1958 and explains how proteins (including enzymes) are produced from deoxyribonucleic acid (DNA). As shown in Figure 2.2, the deoxyribonucleic acid (DNA) in a cell is transcribed by RNA polymerase into ribonucleic acid (RNA). The RNA is then translated in the ribosome into the amino acid sequence that makes up a protein. The whole process is referred to as gene expression. You will notice in Figure 2.2 that the final part of the theory involves the replication of DNA, but this is of less importance to protein production.

As we mentioned above, DNA synthesis technology now allows us to make most genes (DNA) synthetically. This means that we can make DNA from any number of sources, such as bacteria, fungi, plants or animals, and many third parties are available to carry out DNA synthesis on the chemist's behalf. Once we have our synthetic DNA in

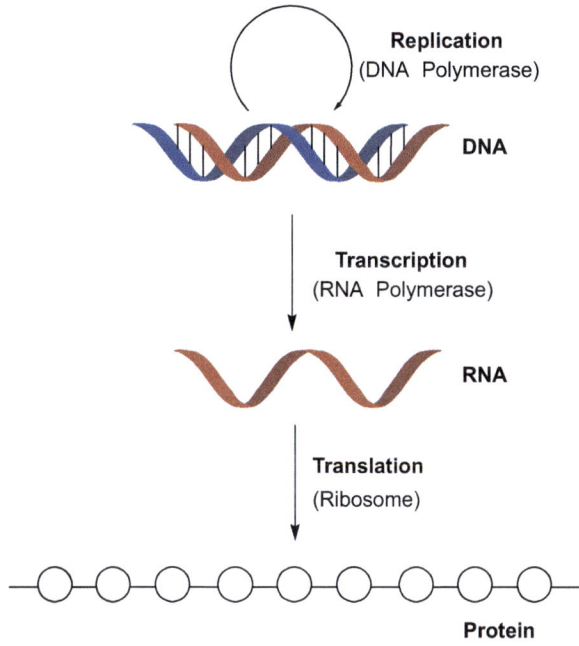

Figure 2.2 The central dogma of biology.

hand, we can then insert it into a "host" microorganism such as *Escherichia coli*. Its job is to make the enzyme we are interested in by using its own cellular machinery in accordance with the central dogma of biology. Proteins produced in this manner are called recombinant proteins. Whilst this method gives chemists the freedom to make any enzyme in which they might be interested, it is not without challenges and so a deal of microbiology knowledge is required for success.

2.3 Advantages of Enzymes over Chemocatalysts

Enzymes have several potential advantages over chemical catalysts that catalyse the same or similar transformations. Due to their complex three-dimensional structures, enzymes typically catalyse reactions with high chemoselectivity, regioselectivity or enantio/diastereoselectivity. In the case of chemoselectivity, enzymes are able to catalyse reactions in the presence of functional groups that might otherwise react under chemical conditions. This is in part due to enzymatic reactions usually taking place at ambient temperature and in aqueous solvent systems. The advantage of this chemoselectivity is that there is less of a requirement for chemists to use protecting groups, which would otherwise make a given synthetic route longer. In the same way, by catalysing reactions with high regioselectivity, enzymes may potentially shorten a route of synthesis, as extra steps to differentiate between two identical groups are not required or regioisomeric products are not produced, increasing the yield. The most value is often gained by using an enzyme with high enantio- or diastereoselectivity. In this case, products are obtained in a pure or almost pure stereoisomeric form, resulting in a higher yield and removing the need for extra steps to separate out enantiomers or diastereomers. Industries, in particular the chemical and pharmaceutical industries, are under increasing pressure to employ or adopt more environmentally benign or sustainable methods of carrying out chemical syntheses. In this regard, enzymes offer several advantages over more traditional chemical methods. This area was developed in 1998 by the chemists Paul Anastas and John Warner, who defined the term green chemistry as: "the utilization of a set of principles that reduces or eliminates the use or generation of hazardous substances in the design, manufacture and applications of chemical products". The twelve principles of green chemistry first proposed by these two chemists are shown in Table 2.2. Beside each principle is a description, where appropriate, of how enzymatic reactions conform or confer an advantage over more traditional chemocatalysed reactions.

In addition to the green chemistry aspects above, enzymes are considered "renewable catalysts" as they are produced by the fermentation of bacteria using sugars and they are biodegradable once they have been used in a process. This is quite different from transition metal catalysts, which have to be mined from the earth and in most cases may not be recovered at the end of the reaction requiring costly disposal if the process is run on an industrial scale. However, despite what appears to be a large number of advantages to using enzymes, there are also some limitations. Due to their highly selective nature, some enzymes may only catalyse a given reaction on a very small number of substrates. In addition, as most enzymes have evolved to work in the mild environments found on the majority of the earth, they may not be stable under the process conditions that chemists may require to run a reaction. Enzyme processes may suffer from low substrate concentrations, leading to low throughput, and many classes of enzymes, as we will see in Part II, require a co-factor

Table 2.2 The impact of biocatalysis on the 12 green chemistry principles (adapted from ref. 1).

Green chemistry principle:	Biocatalysis
1. Prevention (of waste)	Biocatalysis can enable new, more sustainable routes to drugs, reducing the levels of waste
2. Atom Economy	Biocatalysis often enables more efficient synthetic routes due to high selectivity
3. Less Hazardous (less toxic reagents and intermediates) chemical synthesis	Enzymes are generally of low toxicity, replacing more toxic reagents
4. Designing safer (less toxic) chemicals	No impact
5. Safer solvents and auxiliaries	Enzymatic reactions are often preformed in water and organic solvents, if used, are typically of low hazard
6. Design for energy efficiency	Enzymatic reactions are usually performed at or slightly above room temperature, making them energy efficient
7. Use of renewable feedstocks	Enzymes are renewable
8. Reduce derivatives (*e.g.* protecting groups)	Enzymes have high selectivity (chemo-, regio-, and stereo-) typically avoiding the need for protecting groups
9. Catalysis (preferred over stoichiometric reagents)	Enzymes are catalytic
10. Design for degradation (avoid environmental build-up)	No impact on the design of products (although enzymes are biodegradable)
11. Real-time analysis for pollution (& hazard) prevention	No impact
12. Inherently safer chemistry for accident prevention	Enzymatic reactions are generally performed under mild conditions, where the risk of explosions or run-away reactions is very small, making them inherently safe

for activity, which can be expensive and therefore needs to be recycled. We will see how some of these challenges can be addressed in the next section.

2.4 Biocatalytic Technologies and Practical Aspects

We saw in Section 2.2 that we can obtain enzymes from a variety of sources, including a whole microorganism or a recombinant protein that we made from a host organism such as *E. coli*. As a result, enzymes can be used in a variety of different preparations, although in general the most commonly used forms are whole cells or isolated enzymes. As shown in Table 2.3, both of these forms have some advantages and disadvantages.

As a result, chemists can pick the form of enzyme that is most suitable for the application they have in mind, depending upon which factors are most important to the chemist. In addition, both whole cells and enzymes may be immobilised onto some form of solid support. Enzymes may be immobilised *via* a variety of methods, including covalent bonds, physical adsorption, encapsulation and cross-linking, each with its own advantages. Enzyme immobilisation may offer additional advantages, as shown in Table 2.4.

One of the issues associated with the use of isolated enzymes, as shown in Table 2.3, is the need to recycle any co-factors that are present, and we will now look at this in more detail. Co-factors are generally described as inorganic or organic molecules that are required for enzymes to carry out their catalytic function. One of the most commonly used co-factors is nicotinamide adenine dinucleotide (NADH) and its phosphorylated analogue NADPH, shown in Figure 2.3. In their reduced form, these co-factors are typically used as reductants with the red hydrogen atom acting as a hydride equivalent.

Table 2.3 The advantages and disadvantages of running reactions with whole cells or isolated enzymes.

	Pros	Cons
Whole cells	Co-factor recycling enzymes already present Inexpensive and convenient to operate	Low substrate concentrations (typically $1–5 \text{ g L}^{-1}$) Tend to produce more by-products
Isolated enzymes	High substrate concentrations $(100–200 \text{ g L}^{-1})$ Clean and efficient	Requires co-factor recycling, which increases the cost and complexity

Table 2.4 The advantages and disadvantages of running reactions with immobilised enzymes.

	Pros	Cons
Enzyme Immobilisation	Allows recycling of the enzyme Increases stability Easy separation of the enzyme from the product at the end of the reaction May make development of multi-enzyme cascades easier	Adds cost and complexity May affect enzyme activity or selectivity May affect the diffusion of the substrate into the enzyme active site

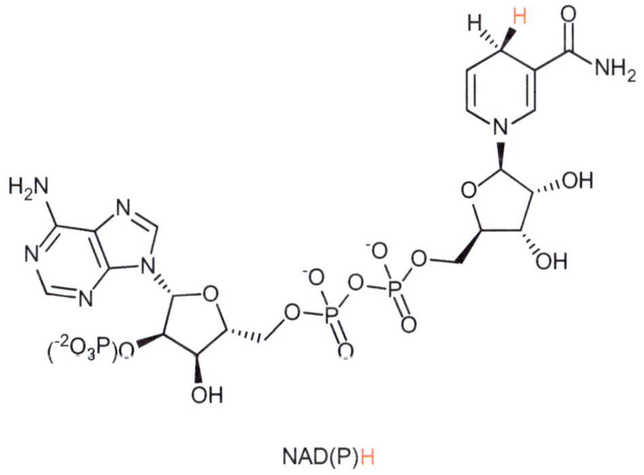

NAD(P)H

Figure 2.3 The reduced form of the co-factor nicotinamide adenine dinucleotide (phosphate), NAD(P)H.

The cost of the nicotinamide adenine dinucleotides NADH and NADPH prohibits us from using stoichiometric amounts of them for reactions with isolated enzymes much above the milligram scale. As a result, chemists have come up with methods to recycle these (and other) co-factors in order to reduce costs. One way to achieve this is to use a second enzyme and a sacrificial substrate to reoxidise the co-factor, as shown in Figure 2.4. If we deliberately choose a sacrificial substrate that is inexpensive, this now becomes a very cost-effective way of running the whole process. In Figure 2.4, we are using a ketoreductase enzyme (which we will meet again in Chapter 5) to carry out the reduction of a ketone to the corresponding alcohol. This reaction requires NADH as the co-factor, and in order to recycle this we

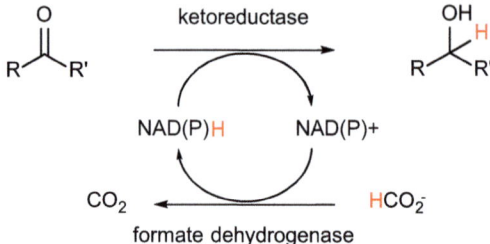

Figure 2.4 An example of a co-factor recycling system for nicotinamide co-factors.

use a second enzyme called formate dehydrogenase and formate as the sacrificial substrate. As we can see, each time the reduction reaction occurs, the oxidised form of the co-factor, NAD^+, is formed. This is then reduced back to the reduced form (NADH) at the expense of formate, which is oxidised to give carbon dioxide.

This system has several advantages, one being that formate is very inexpensive and can be used in excess, and a second being that carbon dioxide is a gas and will leave the reaction mixture helping to drive the reaction equilibrium. This helps produce an efficient recycling system, which in turn ensures that the reaction progresses at the expected rate.

Once we have established whether a co-factor recycling system is required, we also need to think about the conditions that we will need to run the reactions under for optimal enzyme performance. Unless we know these already (such as from literature provided by a commercial supplier) then we will need to screen a number of factors, such as temperature, pH and substrate concentration. Optimisation of these (and other) parameters may be required for success. We have seen above that stability is key to the structure and function of an enzyme, so what do we do if our desired reaction occurs under conditions where our enzyme is not stable? Equally, how can we improve an enzyme's selectivity or activity for a desired reaction? In the section below, we will see how it is possible to improve many different properties of biocatalysts.

2.5 Improving Biocatalysts through Protein Engineering and Evolution

We have seen already in this chapter that the structure of an enzyme is highly instrumental in its biological function and that in turn, the sequence of amino acids that makes up the enzyme is responsible for its structure. As a result, if we change the sequence of amino acids that makes up an enzyme, we will create changes in its structure and therefore its function. It is this idea that forms the basis of strategies to improve many different enzyme properties, from activity and selectivity to pH or temperature stability. Before we look at approaches to achieve this, we first need to consider how many changes we want to make to an enzyme and therefore how many new enzyme "mutants" of the original we would like to produce. If we consider a medium-sized enzyme of around 400 amino acids, then exchanging the amino acid at each position of the starting sequence for all of the other 19 available amino acids gives us a total of 19^{400} variants. This is a huge number, one which we will not even come close to being able to make. Even if we only exchange each position for two other amino acids we would still make 144 000 variants ($400 \times 19 \times 19$). This is known as the sequence-space problem, as there is such as vast about of space in the number of potential sequences that we can't possibly cover all of it. One way around this problem is possible if we already have information about the enzyme's structure, for example from X-ray crystallography. If we know for example the position of the enzyme active site and the residues involved in substrate binding, then we can attempt to make "rational", targeted changes to the protein sequence, only focusing on those key residues. This cuts down hugely the number of mutants that we need to make and screen. If you recall the central dogma of biology above, then in this case, we need to make DNA with a slightly modified sequence, which is then transcribed and translated as before to make the desired variant enzymes. This produces a small collection of mutants that we can then evaluate. This approach is shown in blue in Figure 2.5. If we don't have information about the enzyme crystal structure or if we are looking to make more random changes to an enzyme (for example to improve solvent or pH stability) then we can use an alternative technique called directed evolution. In this case, we now make random changes to the amino acid sequence throughout the entire enzyme resulting in a large library of mutants. At this point, we then subject the entire library to some form of screen or selection with the idea being that just as in the case of Darwinian evolution, the "fittest" enzymes (those most stable or capable of catalysing a reaction) are chosen as the beginning for the next round of the evolution process. If we run several iterations of this process, large changes in enzyme properties (and sequence) become possible. This approach is shown in red in Figure 2.5.

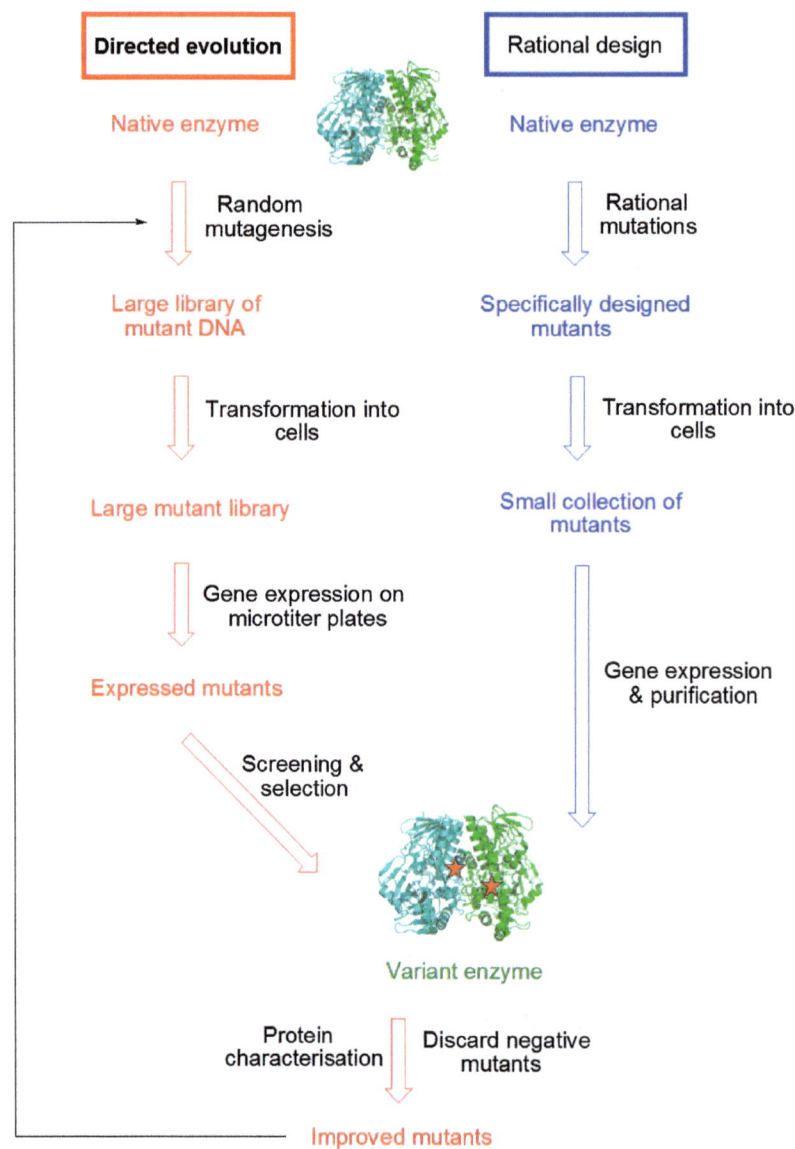

Figure 2.5 Methods of improving biocatalysts through either rational design or directed evolution.

These techniques can be incredibly powerful with very large changes in enzyme function possible. For example, the enantioselectivity of an enzyme can be switched from *R* to *S* selective or the substrate scope of an enzyme can be changed. Increases in enzyme stability to temperature, solvent or pH can be achieved and in some cases even

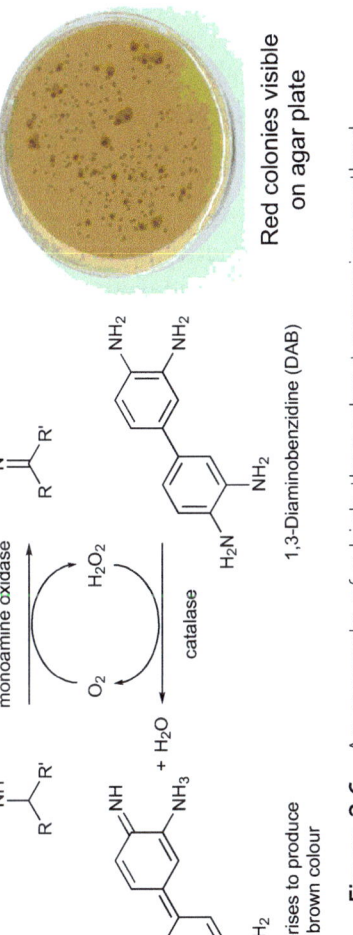

Figure 2.6 An example of a high-throughput screening method.

new reactivity has been created. One of the potential drawbacks of directed evolution is that even taking the sequence-space problem into account, it is relatively easy to generate libraries containing millions or billions of mutants that then need to be screened. Assuming that we would like to screen most, if not all, of these variants, then we need a screening method that will enable us to carry this out as quickly as possible. Fortunately, chemists have created a variety of these "high-throughput" (HPT) screening methods, which allow for up to a million variants to be tested per day. Standard analytical methods such as high pressure liquid chromatography (HPLC) or mass spectrometry (MS) have been modified for use in high-throughput techniques. This typically involves arraying the variants into a 96-well plate format with a liquid handling robot, before the plates are screened using a method with a very short run-time. An alternative to these more traditional approaches is to use a colorimetric screen. In this method, the product (or a by-product) of the desired enzymatic reaction is coupled with a second reaction, which produces a colour change. This allows us to quickly see (visibly or by using fluorescence) which mutant enzymes show the desired characteristics. This approach can be used in solution, in a plate format or in a solid phase such as an agar plate. An example of this approach is shown in Figure 2.6. In this instance, the by-product of the reaction is hydrogen peroxide and this is utilised by a second enzyme (catalase) to oxidise 3,3-diaminobenzidine to form a red polymeric complex. The red colour is then used to identify very quickly which enzymes are the best at catalysing the reaction, either in the liquid phase (*e.g.* a 96-well plate) or as shown in Figure 2.6 in the solid phase on an agar plate. In this case, each bacterial colony that is able to catalyse the reaction appears red in colour.

The screening method is therefore a very important part of any enzyme evolution program. A poor screening method can limit the number of variants that can be investigated, or even worse, if set up incorrectly, can result in mutants with undesired characteristics being chosen. It pays to take heed of the phrase "you get what you screen for" when considering the design of a screening method as part of an enzyme evolution project.

Reference

1. P. T. Anastas and J. C. Warner, *Green Chemistry: Theory and Practice*, Oxford University Press, New York, 1998, p. 30.

3 Hydrolysis

3.1 Introduction

This chapter will cover enzymes that are known in the enzyme classification system as hydrolases. Members of this group of enzymes are given the enzyme classification number 3, followed by a sub-group number depending on which type of bond they hydrolyse. Enzymes in this class are called hydrolases because they use water to break a carbon–heteroatom bond (C–X). However, as this process is reversible, these enzymes can also work in the opposite direction to catalyse the formation of a C–X bond with water produced as a by-product. We will explore this in more detail in Chapter 4. Hydrolases are important enzymes for metabolism and are probably the most widely used biocatalysts within industry. They are used for a wide variety of applications: in detergents they help to hydrolyse fat stains on clothing, and in the cheese industry they help hydrolyse fats and improve the cheese's flavour. They can also be used for the detection and bioremediation of environmental pollutants and most importantly for us, they can be used in organic synthesis to catalyse a variety of reactions. The transformations covered in more detail within this chapter are shown in Figure 3.1 and the hydrolysis of each functional group is discussed in a separate section.

3.2 Hydrolysis of Esters

The hydrolases evolved by Nature for the hydrolysis of esters are versatile enzymes called esterases and lipases. Both of these families of enzymes share a common mechanism, but their structure

Biocatalysis in Organic Synthesis: The Retrosynthesis Approach
By Nicholas J. Turner and Luke Humphreys
© Nicholas J. Turner and Luke Humphreys 2018
Published by the Royal Society of Chemistry, www.rsc.org

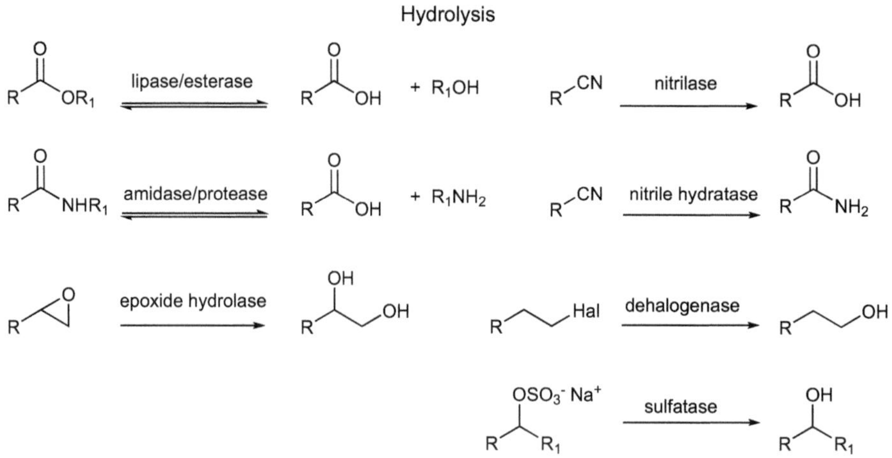

Hydrolysis

Figure 3.1 Overview of transformations presented in this chapter.

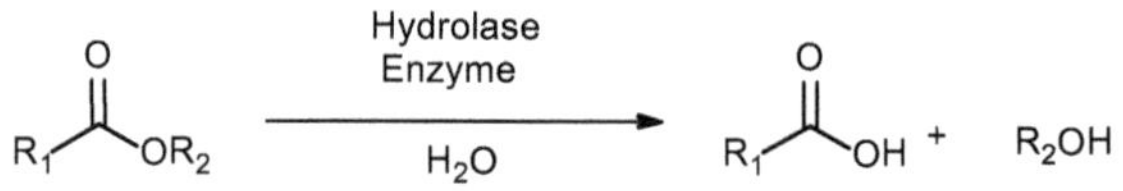

Figure 3.2 General transformation catalysed by hydrolases.

varies slightly depending on whether they have adapted to work in a hydrophobic or hydrophilic environment. As shown in Figure 3.2, if we start with an ester, we can use a hydrolase and water to break the C–O ester bond to give a carboxylic acid and an alcohol.

This is exactly the same as if we had hydrolysed the ester chemically under either acidic or basic conditions. For acidic hydrolysis, we use a dilute acid such as hydrochloric acid to catalyse the reaction, whereas for basic hydrolysis we use hydroxide ions such as sodium hydroxide. These methods are easy to carry out, but we must be careful that the conditions of the hydrolysis reaction are compatible with the rest of the functional groups in the molecule, especially if we need to heat the reaction or use a strong acid or base. By comparison, enzyme-catalysed hydrolysis occurs under much milder conditions.

There is also another advantage to carrying out a hydrolysis reaction using an enzyme: stereoselectivity. If we use a chiral ester as a substrate for the hydrolysis reaction, then the hydrolase enzyme will catalyse the reaction of one enantiomer of the ester faster than the other. This is because the enzyme itself is chiral and so will interact with enantiomers at different rates. As shown in Figure 3.3, if we take a chiral ester

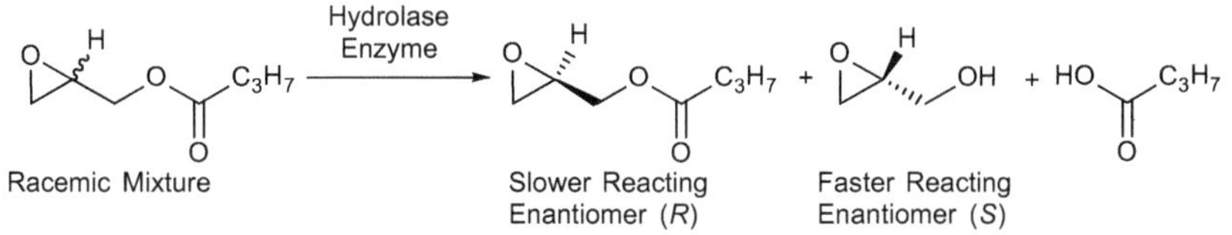

Figure 3.3 Kinetic resolution of an epoxide substrate catalysed by a hydrolase.

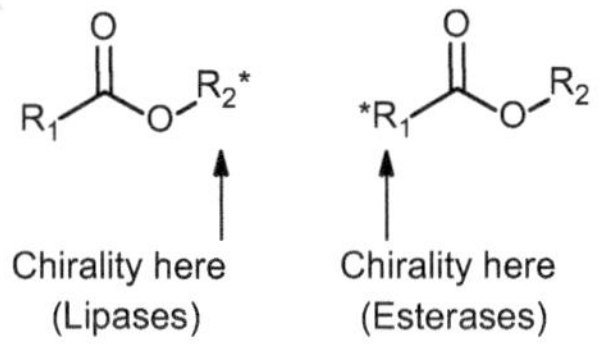

Figure 3.4 The substrate preference of lipases and esterases.

that is a racemic mixture, then the hydrolase enzyme will hydrolyse one ester at a much faster rate, resulting in kinetic resolution.

In this particular example, we are interested in the chiral epoxide that is in the alcohol part of the ester that was hydrolysed. Instead of a racemic mixture of epoxides (which would be difficult to separate), at the end of the reaction we now have an ester and an alcohol, which we can separate easily.

As we have already mentioned, both lipases and esterases will catalyse the hydrolysis of esters. However, when the ester is chiral, the two classes of enzymes prefer different substrates. As shown in Figure 3.4, lipases prefer chirality or prochirality in the alcohol portion and esterases prefer chirality or prochirality in the acid portion.

In general, lipases are more widely used than esterases due to their broad substrate range, high selectivity and ability to hydrolyse water insoluble substrates. This is because lipases have evolved for the hydrolysis of esters, which are not water soluble. As shown in Figure 3.5, lipases typically hydrolyse triglycerides into glycerol and fatty acids.

Lipases are inactive in a purely aqueous medium, but in a lipophilic medium or when micelles form within the aqueous medium, they become active. This is because in aqueous medium the entrance to the active site of lipases is blocked by a "lid". On the inside of the "lid" are predominately hydrophobic residues, such that when placed in a non-aqueous environment the "lid" opens allowing the substrate to access the active site. As a result, lipases are different to most enzymes in that they will often work in typical organic solvents or in organic/aqueous mixtures.

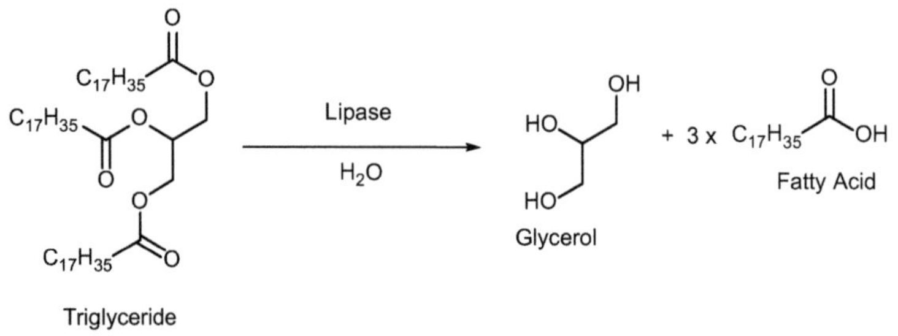

Figure 3.5 Lipase-catalysed hydrolysis of triglycerides into glycerol and fatty acids.

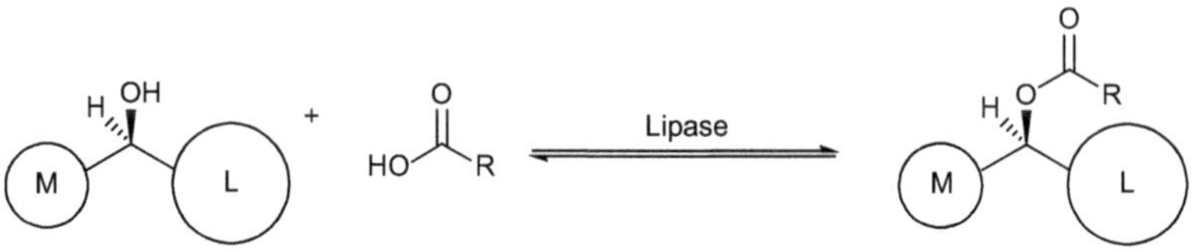

Kazlauskas rule: An empirical rule to predict which enantiomer of a secondary alcohol reacts faster in lipase catalysed reactions. M - medium sized substituent (*e.g.* methyl). L - large sized substituent (*e.g.* phenyl). In acylation reactions, the enantiomer depicted reacts fastest. In hydrolysis reactions, the ester depicted reacts fastest.

Figure 3.6 The Kazlauskas rule for predicting which enantiomer reacts fastest in a lipase-catalysed reaction.

Many different types of substrates have been resolved using lipases, and based upon experimental observations, researchers have suggested a rule to predict which enantiomer from a racemic mixture will react fastest. The Kazlauskas rule (named after the researcher who proposed it) is based on the size of the substituents attached to the ester or, in the opposite direction, the substituents attached to the alcohol. The difference between the sizes of the substituents allows the lipase to differentiate between the enantiomers of the ester. The rule is shown in Figure 3.6 for an ester formed from a secondary alcohol with a lipase. Similar models exist for primary alcohols and secondary amines, but they are not included here.

However, we need to add a note of caution when applying the Kazlauskas rule: the catalytic residues in lipases are often arranged in an inverse manner in esterases leading to the opposite enantiomer reacting faster in esterase reactions to that shown above. As a result, the choice of enzyme will largely influence the outcome of the resolution reaction.

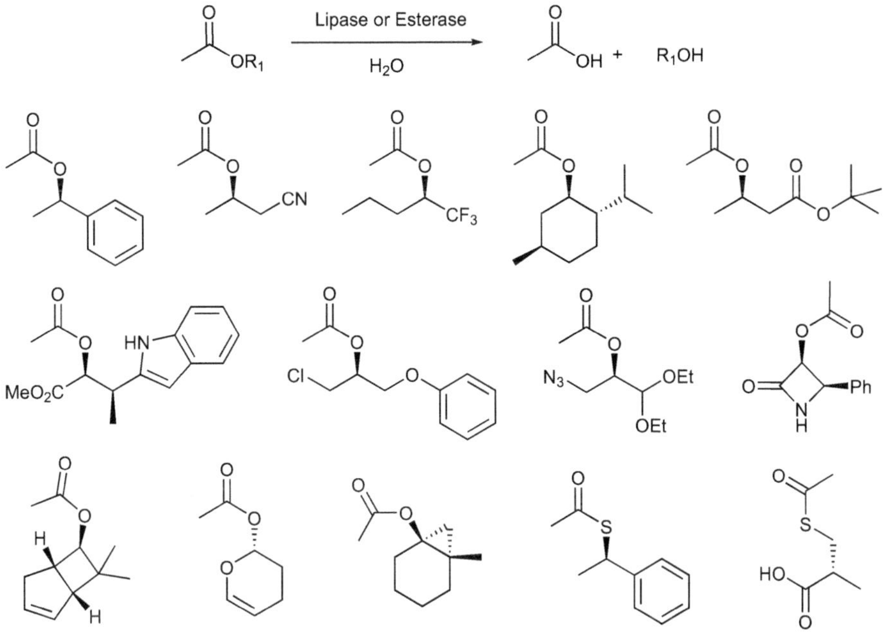

Figure 3.7 Substrate scope of lipase- or esterase-catalysed hydrolysis reactions.

The broad substrate range of lipases includes esters, which are hydrolysed to give primary, secondary and tertiary alcohols and some thiols. For each of the substrates in Figure 3.7, the faster reacting enantiomer is shown. This is the enantiomer which will hydrolyse fastest to give the corresponding alcohol. The substrates in Figure 3.7 also contain a wide variety of functional groups that are stable to the conditions of the lipase reaction, but may not be stable to chemical hydrolysis conditions. A single lipase is unable to accept all of the substrates shown in Figure 3.7; therefore, several enzymes would probably need to be screened to find the best enzyme for each substrate.

Notice that we are able to hydrolyse esters in the presence of other functional groups, such as acetals, nitriles, amides and even another ester. You will also notice that there are no esters made from primary alcohols in Figure 3.7. This is because the selectivity observed is usually lower with these substrates, although it can be improved by optimising the reaction or enzyme evolution.

Carrying out kinetic resolution of a racemic mixture of enantiomers is a useful method of accessing a single enantiomer of the compound we want. However, it is not an ideal process as we only end up using half of the mixture (50% yield) and throwing the unwanted

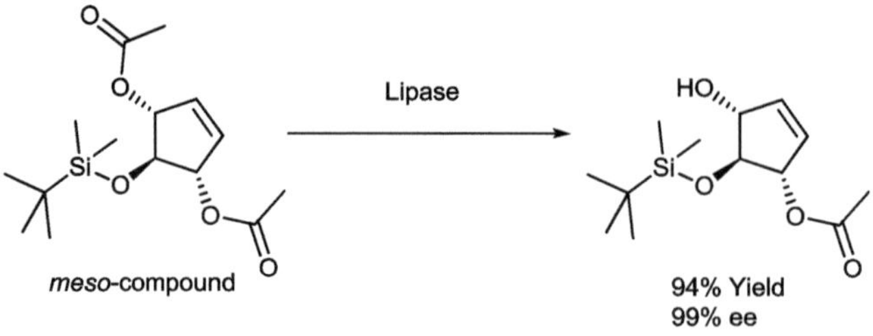

Figure 3.8   Lipase-catalysed desymmetrisation of a *meso*-compound.

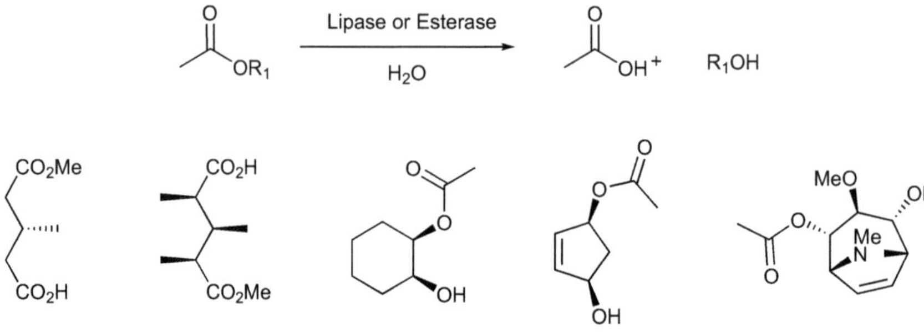

Figure 3.9   Substrate scope of lipase- or esterase-catalysed desymmetrisation reactions.

enantiomer away. An alternative is to start with a compound containing an internal plane of symmetry, a *meso*-compound, and carry out a desymmetrisation reaction. As shown in Figure 3.8, by hydrolysing one of the two esters we "break" the symmetry of the substrate to give one enantiomer. By using this approach, we can now in theory get 100% yield of the desired compound. In this case, the reaction is very selective and high yielding.

A range of *meso*-compounds can be desymmetrised using hydrolase enzymes, as shown in Figure 3.9. In each case, the product enantiomer is shown.

Lipases have been frequently used in the synthesis of pharmaceutical drugs. One particular example is β-adrenergic antagonists, known more commonly as β-blockers, for the treatment of hypertension (high blood pressure) and angina pectoris (pain caused by a lack of blood to the heart). Many compounds in this class of drugs are 3-aryloxy-2-propanolamines and contain a chiral secondary alcohol, such as propranolol, which is shown in Figure 3.10. In this case, the (S)-enantiomer is roughly one hundred times more active than the

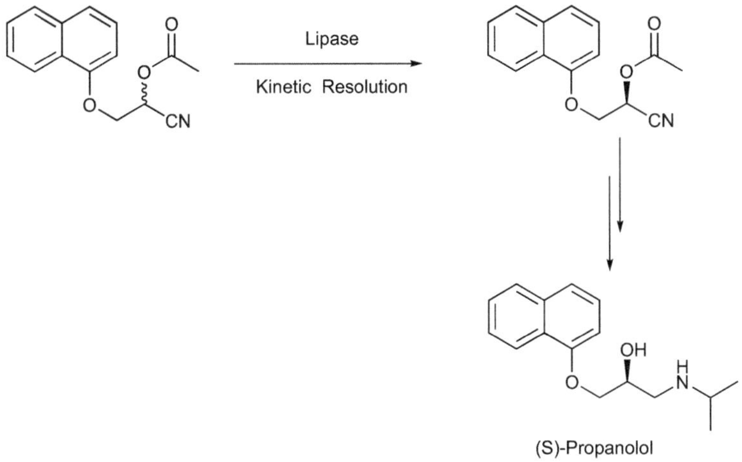

Figure 3.10 Lipase-catalysed kinetic resolution towards (S)-propranolol.

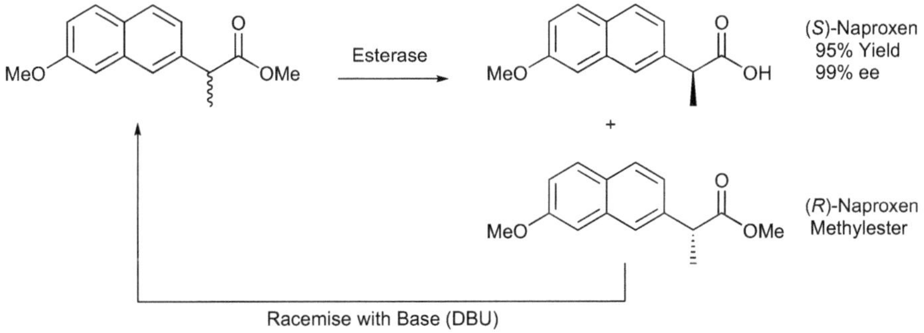

Figure 3.11 Esterase-catalysed resolution for the synthesis of (S)-naproxen.

(R)-enantiomer. This gave the opportunity for chemists to use a lipase in order to selectively produce only the (S)-enantiomer of propranolol.

As we saw in Figure 3.4, in comparison to lipases, esterases prefer substrates where the chiral part is in the acyl portion of the ester. Chemists took advantage of this selectivity for the synthesis of naproxen, a non-steroidal anti-inflammatory drug. In this case, the (S)-enantiomer is around one hundred and fifty times more active than the (R)-enantiomer, which might also cause unwanted side effects. As shown in Figure 3.11, by using an esterase, chemists could selectively produce the (S)-enantiomer through hydrolysis. In addition, they were able to recycle the unwanted (R)-enantiomer through racemisation with a base (DBU). In this way, the chemists were able to get a yield of greater than 50%, the maximum from kinetic resolution.

Lipases and esterases are cofactor- and metal-free enzymes, which means that they are easy to use and no special care needs to be taken when we run reactions with them. The hydrolysis of esters is based on a catalytic triad of glutamic acid, histidine and serine (Glu–His–Ser) and an additional part of the enzyme active site called the "oxyanion hole". It is thanks to these two features that the enzyme is able to catalyse the hydrolysis reaction. A detailed mechanism for ester hydrolysis is shown in Figure 3.12. You do not need to know this mechanism in detail, but it is included here for more advanced readers and those who are interested. Figure 3.12 shows the reversible steps of ester hydrolysis going clockwise starting from the structure at the top of the scheme. By going in the anticlockwise direction, the same steps in reverse give the mechanism for ester synthesis ('reverse hydrolysis', which we will cover in Chapter 4). In the ester hydrolysis direction, the serine residue of the catalytic triad is initially deprotonated by the histidine residue with the glutamic acid helping to delocalise the charge developed on the histidine. This means that the serine hydroxyl group can be deprotonated at neutral or even acidic values, rather than pH 14–15 (remember that the pK_a of alcohols is approximately 15). The deprotonated serine then attacks the ester substrate, generating a tetrahedral intermediate. The charge density on the oxyanion of the tetrahedral intermediate is stabilized by delocalization in the oxyanion hole, allowing this charged species to be formed under neutral reaction conditions. The oxyanion hole is made up of a number of amino acids that have side-chains capable of hydrogen bonding to the oxyanion and therefore stabilising it. This means that the negative charge is delocalised over a wide area and that the oxygen anion shown is not really an accurate representation. Collapse of the tetrahedral intermediate and elimination of the alcohol (in red) then generates an acyl enzyme complex, where the acyl intermediate is covalently bound to the enzyme *via* the serine residue.

Attack of water (in blue), activated by the histidine, then gives a second tetrahedral intermediate. This intermediate then collapses to give the product carboxylic acid and regenerates the enzyme for the next catalytic cycle. As mentioned above, the catalytic cycle can also proceed the other way around, generating esters from alcohols and acids. Consequently, it is also possible to substitute one alcohol group of an ester to another or to replace one acid group by another in a transesterification reaction. It is also useful to note that the catalytic triad is not responsible for any stereoselectivity observed; it is the active site made by the structure of the protein (the protein fold) that

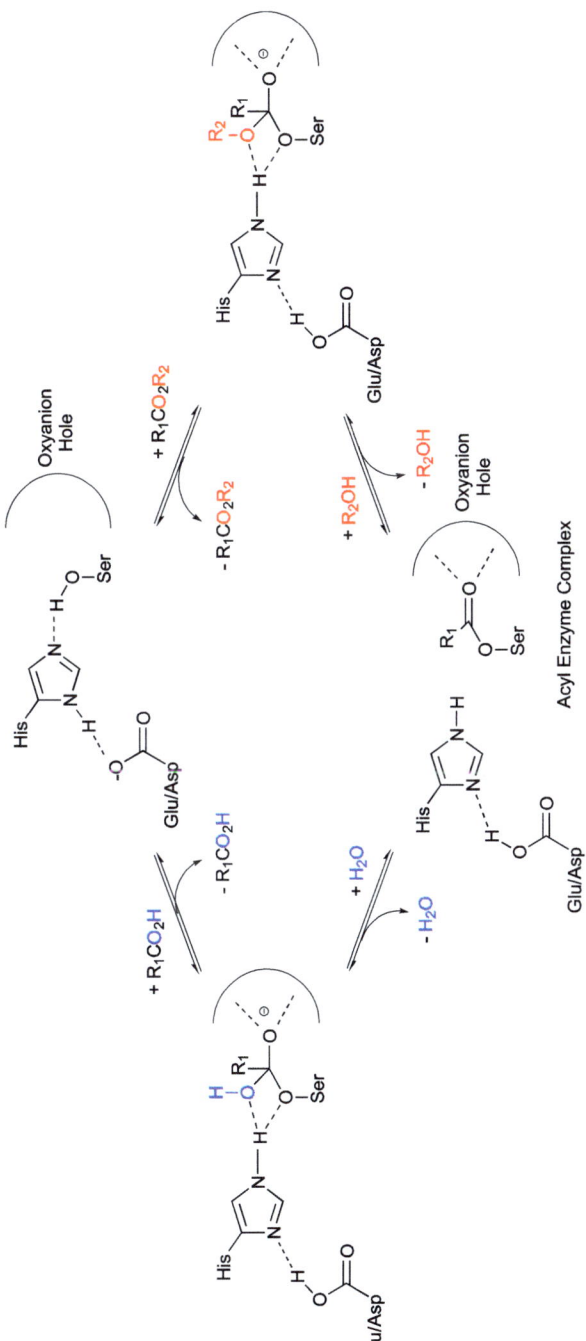

Figure 3.12 The mechanism of ester hydrolysis/synthesis catalysed by lipases and esterases.

is crucial. As a result, two proteins may have the same catalytic triad, but show different stereoselectivity with the same substrate.

3.3　Hydrolysis of Amides

The hydrolysis of amides is carried out by a family of hydrolase enzymes called proteases, which have evolved to hydrolyse polypeptide and protein amide bonds. These enzymes have a variety of functions from digesting proteins found in our food to activating signalling pathways in the body by modifying proteins. These proteins can also hydrolyse esters in the same manner as lipases above, but conversely, lipases are unable to hydrolyse amides. To hydrolyse an amide using chemical conditions typically requires us to use more concentrated acid or base and a higher temperature than we would use to hydrolyse an ester. This is because the amide carbonyl is less electrophilic than the corresponding ester carbonyl, due to the greater electron donating effect of the nitrogen lone pair. As a result, the conditions for chemical hydrolysis of amides may be too harsh for other functional groups present in the molecule. As shown in Figure 3.13, proteases and amidases generally show selectivity towards the acyl portion of a chiral substrate, the same as we saw previously with esterases.

　Amino acid acylases are a class of proteases which, as their name suggests, catalyse the hydrolysis of *N*-acyl-amino acids. As shown in Figure 3.14, this is another kinetic resolution reaction with the natural ʟ-enantiomer of the amino acid of the amide hydrolysed at a much faster rate than the unnatural ᴅ-enantiomer. A range of natural (proteinogenic) and non-natural (non-proteinogenic) amino acids can be resolved.

　Another group of enzymes capable of hydrolysing amide bonds are amidases. As shown in Figure 3.15, these enzymes prefer substrates where the amide is part of the carboxylic acid portion of the amino acid. These enzymes are also useful because they show the opposite

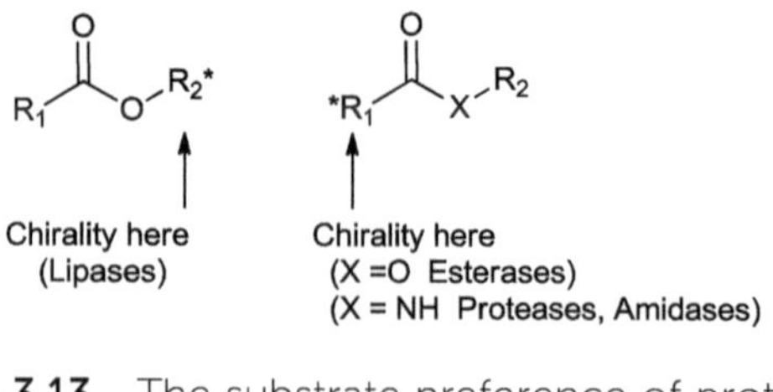

Figure 3.13　The substrate preference of proteases.

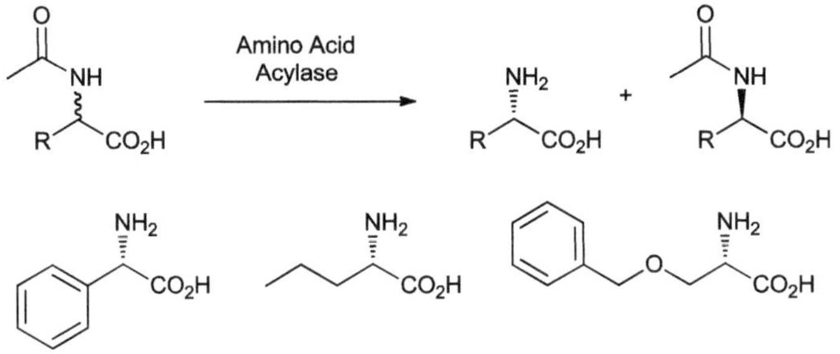

Figure 3.14 Kinetic resolution reactions catalysed by amino acid acylases.

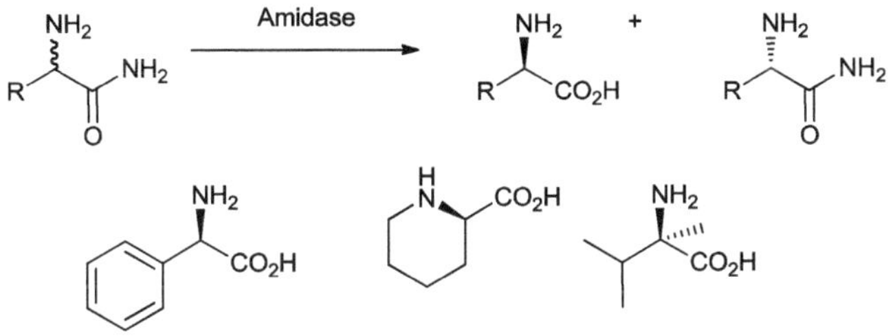

Figure 3.15 Kinetic resolution reactions catalysed by amidases.

selectivity to the amino acid acylases above. This means that they hydrolyse the unnatural D-enantiomer of the substrate faster than the natural L-enantiomer. A range of natural and unnatural amino acids can be resolved using amidases, as shown in Figure 3.15.

There are many different classes of proteases that catalyse amide bond hydrolysis *via* different mechanisms. However, one of the most well understood classes is the "serine proteases". These enzymes catalyse amide hydrolysis by a mechanism that is analogous to the one that we saw above in Figure 3.12. This time, however, we lose ammonia or an amine from the tetrahedral intermediate instead of an alcohol. Another important class of proteases is the metalloproteases. These enzymes typically use a metal ion (hence the name), which is co-ordinated by three amino acid side-chains. These enzymes react by a different mechanism to the serine proteases, which we will not cover here.

Proteases can also hydrolyse substrates other than amino acids. One particularly useful protease is called penicillin G amidase (PGA). As shown in Figure 3.16, its natural function is to hydrolase penicillin G to 6-aminopenicillanate (6-APA). This is an important reaction as the 6-APA produced can be used as a starting material for the synthesis of synthetic penicillins. As shown in Figure 3.16, PGA can also

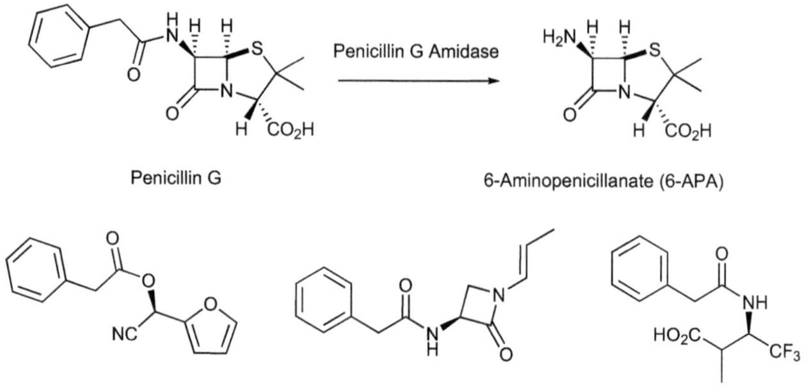

Figure 3.16 Kinetic resolution reactions catalysed by penicillin G amidase.

be used to carry out kinetic resolution reactions on a range of alcohols, amines and amino acids. In this case, the enantiomer shown is the one that is hydrolysed fastest.

The examples shown in Figure 3.16 illustrate an important point: proteases and amidases are also capable of hydrolysing esters as well. Notice that we are able to selectively hydrolyse the amide (or ester) in the presence of other functional groups, such as nitriles, and even in the presence of a lactam (a cyclic amide).

3.4 Hydrolysis of Epoxides

Epoxides are very useful building blocks for organic synthesis in part because they can be opened with a range of nucleophiles. An example of this process is shown in Figure 3.17, with the epoxide being opened by water to give a diol as the product. The epoxide opening is stereospecific because we can think of it as an S_N2 reaction (which occurs with inversion). This means that the nucleophile must attack from the face opposite to that which the epoxide is on. We can see this in Figure 3.17 where we chose the epoxide of cyclohexene as the substrate. If we draw the epoxide in a flattened chair form, then we can see that it is opened to give the diaxial product, putting the two alcohols *trans* to one another.

Because they are useful intermediates in organic synthesis, methods of preparing chiral epoxides are very valuable to chemists. There are several methods of carrying out this transformation, both chemical and enzymatic, which we will see later in Chapter 6. If we want to hydrolyse an epoxide using traditional chemistry, then we

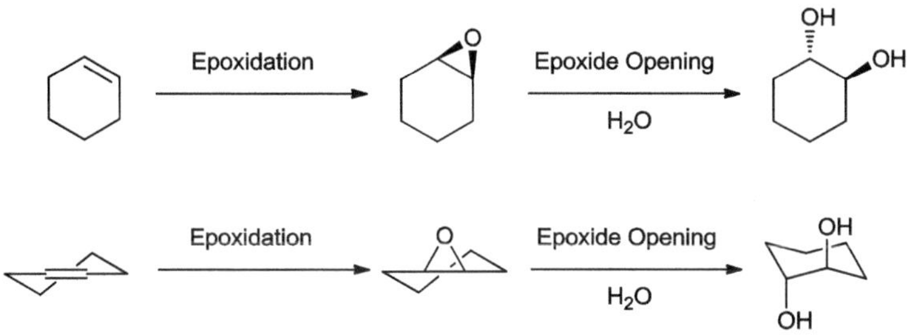

Figure 3.17 Epoxide opening to produce a *trans*-diol.

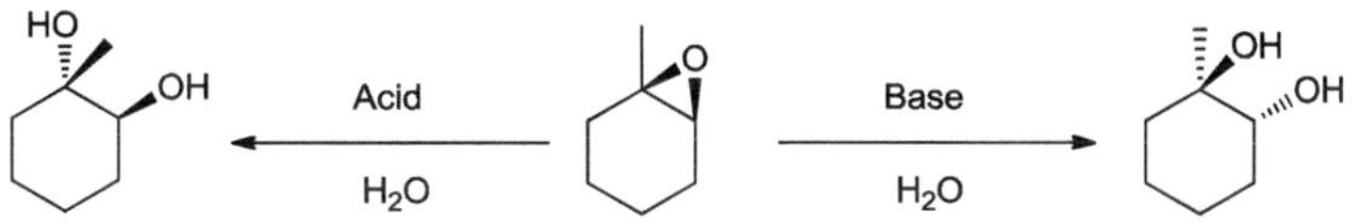

Figure 3.18 Regioselectivity of epoxide opening under acidic or basic conditions.

have two options: acidic or basic conditions. As we can see in Figure 3.18, the choice of conditions will affect which product is produced. Under acidic conditions we protonate the epoxide, creating a build-up of positive charge. In the substrate shown in Figure 3.18, the extra alkyl group at one end of the epoxide helps stabilise the developing positive charge at that end of the epoxide, making it more electropositive than the other epoxide carbon centre. As a result, the nucleophile (water) attacks at this end. However, under basic conditions, there is no charge build-up and now we have a purely S_N2 reaction such that steric hindrance is the controlling factor. Under these conditions, the less hindered and therefore less substituted end is attacked by the nucleophile (hydroxide). As with the previous hydrolysis reactions we have seen, the conditions for hydrolysis of epoxides may not be compatible with acid- or base-sensitive functional groups also present in our molecule.

The enzymes responsible for catalysing this transformation are called epoxide hydrolases. They are found in a wide variety of sources and are catalytically active in a range of solvents. The enzymatic hydrolysis of epoxides can proceed *via* attack at either carbon to give both products, with epoxide hydrolases typically being very regio- and stereoselective. As shown in Figure 3.19, we can take a racemic epoxide and carry out a kinetic resolution reaction, with the epoxide hydrolase reacting selectively with only one enantiomer. The most common form

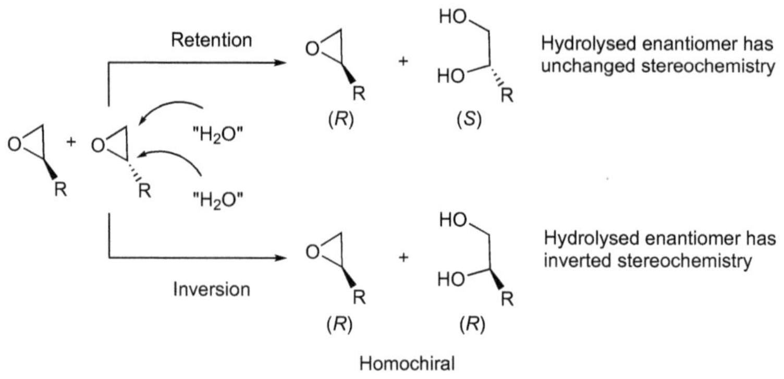

Figure 3.19　Kinetic resolution of epoxides using epoxide hydrolases.

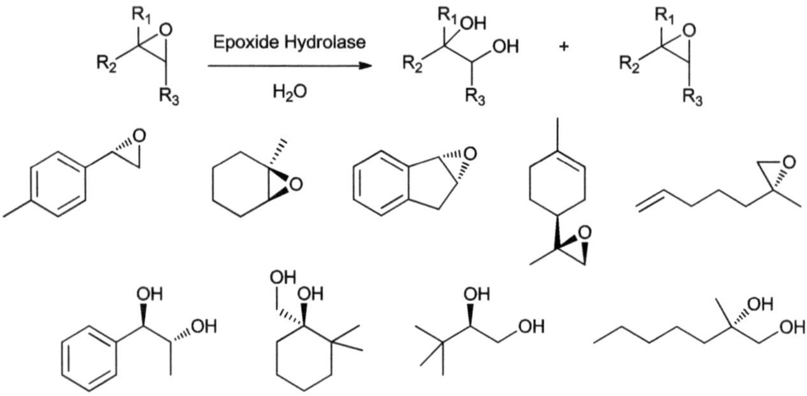

Figure 3.20　Epoxide and diol products from reactions using epoxide hydrolases.

of hydrolysis is at the less hindered end of the epoxide leading to retention of the configuration. However, attack is also possible at the more hindered end of the epoxide resulting in inversion of the configuration. This is useful as we now get two molecules that are homochiral (the same enantiomeric form) from a racemic mixture.

There are two possible methods chemists can use with these enzymes, as shown in Figure 3.20. If we want a chiral epoxide, then we can hydrolyse the unwanted enantiomer to leave behind the desired enantiomer of the epoxide. However, if we want the diol then we need to hydrolyse the desired enantiomer of the epoxide. In Figure 3.20, the desired product, whether a diol or epoxide, is shown to give you some idea of the substrate scope of these enzymes. In many cases, it is possible to use an alternative epoxide hydrolase to obtain the other enantiomer to that shown.

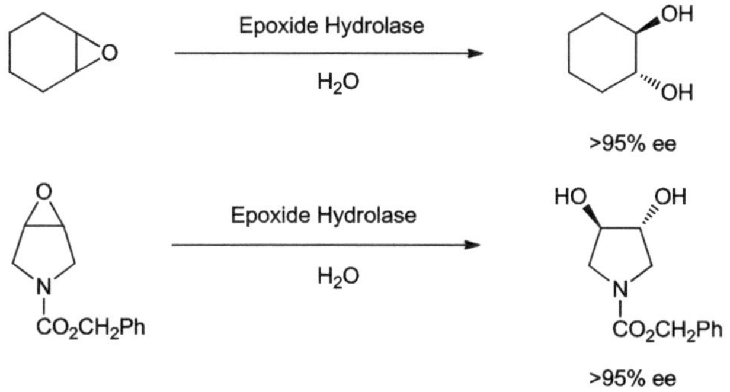

Figure 3.21 Desymmetrisation reactions of *meso*-epoxides using epoxide hydrolases.

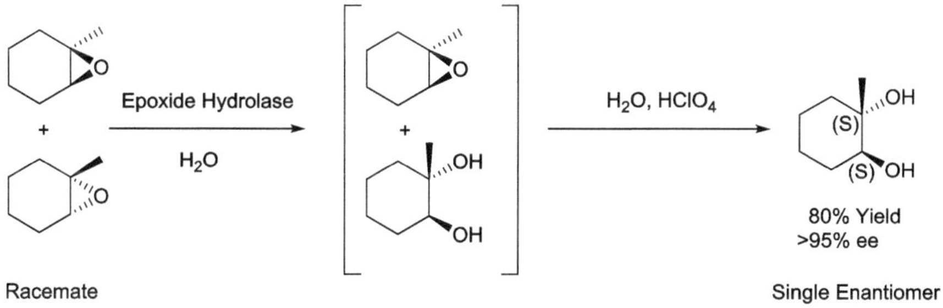

Figure 3.22 Enantioconvergent approach to a single diol enantiomer using an epoxide hydrolase.

As we saw previously with lipases, if we chose a *meso*-epoxide, then we can desymmetrise it to produce a greater than 50% yield of the desired diol. Two examples of this are shown in Figure 3.21.

We can also increase the yield of the desired enantiomer from a kinetic resolution reaction to above 50% without needing to use a *meso*-epoxide substrate. As shown in Figure 3.22, we can use an enantioconvergent approach to obtain a high yield of the desired compound. In Figure 3.22, one enantiomer of the epoxide is opened by the epoxide hydrolase from the least hindered end. At this point, if we are careful and select the right conditions, we can transform the remaining epoxide into the same enantiomer diol produced by the epoxide hydrolase. In the example in Figure 3.22, this is achieved *via* acid-catalysed opening of the remaining epoxide at the more hindered end. The racemic starting material has been transformed into a single enantiomer product.

Like the other hydrolase enzymes in this chapter, epoxide hydro-lases do not require a co-factor. The proposed mechanism of epoxide hydrolases is shown in Figure 3.23 for those who would like more

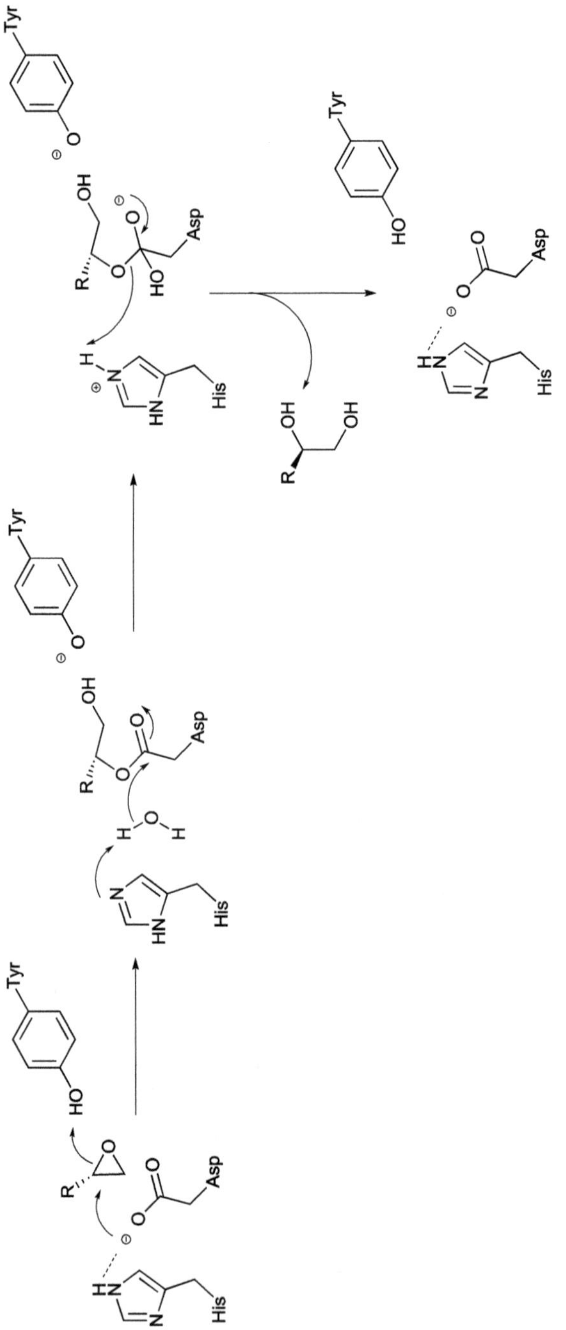

Figure 3.23 Mechanism of epoxide opening catalysed by epoxide hydrolases.

detail on how these enzymes work. The enzymes use a catalytic triad of histidine, aspartic acid and tyrosine. Once the substrate epoxide is bound, the oxygen atom of the epoxide is activated by hydrogen bonding to the tyrosine residue. This is analogous to the acid-catalysed chemical hydrolysis of epoxides. The epoxide is then attacked by the aspartic acid residue, which has been deprotonated by the histidine in order to make it more nucleophilic. This produces a hydroxyester intermediate in which the epoxide is covalently bound to the enzyme. The histidine then deprotonates a molecule of water present in the active site and this then attacks the carboxyl carbon of the hydroxyester to give a tetrahedral intermediate. When the tetrahedral intermediate collapses, the better leaving group (the diol) is released and the catalytic triad is regenerated for the next cycle.

3.5   Hydrolysis of Nitriles

Nitriles are useful synthetic intermediates as they can be thought of as an equivalent of a carboxylic acid but without the acidic proton of that functional group, which may cause problems as it is acidic. However, to convert a nitrile into a carboxylic acid using chemical methods requires high temperatures and either strongly acidic or basic conditions to which sensitive functional groups are not stable. In comparison, nitrile hydrolysing enzymes catalyse this reaction under much milder conditions. As we will also see below, nitrile hydrolysing enzymes often show good regioselectivity, which is an advantage over chemical methods. There are two different pathways *via* which nitrile hydrolysing enzymes are able to hydrolyse nitriles. The first class of enzymes, nitrilases, carry out the direct hydrolysis of a nitrile to the corresponding carboxylic acid as shown in Figure 3.24.

Nitrilases catalyse the hydrolysis of a range of nitriles as shown in Figure 3.25. Nitrilases show good stereo- and regioselectivity (they are able to hydrolyse one nitrile selectively if two different nitriles are present in the substrate) and some examples of this are also shown in Figure 3.25. For each of these examples, we show the product carboxylic acid that came from hydrolysis of the corresponding nitrile.

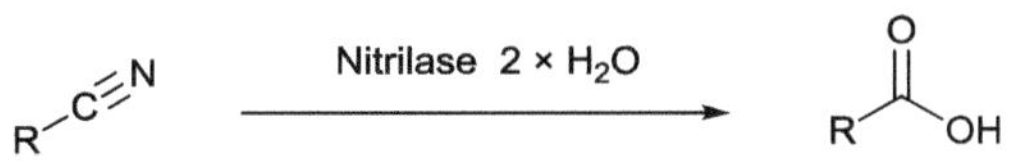

Figure 3.24   General transformation catalysed by nitrilases.

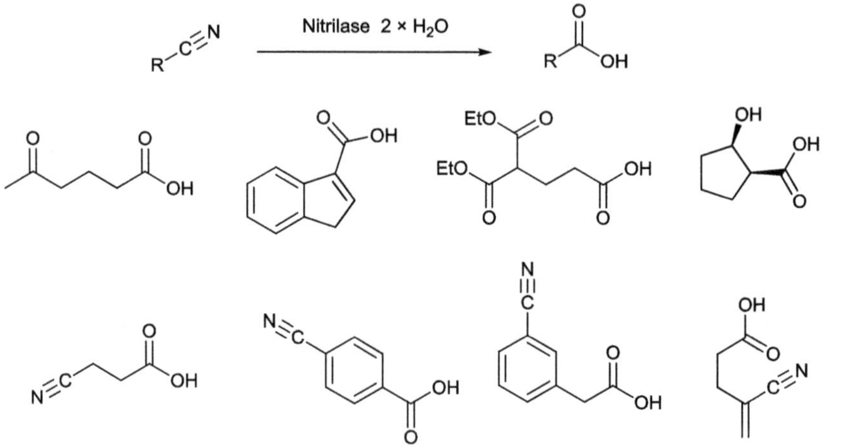

Figure 3.25 Substrate scope of nitriles that are hydrolysed by nitrilases.

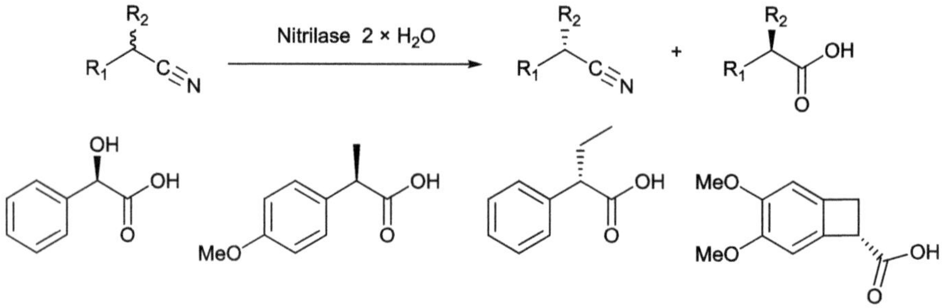

Figure 3.26 Kinetic resolution reactions catalysed by nitrilases.

Notice that we are able to hydrolyse nitriles in the presence of groups that would not usually be stable to acidic hydrolysis conditions, such as esters or alkenes. We can use a nitrilase to carry out the kinetic resolution of nitriles. In Figure 3.26, the faster reacting product enantiomer is shown, although by choosing a different nitrilase we can often make the opposite enantiomer instead.

As we have seen previously with other enzyme classes, we can also carry out the desymmetrisation of *meso*-substrates. Chemists have used this approach to devise a route towards a member of the statin family of drugs, Lipitor. Statins are used to treat high cholesterol levels and prevent heart disease. As shown in Figure 3.27, the chemists took a *meso*-dinitrile substrate and used a nitrilase to carry out a desymmetrisation reaction. The carboxylic acid produced was a single enantiomer and was isolated as the ammonium salt before being reacted further to make Lipitor.

Remember the non-steroidal anti-inflammatory drug naproxen from Figure 3.11? Chemists have also devised a route to this

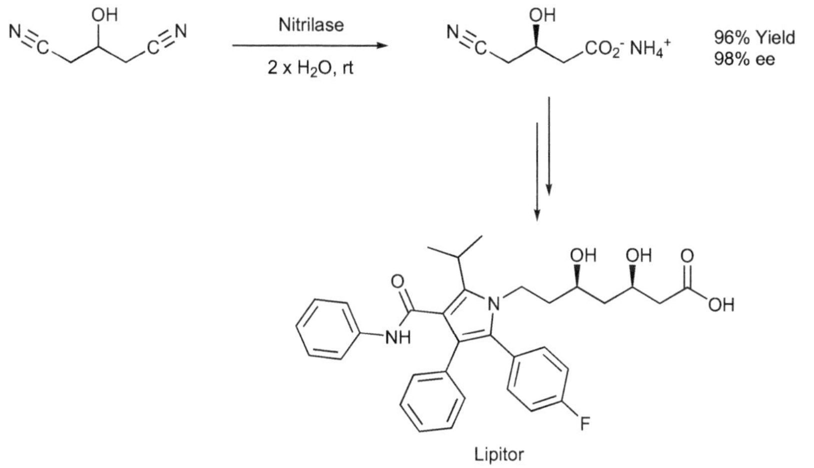

Figure 3.27 Desymmetrisation of a *meso*-dinitrile for the synthesis of Lipitor.

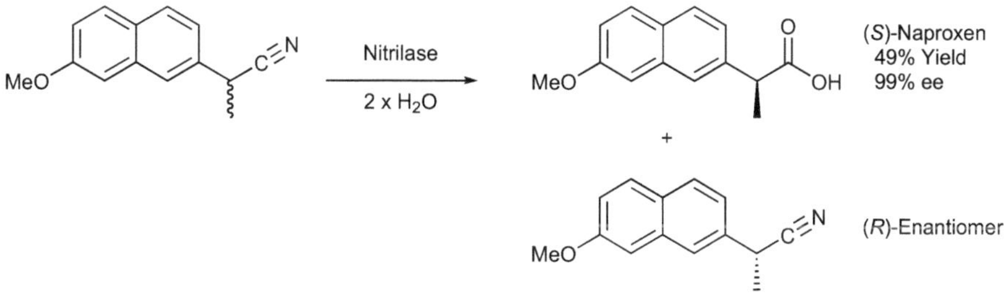

Figure 3.28 Kinetic resolution of a nitrile for the synthesis of (S)-naproxen.

compound using a nitrilase. As shown in Figure 3.28, (*S*)-naproxen can also be synthesised by kinetic resolution of a nitrile substrate.

Nitrilases are oligomeric proteins, typically dimers, and the mechanism is thought to proceed *via* a mechanism involving a catalytic triad of glutamic acid, lysine and cysteine. The conserved cysteine is the most important amino acid in the mechanism and studies have shown that replacing the cysteine with an alanine results in complete loss of activity. As with the other hydrolysis enzymes already discussed in this chapter, nitrilases do not require a co-factor. The mechanism of nitrile hydrolysis, shown in Figure 3.29, bears some resemblance to that used by lipases for ester hydrolysis. In the case of nitrilases, it is the cysteine residue of the triad that initially attacks the nitrile to give a thioimidate intermediate.

At this point the substrate is bound covalently to the enzyme in a very similar manner to the acyl enzyme complex formed in the

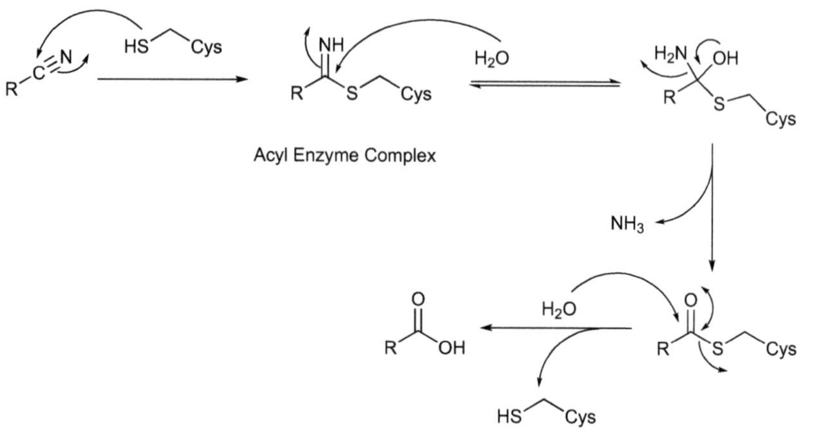

Figure 3.29 The mechanism of nitrile hydrolysis catalysed by nitrilases.

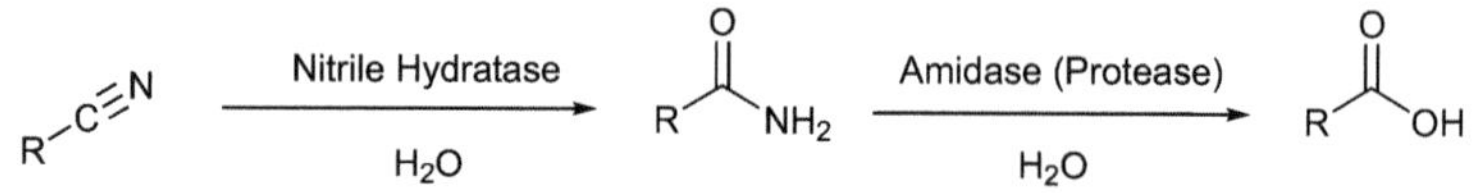

Figure 3.30 General transformation catalysed by nitrile hydratases.

lipase-catalysed hydrolysis of esters. In the hydrolysis of esters, however, it is a serine residue that attacks the ester carbonyl (see Figure 3.11). Water now attacks the acyl enzyme complex to give a tetrahedral intermediate, which then collapses to eliminate ammonia and give a thioester covalently bound to the enzyme. A second molecule of water is now able to attack the thioester, hydrolysing it to the product carboxylic acid and releasing the cysteine residue, which is now able to catalyse the hydrolysis of another nitrile molecule.

There is also a second class of enzymes that carry out the hydrolysis of nitriles. These are nitrile hydratases and they catalyse the hydration of the nitrile to give a primary amide. Using chemical methods, this transformation often requires high temperatures, oxidising reagents and the use of metal catalysts. As shown in Figure 3.30, the primary amide obtained from use of a nitrile hydratase can then be converted into the carboxylic acid by the use of an amidase (which we also saw above in Section 3.2).

As shown in Figure 3.31, a range of nitrile substrates can be converted into primary amides. For example, this process has been used on an industrial scale to make nicotinamide, a vitamin that is added to animal feed as a supplement.

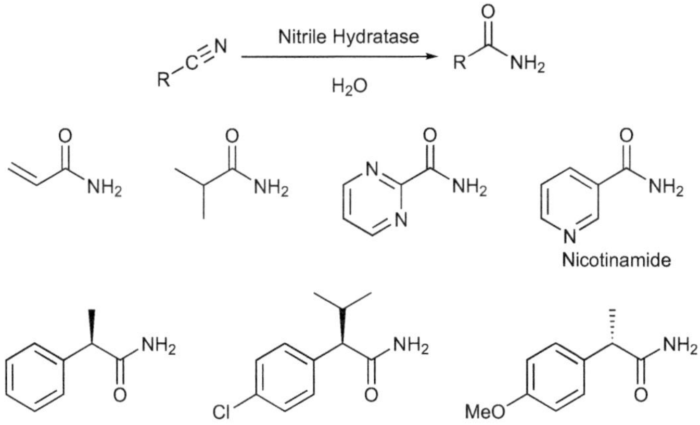

Figure 3.31 Substrate scope of nitriles hydrolysed by nitrile hydratases.

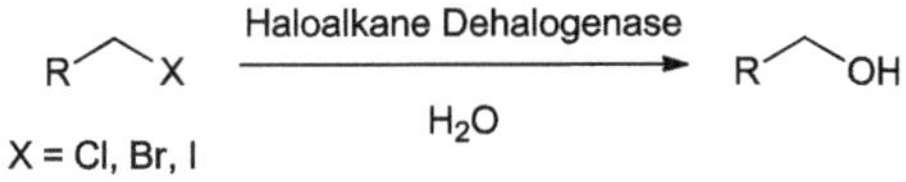

Figure 3.32 General transformation catalysed by haloalkane dehalogenases.

As shown in Figure 3.31, we can also use nitrile hydratases to carry out kinetic resolution of nitrile substrates. In each case above, the faster reacting enantiomer of the substrate is shown. The mechanism of nitrile hydrolysis to the corresponding primary amide with nitrile hydratases is more complex and involves the use of a non-heme iron centre. The mechanism will not be discussed here, but more advanced readers are encouraged to carry out further research.

3.6 Hydrolysis of Carbon–Halogen Bonds

The conversion of an alkyl halide to an alcohol would usually be achieved *via* a chemical substitution reaction with hydroxide as the nucleophile. The exact mechanism (S_N1 or S_N2) is dependent on the substitution pattern of the substrate. However, carrying out a substitution reaction under chemical conditions can result in racemisation if the halogenated substrate is chiral. As shown in Figure 3.32, there are also enzymes that are capable of hydrolysing carbon–halogen bonds, called haloalkane dehalogenases.

Enzymes capable of this transformation have been discovered in microbes living in soil contaminated with alkyl halides. These microbes

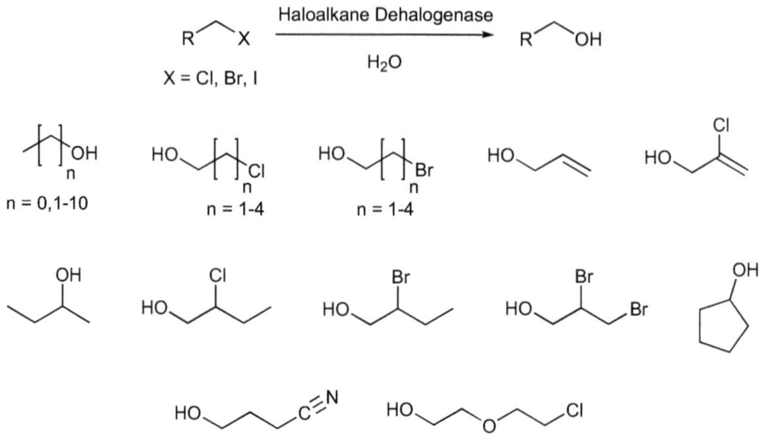

Figure 3.33 Substrate scope of alkyl halides hydrolysed by haloalkane dehalogenases.

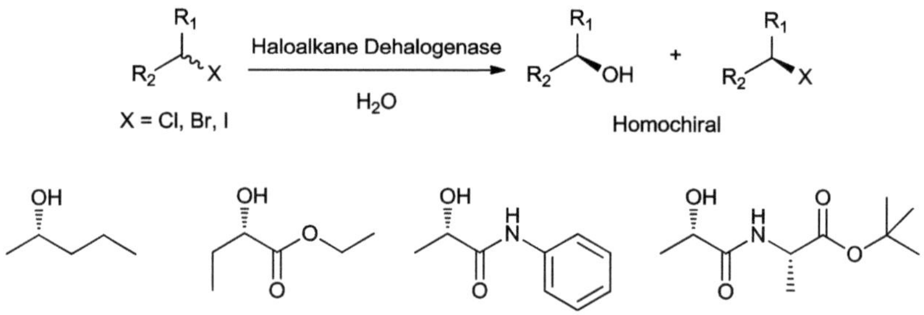

Figure 3.34 Kinetic resolution reactions catalysed by haloalkane dehalogenases.

use halogenated compounds as a carbon source for growth, requiring them to remove the halides from these compounds before utilising them further. Haloalkane dehalogenases have been reported to act on a wide variety of halogenated substrates, as shown in Figure 3.33, although only at sp^3 hybridised carbons. In each case, the product alcohol is shown.

Kinetic resolution reactions can be carried out with haloalkane dehalogenases, as shown in Figure 3.34. In this case, we end up with the product alcohol and the remaining haloalkane being homochiral (the same configuration). This is because the reaction proceeds with inversion of the configuration, just like an S_N2 reaction. So for each example in Figure 3.34, the hydroxy product shown came from the enantiomer of the haloalkane starting material with opposite stereochemistry.

It is also possible to use haloalkane dehalogenases to carry out desymmetrisation reactions if we use a *meso*-haloalkane as the

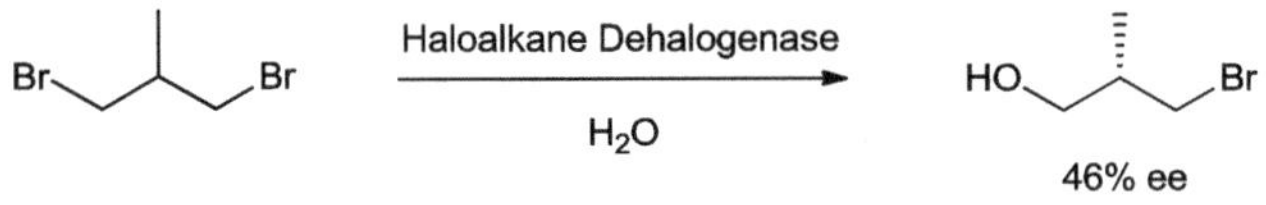

Figure 3.35 Desymmetrisation of a *meso*-dibromide catalysed by a haloalkane dehalogenase.

Ethyl (*S*)-2-hydroxypropionate

(*S*)-Lofexidine

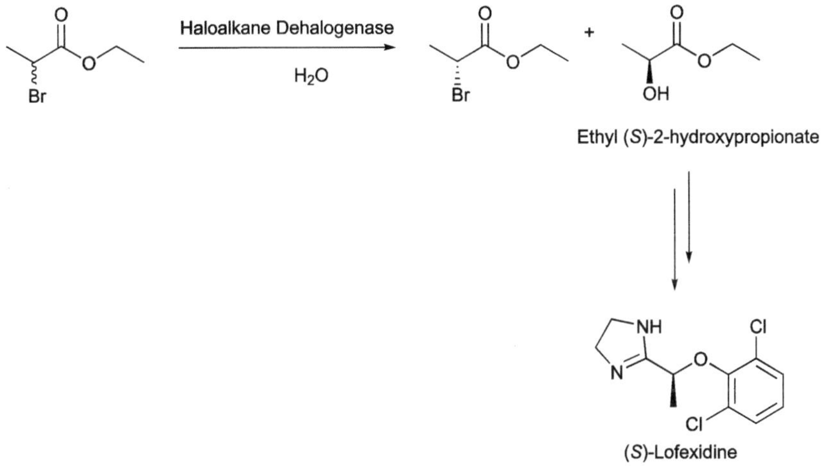

Figure 3.36 Use of a haloalkane dehalogenase in the synthesis of (*S*)-lofexidine.

substrate. Only a few examples of this are known, but one is shown in Figure 3.35. In this example, only one of the two bromides is converted into the corresponding alcohol to make a single enantiomer product, although the selectivity towards one bromide over the other is not particularly high.

Haloalkane dehalogenases are less widely used than the other enzymes discussed in this chapter, but chemists have shown that a haloalkane dehalogenase can be used in a route towards lofexidine, a drug used to treat hypertension (high blood pressure) and for the treatment of symptoms associated with opioid withdrawal. As shown in Figure 3.36, a haloalkane dehalogenase was used to carry out the kinetic resolution of ethyl 2-bromopropionate to produce ethyl (*S*)-2-hydroxypropionate. This substrate has been converted by other chemists into (*S*)-lofexidine, which is roughly 20 times more active than the (*R*)-enantiomer.

The mechanism of haloalkane dehalogenases is included in Figure 3.37. As with the other mechanisms, you don't need to know this, but it is included here for interest. The mechanism itself is similar

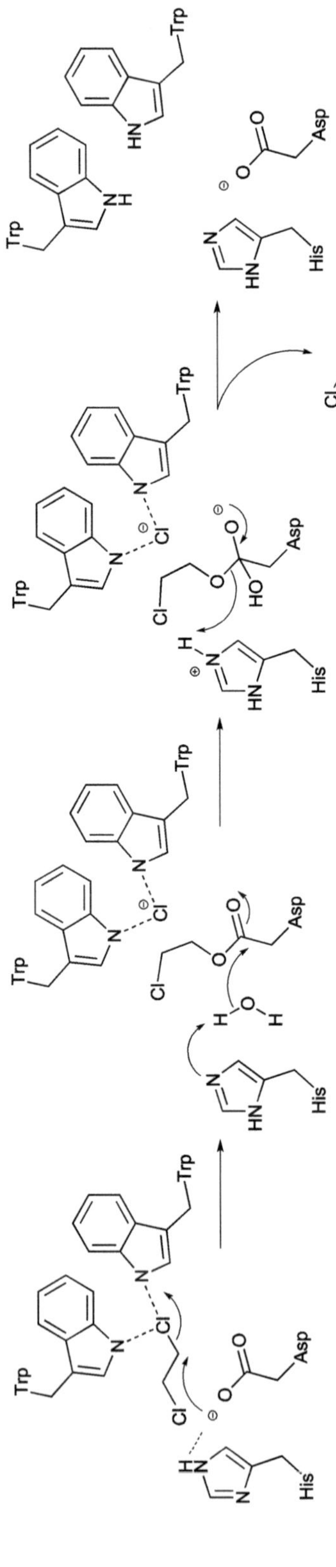

Figure 3.37   The mechanism of alkyl halide hydrolysis catalysed by haloalkane dehalogenases.

to that used by epoxide hydrolases in that an aspartic acid is the key residue. The catalytic triad is made up of aspartic acid, histidine and a second aspartic acid residue. In addition, the enzymes also use "halide stabilising" residues (typically residues such as tryptophan or glutamine) to help stabilise the halide leaving group. As shown in Figure 3.37, the histidine residue deprotonates the aspartic acid, which acts as the nucleophile in a substitution reaction to displace the halide of the starting material. As already discussed, the halide anion is then stabilised by hydrogen bonding to other residues in the active site (in Figure 3.37 we have used two tryptophan residues).

Having displaced the halide from the substrate, we now form an intermediate, which is covalently bound to the enzyme. The histidine then deprotonates a molecule of water in the active site, which attacks the carbonyl of the aspartic acid to form a tetrahedral intermediate. The tetrahedral intermediate then collapses, kicking out the alcohol (which is a better leaving group than the hydroxide) to give the product alcohol and at the same time regenerating the catalytic triad for the next cycle.

3.7 Hydrolysis of Sulfate Esters

Sulfates and their sulfate ester derivatives are found in both natural and manmade compounds. Dialkyl sulfate esters are used as highly reactive alkylating reagents; for example, dimethyl sulfate and cyclic sulfate esters are used in organic synthesis due to their reactivity. The corresponding sulfate monoesters are used less widely in synthesis, but are used as detergents, in particular sodium dodecyl sulfate, which is used in cleaning and cosmetic products. The hydrolysis of sulfate monoesters is catalysed by a family of enzymes called sulfatases. As shown in Figure 3.38, these enzymes can work in a stereocomplementary manner, producing the product alcohol with inversion or retention of the configuration.

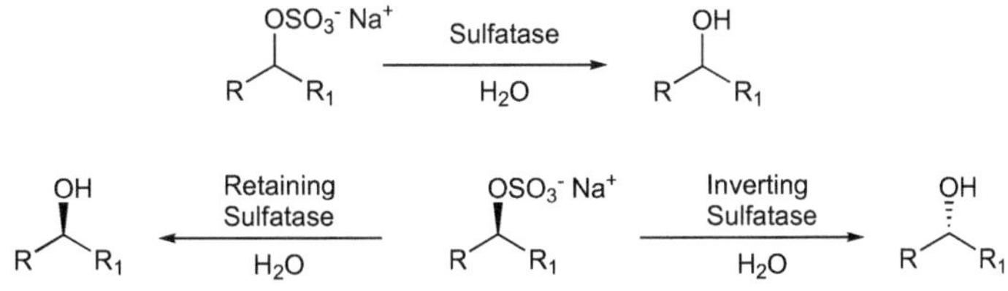

Figure 3.38 Stereocomplementary sulfatases produce either enantiomer of the product alcohol.

Chemical methods of sulfate monoester hydrolysis require the use of strong acids and are able to catalyse the reaction with retention of the configuration; however, they may be unsuitable for more sensitive functional groups present in the molecule. In contrast, under chemical conditions at high pH, the attack of a hydroxide nucleophile on a secondary centre bearing a sulfate is very slow, as sulfate is a poor leaving group. In contrast to chemical methods, sulfatases are able to carry out reactions with inversion or retention with high stereoselectivity under mild conditions without the need for protecting groups. As shown in Figure 3.39, a range of sulfate monoesters undergo reaction to give secondary alcohols. In each case, the substrate is shown.

Chemists have also been able to combine these two different modes of reaction in order to develop a process for the deracemisation of sulfate monoester substrates. Starting from a racemic mixture, one enzyme selectively carries out the reaction with inversion on one of the substrate enantiomers, whilst a second enzyme carries out the reaction with retention on the opposite substrate enantiomer. As shown in Figure 3.40, this process allows access to secondary alcohols in high levels of enantiomeric excess for a small number of substrates.

This reaction requires both enzymes to work under the same reaction conditions and with high stereoselectivity for the particular substrate. An alternative method is to use a chemoenzymatic

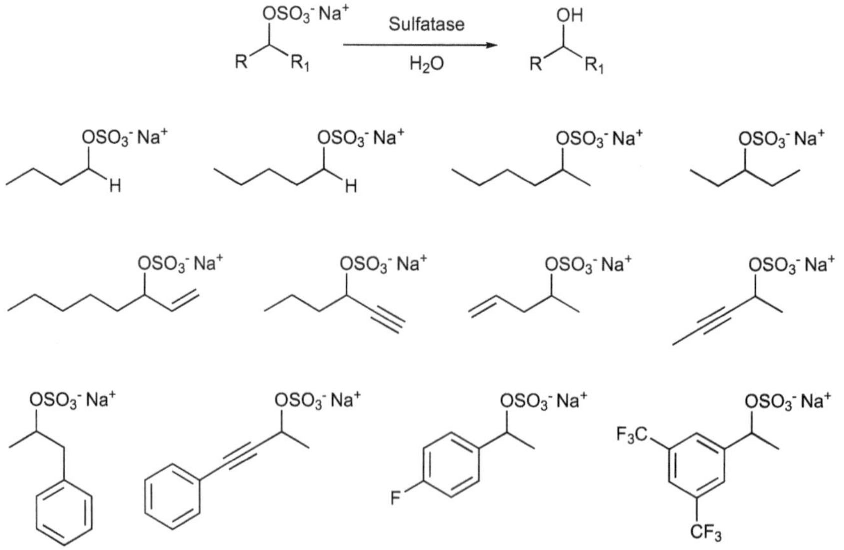

Figure 3.39 Sulfate monoester substrates hydrolysed by sulfatases.

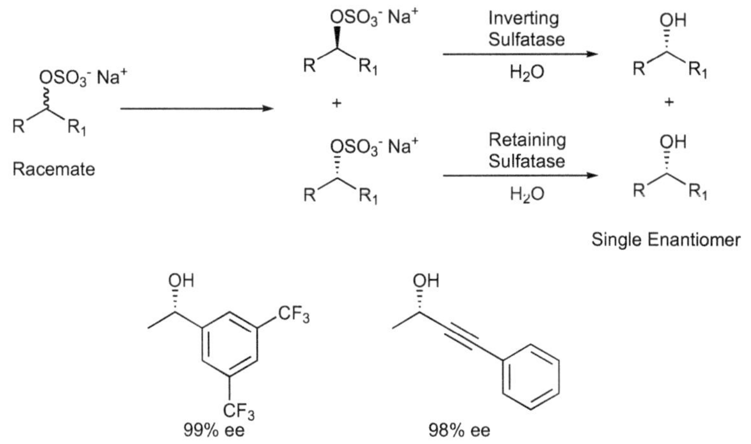

Figure 3.40   Deracemisation of sulfate monoesters to alcohols using a pair of sulfatases.

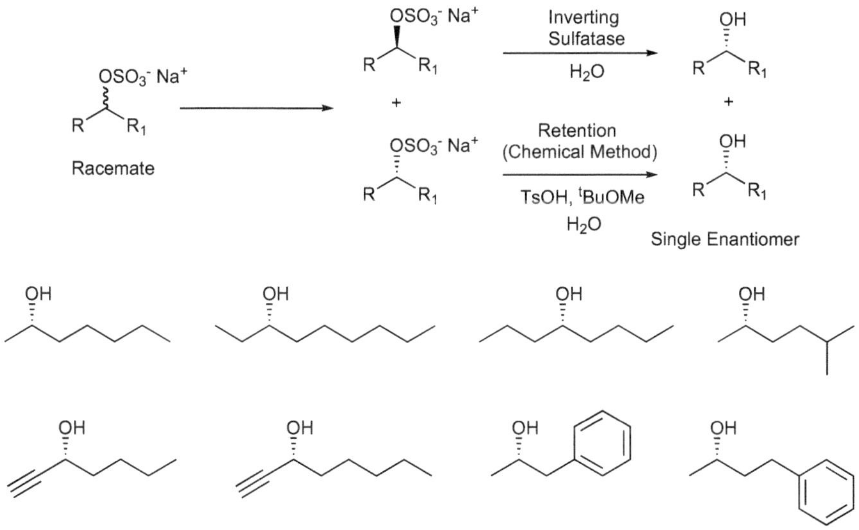

Figure 3.41   Chemoenzymatic deracemisation of sulfate monoesters using sulfatases.

approach with an enzyme carrying out the inversion part of the process before chemically hydrolysing the other enantiomer with retention. Although this is a stepwise process, it allows access to a much wider product scope. In each case, the product alcohol is shown in Figure 3.41, although the conversions may be low in some cases.

This process also has the potential to be very chemoselective, offering a method of producing secondary alcohols in the presence of functional groups that would react under other methods typically used to prepare chiral alcohols, such as the reduction of ketones.

In this chapter, we have seen a range of enzymes that are capable of hydrolysing carbon–heteroatom bonds. They show good stereo- and regioselectivity and have the advantage over chemical reagents of being able to catalyse these transformations under mild conditions. In the next chapter, we will look at some of these transformations in the opposite direction: "reverse hydrolysis".

4 Reverse Hydrolysis

4.1 Introduction

This chapter, like the previous one, covers enzymes that are known in the enzyme classification system as hydrolases. Members of this group of enzymes are given the enzyme classification number 3, followed by a sub-group number depending on which type of bond they hydrolyse. As we saw in Chapter 3, this class of enzymes are called hydrolases because they use water to break a carbon–heteroatom bond (C–X). However, because this reaction is reversible, the same enzymes can also catalyse the formation of a C–X bond, producing water as a by-product. We have already seen that hydrolases are important enzymes in organic synthesis due to their ability to selectively hydrolyse a variety of bonds. In addition, their ability to work in the opposite direction (reverse hydrolysis) makes them even more versatile. The transformations covered in more detail within this chapter are shown in Figure 4.1 and, as the title suggests, they are the reverse of those we saw in Chapter 3. The most widely used "reverse hydrolysis" or synthesis reactions are discussed in more detail in a separate section for each functional group.

4.2 Synthesis of Esters

As we saw in Chapter 3, the hydrolase enzymes responsible for the hydrolysis of esters are lipases and esterases. Therefore, it should come as no surprise that these same enzymes are capable of working in the reverse direction to synthesise esters. To do this, the enzyme

Biocatalysis in Organic Synthesis: The Retrosynthesis Approach
By Nicholas J. Turner and Luke Humphreys
© Nicholas J. Turner and Luke Humphreys 2018
Published by the Royal Society of Chemistry, www.rsc.org

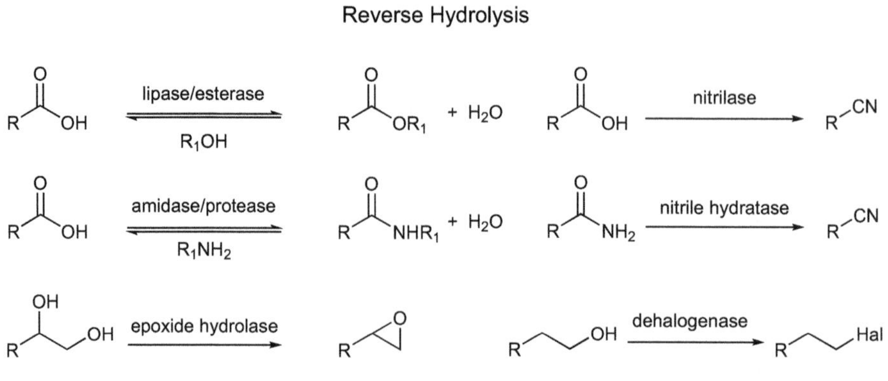

Reverse Hydrolysis

Figure 4.1   Overview of transformations presented in this chapter.

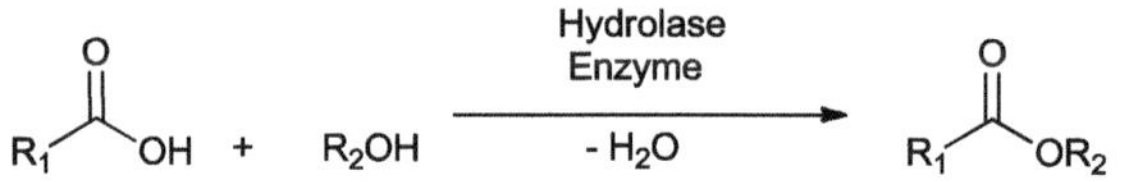

Figure 4.2   General transformation of ester synthesis catalysed by hydrolase enzymes.

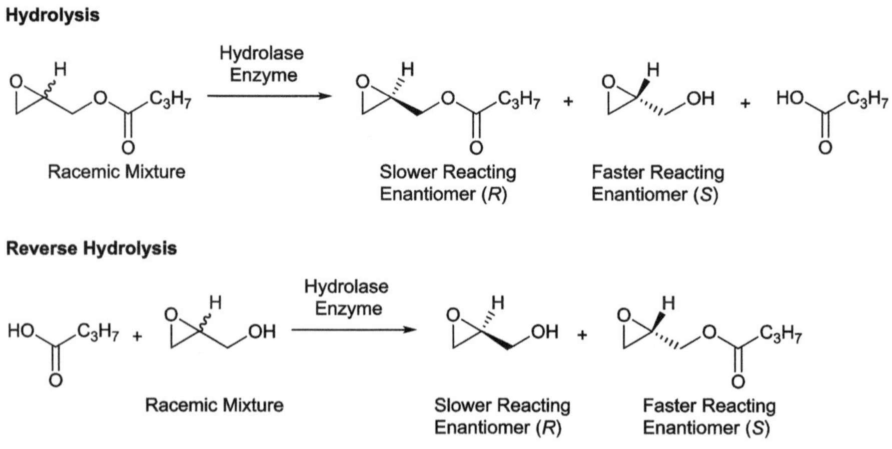

Hydrolysis

Racemic Mixture   Slower Reacting   Faster Reacting
                                    Enantiomer (R)   Enantiomer (S)

Reverse Hydrolysis

Racemic Mixture   Slower Reacting   Faster Reacting
                                    Enantiomer (R)   Enantiomer (S)

Figure 4.3   Kinetic resolution of an epoxide using a hydrolase enzyme in the reverse hydrolysis direction.

catalyses the reaction of an alcohol with a carboxylic acid to form an ester, as shown in Figure 4.2.

Hopefully you can remember the hydrolysis reaction of the epoxide in Chapter 3, Figure 3.3, which we used as an example of kinetic resolution. Let's imagine now hypothetically that we can run this reaction in reverse. The reverse hydrolysis reaction (and the original hydrolysis one) are shown in Figure 4.3.

In the hydrolysis direction, we start with a racemic mixture of the ester and the (S)-enantiomer reacts fastest, being hydrolysed to produce the (S)-alcohol and butanoic acid as a by-product. By comparison, in the reverse hydrolysis direction, we would have to start with butanoic acid (this is the substrate we wish to make the ester of) and a racemic mixture of the alcohol (which will become the alkoxyl part of the ester). If we were able to perform this reaction, we would now carry out a kinetic resolution reaction of the alcohol with one enantiomer (in this case the (S)-enantiomer) reacting faster than the other. The faster reacting enantiomer produces the corresponding ester, leaving behind the slower reacting enantiomer of the alcohol.

At this point, we also need to remember from Chapter 3 that if the substrate is chiral, then lipases prefer chirality or prochirality in the alcohol portion, whilst esterases prefer chirality or prochirality in the acid portion, as shown in Figure 4.4.

In theory, this means that if we have a chiral alcohol, then it would best to use a lipase to resolve it, but if we have a chiral acid, then an esterase would be preferable. We will see examples of both of these reactions next, although in practice both lipases and esterases are capable of resolving both types of chiral substrate. Many different types of substrates have been resolved using lipases, although perhaps the most well studied are secondary alcohols. In a similar manner to the previous chapter, we can use the Kazlauskas rule to predict which alcohol will react fastest. This rule is shown again in Figure 4.5.

As with all rules, there are some exceptions (for both substrates and enzymes), but in general it is a good starting point for assessing which enantiomer of a secondary alcohol will react fastest in a lipase-catalysed resolution. A wide range of alcohols have been resolved using a variety of lipase enzymes, some of which are shown in Figure 4.6. As with many of the other figures in this chapter, one lipase or esterase is unlikely to be active on all of the substrates shown. However, with many lipase or esterase enzymes to choose from, screening several enzymes is likely to identify one which works for the desired substrate. Esterases also catalyse this transformation,

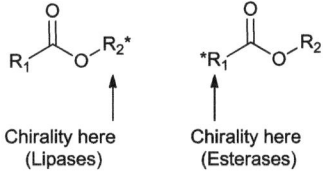

Figure 4.4 The substrate preference of lipases and esterases.

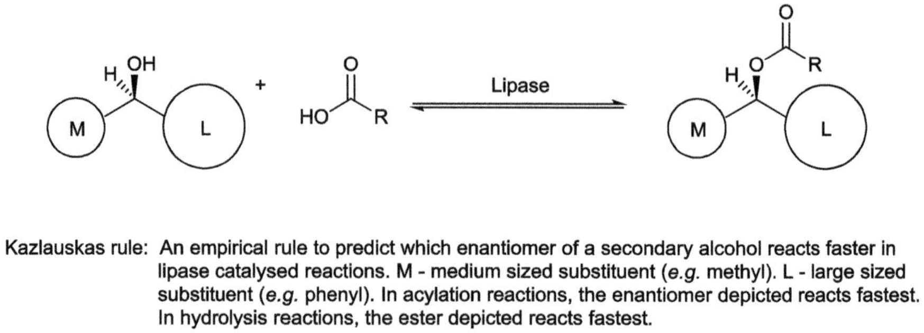

Kazlauskas rule: An empirical rule to predict which enantiomer of a secondary alcohol reacts faster in lipase catalysed reactions. M - medium sized substituent (*e.g.* methyl). L - large sized substituent (*e.g.* phenyl). In acylation reactions, the enantiomer depicted reacts fastest. In hydrolysis reactions, the ester depicted reacts fastest.

Figure 4.5 The Kazlauskas rule for predicting which enantiomer reacts fastest in a lipase-catalysed reaction.

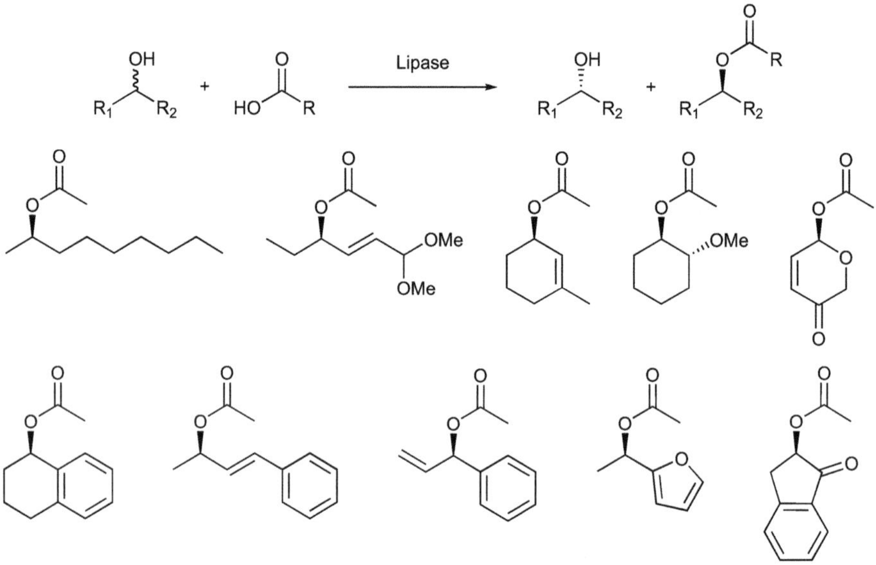

Figure 4.6 The substrate scope of esters produced using lipases.

but there are far fewer examples and remember, esterases often react fastest with the opposite enantiomer to lipases. In each case, the ester shown is formed from the faster reacting enantiomer of the alcohol.

As shown in Figure 4.6, a wide range of functional groups are compatible with the reaction conditions. As with the hydrolysis reaction, we can also carry out desymmetrisation reactions in the reverse hydrolysis direction if we use a *meso*-substrate. Some examples are shown in Figure 4.7, and in each case the product enantiomer is shown. These reactions are typically very stereoselective and high-yielding.

Primary alcohols can also be resolved using lipases and esterases, whereas for tertiary alcohols a much slower rate of reaction is observed

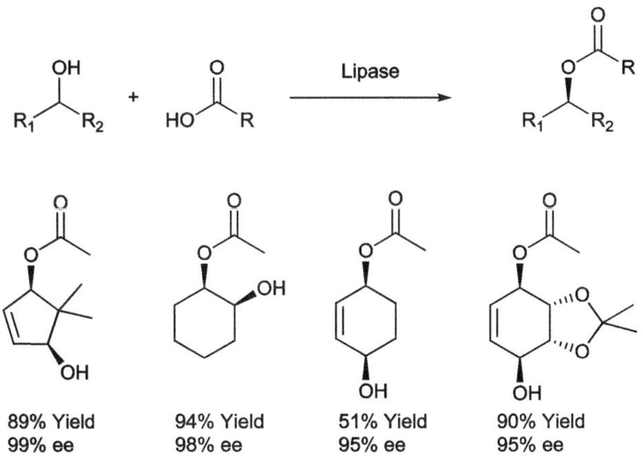

Figure 4.7 The desymmetrisation of *meso*-diols using lipases.

due to the extra steric hindrance present. Although an empirical rule for primary alcohols similar in that shown in Figure 4.5 exists for some lipases, it is not as reliable and is not dealt with here. Readers are encouraged, however, to seek it out if they are interested, as it can be useful in some cases. As might be expected, it is also possible to carry out lipase-catalysed desymmetrisation of primary alcohols that are *meso*-substrates. Lipases can also differentiate between the enantiomers of primary alcohols with an α-quaternary stereocentre. Examples for each of these substrate types are given in Figure 4.8 and, as previously, the enantiomer shown is the one that reacts fastest.

You will notice that in the previous three schemes, we only made acetate esters of the alcohol substrates. It is of course possible to make other esters depending on what we chose to react with our alcohol and what we would like our product to be. The compound that we react with the alcohol is called the acyl donor. Ideally, it should be inexpensive, lead to a fast, irreversible enzyme-catalysed reaction and be unreactive in the absence of the enzyme. In practice, it isn't possible to have all of these things, so chemists make their choice based on what is important for a specific reaction, as shown in Figure 4.9. For inexpensive chemicals where cost is the most important issue, then chemists pick simple esters, *e.g.* methyl esters. However, acylations with these acyl donors are slow and reversible, so to drive the reaction to completion chemists remove the water formed by distillation or with chemical drying agents such as molecular sieves. Another option is to crystallise the product under the reaction conditions in order to drive the equilibrium in the desired direction. For more high-value products, chemists use more

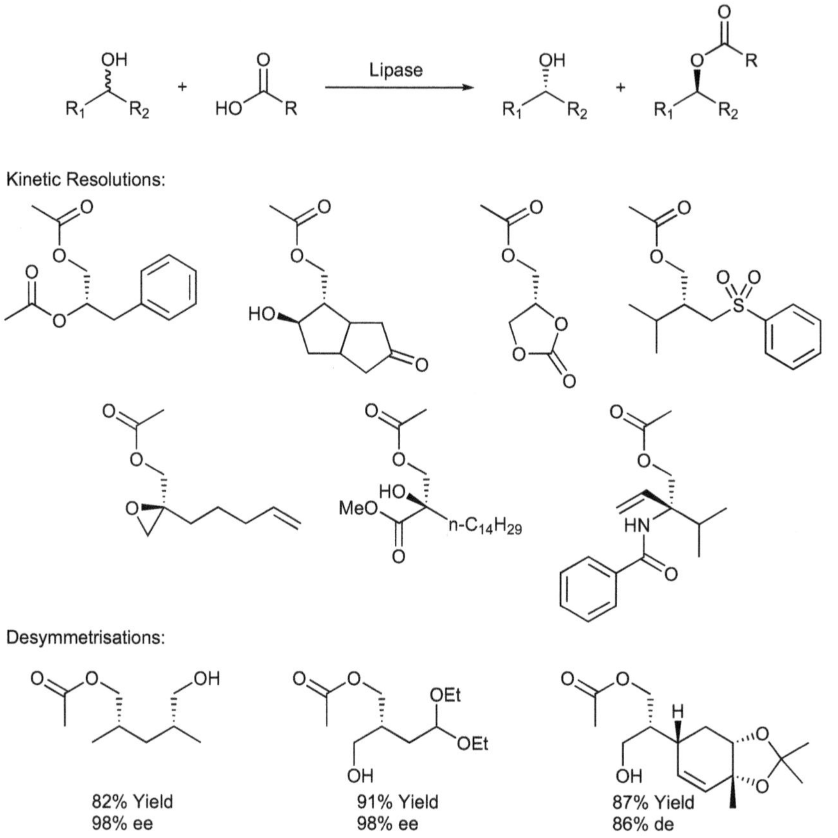

Figure 4.8 Further substrate scope of kinetic resolution and desymmetrisation reactions catalysed by lipases.

active acyl donors (such as vinyl acetate, as shown in Figure 4.9). These
allow much faster reactions and help move the equilibrium in favour of
the acylation reaction. In the case of an activated ester, chemists pick a
leaving group that is better than the alcohol they are reacting the acyl
donor with. This will ensure that the equilibrium will lie on the side of
the products. In the case of enol esters, the by-product of the acylation
reaction can tautomerise from the enol to keto form, again shifting
the equilibrium. In the case vinyl acetate, as shown in Figure 4.9, the
by-product formed (vinyl alcohol) tautomerises to form acetaldehyde,
which is very easily removed from the reaction.

So far, we have looked at substrates where the alcohol contains a
chiral centre. However, it is also possible that the acid portion of the
molecule could contain a chiral centre instead, or indeed that both
portions of the molecule could be chiral. As we might expect from

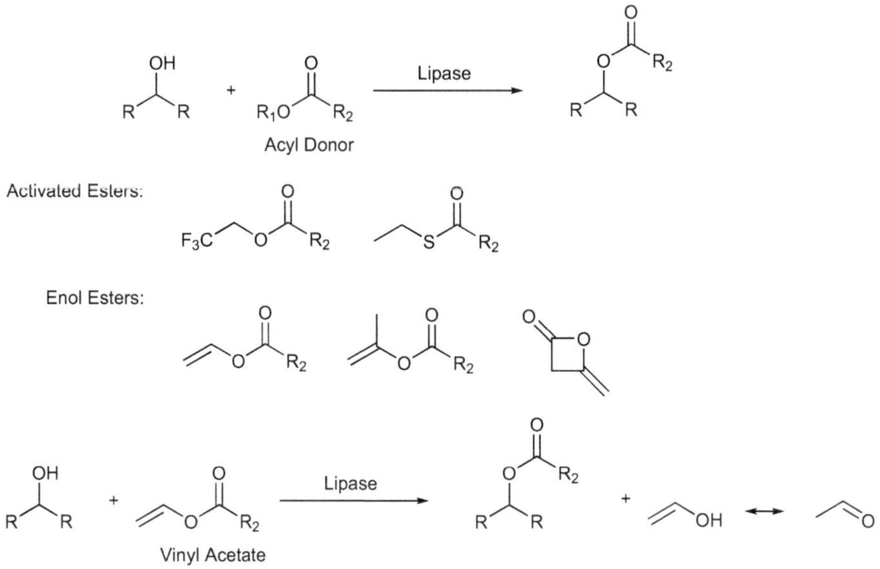

Figure 4.9 Acyl donors that may be used in lipase-catalysed acylation reactions.

Figure 4.4, some lipases tend to show less stereoselectivity with these substrates. In general though, both lipases and esterases are capable of resolving chiral carboxylic acids that have a stereocentre at the α or β position relative to the carboxylic acid. The substrate for these reactions can be the carboxylic acid or an ester, which then undergoes a transesterification reaction. In addition, esterases are also able to carry out the desymmetrisation of dicarboxylic acid substrates. For each of these reactions, a range of substrates is shown in Figure 4.10. In each case, the product of the reaction with the faster reacting enantiomer is shown. We can also see in Figure 4.10 that a range of different alcohols can be used in these reactions depending on which enzyme we chose. Selection of the appropriate alcohol may depend on factors such as the enzyme's preference, steric effects or how easy it is to remove the excess alcohol from the product at the end of the reaction.

In addition to resolving carboxylic acids, lipases and esterases are also capable of catalysing the formation of lactones and the ring opening of cyclic anhydrides. Lactones are important synthetic intermediates and are often found in flavour compounds. Rather than starting from the hydroxy acid as you might expect, to carry out these reactions we start from the hydroxy ester. This is because the hydroxy acid may be prone to spontaneous lactonisation without the involvement of the enzyme. As a result, these reactions are more

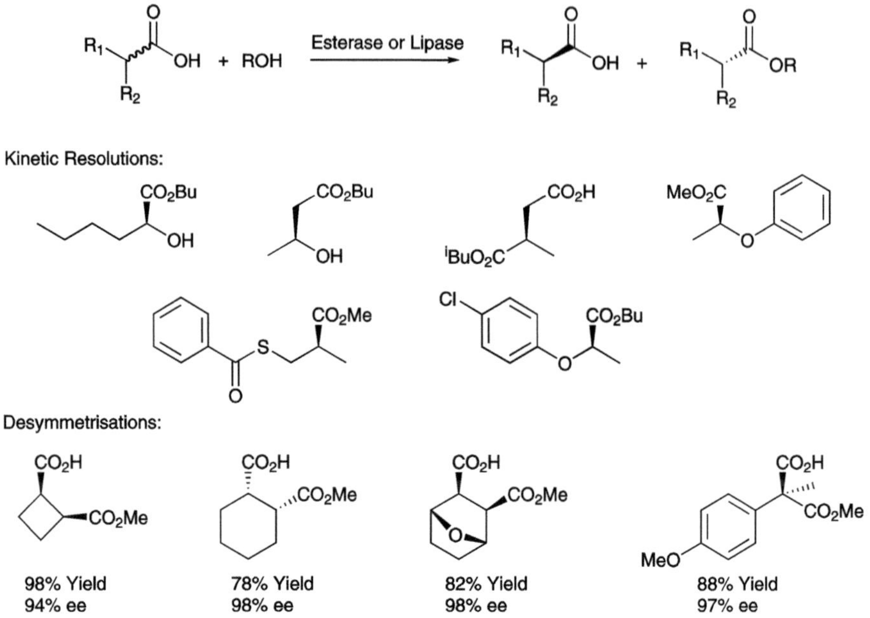

Figure 4.10 Kinetic resolution and desymmetrisation of carboxylic acids using lipases and esterases.

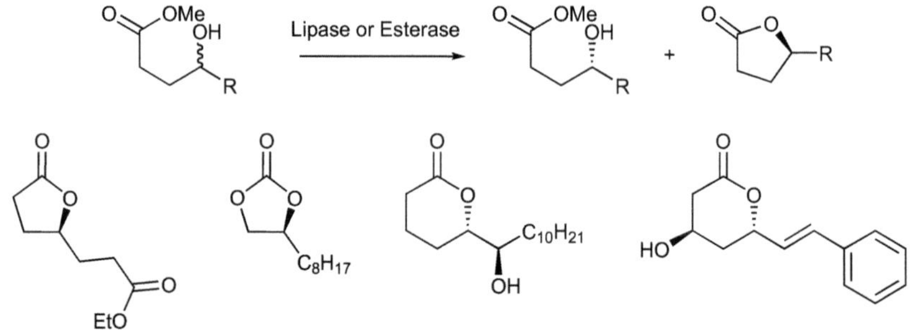

Figure 4.11 The synthesis of lactones catalysed by lipases and esterases.

like transesterification reactions. Some examples of this process are shown in Figure 4.11, with the enantiomer shown being the one that reacts fastest. Note that although we are resolving a secondary alcohol in these reactions, we can't use the Kazlauskas rule to predict which enantiomer will react fastest. This is because the conformational difference between acyclic esters and lactones affects how these substrates bind in the enzyme active site.

As mentioned above, lipases have also been shown to open cyclic anhydrides. Although the regioselectivity and enantioselectivity of

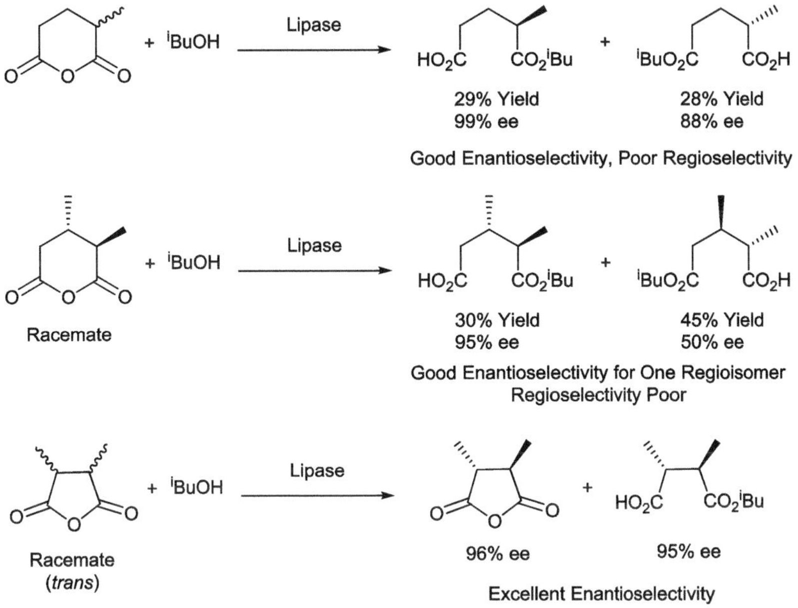

Figure 4.12 The ring opening of cyclic anhydrides catalysed by lipases.

these reactions can be good, enzymes tend to display either high regio- or enantioselectivity, rather than both. The substrate scope is not as wide as that for other reactions we have seen, with 5- and 6-membered cyclic anhydrides being most common, as shown in Figure 4.12.

At this point, we have only discussed the use of lipases and esterases for the synthesis of esters. However, this reaction can also be catalysed by proteases. Remember from Chapter 3 that proteases are enzymes that, in the hydrolysis direction, cleave amide bonds, but they can also make esters from an alcohol and an acid in the same way that lipases and esterases do.

We have now seen a variety of methods for making or breaking esters and how we can use these reactions to resolve a chiral alcohol, a chiral acid, or a chiral ester. As chemists, we can now evaluate all of these approaches when faced with a retrosynthetic challenge. Which option we chose will depend on the product we want, but also on which is the shortest route and the inherent selectivity of the enzymes we have in our toolbox. Let's consider again the drug propranolol, which we met in the last chapter. As shown in Figure 4.13, we can think of different ways of making the desired (S)-enantiomer.

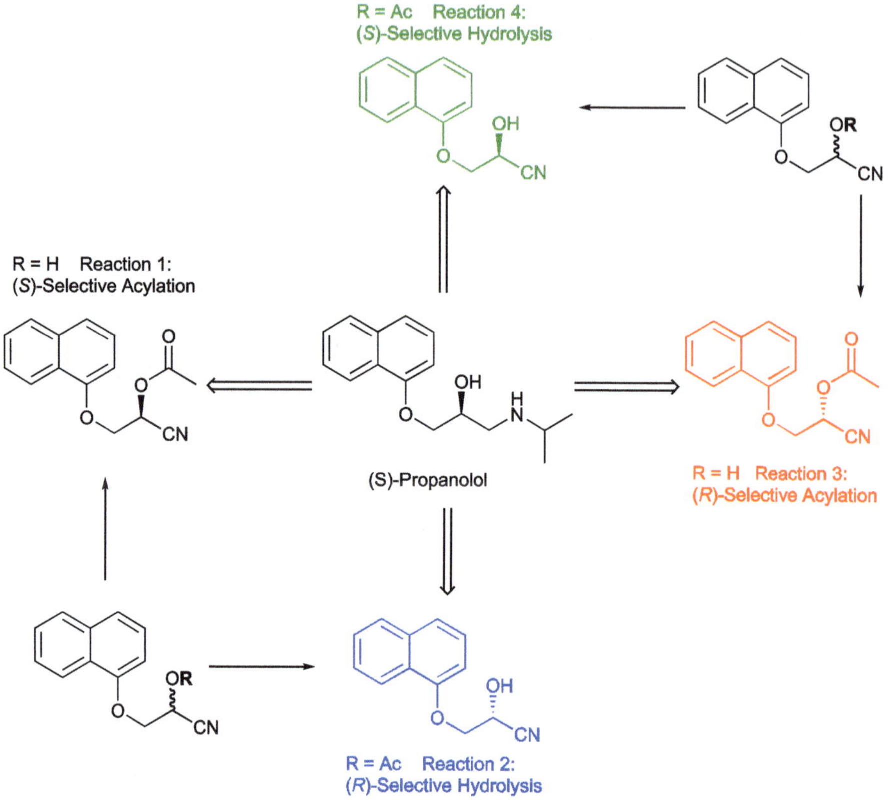

Figure 4.13 Retrosynthetic analysis of different lipase-catalysed approaches to (S)-propranolol.

If we disconnect propranolol back to the cyanohydrin compound we saw in the previous chapter, then we can potentially get back to four compounds: the two enantiomers of the alcohol and the two enantiomers of the ester. In theory, we could convert all of these compounds into (*S*)-propranolol, but some would require more steps than others so let's look at each individually. If we start from the racemic alcohol (R = H) in the bottom left of Figure 4.13, then we could carry out a kinetic resolution reaction using an (*S*)-selective lipase to give us the (*S*)-ester in black and the (*R*)-alcohol in blue as the other product. Alternatively, if we took the racemic ester and used an (*R*)-selective lipase to carry out a hydrolysis reaction, we would end up with the same products. Looking at those two products, the (*S*)-ester looks like it could be more straightforward to convert into (*S*)-propranolol as the stereocentre is already set in the correct configuration. However, at some point we would have to cleave the ester again to make the product, adding an extra step to the synthesis. Another alternative is to start from the racemic alcohol and use an (*R*)-selective acylation reaction to give the (*R*)-ester in red and the (*S*)-alcohol in green. We can also make these two materials by an (*S*)-selective hydrolysis of the racemic ester. Looking at these two routes, the (*S*)-alcohol in green looks like it should be the most straight forward to convert into the desired (*S*)-propranolol, but we still have to choose between the hydrolysis or acylation resolution in order to make it. This is where the difficulty of making either of the racemic starting materials or their stability might help us choose. For example, if we had to make the racemic ester from the alcohol before carrying out the kinetic resolution reaction, then this would add an extra step to the synthesis. Therefore, the shortest route might be to make the racemic alcohol and then use an (*R*)-selective acylation reaction to acylate the (*R*)-enantiomer. This in turn would leave behind the (*S*)-alcohol in green as the unacylated product. The (*S*)-alcohol has the correct stereocentre for (*S*)-propranolol and does not need to be deprotected. Being able to look at all possible options and then choose the best strategy will be covered in more detail later in Chapter 11.

Remember that the mechanism of these acylation reactions is the same as for the hydrolysis reactions but in reverse. The mechanism was given in detail in Chapter 3, Figure 3.11. To go in the acylation direction, we need to start at the top and go anti-clockwise rather than clockwise, which is what we did for the hydrolysis reactions. This is a useful exercise to help familiarise yourself with the mechanism.

4.3 Synthesis of Amides

Amides are a very common functional group found in peptides and proteins, natural products and pharmaceutical compounds. This makes methods of their synthesis very valuable. If we try to carry out this reaction chemically by just mixing together the carboxylic acid and amine, then we end up with no reaction as the amine deprotonates the carboxylic acid to give a salt. To overcome this, chemists have devised many ways of converting the hydroxyl group of the carboxylic acid into a better leaving group, such as using thionyl chloride to make the acid chloride or carbonyl diimidazole to make an imidazolide. Depending on the substrate, these approaches may racemise chiral centres present due to the reaction conditions and/or require the use of protecting groups. Additionally, some of these reagents also suffer from poor atom economy. By comparison, proteases or amidases operate under mild reaction conditions and therefore are able to catalyse the reaction between a carboxylic acid and amine without racemising chiral centres, such as those found in amino acids. In addition, enzyme-catalysed processes often have high enantioselectivity and rarely require the use of protecting groups. However, some issues still need to be overcome, and so chemists using proteases or amidases take one of two approaches to these reactions, as shown below. Hydrolysis of peptides is favoured due to the formation of the carboxylate salt and its solvation in water. Therefore, to overcome this, chemists shift the equilibrium by running the reactions in an organic solvent, reducing the concentration of water and lowering salt formation of the two starting materials (as they cannot be as efficiently solvated). This is called "thermodynamic control" and is illustrated in Figure 4.14.

Chemists have also made use of an alternative approach that enables esters to be reacted with amines. This method only works if we use a serine protease (or lipase as we will see below), so metalloproteases will not work with this approach. The method uses an ester as the starting substrate, which initially reacts with the serine residue of the protein as per the mechanism in the previous chapter to give an acyl-enzyme intermediate. The acyl-enzyme intermediate can now undergo one of two reactions, as shown in Figure 4.15. In the desired pathway, the amine reacts with the acyl-enzyme intermediate to make the product amide. However, water present can also react with the acyl-enzyme intermediate to give the carboxylic acid. The ratio of these two processes (synthesis *versus* hydrolysis) can be an important

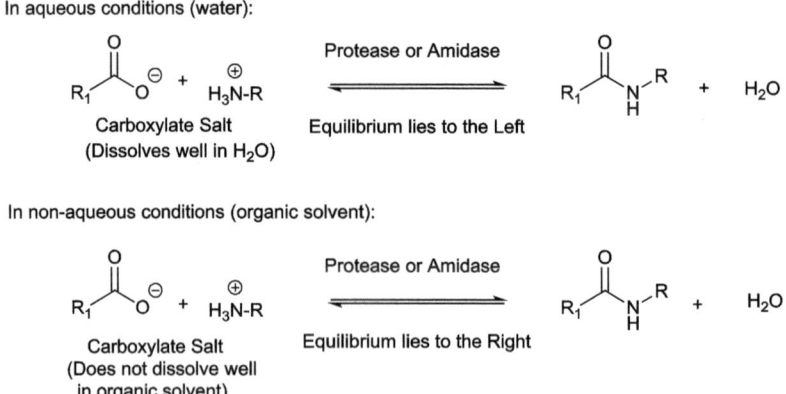

Figure 4.14   Using thermodynamic control to favour amide formation.

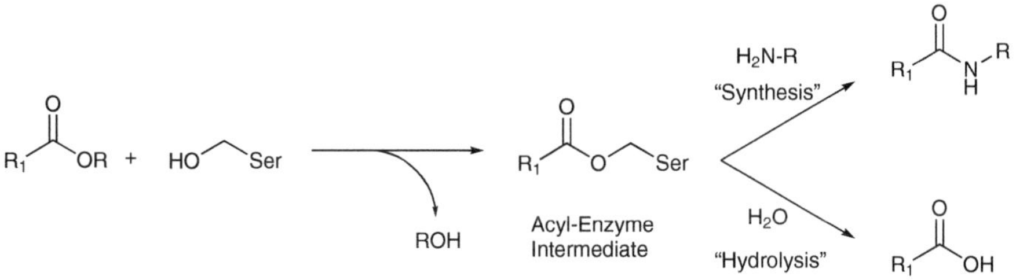

Figure 4.15 Using kinetic control to favour amide formation.

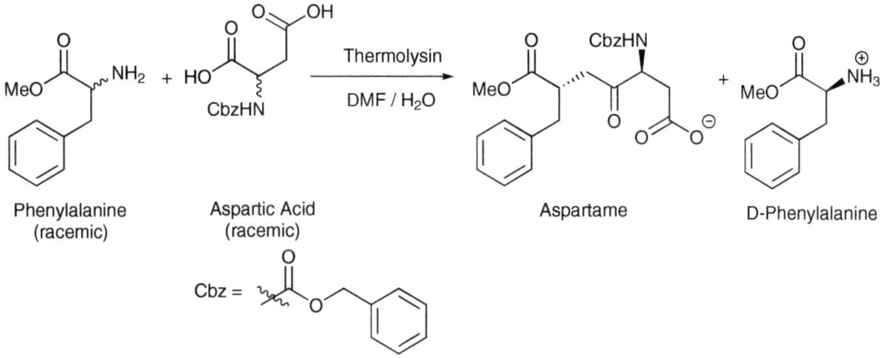

Figure 4.16 Using thermodynamic control for the thermolysin-catalysed synthesis of aspartame.

factor in choosing which enzyme to carry out the reaction with a particular substrate. This approach is called "kinetic control" and is more commonly used than "thermodynamic control" shown above.

Proteases and amidases have mostly been used to make peptide bonds using natural L-amino acids, unnatural D-amino acids and non-amino acid substrates. They usually show the highest selectivities towards the acyl part of an ester or amide (like esterases). One of the most well-known examples is the industrial production of aspartame, an artificial sweetener. As shown in Figure 4.16, a protease (thermolysin) catalyses the reaction of a protected aspartic acid and phenylalanine methyl ester to make the product. Both starting materials are racemic with the phenylalanine used in excess, the reason for which we shall see shortly. The protease enzyme is very selective, with only the L-enantiomer being coupled. In addition, the α-carboxylic acid of the aspartic acid is selectively coupled in the presence of the β-carboxylic acid. This reaction is run under thermodynamic control because the aspartame produced forms a salt with phenylalanine methyl ester, which is insoluble and precipitates from the reaction mixture, driving the equilibrium in the synthesis direction. This is why an excess of the phenylalanine methyl ester is required.

Acylases and amidases are also able to catalyse the formation of amide bonds. Penicillin *G*-acylase (PGA) in particular has been used to make a number of semi-synthetic penicillin and cephalosporin antibiotics, as shown in Figure 4.17.

As we have already mentioned, if we use a kinetic control approach, then we should be able to use lipases to make amide bonds as well. A range of benzyl esters can be reacted with a variety of amines to make amides, as shown in Figure 4.18. The reaction takes place in iso-propyl ether and is catalysed by a lipase from *Pseudomonas cepacia*. This is marketed as "Amano PS-30" and in each example in Figure 4.18, the product amide is shown.

This is because lipases use a catalytic serine residue in the same way that serine proteases do. Chemists have used this reactivity of lipases to catalyse the kinetic resolution of a range of racemic amines, as shown in Figure 4.19. In each case, the slower reacting amine is shown.

The empirical rule for determining which enantiomer of an alcohol reacts fastest (Figure 4.5) can also be applied to amine substrates. However, a word of caution is required here as proteases (such as subtilisin) often show the opposite selectivity to lipases!

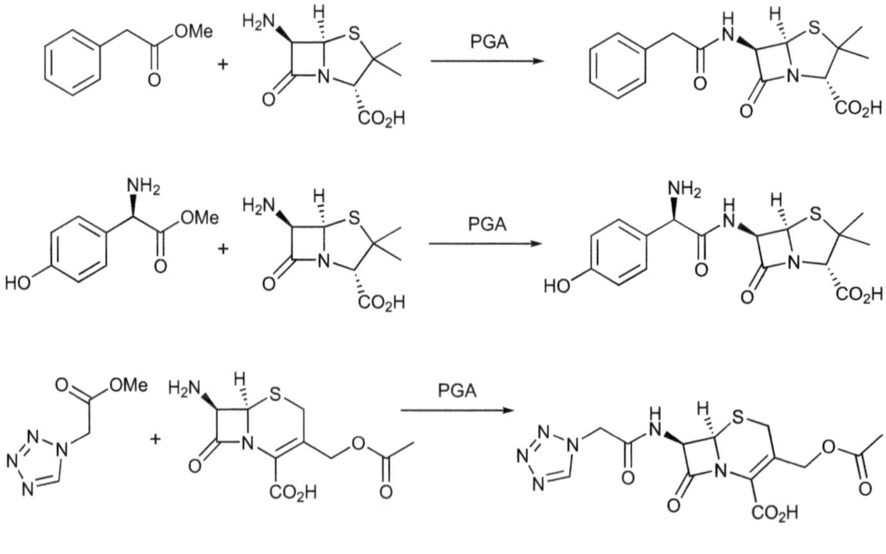

Figure 4.17 The synthesis of amides catalysed by penicillin G acylase.

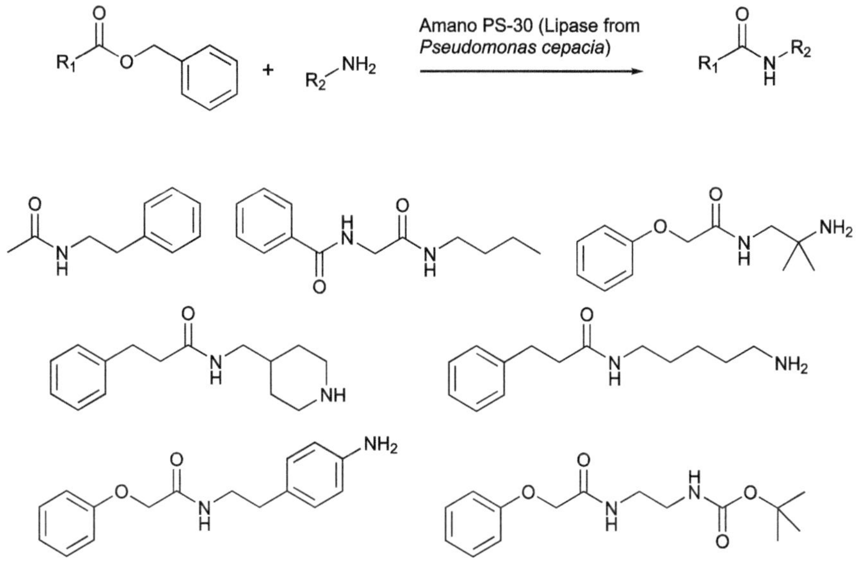

Figure 4.18 Substrate scope of amide bond synthesis catalysed by Amano PS-30 lipase.

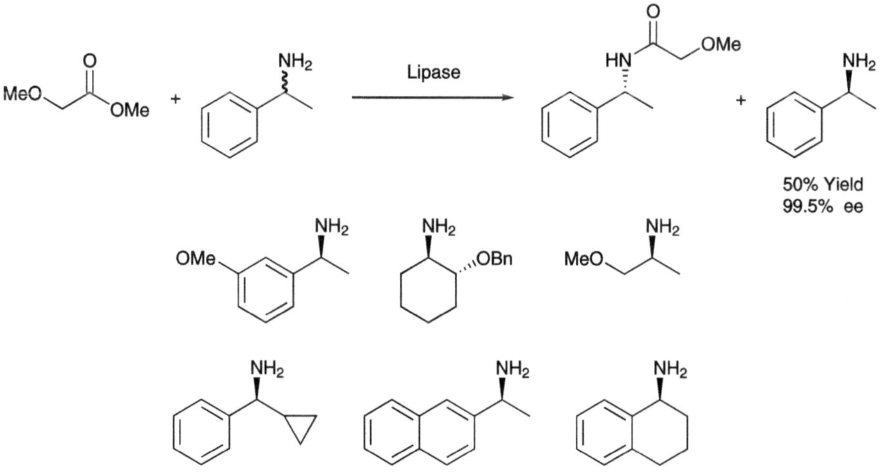

Figure 4.19 The kinetic resolution of amines catalysed by lipases.

In addition to stereoselectivity, hydrolases also show regio-selectivity in amide bond forming reactions. This can occur when we have two esters that are not equivalent and one is reacted in the presence of the other. As shown in Figure 4.20, chemists at AstraZeneca wanted to make a seven-membered lactam. To do this,

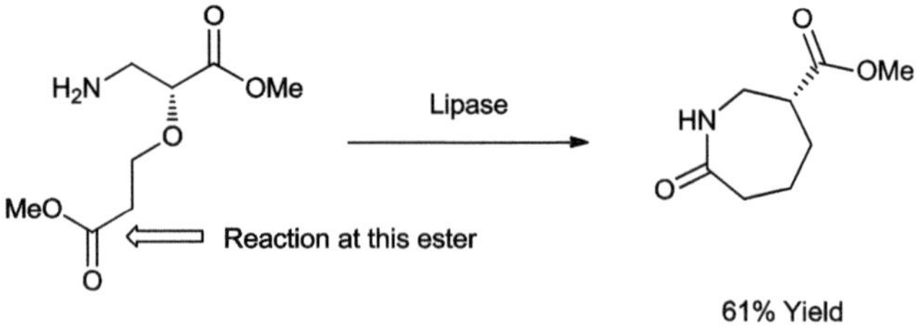

Figure 4.20　Regioselective lactam formation catalysed by a lipase.

they took a diester starting material and used a lipase to make the lactam from only one of the two esters present.

4.4　Reverse Hydrolysis Using other Hydrolase Enzymes

Although we included the reverse hydrolysis reactions for epoxide hydrolases, nitrilases and haloalkane dehalogenases in Figure 4.1 of this chapter, we will not cover these transformations in detail here. Epoxide hydrolases are not typically used to make epoxides, as to make a chiral epoxide we would normally require the chiral diol, which may be difficult to selectively synthesise. The most common way of obtaining a chiral epoxide is to carry out kinetic resolution of a racemic epoxide. There are also other enzymes capable of carrying out epoxidations to make chiral epoxides, as we will see later. Nitrilases have not been reported to carry out the reverse hydrolysis process shown in Figure 4.1. Very recently, an enzyme has been discovered that catalyses the conversion of a carboxylic acid into a nitrile, but its mechanism is different to that of nitrilase enzymes and so far, its substrate range is too narrow to be of synthetic use. As with nitrilases, haloalkane dehalogenase enzymes have not been reported to work in the reverse direction shown in Figure 4.1. However, there are other enzymes that are capable of adding halogens to substrates, which we will see in Chapter 7.

5 Reduction

5.1 Introduction

This chapter will cover a range of different reduction reactions that are catalysed by several different types of enzymes. Despite catalysing different transformations, these enzymes are all from the oxidoreductase group within the enzyme classification system. Oxidoreductases are given the enzyme classification number 1 and are defined as enzymes that catalyse oxidation/reduction reactions *via* the transfer of H and O atoms or electrons from one substance to another. Reductions are an important reaction class for the synthetic chemist. If a particular functional group is sensitive to certain reaction conditions, then we may be able to carry it through the synthesis at a higher oxidation level until it is required. A good example of this is an aldehyde, as shown in Figure 5.1. We need to protect the aldehyde due to its unstable nature, but rather than use a protecting group, we can use an oxidised aldehyde instead: the ester. This way we bring the aldehyde through our synthesis until we need to react it, such as if we need to carry out a reductive amination. At this point, we see the benefit of reduction reactions as we can reduce the ester to the aldehyde ready for the desired reaction.

The transformations covered within this chapter are shown in Figure 5.2. Each transformation and its synthetic use are discussed in more detail in a separate section.

Biocatalysis in Organic Synthesis: The Retrosynthesis Approach
By Nicholas J. Turner and Luke Humphreys
© Nicholas J. Turner and Luke Humphreys 2018
Published by the Royal Society of Chemistry, www.rsc.org

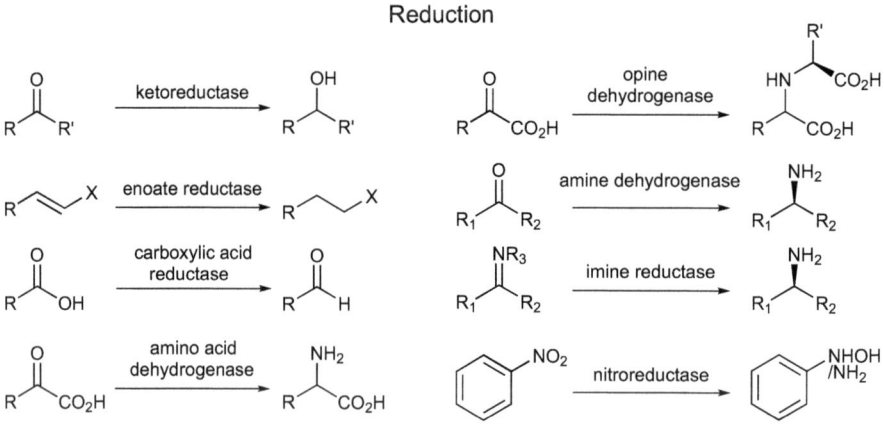

Figure 5.1 Use of a higher oxidation state to protect a reactive functional group.

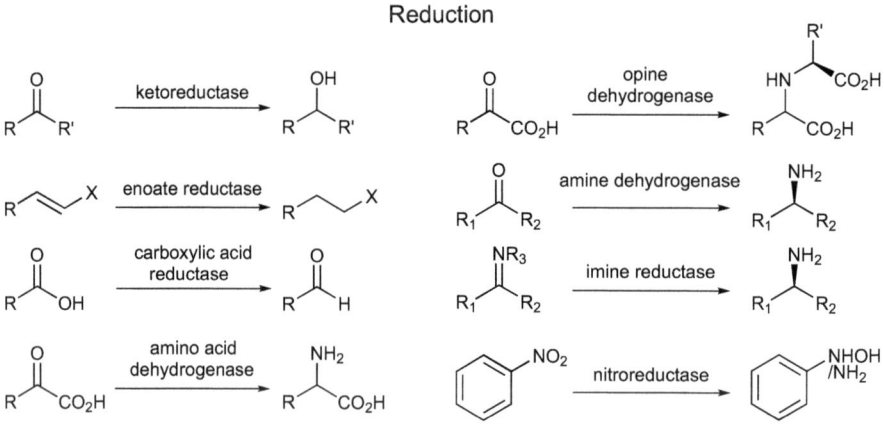

Figure 5.2 Overview of the transformations presented in this chapter.

5.2 Reduction of Ketones and Aldehydes

Ketones and aldehydes are widespread in Nature, often in combination with other functional groups. The oxygen atom of the carbonyl group in ketones and aldehydes is capable of forming hydrogen bond interactions with other functional groups and the carbonyl itself is a useful handle that can be interconverted into a range of other functional groups. Ketones and aldehydes are one oxidation level above alcohols without containing the acidic proton of the alcohol. Therefore, they can be used as a substitute for an alcohol and reduced to the alcohol at an appropriate point in a synthesis, as shown in Figure 5.3. The reduction of unsymmetrical ketones has the potential to produce chiral alcohols, which are valuable intermediates and building blocks

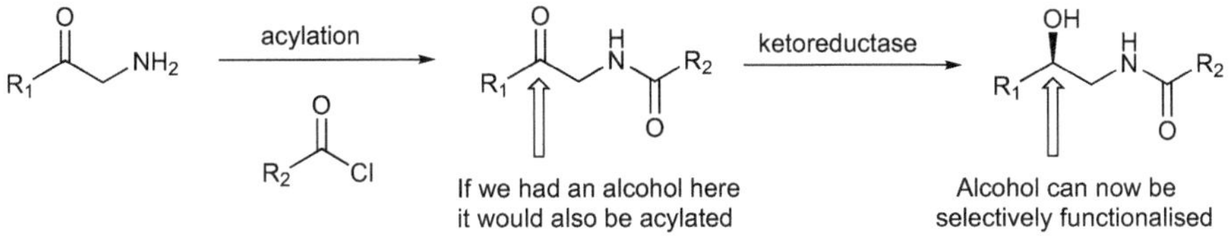

Figure 5.3 Reduction using a ketoreductase.

in synthesis. One method of reducing ketones and aldehydes is to use a ketoreductase (Kred) enzyme, as shown in Figure 5.3.

There are several other chemical methods for the reduction of ketones and aldehydes, such as traditional metal hydride reducing agents like sodium borohydride and lithium aluminium hydride. However, these reagents can be difficult to use on a large scale due to the danger of flammable hydrogen gas being generated and the production of metal waste that must be disposed of. These reagents are also achiral and therefore produce racemic products. Chiral reduction can be carried out with borane using a chiral catalyst named after its inventors: the Corey–Bakshi–Shibata (CBS) catalyst. This system is highly enantioselective, but requires the synthesis of the ligand and care needs to be taken not to produce hydrogen gas. Another alternative is the use of metal-catalysed hydrogenation with a chiral ligand. The enantioselectivities are high but the cost of the metal and ligand need to be taken into account. These and other chemical methods for the reduction of ketones are shown in Figure 5.4.

In contrast, ketoreductase enzymes display high stereo- and regioselectivity and can be used under aqueous conditions, which are not compatible with hydride reagents or borane. No metal or chiral ligands are required and the enzyme is easily disposed of at the end of the reaction without any specialised treatment. A huge number of ketoreductase enzymes have been reported and characterised and they are the most frequently used biocatalysts after lipases/esterases. Ketoreductases are co-factor-dependent enzymes that use a nicotinamide adenine dinucleotide co-factor. As shown in Figure 5.5, the co-factor exists in two forms: the phosphorylated version is abbreviated to NADPH, whilst the non-phosphorylated version is NADH. The nicotinamide co-factor is the source of the "hydride" (in red), which the enzyme uses to carry out the reduction of the substrate. As shown in Figure 5.5, the by-product of the reduction is the oxidised co-factor, which is abbreviated to NAD^+ or $NADP^+$ depending upon whether it is phosphorylated or not.

These enzymes are also capable of catalysing the reverse process, *i.e.*, oxidising an alcohol to a ketone or aldehyde, as we will see in the

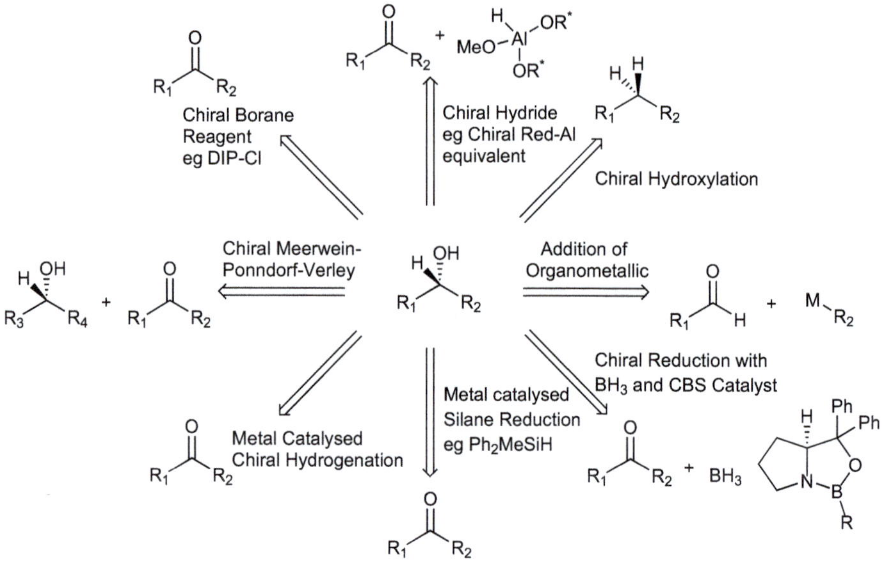

Figure 5.4   Selected chemical methods of reducing ketones.

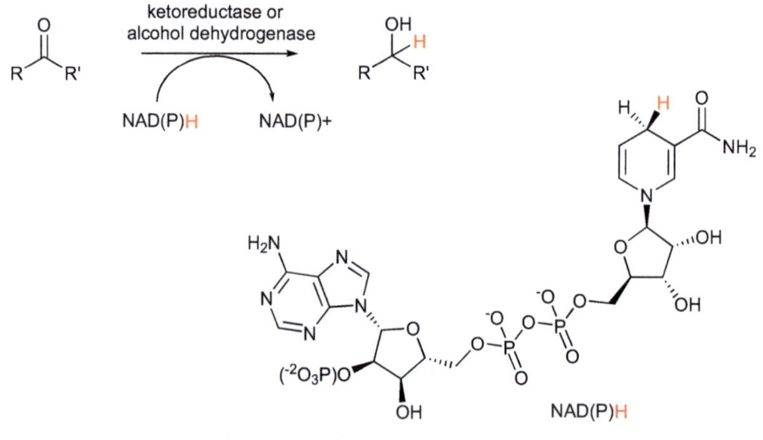

Figure 5.5   General transformation of ketone reduction catalysed by ketoreductases.

next chapter. As a result, these enzymes are often also called alcohol dehydrogenases (ADHs). The enzymes are the same but the names are used somewhat interchangeably, so remember when you see ADH that it is the same as Kred and *vice versa*.

The chiral environment of the enzyme active site determines which face of the ketone is attacked by the hydride. To predict which product is produced, we can use a rule that was proposed in order to predict which

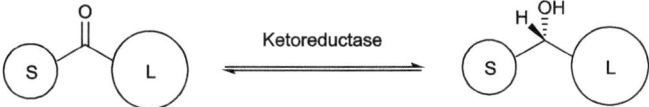

Prelog rule: A rule to predict which face of a prochiral ketone is reduced in ketoreductase catalysed reactions. S - small substituent (*e.g.* methyl). L - large substituent (*e.g.* phenyl). In reductions obeying the rule only the *re* face of the prochiral substrate ketone is reduced.

Figure 5.6   The Prelog rule for predicting which face of a prochiral ketone is reduced.

face of an unsymmetrical ketone is reduced. As shown in Figure 5.6, the Prelog rule (named after the researcher who suggested it) states that for a prochiral ketone, only the *re* face is reduced. Enzymes can therefore be characterised as showing Prelog selectivity (*e.g.* yeast ADH, horse liver ADH (HLADH) and *Thermoanaerobium brockii* ADH (TBADH)) or anti-Prelog selectivity (*i.e.* they reduce the *si* face of a prochiral ketone).

A wide range of ketones have been shown to undergo reduction with ketoreductase enzymes, a selection of which are shown in Figure 5.7. Aromatic, aliphatic and cyclic ketones can be reduced, along with bulkier diaryl substrates. In each case, the product enantiomer is shown but the other enantiomer can often be accessed through the choice of an alternative enzyme.

As shown in Figure 5.7, ketoreductases are also highly chemoselective. Reduction reactions can be carried out in the presence of other groups such as esters, nitriles and alkenes, all of which could otherwise be reduced under the conditions of some chemical reduction methods. Due to the number of enzymes available and their wide substrate range, ketoreductases have been used to develop biocatalytic routes for the synthesis of a number of pharmaceutical drugs. Rivastigmine is a drug used to treat Alzheimer's or Parkinson's disease. As shown in Figure 5.8, we can use a ketoreductase to make a chiral alcohol intermediate, which can then be transformed into the drug compound.

We have already seen one enzymatic route (using a nitrilase in Chapter 3) that chemists have used towards a member of the statin family of drugs, Lipitor. The statin family of drugs are used to treat high cholesterol levels and prevent heart disease. The side-chain of Lipitor contains two stereocentres with only the *R,R* diastereoisomer used in the active ingredient. Chemists have developed a three-enzyme route to make the Lipitor side-chain, in which two of the steps are catalysed by ketoreductases (the third enzyme we will meet in Chapter 8). As shown in Figure 5.9, both chiral alcohols are set by using ketoreductases. The diol product is then reacted further to give Lipitor.

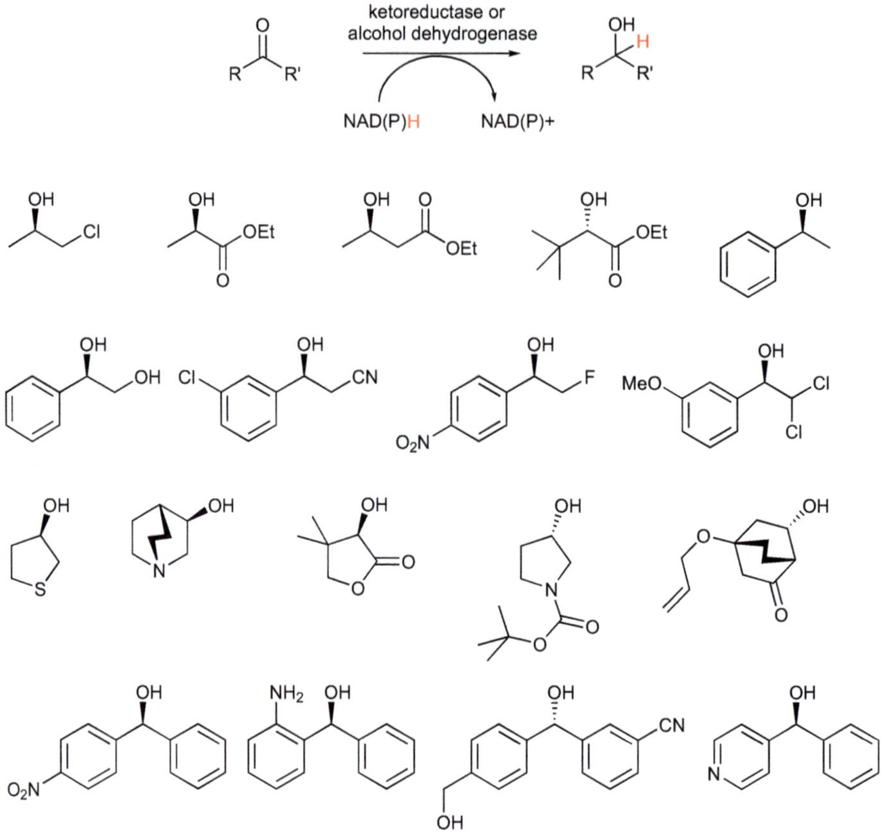

Figure 5.7 The scope of product alcohols produced by ketoreductases.

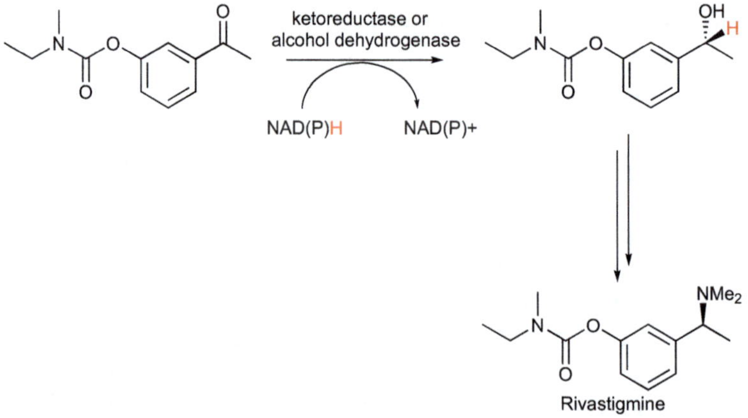

Figure 5.8 The synthesis of rivastigmine using a ketoreductase.

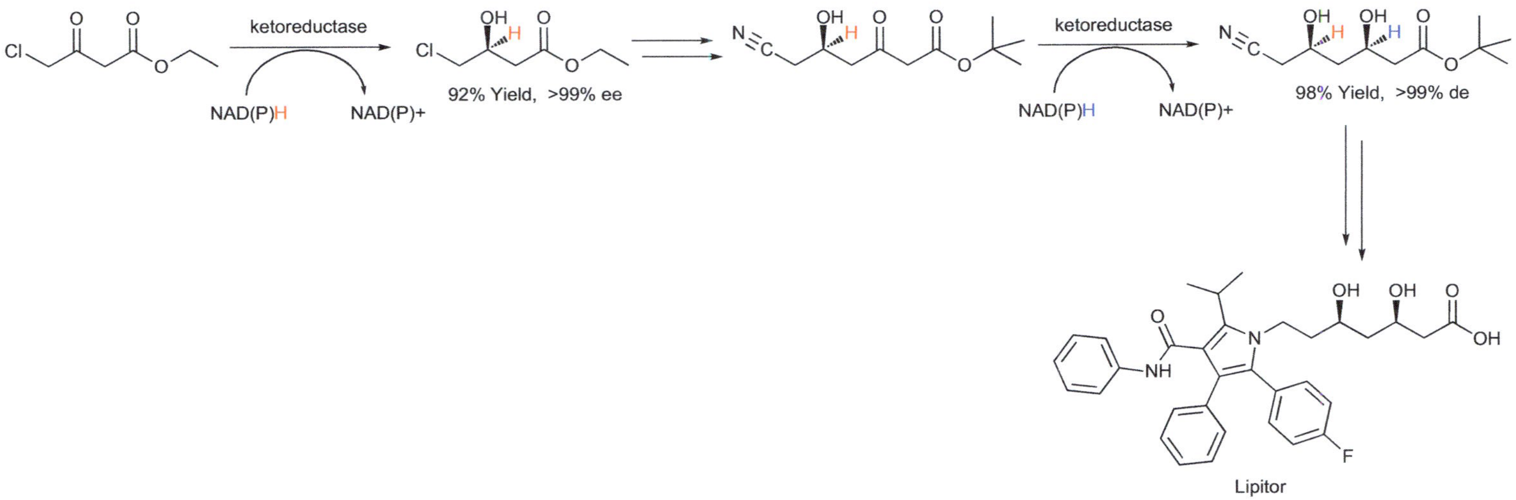

Figure 5.9 The synthesis of Lipitor using a ketoreductase.

Ketoreductases have also been used for the synthesis of montelukast, a pharmaceutical drug used for the treatment of asthma. It was originally made *via* a chemical reduction with a chiral reducing reagent, (−)-β-chlorodiisopinocamphenylborane. As well as being a bit of a mouthful to pronounce, this reagent is flammable (making it difficult to handle), relatively expensive and needs special treatment of the waste produced during the reaction. By using a ketoreductase instead, we should be able to develop a more sustainable, more selective and higher yielding process. As shown in Figure 5.10, the ketoreductase-catalysed reduction works very well giving the desired alcohol in high yield and enantiomeric excess.

As well having high stereoselectivity, ketoreductases can also display high regio- and chemoselectivity. We saw in Figure 5.7 how ketoreductases are highly chemoselective, reducing ketones in the presence of other functional groups that are also prone to reduction. Chemists have taken advantage of this for the synthesis of L-carnitine, an animal feed supplement. As shown in Figure 5.11, the ketoreductase selectively reduces the ketone without reacting with the ester that is also present.

An example of the high regioselectivity possible with ketoreductases has been reported by chemists from Merck. As shown in Figure 5.12, the diketone substrate can be selectively reduced at either of the two ketones by choosing a particular ketoreductase. For the trifluoromethyl ketone, chemists found enzymes that would also produce either the (*R*)- or (*S*)-enantiomer selectively.

Ketoreductases have also been used to set two stereocentres at the same time using dynamic kinetic resolutions. In these processes, the ketoreductase carries out an enantiospecific and diastereoselective reduction of one enantiomer of the starting material, whilst the unreactive enantiomer is racemised *in situ*. This enables up to 100% yield of a single diastereoisomer from the reaction mixture. As shown in Figure 5.13, this process is used in the synthesis of a hydroxy amino ester. Here, the ketoreductase selectively reduces the ketone carbonyl of only one substrate enantiomer, whilst the other enantiomer is racemising under the reaction conditions (pH 7 and 30 °C). The product is one of four possible diastereoisomers and is formed in very high yield and enantiomeric excess.

Although less commonly reported than for ketones substrates, ketoreductases are also capable of reducing aldehydes to the corresponding primary alcohols. A range of aldehydes can be reduced, as shown in Figure 5.14. In each case, the substrate aldehyde is shown.

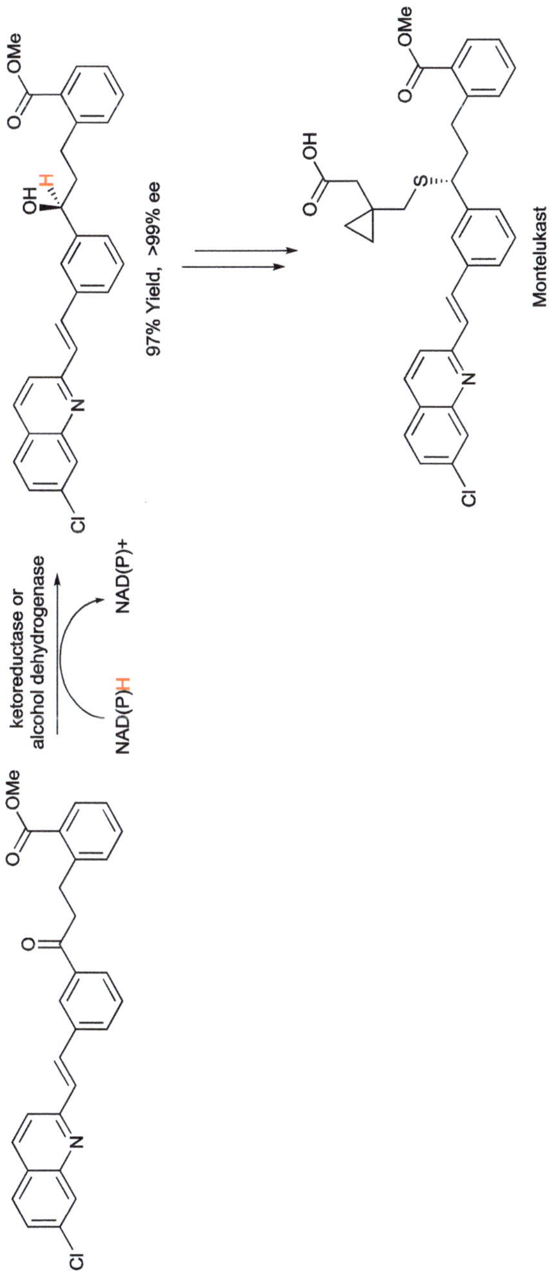

Figure 5.10 The synthesis of montelukast using a ketoreductase.

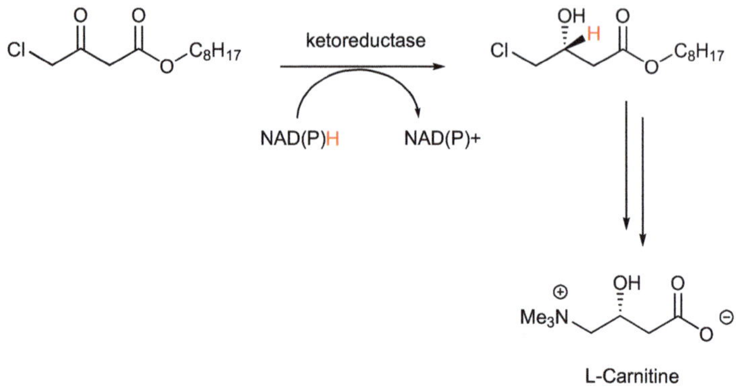

Figure 5.11   The synthesis of L-carnitine using a ketoreductase.

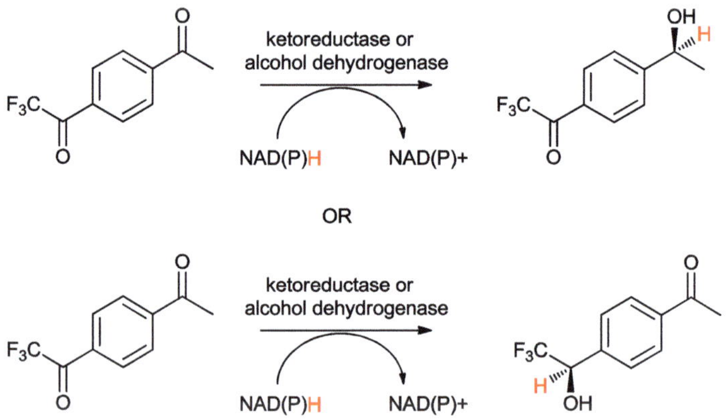

Figure 5.12   Regioselective reduction of a diketone using ketoreductases.

As we have already seen, ketoreductase enzymes require a nico-tinamide co-factor in order to carry out the reduction of ketones and aldehydes. Nicotinamide co-factors are relatively expensive to buy, so if we plan on carrying out a ketoreductase-catalysed reaction on anything other than a small scale, the cost of using a stoichiometric amount of the co-factor could be prohibitive. To address this issue, chemists have come up with ways of recycling the co-factor, so that we need much less than one equivalent. There are two approaches that chemists use for co-factor recycling. The first method, as shown in Figure 5.15, uses an alcohol to recycle the nicotinamide co-factor. Therefore, in the course of the reaction, the substrate ketone gets reduced to the product alcohol, whilst the sacrificial alcohol is oxi-dised to the corresponding ketone. This is called a substrate-coupled recycling system because the sacrificial alcohol must also be a

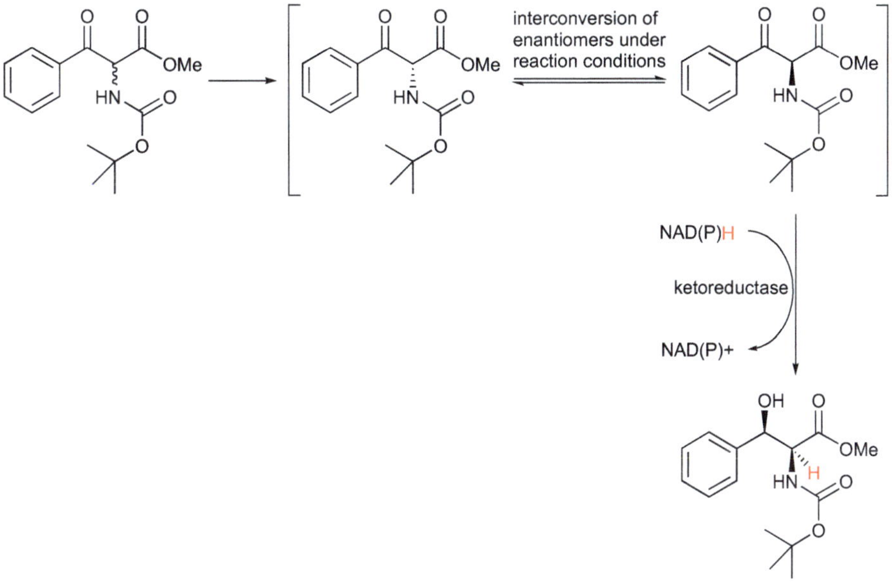

Figure 5.13 A dynamic kinetic resolution reaction using a ketoreductase.

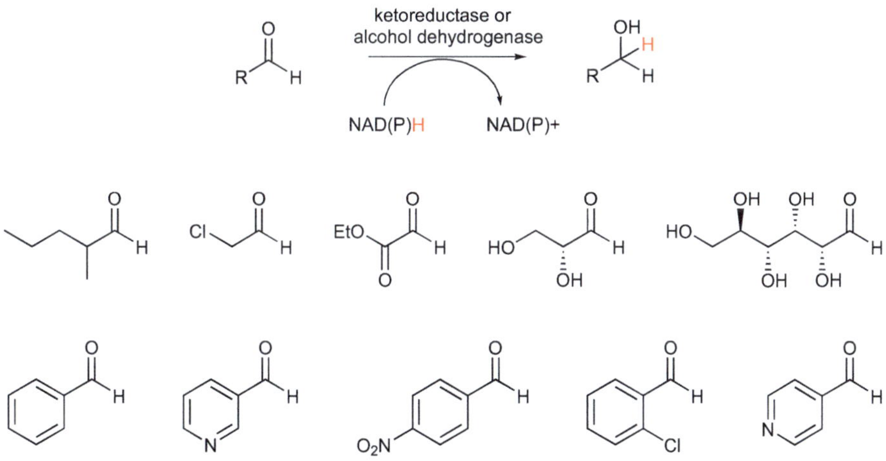

Figure 5.14 The substrate scope of aldehydes reduced by ketoreductases.

substrate for the ketoreductase enzyme. The advantage of this approach is that we only use one enzyme in the process. The disadvantages are that the enzyme must accept the sacrificial alcohol and that because we now have two ketones and two alcohols, an equilibrium is set up between the starting materials and the products, which may prevent us from achieving high yields. To overcome any

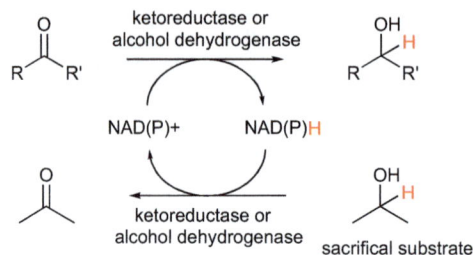

Figure 5.15 A substrate-coupled co-factor recycling system for ketoreductases.

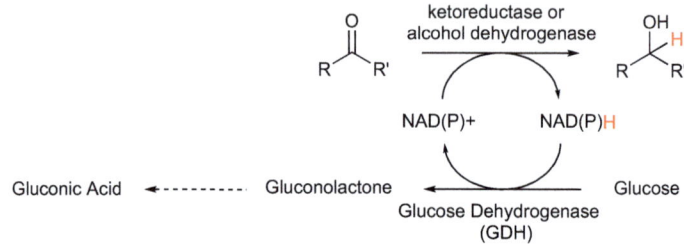

Figure 5.16 An enzyme-coupled co-factor recycling system for ketoreductases.

equilibrium problems, chemists often use isopropyl alcohol as the sacrificial alcohol. This is because it is oxidised (removing the red hydrogen, which in turn is transferred to the nicotinamide co-factor) to acetone, which can be removed by distillation to drive the equilibrium of the reaction.

The second option is to use a second enzyme (and substrate) to recycle the nicotinamide co-factor. There are many systems capable of achieving this, with one of the most commonly used being glucose dehydrogenase, as shown in Figure 5.16. In this case, the glucose dehydrogenase enzyme recycles the nicotinamide co-factor at the expense of glucose, which is oxidised to gluconolactone. This is called an enzyme-coupled recycling system because we use a separate enzyme and substrate to recycle the co-factor. The advantage of this system is that by using a second enzyme, we have an irreversible recycling reaction and so don't need to worry about an equilibrium. The disadvantage of a two-enzyme system is that we need to find reaction conditions that are suitable for both enzymes and we need to make sure that the substrate and product of one enzyme don't inhibit the other enzyme.

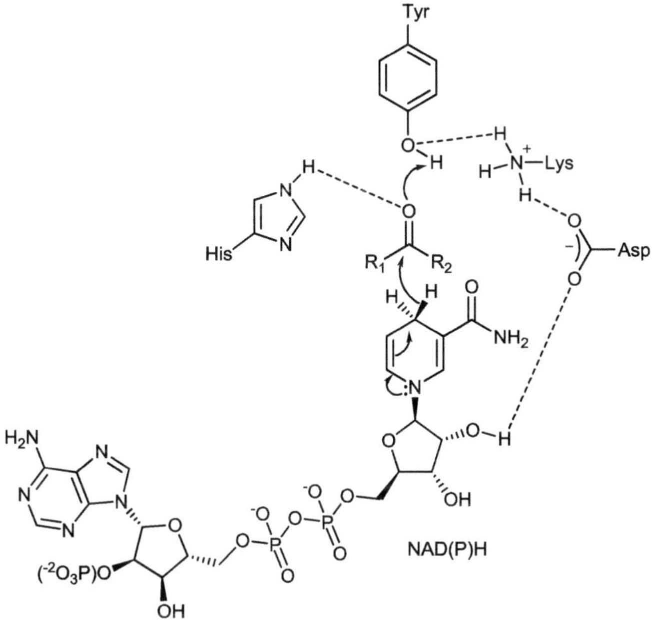

Figure 5.17 The mechanism of ketone reduction catalysed by ketoreductases.

A more detailed mechanism of ketoreductase-catalysed reduction of ketones is shown in Figure 5.17. You do not need to know the mechanism in this level of detail, but it is included here for those who wish to see a more in-depth explanation. As mentioned previously, the enzyme active site controls the facial selectivity of the hydride attack from the nicotinamide co-factor to the ketone (or aldehyde). The important residues are a tetrad of tyrosine, lysine, aspartic acid and histidine. As shown in Figure 5.17, the carbonyl of the substrate is activated by the histidine and tyrosine residues. The tyrosine acts as the proton donor and is stabilised by an extended hydrogen bond network involving the lysine and aspartic acid residues. The nicotinamide co-factor has its own binding site within the protein in order to bring it into position to carry out the reduction reaction. However, the aspartic acid of the tetrad is also involved in binding the co-factor. The nitrogen lone pair of the nicotinamide is electron donating, resulting in hydride transfer from the co-factor to the ketone, which picks up a proton from the tyrosine residue. The two hydrogens of the prochiral nicotinamide are called pro-*R* and pro-*S* as reduction with one leads to the (*R*)-enantiomer whilst the other leads to the (*S*)-enantiomer. As mentioned previously, this is controlled by the orientation of the enzyme active site.

5.3   Reduction of Carboxylic Acids

Like ketones and aldehydes, carboxylic acids are widespread in both natural and man-made compounds. Many important molecules contain carboxylic acid groups, such as amino acids, the building blocks of proteins, pyruvic and citric acid, vital to central cell metabolism, and fatty acids, which are involved in metabolic processes. Carboxylic acids are an oxidation state above ketones and aldehydes and (if protected as esters) can be useful in synthesis. They can be taken through a series of synthetic steps that the more sensitive aldehyde might not survive, and then reduced to the aldehyde when it is needed for a particular reaction as we saw in Figure 5.1. However, in order for this strategy to work, we must have an effective method of reducing the acid (or ester) to the corresponding aldehyde. One way of achieving this transformation is to use a carboxylic acid reductase (CAR) enzyme, as shown in Figure 5.18. These enzymes are large, multi-domain proteins that utilise both adenosine triphosphate (ATP) and nicotinamide (NADH or NADPH) co-factors.

A range of chemical alternatives also exist for carrying out this transformation. The reducing reagent borane (BH_3) selectively reduces carboxylic acids in the presence of groups such as ketones. However,

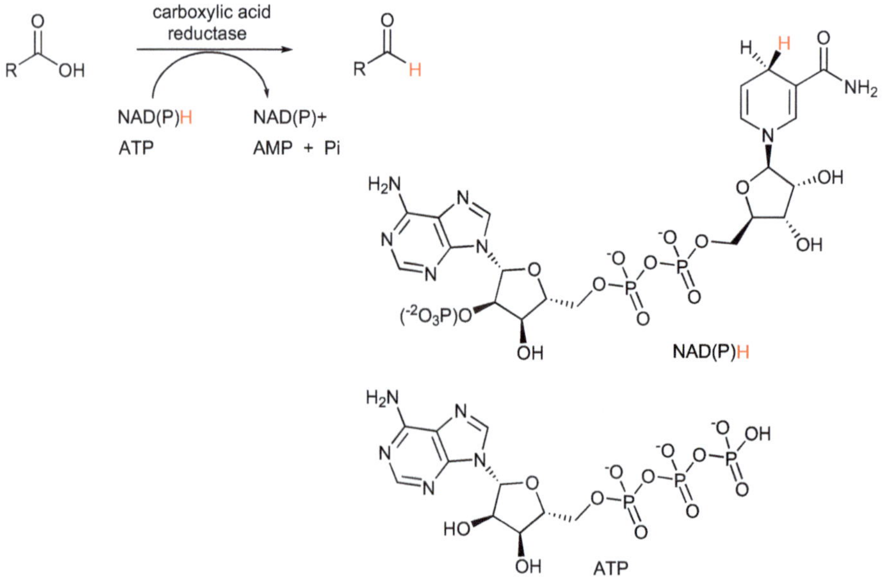

Figure 5.18   General transformation for the reduction of carboxylic acids using carboxylic acid reductases.

the product of the initial reduction, the aldehyde, is more electrophilic than the carboxylic acid and so over-reduction occurs to give the alcohol. This then requires a second step to oxidise the alcohol back up to the aldehyde. Chemists have come up with a number of ways around this fundamental problem of reactivity, as shown in Figure 5.19. The downside to these alternative processes is that they often start from a carboxylic acid equivalent, which requires an extra synthetic step to make. For example, the Rosenmund reaction to produce an aldehyde starts from the acid chloride, whilst the reducing reagent diisobutylaluminium hydride (DIBAL) works only on esters. In addition, these reducing reagents are often incompatible with aqueous conditions and use metals, which produce waste that can be expensive to dispose of. Alternatives to more traditional reducing reagents often use expensive metals and chiral ligands and may require high temperature or pressure to carry out the reduction. As a result, many of the chemical methods are incompatible with sensitive functional groups and therefore protecting groups may have to be used.

In contrast, carboxylic acid reductase-catalysed reactions occur under mild conditions in aqueous solvent (or mixtures) removing the need for protecting groups. These enzymes are highly regio- and stereoselective and, unlike many chemical reductants, the product aldehyde is not

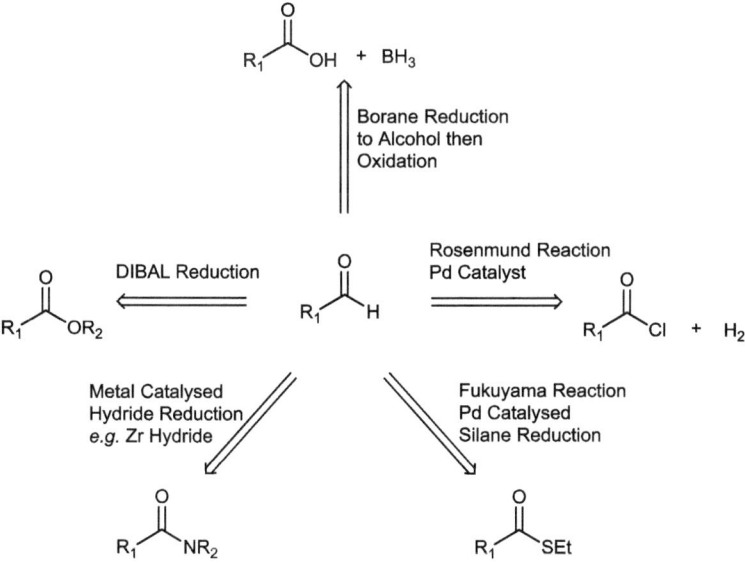

Figure 5.19 Selected chemical methods of aldehyde synthesis.

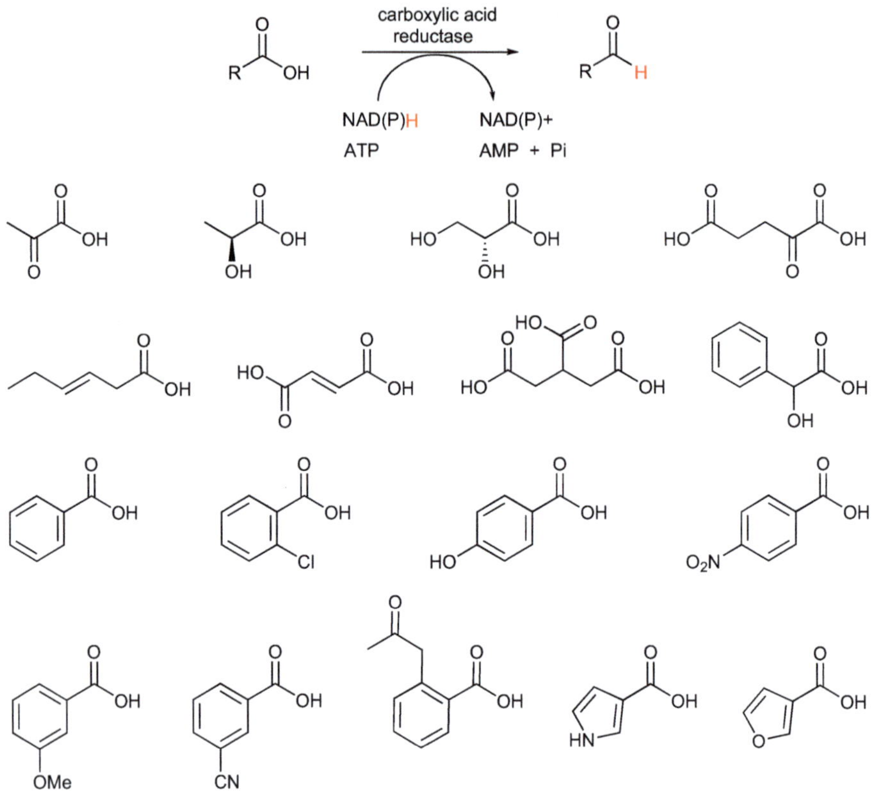

Figure 5.20 Substrate scope of carboxylic acids reduced by carboxylic acid reductases.

reduced further. Carboxylic acid reductases catalyse the reduction of a range of carboxylic acids although these tend to be limited in general to aromatic (or heteroaromatic) and aliphatic acid substrates. A range of acid substrates that are reduced by these enzymes are shown in Figure 5.20. In each case, the substrate carboxylic acid is shown.

Aldehydes and alcohols are prevalent in compounds that are important for the food, flavour and fragrance industries. An example of using a CAR enzyme is the production of vanillin, a compound used to flavour sweet foods such as ice cream or chocolate. As shown in Figure 5.21, the starting material for this process is vanillic acid, which is reduced by the carboxylic acid reductase to give the corresponding aldehyde, vanillin, in very high yield. No by-products are formed in the reaction, making the isolation of the product easy.

As mentioned previously, carboxylic acid reductases are co-factor-dependent enzymes requiring both ATP and a nicotinamide co-factor.

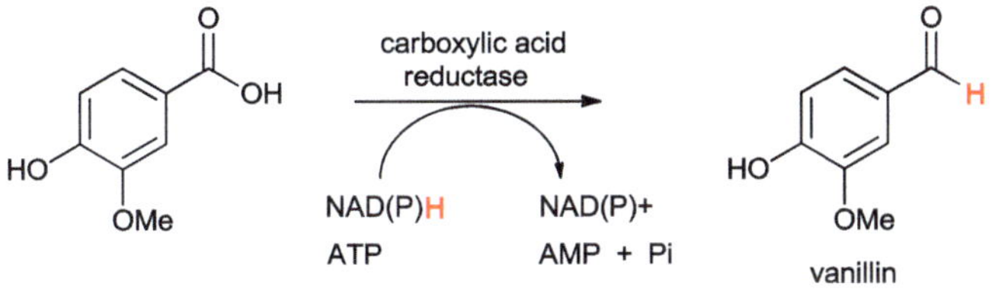

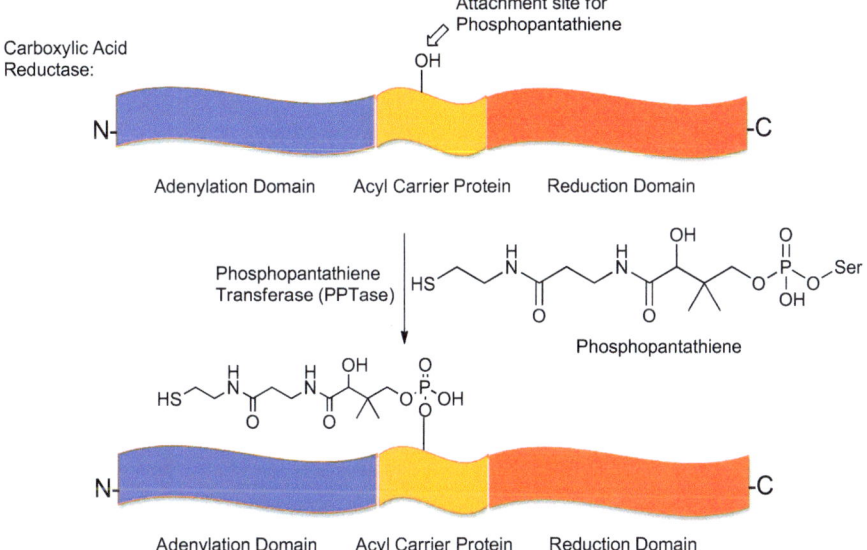

Figure 5.21 The synthesis of vanillin using a carboxylic acid reductase.

Figure 5.22 Attachment of the phosphopantatheinyl "arm" to the carboxyl acid reductase.

A simplified version of the mechanism showing the use of these co-factors is outlined in Figures 5.22 and 5.23. You don't need to know this, but it is included here for advanced readers and those who are curious as to how these enzymes work. Carboxylic acid reductase enzymes are made up of two domains linked together by an acyl carrier protein (which is important, as we shall see shortly). For the enzyme to be active, it requires attachment of a prosthetic group called phosphopantetheine. This group is attached at a specific site in the protein by another enzyme called phosphopantetheine transferase (PPTase), as shown in Figure 5.22.

Once the prosthetic group is added, the mechanism begins in the "adenylation domain" of the enzyme, as shown in Figure 5.23. The carboxylic acid (in red) initially reacts with adenosine triphosphate

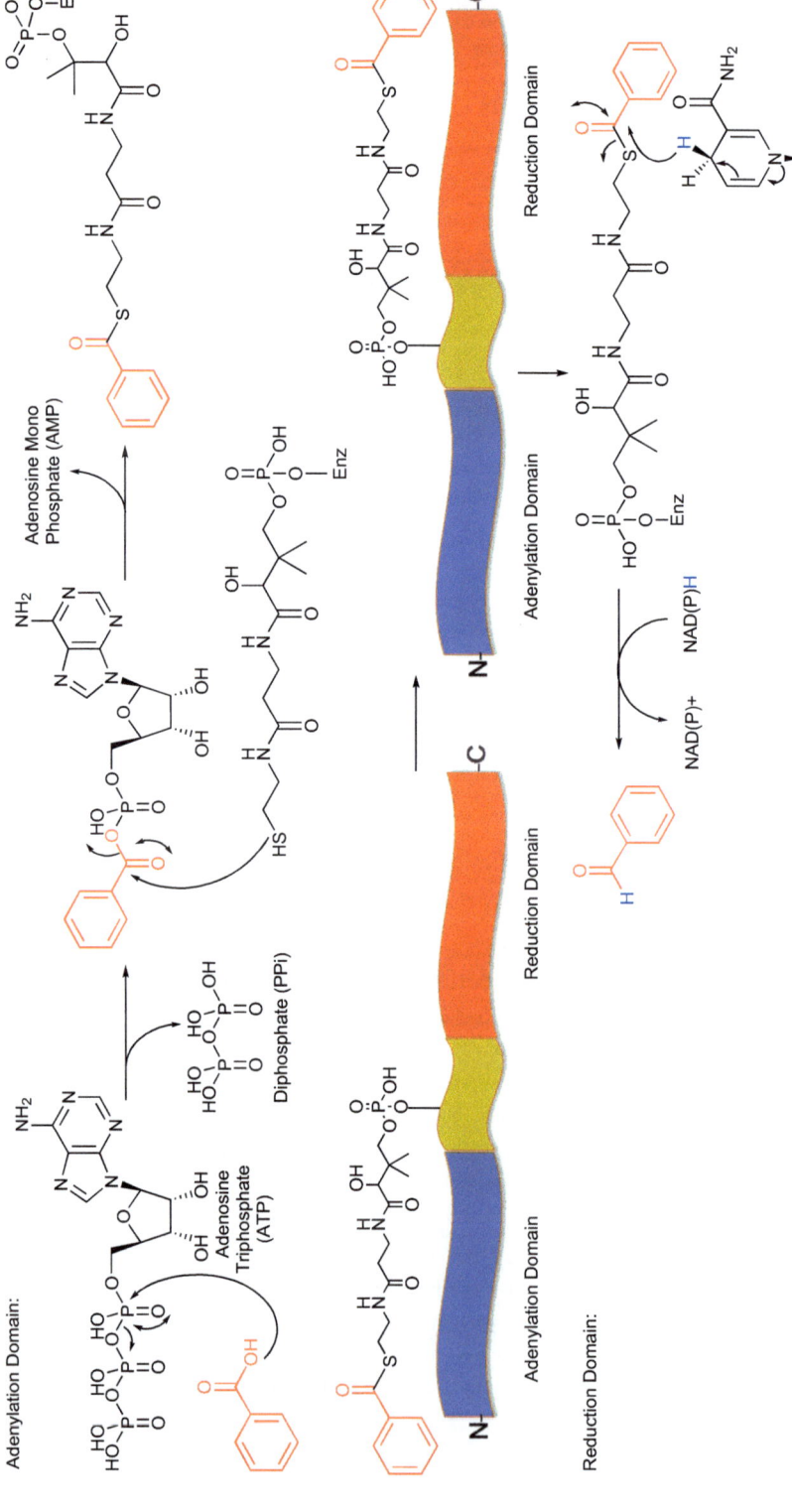

Figure 5.23 A simplified version of the mechanism of carboxylic acid reductases.

(ATP) to give the adenosine monophosphate ester (AMP ester) of the carboxylic acid and diphosphate (PPi) as a by-product. The adenylated substrate is then attacked by the sulfur atom at the end of the phosphopantetheine arm to give a thioester and adenosine mono-phosphate (AMP) as a by-product. Now that the substrate has been loaded onto the phosphopantetheine, the arm swings around to carry the substrate to the "reduction domain" of the enzyme. Here, the thioester is reduced by a hydride (in blue) from the nico-tinamide co-factor to give the aldehyde product, releasing the phos-phopantetheine arm, which can then move back to the adenylation domain to react with another adenylated substrate.

5.4 Reduction of Alkenes

Carbon–carbon double bonds or alkenes are versatile functional groups in organic chemistry, which can be transformed in a number of different ways. As we will see in Chapter 7, depending on the reaction conditions, we can add electrophiles or nucleophiles to an alkene. However, we may not need to add anything to the double bond and instead reduce the alkene to a saturated alkane. This can often be the case in organic synthesis when we join two molecules together to create an alkene, which then needs to be reduced. There are many useful methods of alkene synthesis, such as Wittig or Wadsworth–Horner–Emmons reactions, metathesis reactions or con-densation and elimination processes. In addition, if the alkene is di-substituted at one end, then we can generate a chiral centre in the reduction reaction. One method of alkene reduction is to use an enoate reductase (Ered) enzyme, as shown in Figure 5.24. Enoate reductases are flavoproteins, as they require a co-factor that is a nucleic acid de-rivative of riboflavin. Ereds typically use flavin mononucleotide (FMN) as the riboflavin co-factor. They also require a nicotinamide co-factor, which is the source of the hydride required for reduction. As can be seen in Figure 5.24, enoate reductases work best if the alkene is "ac-tivated" by having an electron withdrawing group (EWG) substituent.

As we can see from Figure 5.24, the reaction results in the *trans* addition of hydrogen across the double bond. As we have mentioned, the hydride (in red) comes from the nicotinamide co-factor *via* the flavin, which attacks in a 1,4-Micheal fashion at the β-carbon of the substrate. The proton (in blue) comes from the solvent or a tyrosine residue on the opposite face of the double bond, and is added at the α-carbon.

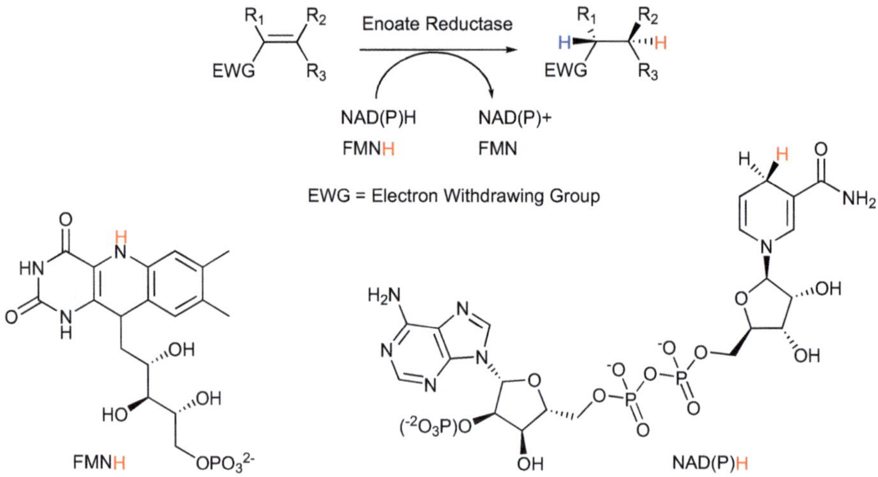

Figure 5.24 General transformation of alkene reduction catalysed by enoate reductases.

A number of chemical methods exist for the reduction of alkene double bonds. As with ketones and aldehydes, it is possible to reduce alkenes with hydride reagents such as sodium borohydride. As well as the potential difficulties of handling these reagents on a large scale (due to the potential for evolution of flammable hydrogen gas), these reagents are not very chemoselective. This means that we often get reduction of other groups present in the molecule, such as ketones. This is particularly problematic if we try to reduce an enone, such as cyclohexenone, as shown in Figure 5.25. The other classical method of alkene reduction is hydrogenation using a metal catalyst and a hydrogen source. Although a wide range of catalysts and conditions exist, the metal catalyst can be expensive and flammable hydrogen gas is generated. Alternative reductants are available, such as silanes or even sodium metal and a proton source, but these methods also tend to use transition metal catalysts as well. Another option is to use an organocatalyst and hydride source. As shown in Figure 5.25, chemists have used a dihydropyridine (called the Hantzsch ester) that looks very much like the nicotinamide co-factor that an Ered would use. The chiral phosphoric acid is used to activate the enone and determine the enantioselectivity of the reaction.

In contrast to the chemical methods discussed above, enoate reductases are able to catalyse the reduction of activated alkenes with high stereoselectivity under mild aqueous conditions. They are able to

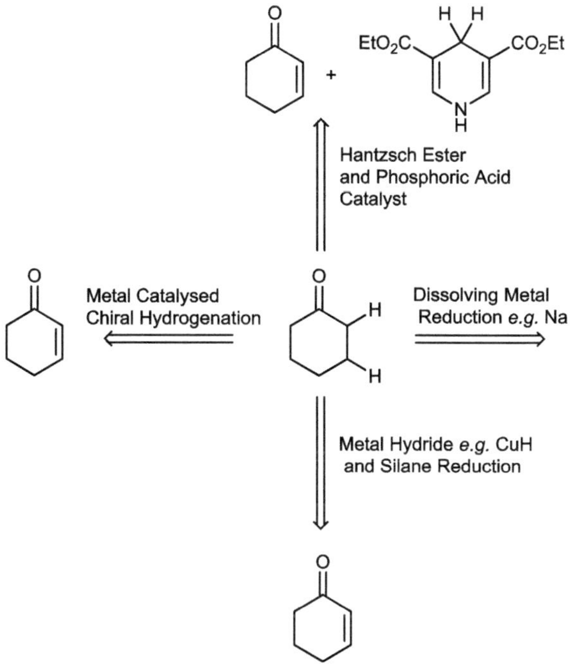

Figure 5.25   Selected chemical methods for the reduction of alkenes.

reduce double bonds chemoselectively in the presence of other groups, avoiding the need for metal catalysts and ligands, difficult to handle hydrogen gas or hydride reagents. These factors make them a good alternative option to more traditional chemical methods. A range of activated alkene substrates reduced by enoate reductases are shown in Figure 5.26. In each case, the product is shown and in many cases either enantiomer of the product can be produced by choosing the appropriate enzyme.

Although the wide diversity of enoate reductase enzymes often means that the choice of enzyme will determine which enantiomer we make, it is also possible to change the configuration of the product by other means. In particular, it may be possible to change the enantioselectivity of the product by changing the geometric configuration of the double bond in the substrate. In some cases, changing the configuration of the alkene from *trans* (*E*) to *cis* (*Z*) can result in a switch in the stereochemical outcome of the reduction reaction. This is because the alternate configuration of the alkene may result in an alternate binding mode of the substrate. This alternate, "flipped" mode can then result in production of the other product enantiomer, as shown in Figure 5.27.

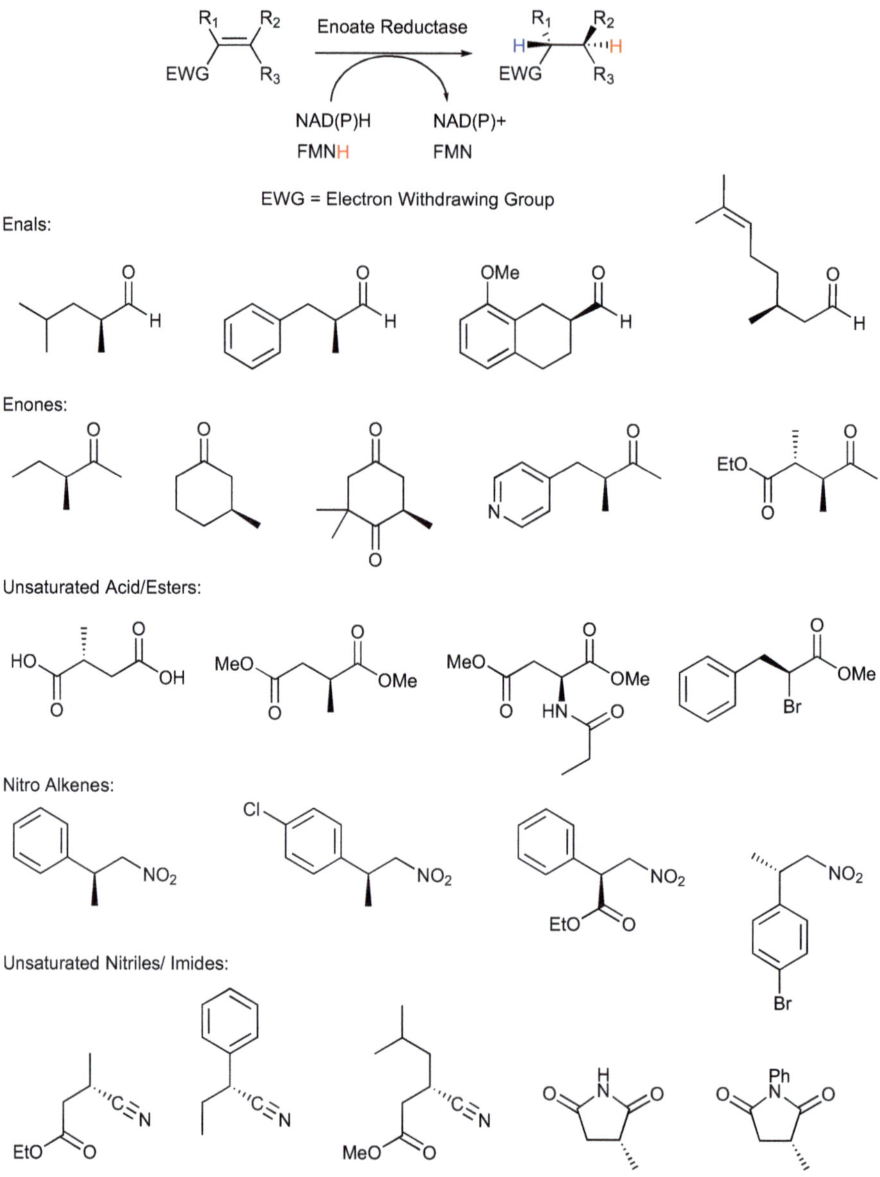

Figure 5.26 The substrate scope of products produced by the reduction of alkenes catalysed by enoate reductases.

Due to the wide number of enzymes available and their advantages over some chemical reductants, enoate reductases have been used to synthesise a range of pharmaceutical drugs or intermediates. Baclofen is a drug used to treat disorders of the central nervous *trans (E)*-Alkene:

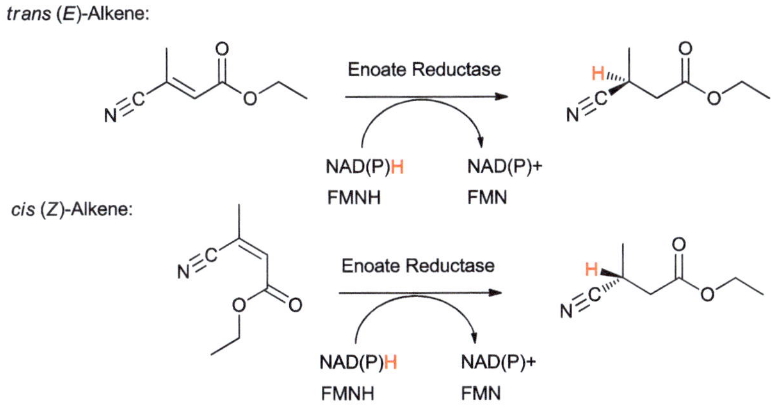

cis (Z)-Alkene:

Figure 5.27 Stereochemical outcome of reducing *cis* or *trans* bonds with enoate reductases.

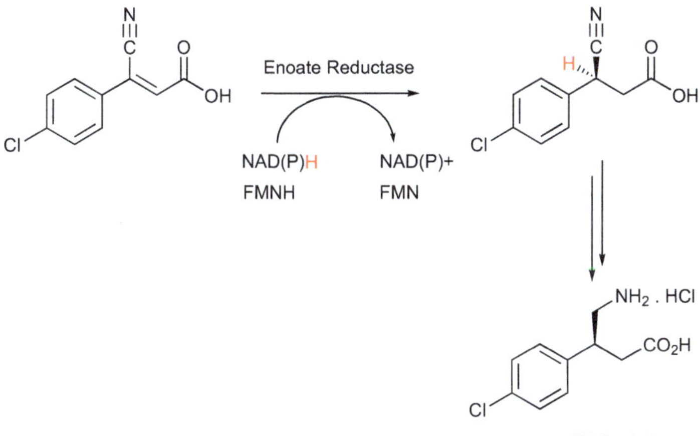

(*S*)-Baclofen

Figure 5.28 The synthesis of baclofen using an enoate reductase.

system, such as spinal cord injury, cerebral palsy and multiple sclerosis. Chemists have designed a route to selectively produce the (*S*)-enantiomer of baclofen by using an enoate reductase to generate an intermediate compound, as shown in Figure 5.28.

Enoate reductases have also been used in the synthesis of flurbiprofen, a non-steroidal anti-inflammatory used to treat dental pain and arthritis. In this case, the enzyme-catalysed reduction using an enoate reductase selectively gave the (*S*)-enantiomer of the intermediate, as shown in Figure 5.29.

We have seen that enoate reductases are co-factor-dependent, with both flavin and nicotinamide co-factors. As we mentioned previously

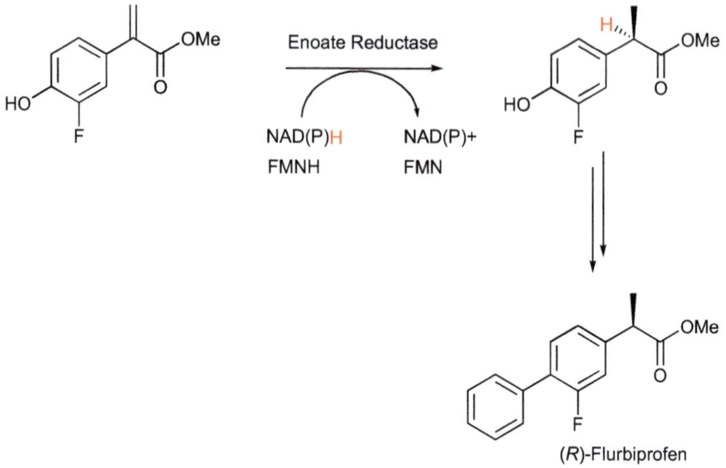

Figure 5.29 The synthesis of flurbiprofen using an enoate reductase.

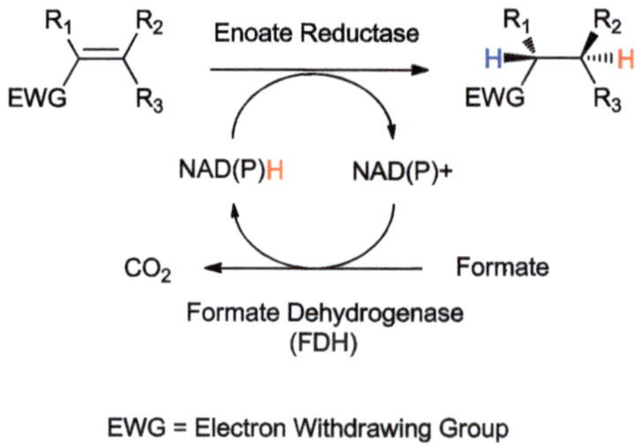

EWG = Electron Withdrawing Group

Figure 5.30 An enzyme-coupled co-factor recycling system for enoate reductases.

in the section on ketoreductases, the cost of a stoichiometric amount of the co-factor is prohibitive on a large scale. As a result, chemists have developed nicotinamide recycling systems that are similar to those used with ketoreductases. Most enoate reductase reactions use the enzyme-coupled approach to nicotinamide co-factor recycling with a number of systems being reported. In Figure 5.30, we have shown a typical enzyme-coupled recycling system using formate dehydrogenase, which recycles the nicotinamide at the expense of formate, which in turn is oxidised to carbon dioxide.

A simplified version of the mechanism of enoate reductases showing the use of both flavin and nicotinamide co-factors is outlined

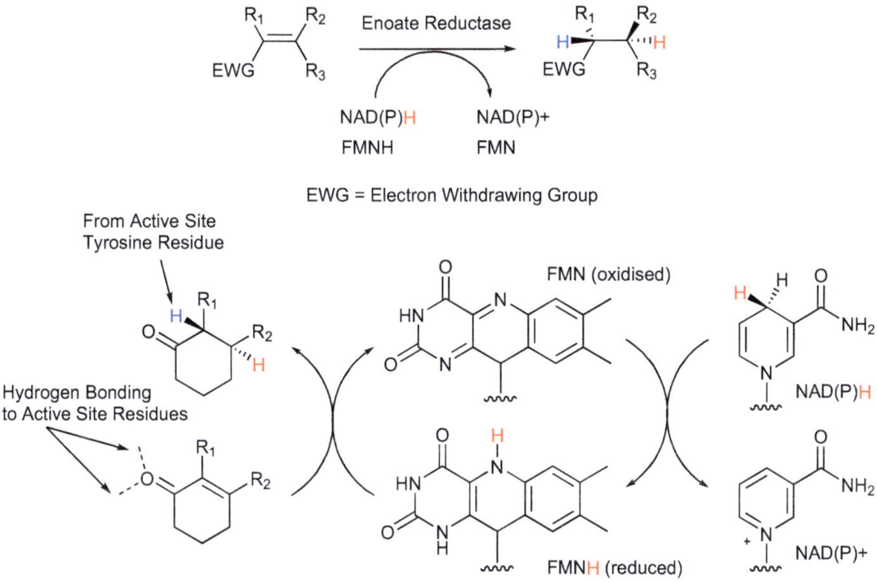

Figure 5.31 A simplified mechanism for the reduction of alkenes using enoate reductases.

in Figure 5.31. As with the other more detailed mechanisms within this chapter, you don't need to know this, but it is included here in case you would like more information on how these enzymes work. The mechanism starts with the substrate binding to the active site of the enzyme. The electron withdrawing group on the alkene is important here as it helps bind the substrate in the active site *via* hydrogen bonding to a histidine/aspartic acid pair of residues. The alkene then undergoes reduction initially by attack of a hydride from the reduced form of the flavin molecule. As mentioned previously, the hydride (in red) from the flavin adds to the β-carbon of the enone. The reduction is completed by the addition of a proton (in blue) from a tyrosine residue on the side opposite to that of the hydride attack. This generates the product and gives the oxidised form of the flavin co-factor as a by-product. The oxidised flavin is then reduced back to the form required for the reduction of the next substrate molecule by the nicotinamide co-factor. This is turn is oxidised to give the oxidised form of the nicotinamide, which, if using co-factor recycling, must then be recycled back to the reduced form for the catalytic cycle to continue. As we can see in the product, the stereochemical outcome is from the *trans* addition of hydrogen across the carbon–carbon double bond.

5.5   Reduction of Imines

Chiral amines are of high importance and therefore have high value as pharmaceutical and agrochemical intermediates. A wide range of methods are available for the synthesis of chiral amines, but one of the most well used is the reduction of an imine, thus making imines key functional groups. Imines are also key functional groups in Nature, as demonstrated by their use in many biosynthetic pathways. For example, they are one of the methods Nature uses to make carbon–nitrogen bonds through the condensation of an aldehyde or ketone with an amine. Several chemical procedures for the reduction of imines have been reported, as shown in Figure 5.32. These typically include: hydrogenation with a metal catalyst, hydrosilylation with a Lewis acid (or base) catalyst, or organocatalysed reduction with the Hantzsch ester. We have seen methods equivalent to these processes for the reduction of other functional groups previously in this chapter. These methods can be very stereoselective, but may use environmentally unfriendly solvents and expensive transition metal catalysts and ligands. An additional disadvantage is that a protecting group is usually required for the imine nitrogen, adding extra steps to the synthesis of

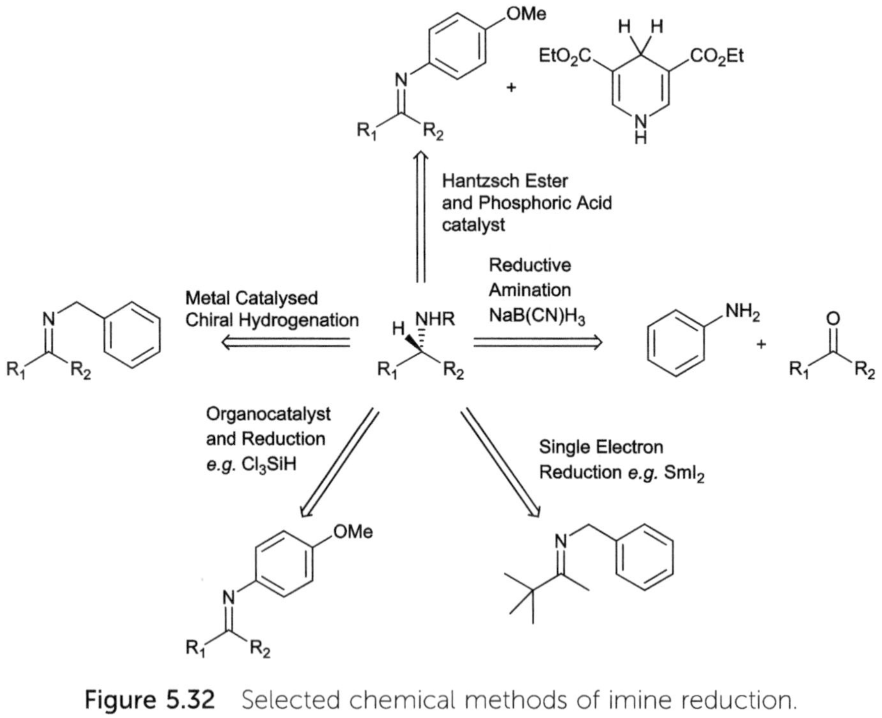

Figure 5.32   Selected chemical methods of imine reduction.

the chiral amine. In some cases, the reaction conditions may also be incompatible with other functional groups present in the molecule.

As an alternative method, an enzyme-catalysed process has the advantage of being carried out in much greener solvents under milder reaction conditions; furthermore, it does not need chiral ligands or metals and can still be highly stereoselective. In addition, the enzymes detailed below can also display high chemo- and regioselectivities. Enzymatic processes don't require the use of protecting groups, which makes the synthesis of amines simpler and often more efficient.

5.5.1 Amino Acid Dehydrogenases and Opine Dehydrogenases

α-Amino acids are not only the building blocks of proteins, but also important biological compounds that play critical roles in organisms. In humans, amino acids are important for neurotransmission, red blood cells, lipid transport and organ function, to name just a few of the roles that they play in the body. They are also used in nutritional supplements, fertilisers, and pharmaceutical drugs and as chiral catalysts. α-Amino acids are typically synthesised in the body by enzymes called amino acid dehydrogenases. As shown in Figure 5.33, these enzymes catalyse the reductive amination of an α-keto acid with ammonia. It is also possible to use these enzymes in the reverse direction to deaminate α-amino acids to the corresponding α-keto acids. As shown in Figure 5.33, these enzymes are co-factor-dependent using a nicotinamide co-factor (NADH).

Many methods for α-amino acid synthesis exist, of which a small selection is detailed in Figure 5.34. Some of these we will see again in Chapter 8, but the equivalent process to the amino acid dehydrogenase above is that of reductive amination. As already mentioned, chemical methods for reductive amination generally use hydride reagents such as sodium cyanoborohydride or metal-catalysed hydrogenation to reduce the imine once it is formed. Imine formation can

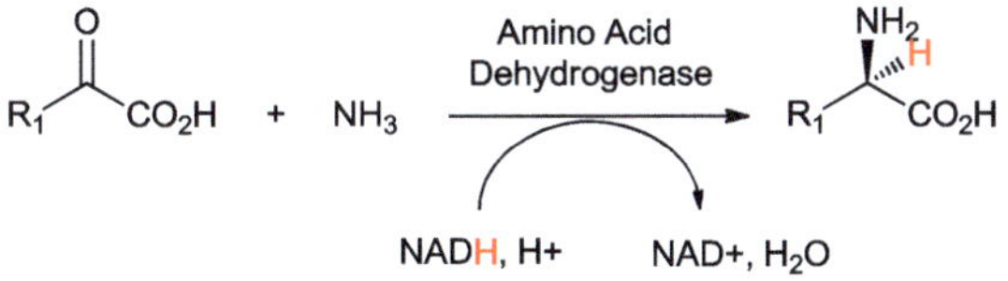

Figure 5.33 General transformation to make amino acids catalysed by amino acid dehydrogenase.

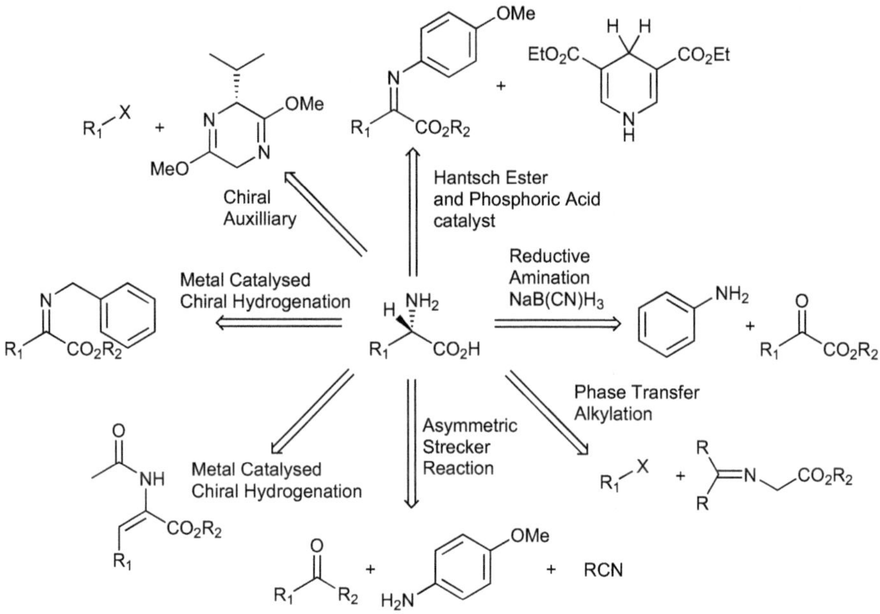

Figure 5.34 Selected chemical methods of α-amino acid synthesis.

be difficult, requiring the use of strong Lewis acids and a protecting group on the imine nitrogen, which must be removed later.

In comparison, amino acid dehydrogenases work directly on the α-keto acid with ammonia under mild conditions to produce the amino acid with high stereoselectivity and without the need for a protecting group. A number of amino acid dehydrogenases have been reported with each enzyme having a different substrate range. This means that we can make a range of amino acids by choosing the appropriate enzyme. Amino acid dehydrogenases are ʟ-selective, producing natural amino acids, although enzymes with the opposite selectivity capable of producing unnatural ᴅ-amino acids have been reported through the use of enzyme evolution. As shown in Figure 5.35, a range of α-keto acids undergo amination to give the product α-amino acids.

Because chiral amines are such important chemicals, amino acid dehydrogenases have been used for the synthesis of important α-amino acids such as non-proteinogenic amino acids. These amino acids (which are not used in protein biosynthesis), have a wide range of activities and are found in a range of pharmaceutical compounds. An example is *tert*-leucine, a non-proteinogenic amino acid used as a building block for a number of pharmaceutical compounds, such as telaprevir, an antiviral drug used for the treatment of hepatitis C. As shown in Figure 5.36,

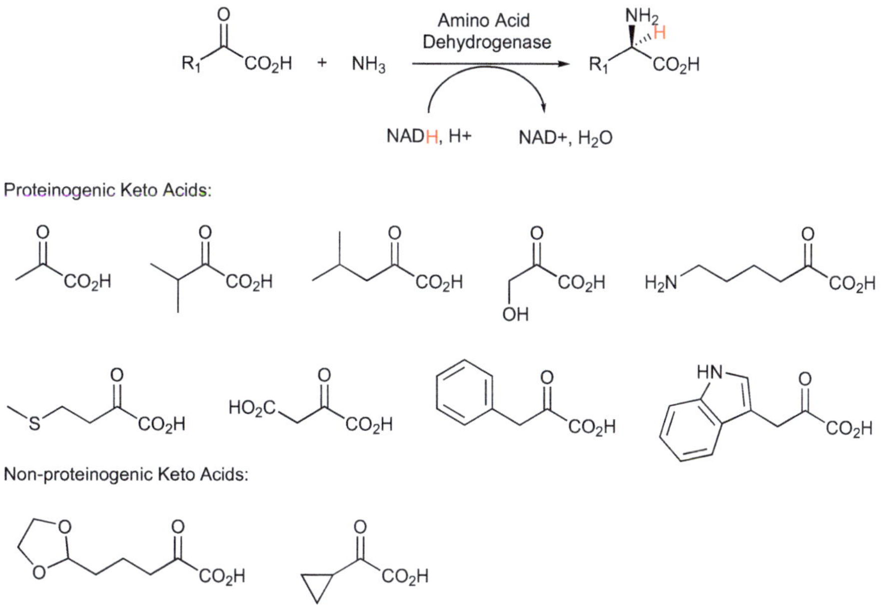

Figure 5.35 Substrate scope of α-keto acids used by amino acid dehydrogenases.

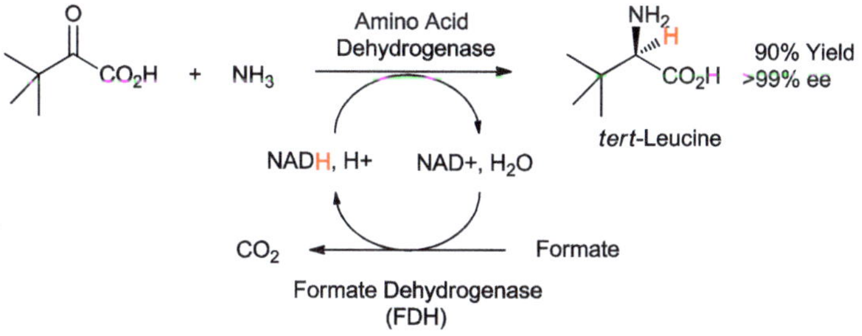

Figure 5.36 The synthesis of *tert*-leucine using an amino acid dehydrogenase.

reductive amination of trimethylpyruvate with leucine dehydrogenase gives (S)-*tert*-leucine in very high yield and stereoselectivity. As with the other nicotinamide co-factor-dependent enzymes we have already seen in this chapter, the nicotinamide can be readily recycled through the use of a second enzyme (an enzyme-coupled recycling system). In this case, chemists use formate dehydrogenase to recycle the nicotinamide, oxidising the formate to give carbon dioxide as a by-product.

Amino acid dehydrogenases have also been shown to work on non-natural substrates. Chemists at Bristol-Myers-Squibb used

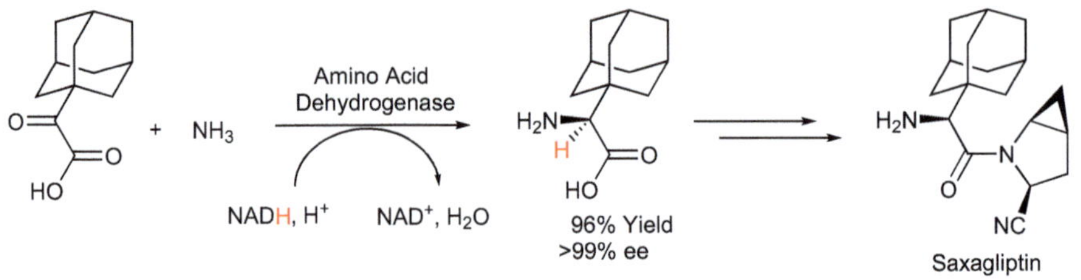

Figure 5.37   The synthesis of saxagliptin using an amino acid dehydrogenase.

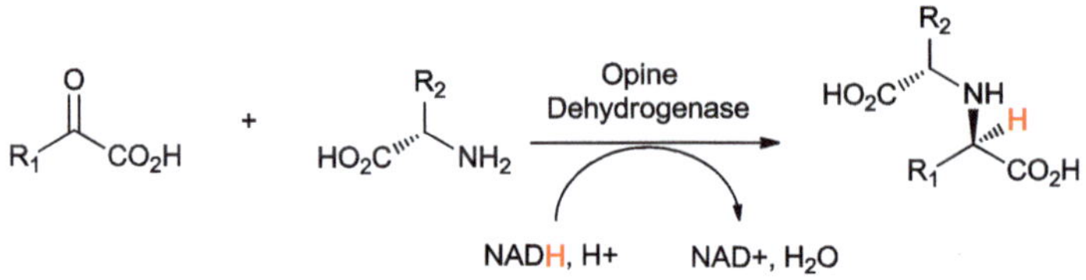

Figure 5.38   The general transformation catalysed by opine dehydrogenases.

phenylalanine dehydrogenase to carry out a reductive amination on a keto acid with a bulky adamantane group attached. As shown in Figure 5.37, the product of the enzyme-catalysed transformation was then reacted further to give saxagliptin, a drug used to treat diabetes.

Opine dehydrogenases are a related family of enzymes that catalyse a very similar reaction to α-amino acid dehydrogenases. Opine dehydrogenases catalyse the reductive amination of an α-keto acid substrate (the same as amino acid dehydrogenases), but rather than using ammonia as the nucleophile, opine dehydrogenases use an amino acid to produce an opine, as shown in Figure 5.38. These proteins are found in bacteria and marine cephalopods such as scallops.

The substrate scope of these enzymes is also rather limited, as shown in Figure 5.39. As a result, chemists have carried out protein engineering on this class of enzymes in an attempt to increase the substrate scope.

The success of the protein engineering program was demonstrated with the synthesis of vernakalant, a potential drug for irregular heart rhythms (atrial fibrillation). The power of the enzyme evolution was such that both the amine and ketone were changed quite significantly from the enzyme's original (natural) substrates. Using an engineered opine dehydrogenase, vernakalant was produced with very high stereoselectivity, as shown in Figure 5.40.

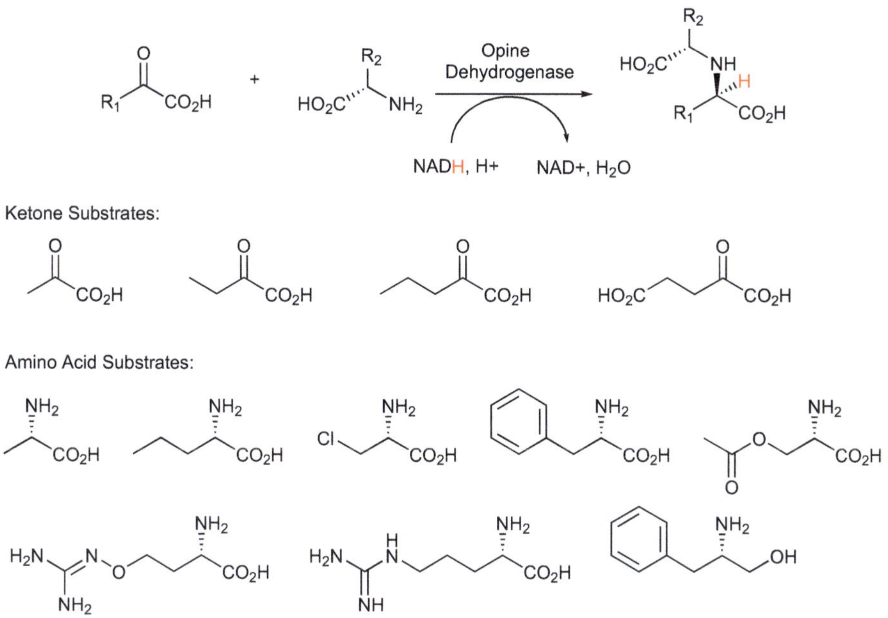

Figure 5.39   The substrate scope of opine dehydrogenases.

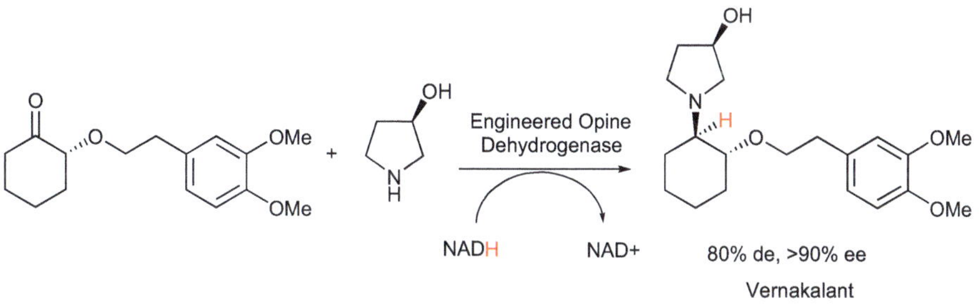

Figure 5.40   The synthesis of vernakalant using an opine dehydrogenase.

5.5.2   Amine Dehydrogenases

Although they are useful enzymes for the synthesis of α-amino acids, amino acid dehydrogenase enzymes are limited in the fact that they only work on α-keto acid substrates. To overcome this limitation, chemists have carried out protein engineering on amino acid dehydrogenases in order to widen their substrate specificity. The resulting enzymes no longer require an α-keto acid substrate and are able to work on isolated ketones. These enzymes are called amine dehydrogenases and catalyse the transformation shown in Figure 5.41.

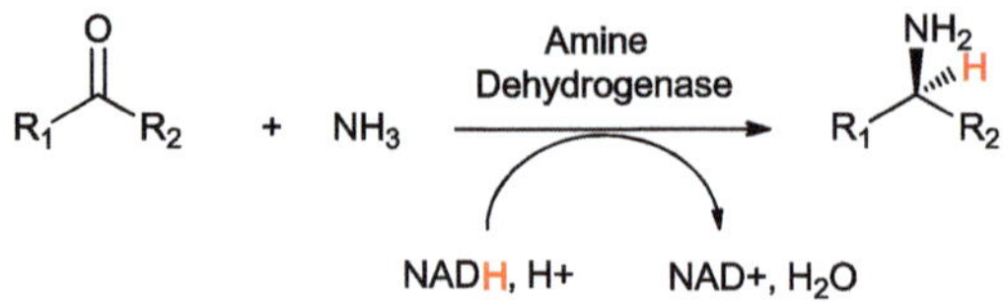

Figure 5.41 General transformation catalysed by amine dehydrogenases.

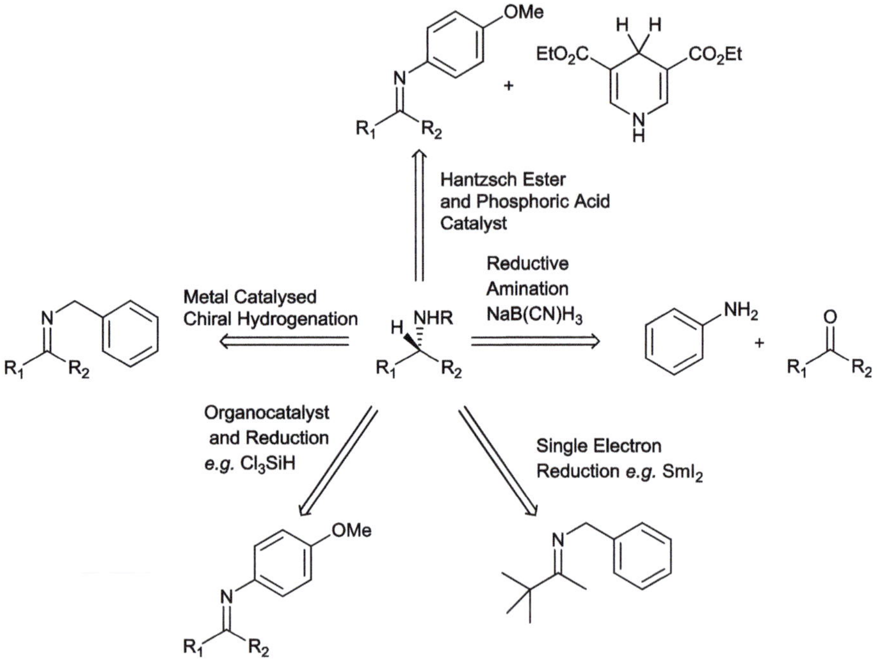

Figure 5.42 Selected chemical methods of amine synthesis.

As with the amino acid dehydrogenases discussed in Section 5.4.1, amine dehydrogenases effectively catalyse the reductive amination of ketones. These enzymes are co-factor-dependent, requiring a nicotinamide co-factor (NADH) in order to catalyse the reaction. We have already discussed in this chapter the various chemical methods for carrying out reductive aminations and some of their potential disadvantages. Chemical methods of reductive amination are shown again for you in Figure 5.42.

Amine dehydrogenases offer an alternative to chemical methods without a number of the drawbacks associated with them. Amine dehydrogenases work under mild, aqueous reaction conditions with high stereoselectivity and without the need for metal catalysts, chiral ligands or protecting groups. Amine dehydrogenases also offer a complementary

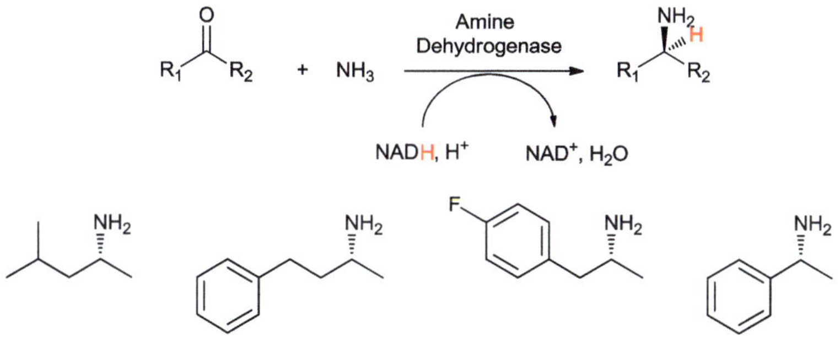

Figure 5.43 The scope of product amines produced by amine dehydrogenases.

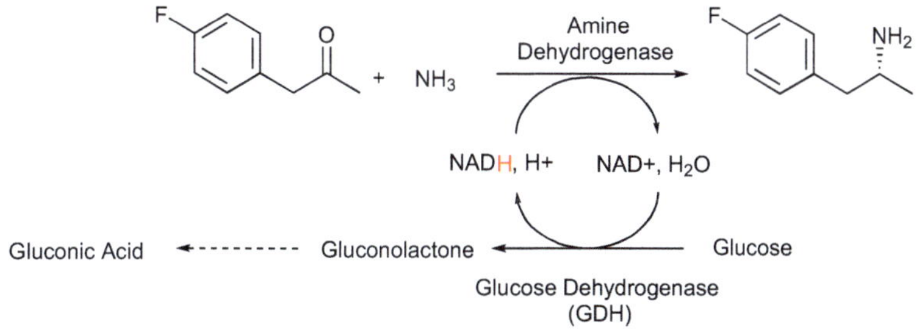

Figure 5.44 An enzyme-coupled co-factor recycling system for amine dehydrogenases.

method of amine synthesis to another class of enzymes, transaminases, which we will look at in more detail in Chapter 7. As this class of enzymes is still relatively new, the range of substrates that undergo reaction is quite limited, as shown in Figure 5.43. In each case the product enantiomer is shown and access to the opposite enantiomer may also be possible through the choice of an alternate enzyme or enzyme engineering.

We mentioned earlier that amine dehydrogenases are co-factor-dependent enzymes requiring a nicotinamide co-factor in order to carry out the reaction. Like the other nicotinamide co-factor-dependent enzymes we have already discussed in this chapter, it is possible to recycle the co-factor through the use of a second enzyme such as glucose dehydrogenase, as shown in Figure 5.44.

5.5.3 Imine Reductases

Imine reductases (Ireds) are a class of enzymes related to amino acid, opine and amine dehydrogenases. Instead of making and then reducing an imine, as their name suggests, imine reductases only carry out the

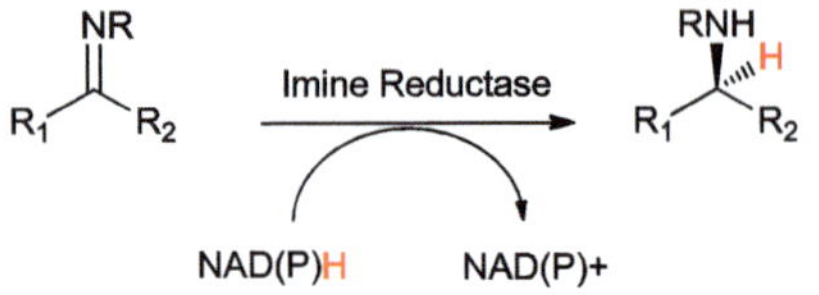

Figure 5.45   General transformation catalysed by imine reductases.

reduction of a preformed imine. As shown in Figure 5.45, this offers an alternative method of imine reduction that is complementary to methods that have already been discussed in this chapter. These enzymes have been found in a diverse range of organisms including bacteria, fungi and animals. Like the previous enzyme classes, imine reductases are co-factor-dependent enzymes and use a nicotinamide co-factor.

This class of enzymes is relatively new and unstudied and therefore the extent of imine reduction in Nature beyond a few reported examples is unknown. Only a handful of imine reductases have been described so far, but this number will inevitably grow as the usefulness of these enzymes is demonstrated. We have already covered a range of chemical methods that are used for the reduction of imines in Figure 5.32 and 5.42. As previously mentioned, the use of flammable hydrogen gas or expensive metal catalysts and ligands can be avoided by using an enzyme-catalysed reduction instead. Imine reductases are highly stereoselective and carry out the reduction of imines under mild aqueous conditions. They are also highly chemoselective, only reducing imines even in the presence of other functional groups that are prone to reduction, such as ketones.

Before using an imine reductase, we need to first synthesise the imine we want to use as a substrate for the reaction. The most popular method of imine synthesis is the condensation between a carbonyl compound (aldehyde or ketone) and an amine, as shown in Figure 5.46. The equilibrium position of this reaction lies in favour of the thermodynamically stable ketone, meaning that imines are prone to hydrolysis with water. This is turn presents us with a problem: because imine reductases work under aqueous conditions we need to find ways of stabilising our imine substrate, otherwise it will hydrolyse before it can be reduced by the enzyme. Chemists have overcome this problem by using more stable cyclic imines as substrates for these enzymes. Methods of cyclic imine substrate synthesis are shown in Figure 5.46.

Imine reductases have a reasonable substrate scope, which is limited by the range of imines available that are hydrolytically stable to the aqueous reaction conditions. Due to this issue, the number of acyclic imines reported to undergo successful reduction is rather

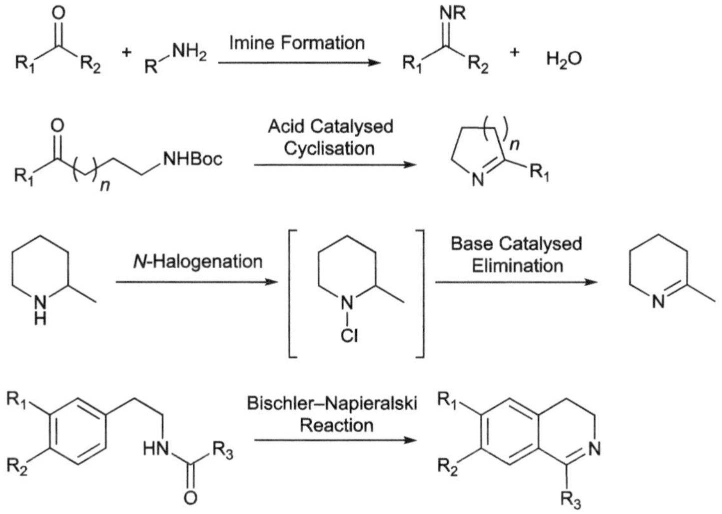

Figure 5.46 Selected examples of cyclic imine synthesis.

small. On the other hand, a wide range of cyclic imines can be successfully reduced, as shown in Figure 5.47. In each case, the substrate imine is shown with either enantiomer of the product often accessed by the choice of an appropriate enzyme. Imine reductases are also very chemoselective making them useful catalysts for synthesis.

Like the other nicotinamide-dependent enzymes already discussed in this chapter, it is possible to recycle the nicotinamide co-factor used by imine reductases through the use of a second enzyme, such as glucose dehydrogenase, as shown in Figure 5.48.

Imine reductases are a more recently discovered class of enzymes and their mechanism is yet to be fully elucidated. Nonetheless, amino acids that are potentially relevant to the mechanism have been postulated based on conserved residues from alignment of proteins from the superfamily. Further studies into the mechanism are ongoing.

5.6 Other Reductions

A number of other enzyme-catalysed reduction reactions have been reported, which are less well developed than those already covered. In each case, fewer enzymes have been reported to carry out the transformation and the range of substrates is not as wide. They are included here though for completeness and to give you an idea of other enzyme-catalysed reductions that could be useful in synthesis.

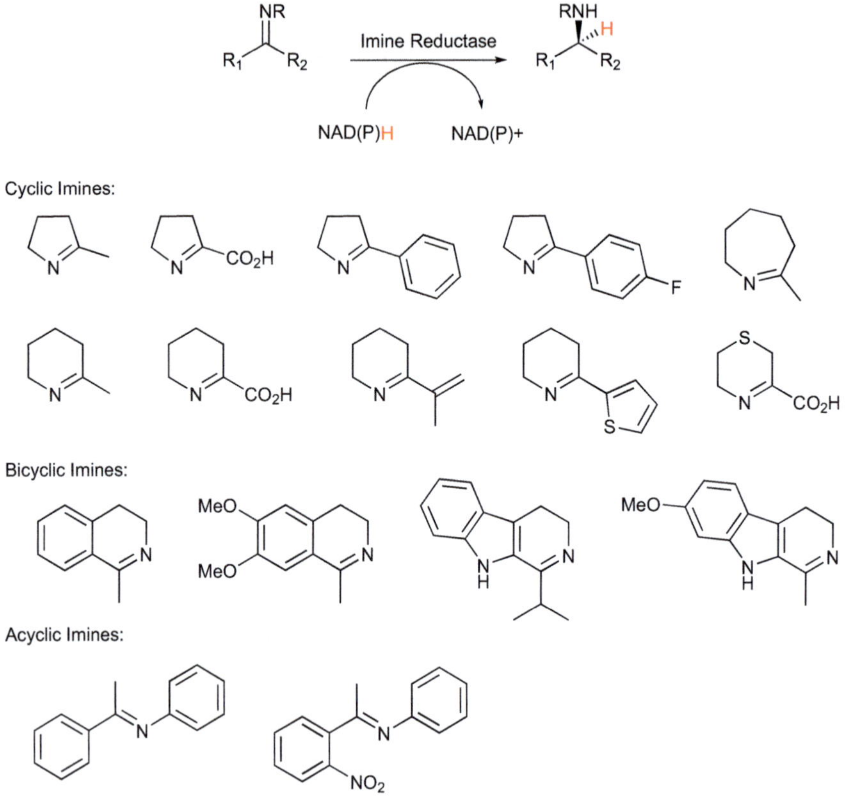

Figure 5.47 The scope of substrate imines reduced by imine reductases.

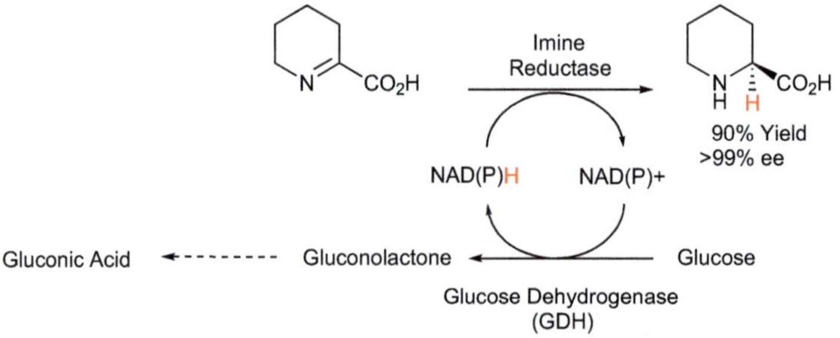

Figure 5.48 An enzyme coupled co-factor recycling system for imine reductases.

5.6.1 Reduction of Nitroaromatics

Nitroreductases are a family of enzymes that reduce nitrogen-containing compounds, in the form of a nitro, nitroso or hydroxylamine group. Although the biological function of nitro-reductases is not entirely known, they are thought to be involved in oxidative stress mechanisms, in some cases biosynthetic pathways and even bioluminescence. They have been used for bioremediation of nitro compounds, as biological sensors and in cancer therapies where, in affected cells, they are used to activate prodrugs into therapeutic agents. These enzymes catalyse the general reduction shown in Figure 5.49. Typically, the nitro group is reduced to the corresponding hydroxylamine, which may then be further reduced or undergo disproportionation to the amine and nitroso species.

Unlike the chemical reduction of nitro groups, which often involves the use of metal catalysts and flammable hydrogen gas, nitroreductase enzymes work under mild ambient conditions without the use of transition metals. A small range of nitro compounds can be reduced, as shown in Figure 5.50.

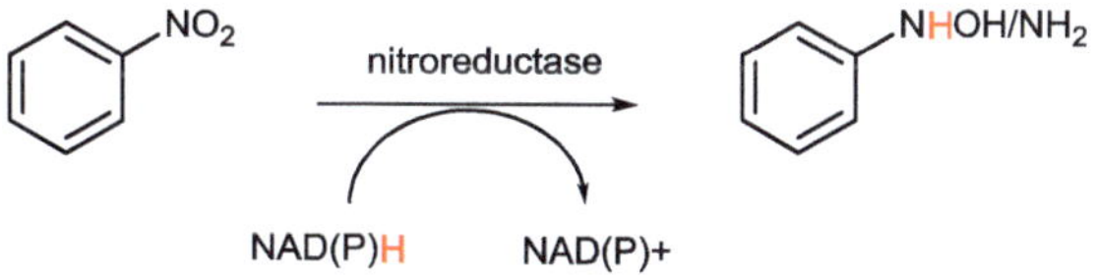

Figure 5.49 General transformation catalysed by nitroreductases.

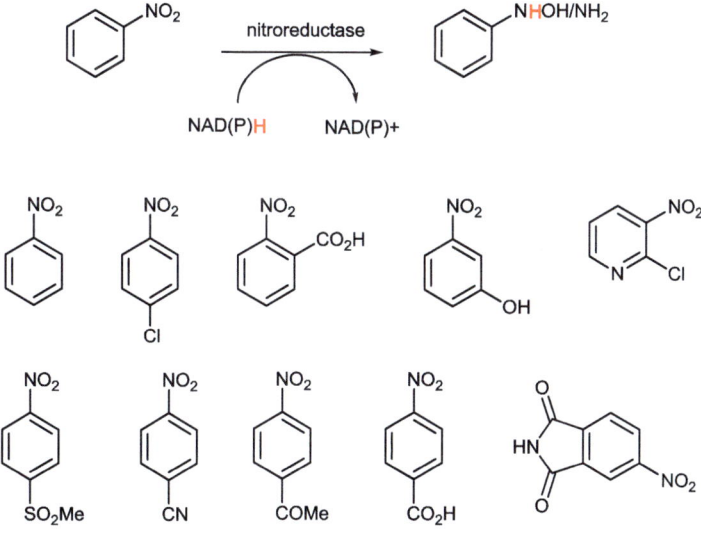

Figure 5.50 The scope of nitro compound substrates reduced by nitroreductases.

5.6.2 Reduction of Sulfoxides

As we will see again in Chapter 6, sulfur is an abundant naturally occurring substance that is found in its elemental form or in organo and metal sulfides. It is an essential element for all life, and is also found in co-factors such as glutathione and thioredoxin. It is required for the activity of iron–sulfur proteins, whilst disulfide bridges in proteins confer extra stability and strength. Whilst elemental sulfur is a product from and an oxidant for various species of bacteria, the more oxidised forms of sulfur (sulfoxides and sulfones) are found in a variety of natural products and man-made compounds, such as pharmaceuticals, agrochemicals, chiral auxiliaries and ligands. Sulfoxides, in particular, are of interest as the sulfur and oxygen atoms do not share a typical p-orbital π-bond like a carbonyl group. Instead, the sulfur assumes an sp^3 hybridisation with a lone pair of sulfur electrons occupying one of the orbitals. Sulfoxides are conformationally stable at room temperature and can be separated into enantiomers. As a result, the enzymatic reduction of sulfoxides has been reported as a method of carrying out the kinetic resolution of racemic sulfoxides. As shown in Figure 5.51, a small number of substrates can be resolved using this method. In each case, only the *S*-enantiomer was reduced leaving behind the *R*-sulfoxide.

5.6.3 Reduction of Azides

Azides are useful compounds in organic synthesis, as they can be reacted to make a heterocyclic ring or used as a leaving group in substitution reactions. They can also be used as a protected amine, which can be produced by reducing the azide. The reduction of azides to amines has been reported using baker's yeast. A small number of substrates have been reported, as shown in Figure 5.52.

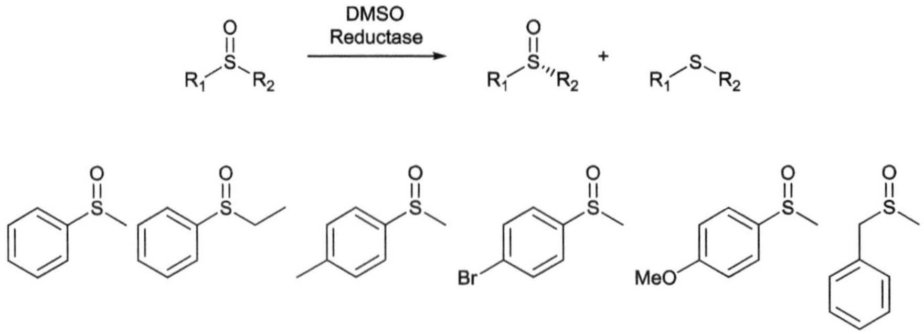

Figure 5.51   The scope of sulfoxide substrates reduced by DMSO reductase.

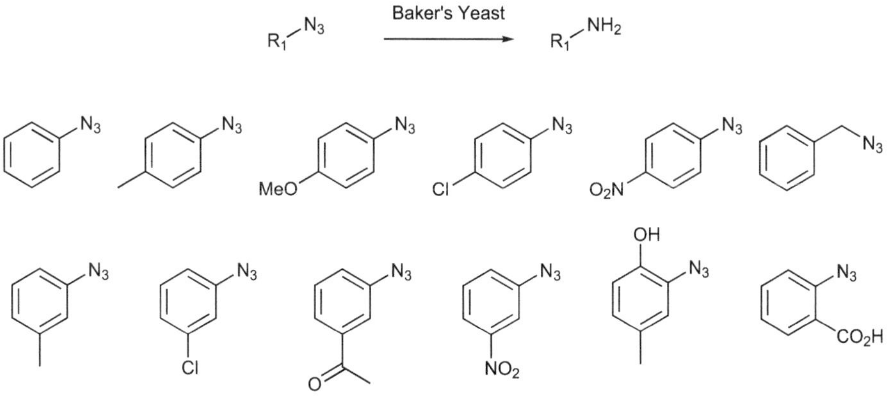

Figure 5.52 The scope of azide substrates reduced by baker's yeast.

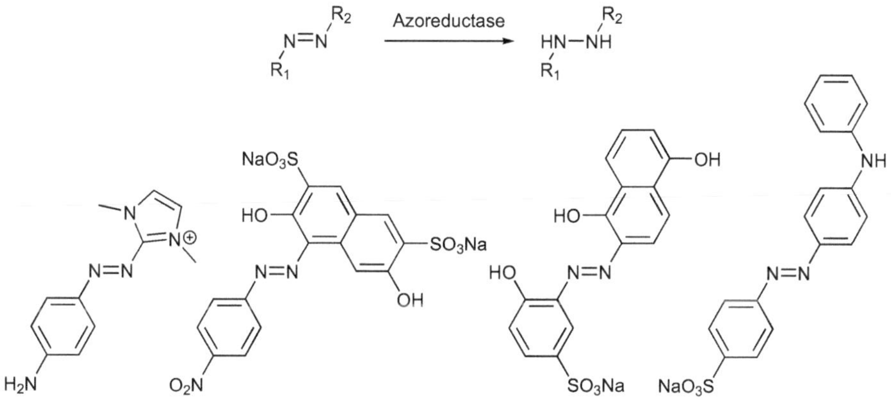

Figure 5.53 The scope of azo substrates reduced by azoreductases.

5.6.4 Reduction of Azo Compounds

Azo compounds are important compounds as the azo functional group often extends the *pi*-system in a molecule with the result that these compounds, in particular aromatic azo compounds, are often highly coloured. This property makes them very useful in the textile industry, where they are used as colourants and dyes. As a result, waste from the textile industry is often coloured and environmental rules limit the appearance of colour in any water discharges. As one solution to this problem, a number of enzymes have been reported to reduce azo dyes and hence remove the colour from these waste streams. These enzymes reduce a variety of azo compounds to the corresponding diamines. Many dyes exist, so to give you an idea of the types of substrates for these enzymes, a few dye compounds are shown in Figure 5.53.

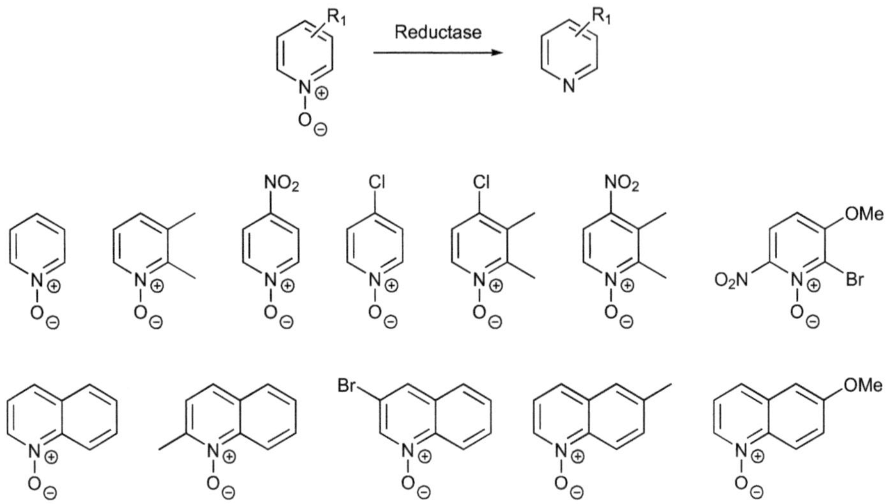

Figure 5.54 The scope of N-oxide substrates reduced by a number of reductase enzymes.

5.6.5 Reduction of N-oxides

N-oxides are highly polar, hydrophilic functional groups, which as their name suggests are one oxidation state higher than amines. They are used as protecting groups for amines in organic synthesis and more generally as surfactants in a number of products such as surface cleaners, shampoos and detergents. A number of enzymes have been reported to reduce N-oxides to the corresponding amines. As shown in Figure 5.54, a number of pyridine and quinolone N-oxides can be reduced.

6 Oxidation

6.1 Introduction

This chapter will cover a number of different oxidation reactions catalysed by a range of enzymes. Despite catalysing different transformations, these enzymes are all from the oxidoreductase group within the enzyme classification system. Oxidoreductases are given the enzyme classification number 1 and are defined as enzymes that catalyse oxidation/reduction reactions *via* the transfer of H and O atoms or electrons from one substance to another. Just like the reductions in the previous chapter, oxidations are an important methodology for the synthetic chemist. As we saw in Chapter 5 (Reduction), if a particular functional group is sensitive to certain reaction conditions, then we may be able to carry it through the synthesis at a different oxidation level until it is required. This time, if we carry a functional group through the synthesis in a lower oxidation state in order to prevent it from reacting, then we will need to make use of an oxidation reaction to oxidise it prior to carrying out the desired reaction upon it. This is the opposite of the strategy that we covered in Chapter 5.

The transformations covered within this chapter are shown in Figure 6.1, with each transformation and its synthetic use discussed in more detail in a separate section. Often, more than one class of enzyme is capable of carrying out the same transformation and therefore the sections in this chapter will contain a section for each enzyme capable of catalysing a specific transformation.

Biocatalysis in Organic Synthesis: The Retrosynthesis Approach
By Nicholas J. Turner and Luke Humphreys
© Nicholas J. Turner and Luke Humphreys 2018
Published by the Royal Society of Chemistry, www.rsc.org

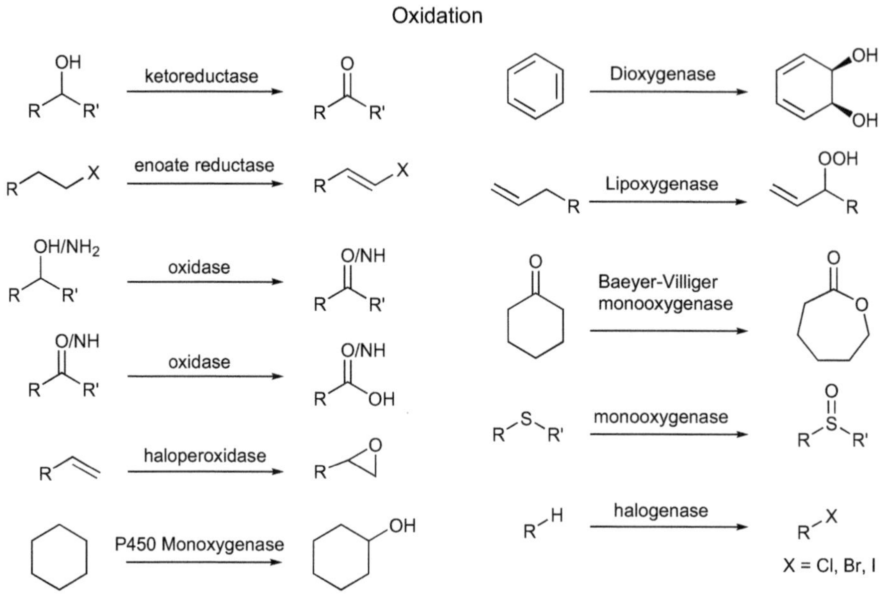

Oxidation

Figure 6.1 Overview of transformations presented in this chapter.

6.2 Oxidation of Alkanes

Alkanes are saturated hydrocarbons, which makes them the lowest oxidation level of carbon atoms. They are highly important molecules, forming the basis of the chemical industry itself. They can be obtained from natural resources such as gas and crude oil and are also found in algae, bacteria and plants. They are used as a fuel to produce energy and converted into alkenes, which in turn are raw materials for products such as polymers, adhesives and detergents, as well as chemical synthesis. Chiral alcohols, formed by the oxidation of alkanes, are also valuable and are precursors for compounds used in agrochemicals, pharmaceuticals or liquid crystals. This makes the selective oxidation of alkanes to alcohols an important reaction for chemical synthesis. However, this reaction is challenging as carbon–hydrogen bonds are relatively inert with high activation energy, and for a substrate with many C–H bonds, obtaining high chemo-, regio- and stereoselectivity is a problem. In addition, the product alcohols are more easily oxidised than the starting material, so these reactions are difficult to control. In addition to chemical approaches, enzymes have been reported to catalyse the selective oxidation of alkanes to alcohols. These methods use mild conditions and green solvents, and the products can be obtained with high chemo-, regio- or ste-reoselectivity. In the next two sections, we will look at families of enzymes that are capable of performing these reactions.

6.2.1 Oxidation of Alkanes with P450 Monooxygenases

One of the families of enzymes capable of catalysing the hydroxylation of alkanes is the P450 monooxygenases. As shown in Figure 6.2, these enzymes use molecular oxygen and two electrons from a nicotinamide co-factor, which is the terminal reductant to carry out the hydroxylation.

There are many methods of producing alcohols, some of which are detailed in Figure 6.3. Although the direct functionalisation of

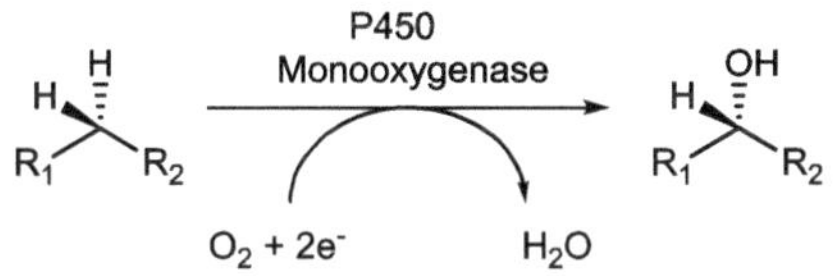

Figure 6.2 General transformation of alkane oxidation catalysed by P450 monooxygenases.

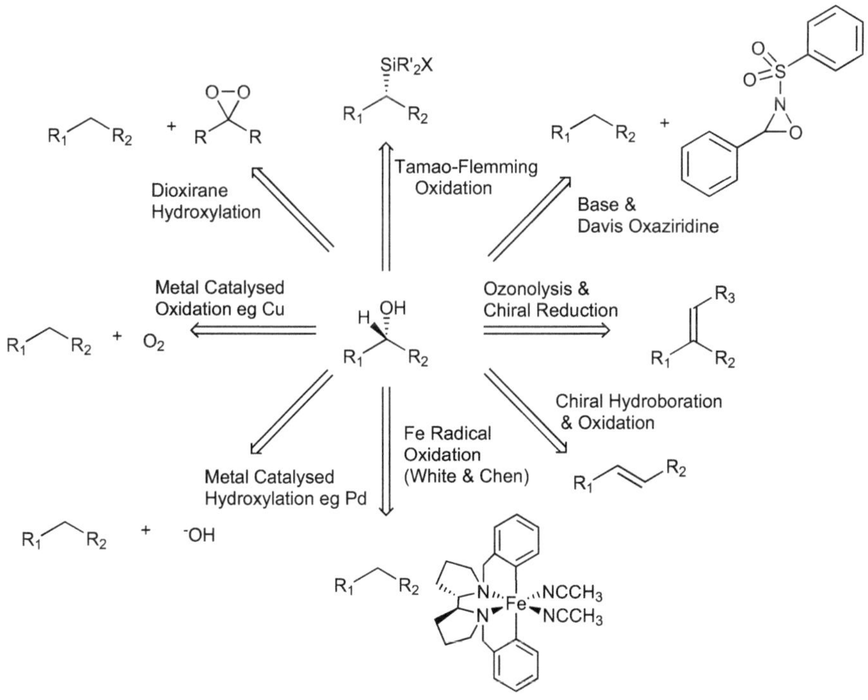

Figure 6.3 Selected chemical methods of alkane oxidation or equivalent.

alkenes has been known for a long time, the harsh conditions and lack of selectivity have prevented its use on more complex substrates until more recently. Traditional methods used in organic synthesis on complex intermediates often use a starting material at the alcohol oxidation state (or higher) and may require more than one step to put the alcohol group in place. Many of these methods require the use of transition metal catalysts and chiral ligands and may also use harsh reaction conditions. Some catalytic methods may also use terminal oxidants that are difficult to use on a large scale, such as peroxides. Even with more modern state of the art systems, regioselectivity and low yields may be a problem with some substrates.

In contrast, P450 monooxygenase (or cytochrome P450) enzymes catalyse the hydroxylation of a wide range of substrates under very mild aqueous reaction conditions using oxygen as the terminal oxidant. The chemo-, regio- and stereoselectivities are typically very high and the yields are often good. This makes P450 monooxygenases a useful, environmentally benign, non-toxic alternative to chemical methods. Due to the presence of alkanes in the environment, a wide variety of microbes have evolved to oxidise alkanes. As a result, P450 monooxygenases are found in a large number of organisms, from bacteria to plants and humans. P450 monooxygenases are complex proteins which, due to their mechanism (which we will see shortly), require additional proteins known as redox partners to facilitate the reaction. As a result, P450 monooxygenases can be divided into four different classes depending on the relationship with their redox partners. Two of the four classes are shown in Figure 6.4. You do not need to know the details of how the electrons required in the reaction are transported, but it should give you an appreciation of these systems and help in choosing which might be best to try first. Class I P450 monooxygenases are found in bacteria and are soluble proteins, unlike the other classes, which are found bound to membranes. As can be seen in Figure 6.4, in order for electrons to be shuttled from the nicotinamide co-factor NAD(P)H to the P450 monooxygenase, we need two additional proteins: a ferrodoxin and a ferrodoxin reductase. In class I systems, these two extra proteins are not connected to the P450 monooxygenase and so must be added in addition to the P450. They must also be able to "talk" to one another and the P450 in order for the electrons to be shuttled efficiently. As a consequence, when we want to run a reaction in the laboratory, we have to make (or buy) all three proteins and ensure that they are compatible in order to produce an active system.

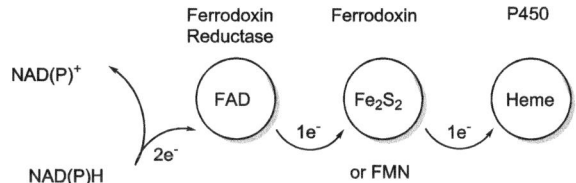

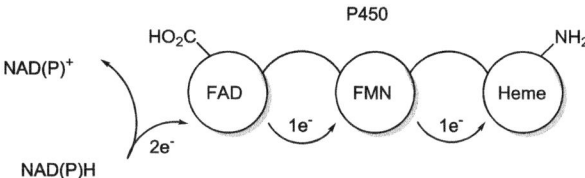

Figure 6.4 The enzyme structure of class I and class IV P450 monooxygenases.

In comparison, class IV systems contain all three components (P450, ferrodoxin and ferrodoxin reductase) linked together to form one large protein. This makes the system "self-sufficient" in that we only need to make one (albeit big) protein for the system to be active. We also remove the problem of the components not being compatible with one another as self-sufficient P450s have evolved to include compatible redox partner proteins. This system is much easier to use in the lab as we now only need to make or buy one protein. Just in case you are wondering, class III systems are also self-sufficient, but differ slightly in the active site of the ferrodoxin. Class II systems are slightly different as they are bound to membranes and the two redox partner proteins are linked, whilst the P450 monooxygenase is separately bound to the membrane.

Thousands of P450 monooxygenase genes have been reported and classified into families according to their sequence. In general, the substrate scope of P450 monooxygenases as a whole is very varied, with a wide range of substrates undergoing hydroxylation. However, for each particular enzyme, the range of substrates can in some cases be rather narrow and this is, in part, the reason for their excellent selectivity. As a result, several P450 monooxygenases may need to be screened in order to find the ideal enzyme. A similar scenario is employed by our livers, where a number of P450s, each with their own substrate scope, are used in xenobiotic metabolism.

As mentioned previously, P450 monooxygenases display high chemo- and regioselectivity, such that hydroxylations can be carried out in the presence of other functional groups such as alkenes. In each case in Figure 6.5, the product of the oxidation is shown. Where more than one alcohol is present, the one incorporated by the P450 monooxygenase is coloured red. In some cases, alternative regio- or stereoisomers can be produced through the use of an alternative enzyme.

The mechanism of P450 monooxygenases is rather complex, although you do not need to know the details in order to be able to

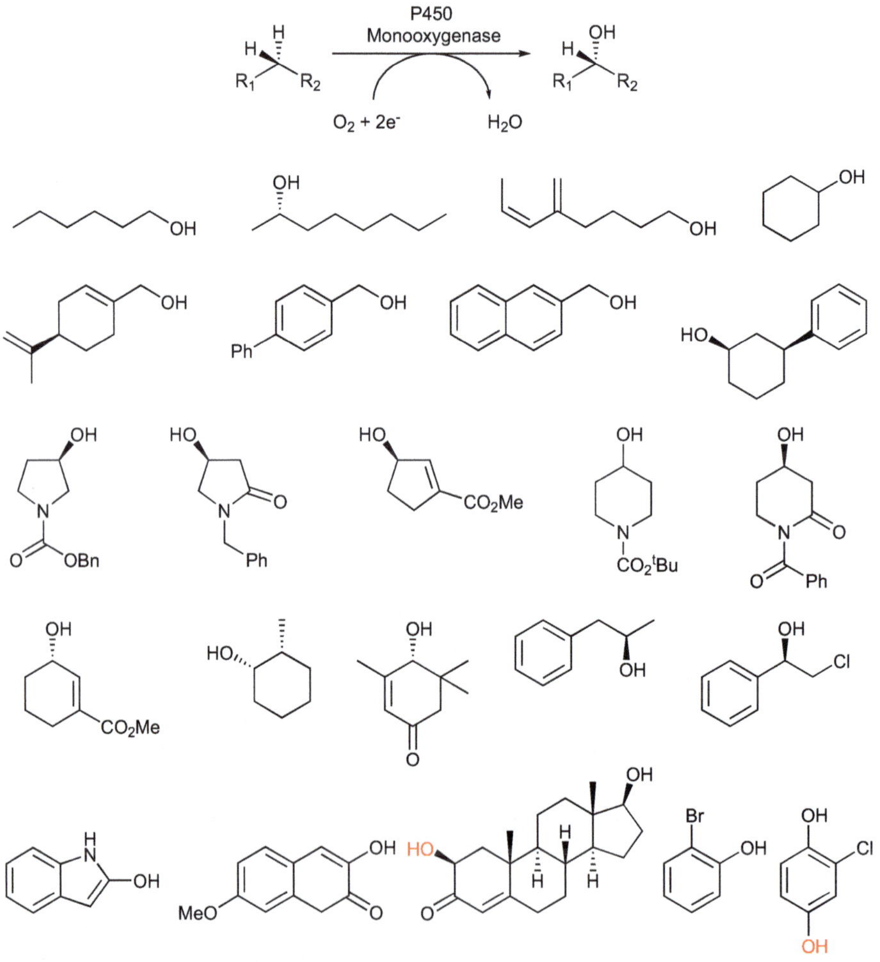

Figure 6.5 The scope of alcohol products produced by oxidation reactions catalysed by P450 monooxygenases.

suggest when it might be applicable to use these enzymes in synthesis. However, like the other enzyme mechanisms in this section, we have included it here for your information and for those who would like greater detail. The overall reaction catalysed by P450 monooxygenases is shown in Figure 6.6. As we can see, the substrate is hydroxylated using the nicotinamide co-factor and oxygen, with water generated as a by-product.

We have seen that electrons are transported to the P450 monooxygenase *via* two redox partner proteins, and this system is shown again in Figure 6.7 for the self-sufficient class IV system. The key thing to remember here is that the terminal oxidant is the nicotinamide NAD(P)H, which gets oxidised to $NAD(P)^+$, and this is where the electrons are supplied from in the mechanism. P450

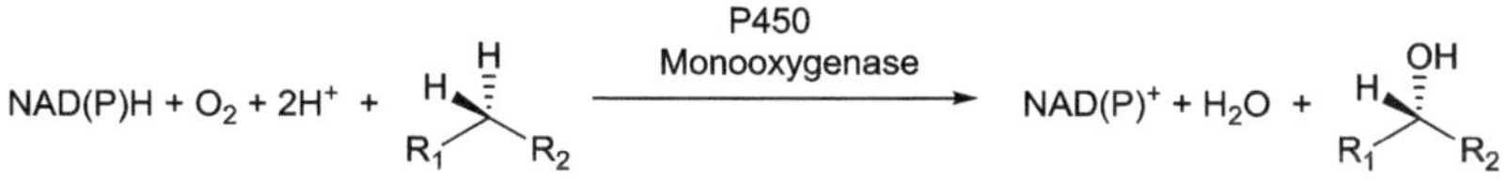

$$NAD(P)H + O_2 + 2H^+ + \quad \underset{R_1 \quad R_2}{\overset{H}{\diagdown}} \quad \xrightarrow[\text{Monooxygenase}]{\text{P450}} \quad NAD(P)^+ + H_2O + \quad \underset{R_1 \quad R_2}{\overset{OH}{\diagdown}}$$

Figure 6.6 A more detailed equation of P450 monooxygenase-catalysed oxidation of alkanes.

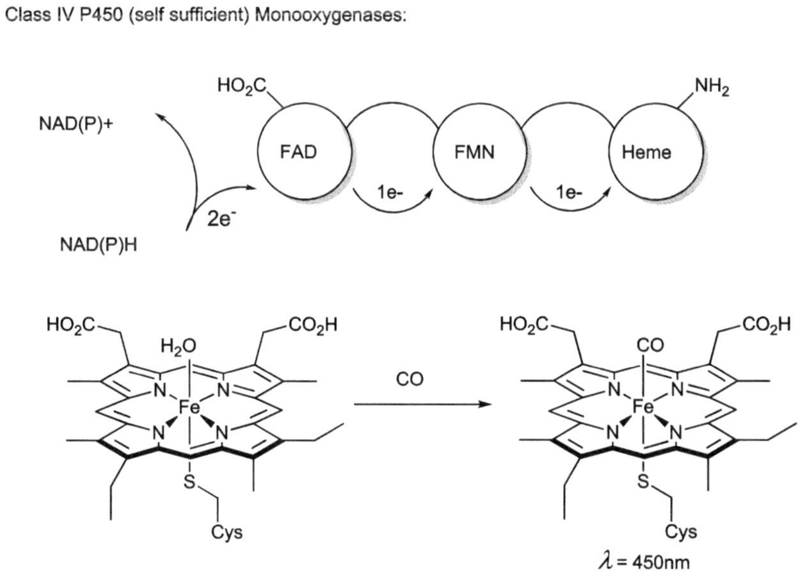

Class IV P450 (self sufficient) Monooxygenases:

Figure 6.7 The heme group found in the active site of P450 monooxygenases.

enzymes are actually a subclass of a larger family of heme-containing oxidoreductases. As the name suggests, in the active site of P450 enzymes, there is a heme group consisting of an iron(III) atom complexed by a porphyrin ring. Typically, the iron is also bound by an active site cysteine residue and a water molecule, as shown in Figure 6.7. When the water ligand is replaced by carbon monoxide in the active site, the absorption spectra of these proteins show a large characteristic peak at 450 nm, which is why they are called P450s.

The hydroxylation mechanism of P450 monooxygenases starts from the heme complex with water bound, as shown in Figure 6.8. This can be considered to be the resting state of the enzyme. The mechanism starts with the substrate being bound into the active site, substituting the water molecule previously bound to the heme and changing the heme iron from a low to high spin state. At this point, the first electron transfer occurs from the redox partner enzymes, reducing the iron from oxidation state III to II. In this oxidation state the iron is now able to bind molecular oxygen, forming a superoxo complex with the iron being oxidised back to iron(III). The second electron transfer now occurs to generate a peroxo anion, which is protonated very rapidly resulting in an iron(III) hydroperoxy complex. Further protonation of the same oxygen atom leads to the loss of water. In order to stabilise the remaining oxygen atom, the iron is oxidised to form a high-valence iron oxo complex, which is highly reactive. This can either be represented as a neutral structure with iron in the (V) oxidation state, or as an iron(IV) complex that overall is a radical cation with the radical stabilised by the heme. Irrespective of how this is represented, the iron oxo complex is the key catalytic complex if we want to carry out a hydroxylation reaction. To complete the catalytic cycle, the substrate is hydroxylated and then released by exchange with a water molecule, getting us back to the resting state.

Having spent this much time discussing the catalytic cycle, it is also worth studying the hydroxylation reaction in a little more detail. In fact, it is probably more important to know how the iron oxo complex reacts with the substrate than all of the steps involved in generating the iron oxo complex itself. Therefore, if you only want to remember one part of the mechanism of P450 monooxygenases, Figure 6.9 will be most useful. The hydroxylation part of the mechanism is known as a "radical rebound" mechanism, as we shall see below. We start with the iron oxo complex, which we have represented as the iron(IV) radical cation. In the first step of the mechanism, the oxygen atom

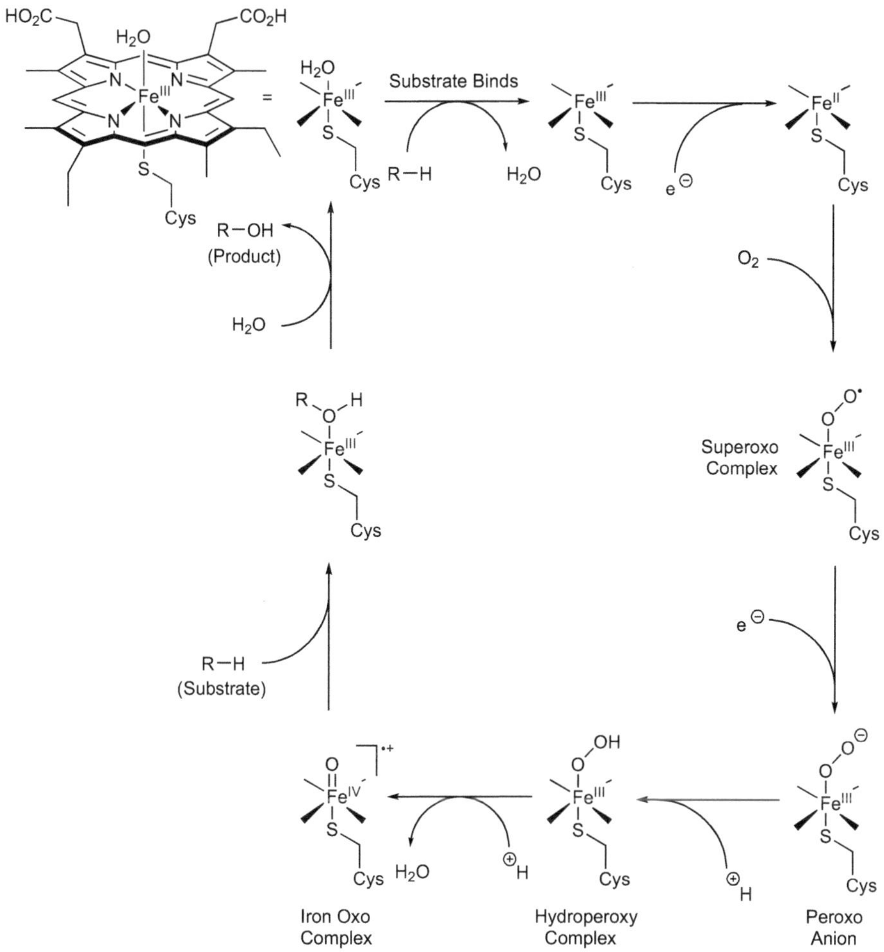

Figure 6.8 General mechanism of the oxidation of alkanes catalysed by P450 monooxygenases.

abstracts a hydrogen radical from the substrate and at the same time the heme radical cation is reduced to give a neutral iron(IV) complex. The substrate radical now "rebounds", attacking the oxygen atom to give the hydroxylated product and further reducing the iron to a more stable iron(III) oxidation state.

6.2.2 Oxidation of Alkanes with Lipoxygenases

In addition to the P450 monooxygenases discussed above, the oxidation of alkanes is also catalysed by a group of enzymes called

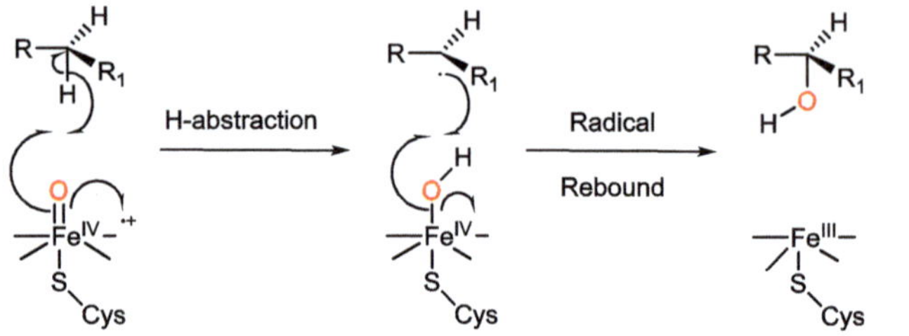

Figure 6.9 The radical rebound mechanism used by P450 monooxygenases for the oxidation of alkanes.

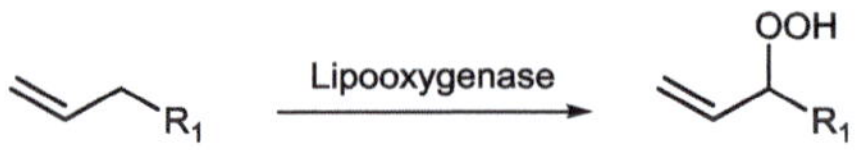

Figure 6.10 General transformation for the oxidation of alkanes catalysed by lipoxygenases.

lipoxygenases. These enzymes are found widely in plants, fungi and animals, where they react with polyunsaturated fatty acid substrates that contain *cis* double bonds. These are essential fatty acids in humans, but are not present in yeast and the majority of bacteria. As a result, lipoxygenases are absent from yeast and typical prokaryotes. Lipoxygenases catalyse the oxidation of fatty acids as shown in Figure 6.10.

As can be seen from Figure 6.10, lipoxygenases catalyse the oxidation of substrates in the allylic position (the position α to a double bond) to produce a hydroperoxide. The hydroperoxide can undergo a number of reactions including reduction to the alcohol, which is formally equivalent to an allylic oxidation. A number of chemical methods exist for the allylic oxidation of alkanes or the oxidation of alkanes to form hydroperoxides. Some of these methods are shown in Figure 6.11 and they usually involve the use of metals or organic peroxides, which can be difficult to handle, particularly on a large scale.

Lipoxygenase enzymes offer a complementary alternative to the chemical methods shown above. They catalyse the oxidation of the allylic position of alkanes under mild reaction conditions and in high regio- and stereoselectivity. Lipoxygenases have a relatively narrow substrate range, which is partly why they are so selective. They oxidise

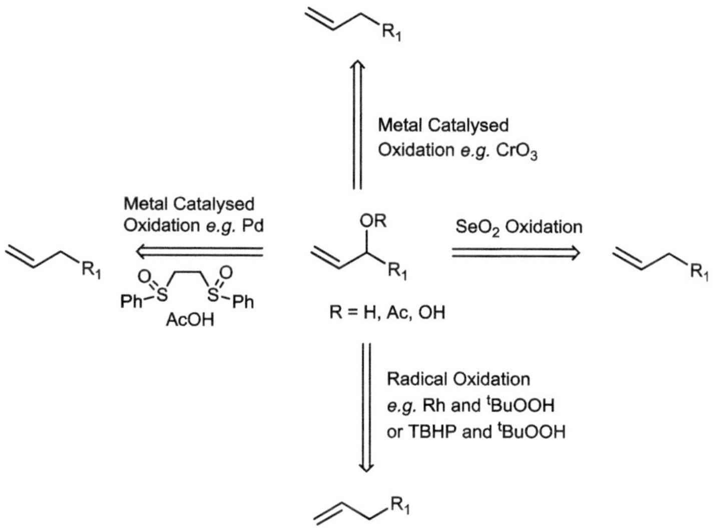

Figure 6.11 Selected chemical methods of alkane oxidation.

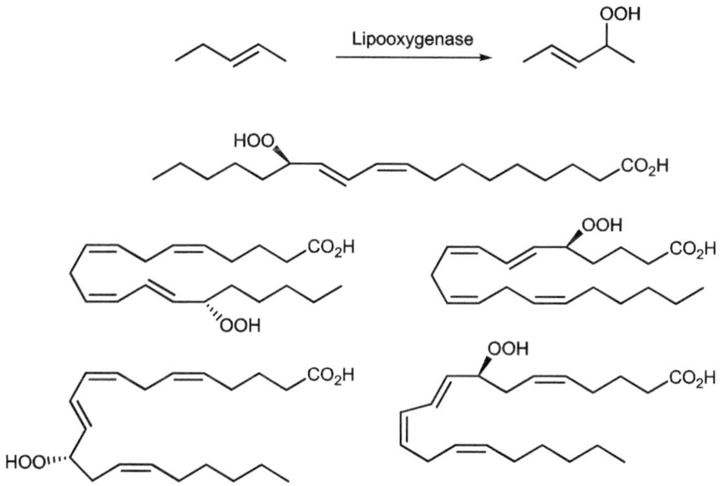

Figure 6.12 The substrate scope of hydroperoxides produced by lipoxygenases.

a range of fatty acid substrates as shown in Figure 6.12. In each case the product hydroperoxide is shown, although the products can undergo further reaction to produce either the alcohol or epoxide in some cases.

Lipoxygenases display some similarities to the P450 mono-oxygenase enzymes we discussed previously. Like P450 mono-oxygenases, lipoxygenases use an active site iron atom to catalyse the oxidation. In P450 monooxygenases, the iron is ligated by a heme ligand, whereas in lipoxygenases the iron is ligated by four residues of which three are histidines. Both types of enzymes also use molecular oxygen as the oxidant.

6.3   Oxidation of Alkenes

Alkenes are useful functional groups as they can be transformed into a range of other functional groups through a number of reactions. This makes them versatile intermediates due to the array of functional group interconversions they can undergo, as shown in Figure 6.13. With respect to oxidation, alkenes may be converted to epoxides and diols or into ketones by transition metal-catalysed oxidation. The alkene may also be oxidatively cleaved to give an aldehyde or ketone depending upon the substitution of the alkene.

The functional group interconversions shown in Figure 6.13 are, for the most part, carried out using classical organic chemistry

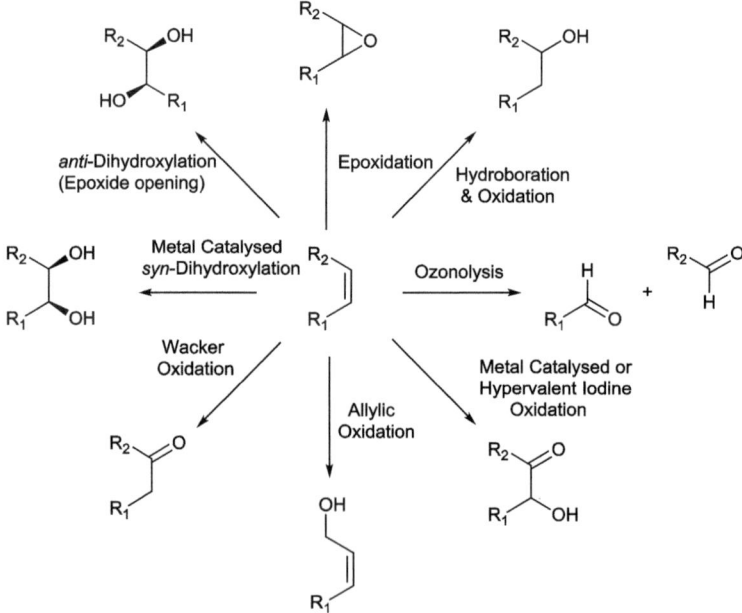

Figure 6.13   Selected chemical methods of derivatising an alkene.

reagents. Although these are extremely useful reactions, the reagents required may be toxic and typically generate metal waste, which has to subsequently be disposed of. Stereoselectivity is achieved through the use of chiral ligands, which may need several steps to be synthesised. In addition, some of the oxidants used may not be compatible with other sensitive functional groups present. Enzymatic methods are available for most of the transformations shown in Figure 6.13 and, in contrast to more traditional chemistry approaches, they do not require the use of metals or chiral ligands and thus generate much less waste. The mild reaction conditions are likely to be compatible with more sensitive functional groups and they typically display high regio- and stereoselectivity. Each of the following sections details a family of enzymes that are capable of oxidising alkenes to some of the products shown in Figure 6.13.

6.3.1 Oxidation of Alkenes and Aromatic Rings with Dioxygenases

Dioxygenase enzymes were first identified in soil bacteria over 50 years ago and are involved in environmental degradation of aromatic compounds. They have the ability to catalyse the *cis*-dihydroxylation of alkenes and, more importantly, aromatic hydrocarbons as shown in Figure 6.14. The equivalent *cis*-dihydroxylation of alkenes is traditionally carried out with osmium tetroxide, which is a particularly toxic compound. Although modern methods use catalytic amounts of this chemical and a less toxic reagent to re-oxidise the osmium, these methods still require careful disposal of the metal residues at the end of the reaction.

Importantly, dioxygenase enzymes catalyse the same transformation under mild aqueous reaction conditions without the use of toxic metals such as osmium. The dihydroxylation of aromatic rings can be carried out on a range of substituted aromatic and heteroaromatic substrates. A range of *cis*-diol products are shown in Figure 6.15.

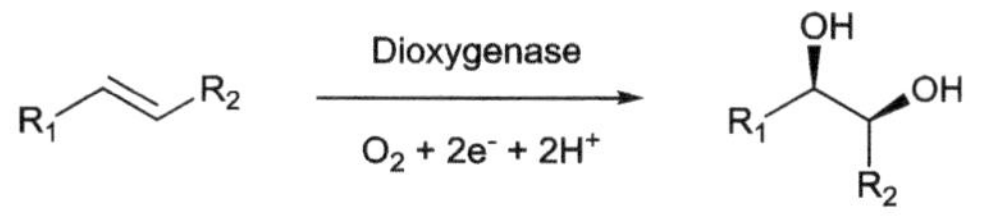

Figure 6.14 General transformation for the oxidation of alkenes using dioxygenases.

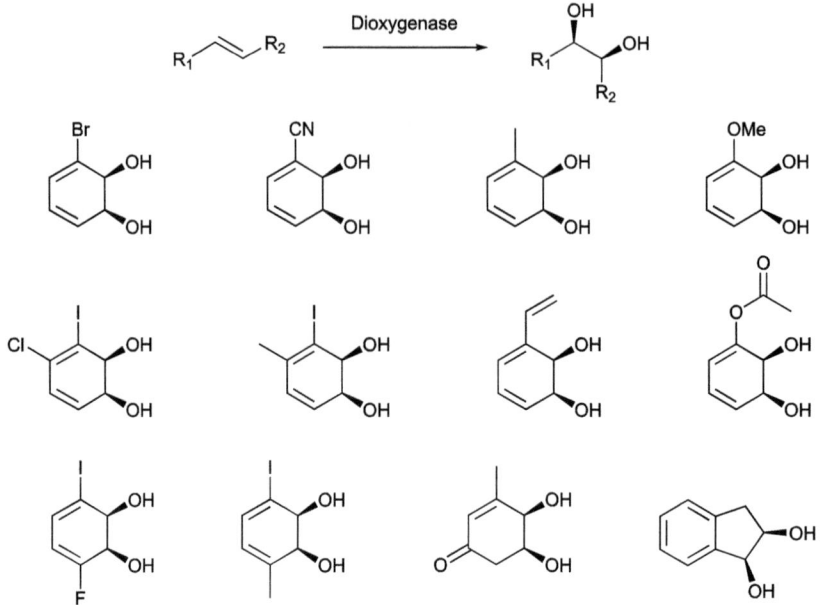

Figure 6.15 The substrate scope of product diols produced by dioxygenases.

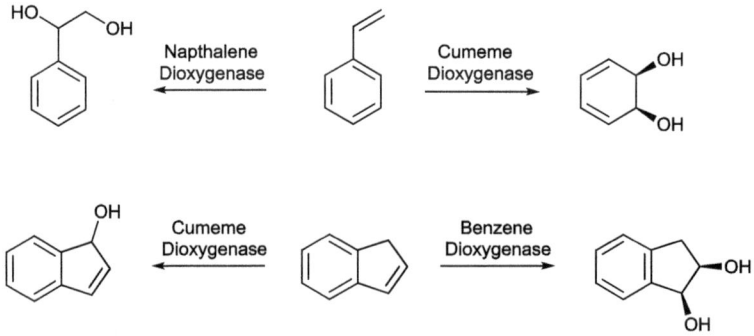

Figure 6.16 Examples of the regioselectivity obtained by using different dioxygenases.

In addition, bicyclic or benzylic alkenes can also be hydroxylated, usually with high regio- and stereoselectivities. Complementary selectivities can often be achieved through the choice of an appropriate enzyme, as shown in Figure 6.16.

Due to the similarity of the transformations that they catalyse, it may come as no surprise to learn that P450 monooxygenases and dioxygenases also have some similarities in their structure and mechanism. You do not need to know the more detailed information about the structure and mechanism that follows, but we have included it here for those who would like a more in depth understanding of these enzymes. The exact mechanism of dioxygenases has not been fully elucidated yet and several possibilities exist, so instead we will discuss some of the details to give you an idea of how these enzymes work. Like P450 monooxygenases, dioxygenases use an electron transport system made up of redox partners. However, unlike P450 monooxygenases, which contain a heme ligand for the active site iron, dioxygenases use an iron centre stabilised by histidine or aspartic acid side-chains. As shown in Figure 6.17, they also use an iron sulfur redox partner rather than the flavin-dependent one favoured by P450 monooxygenases.

One possible mechanism is shown in Figure 6.18, with the dioxygenase starting with iron(III) bound in the active site. Upon substrate binding, this is reduced by electron transfer to an iron(II) complex. As with P450 monooxygenases, the active site iron(II) complex is able to bind molecular oxygen to initially give an analogous iron(III) superoxo complex. However, when the dioxygenase superoxo complex undergoes reduction and protonation, rather than losing water (like the P450 monooxygenase), it forms an iron(V) intermediate instead.

Class IV P450 (self sufficient) Monooxygenases:

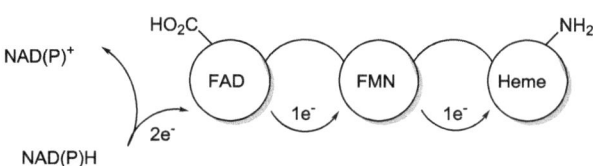

Dioxygenases:

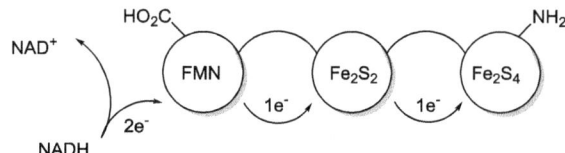

Figure 6.17 A comparison of P450 monooxygenase and dioxygenases.

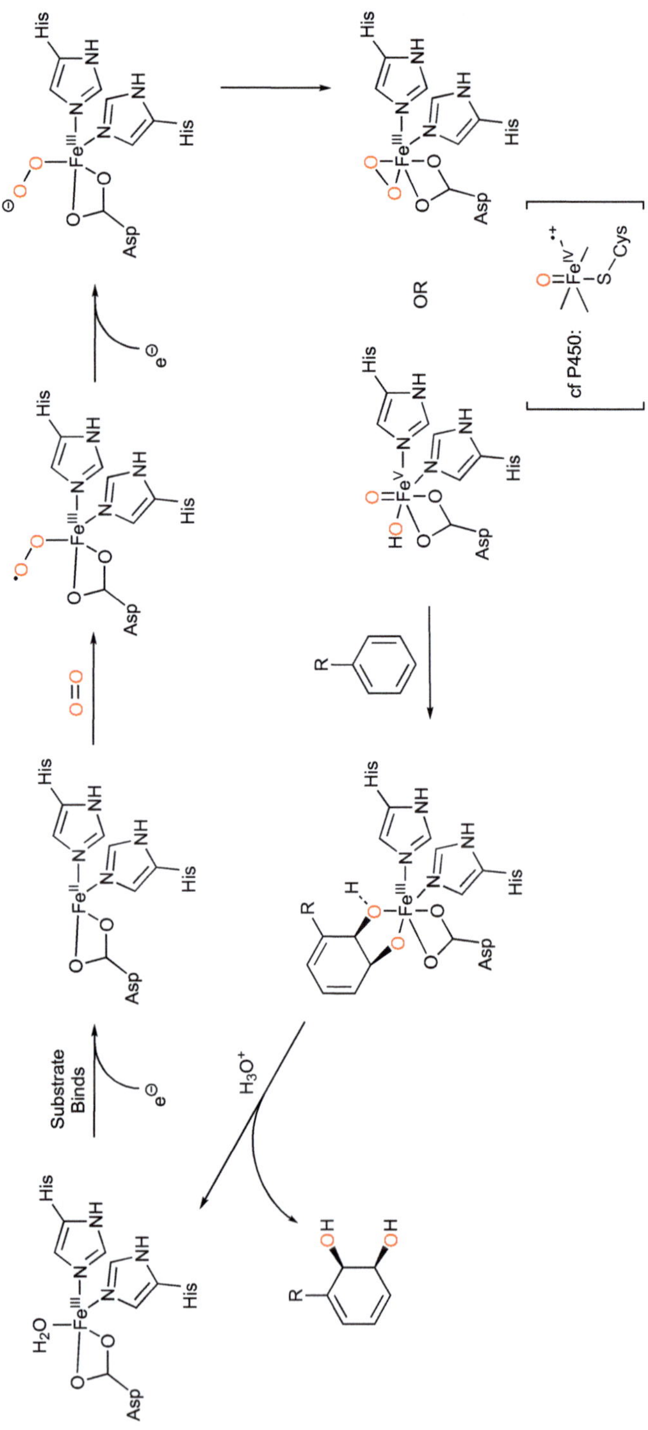

Figure 6.18 General mechanism for the oxidation of alkenes catalysed by dioxygenases.

This is the equivalent to the P450 iron(IV) oxo complex and as shown in Figure 6.18 can be represented as the dioxirane or an oxo species. The dioxygenase then adds both oxygen atoms to the substrate (in this case a benzene ring) to give an iron(III) substrate complex. Finally, with the addition of water, the product is protonated and released from the active site, regenerating the original iron(II) catalyst resting state ready for another catalytic cycle to begin.

6.3.2 Oxidation of Alkenes with Monooxygenases (Including P450 Monooxygenases)

Enantiomerically pure epoxides are valuable intermediates for the synthesis of both natural products and industrial compounds. As we saw in Figure 6.13, epoxides can be easily accessed through the epoxidation of an alkene. A number of excellent chemical methods exist for asymmetric epoxidation, such as the Sharpless and Jacobsen methods, which produce the desired epoxides with high enantio-selectivity. However, these methods (and others) use metals and chiral ligands, which often need to be synthesised and generate waste that must be disposed of at the end of the reaction. Although these methods work well for certain substrates, such as allylic alcohols and isolated alkenes, asymmetric epoxidation of some alkenes remains challenging. In contrast, monooxygenase enzymes are a complementary methodology catalysing the epoxidation of alkenes. These enzymes work under mild conditions without generating potentially hazardous metal waste or the need to use chiral ligands. Furthermore, monooxygenase enzymes display high levels of regio-, chemo- and stereoselectivity.

One of the most well-used enzymes to carry out this transformation is styrene monooxygenase. This protein is a flavin-dependent mono-oxygenase that uses a flavin co-factor to carry out the epoxidation and an additional nicotinamide co-factor to reduce the flavin prior to the start of the catalytic cycle. As the name suggests, styrene mono-oxygenase works predominately on styrene derivatives as shown in Figure 6.19. In each case the product epoxide is shown.

A number of other monooxygenases also carry out these reactions, but they are not included here. Instead we will look at P450 mono-oxygenases, which, in addition to carrying out the hydroxylation reactions described above, are also able to catalyse the epoxidation of alkenes as shown in Figure 6.20.

In addition to styrenes, P450 monooxygenases also react with aliphatic and cyclic alkenes. As can be seen in Figure 6.21, a range of

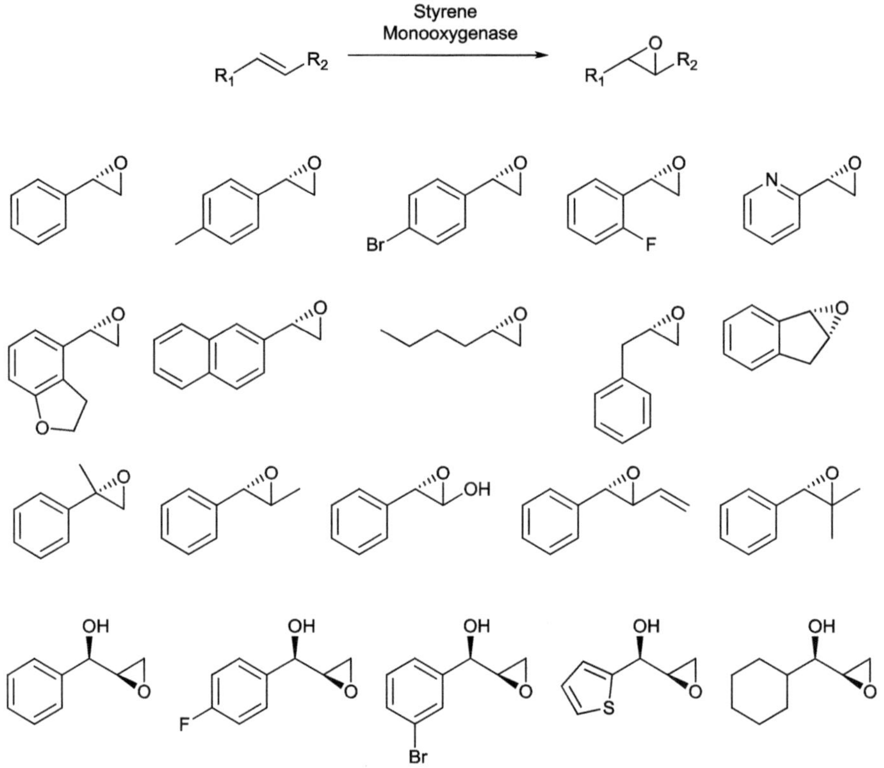

Figure 6.19 The scope of epoxide products produced by the styrene monooxygenase-catalysed oxidation of alkenes.

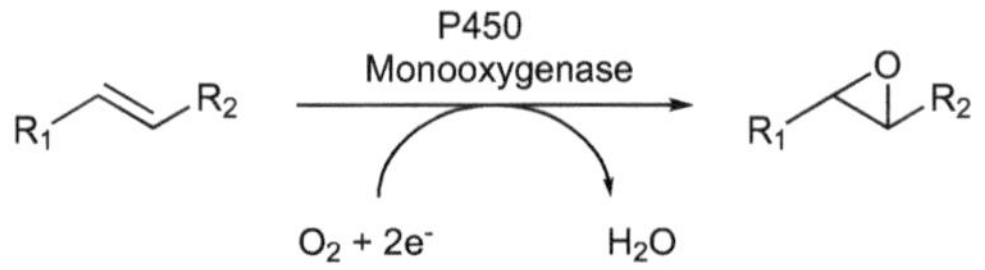

Figure 6.20 General transformation for the oxidation of alkenes catalysed by P450 monooxygenases.

substrates undergo reaction often in high regio- and stereoselectivity. In some cases, the opposite stereoisomer can be produced by the choice of an alternative enzyme.

6.3.3   Oxidation with Peroxygenases and Peroxidases

Peroxygenases are a diverse group of proteins that are found only in fungi. These enzymes are secreted outside of the organism in order to

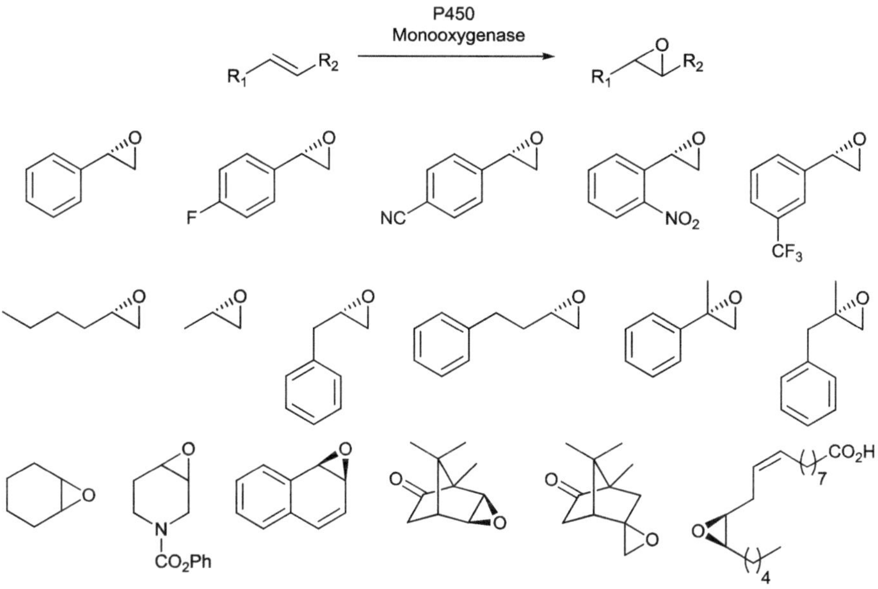

Figure 6.21 The scope of epoxide products produced by the P450 monooxygenase-catalysed oxidation of alkenes.

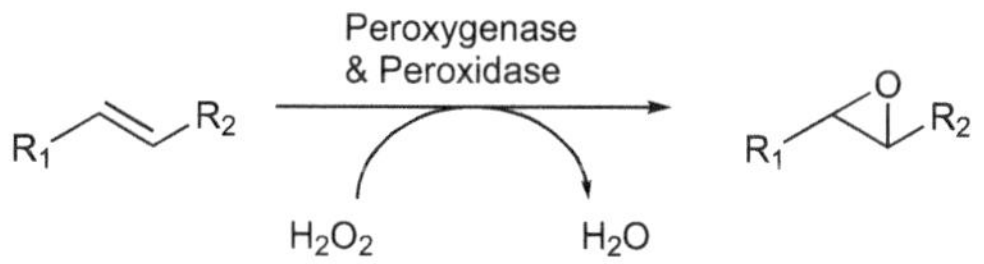

Figure 6.22 General transformation for the peroxygenase- and peroxidase-catalysed oxidation of alkenes.

break down chemicals present in the environment, such as lignin in wood. These proteins are similar to P450 monooxygenases in that they usually contain a heme ligand, but differ because, as the name suggests, they use hydrogen peroxide as the oxidant. Like P450 monooxygenases, peroxygenases are also able to carry out the epoxidation of alkenes, as shown in Figure 6.22.

Peroxygenases have a reasonable substrate scope and have been reported to react with linear, branched and cyclic alkenes. In some cases, good regio- and stereoselectivity is observed for the product epoxide. Figure 6.23 shows a range of product epoxides that can be made with these enzymes.

Peroxygenases are also able to catalyse a number of other "P450" like transformations, such as hydroxylation, but fewer examples have been reported. Another related family of enzymes capable of carrying

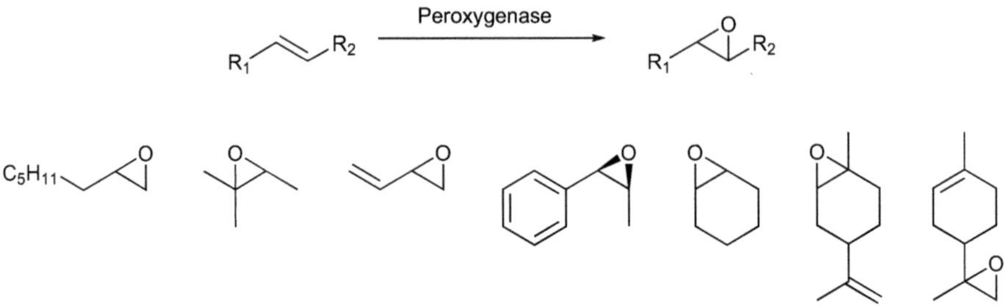

Figure 6.23 The scope of epoxide products produced by the peroxygenase-catalysed oxidation of alkenes.

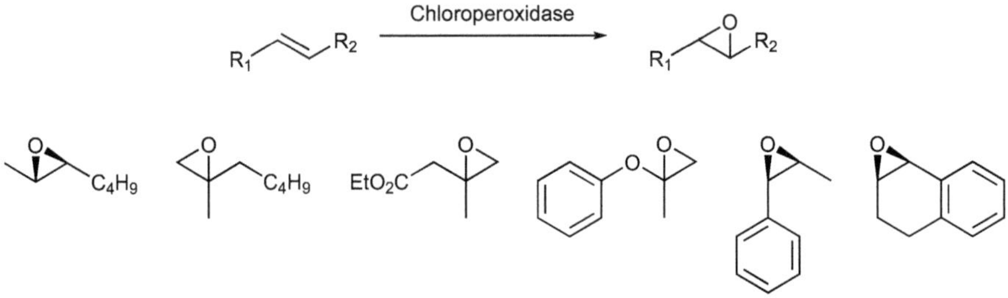

Figure 6.24 The scope of epoxide products produced by the chloroperoxidase-catalysed oxidation of alkenes.

out epoxidation reactions is the peroxidases. This family of enzymes is found much more widely in bacteria, plants and animals, including humans. These proteins share the same heme ligand as P450 and peroxygenase enzymes, but unlike the latter two families they are unable to catalyse the insertion of oxygen into a C–H bond. However, they are still capable of catalysing the epoxidation of a variety of alkenes, as shown in Figure 6.24 for the enzyme chloroperoxidase.

One potential drawback to using peroxygenases and peroxidases is that hydrogen peroxide is required for the reaction. If this accumulates to sufficient levels, it may prevent the enzyme from working or form unwanted side products. The most practical method of dealing with this problem is to add hydrogen peroxide slowly to the reaction mixture or use a second enzyme that generates hydrogen peroxide *in situ*. As they have very similar names, it's easy to get confused between P450 monooxygenases, peroxygenases and peroxidases, especially as there is some overlap in the reactions they catalyse. To help avoid confusion, a summary of the features and reactions catalysed by each of these enzymes is shown in Figure 6.25.

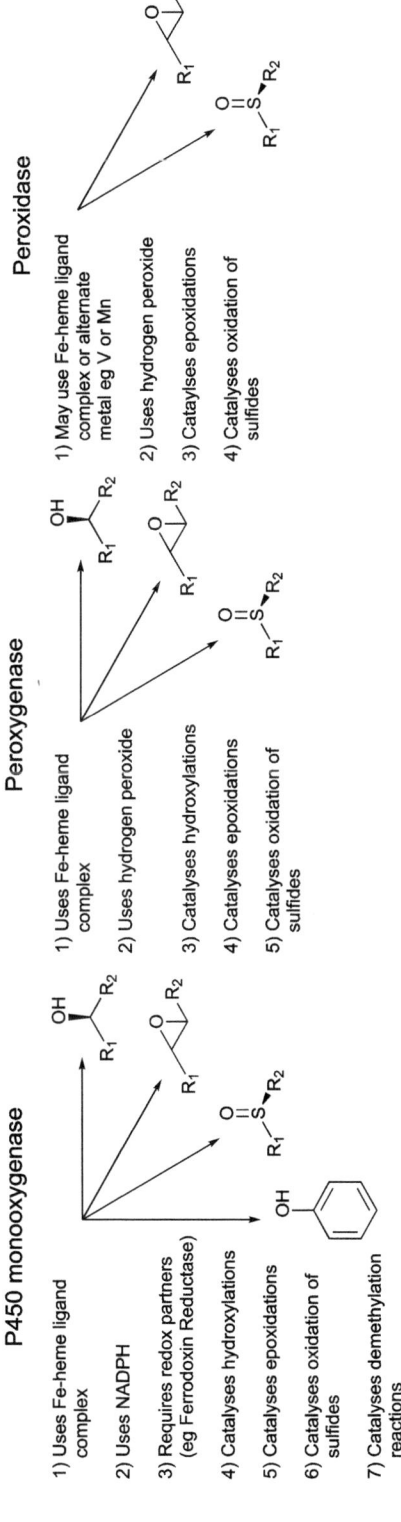

Figure 6.25 A comparison of the attributes of and reactions catalysed by P450 monooxygenase, peroxygenase and peroxidase enzymes.

6.4   Oxidation of Alcohols

In Chapter 5, we talked about the importance of ketones and aldehydes. They are widespread in Nature and found in many industrial
compounds. The oxygen atom of the carbonyl group in ketones and
aldehydes is capable of forming hydrogen bond interactions with
other functional groups and the carbonyl itself is a useful handle that
can be interconverted into a range of other functional groups.
Ketones and aldehydes are one oxidation level above alcohols and do
not contain the acidic proton of the alcohol. Therefore, one of the
more obvious methods to produce ketones and aldehydes is to oxidise
alcohols. A huge number of chemical methods exist for the oxidation
of alcohols with a list (by no means exhaustive) of reagents shown in
Figure 6.26.

Despite many useful synthetic methods, the oxidation of alcohols
often uses metal catalysts that need to be disposed of at the end of the
reaction or conditions that may be unsuitable for more sensitive
functional groups. This in turn may require the use of protecting
groups, which adds extra steps to a synthesis. It can also be difficult to
stop the oxidation at one level above the alcohol, with over-oxidation
to give carboxylic acids or esters often observed. As a complementary
technology, the oxidation of alcohols using enzymes offers the advantage of mild reaction conditions and high selectivity. This could be
regioselective oxidation of a polyol substrate or selective oxidation of
one enantiomer of a racemate. Each of the following sections details a
family of enzymes that is capable of oxidising alcohols to aldehydes
and ketones.

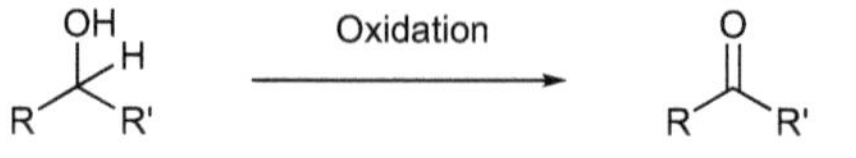

Chemical Reagents:

1) H_2SO_4, CrO_3 - Jones Oxidation

2) Pyridinium Dichromate - anhydrous conditions

3) Me_2SO, $(COCl)_2$, NEt_3 - Swern Oxidation

4) Dess-Martin Periodinane - Hypervalent Iodine

5) Tetrapropylammonium Perruthenate (TPAP), NMO

6) $KHSO_5$ - Oxone

7) TEMPO, HOCl - radical conditions

8) $R_1R_2C=O$, $Al(OR_3)$ - Oppenhauer Oxidation

9) V_2O_5, O_2 - Metal catalysed aerobic oxidation

10) Ag, N-heterocyclic Carbene

11) RuCp Complex - Shvo Catalyst

12) $BiBr_3$, H_2O_2

Figure 6.26   Selected methods for the chemical oxidation of alcohols to
ketones.

6.4.1 Oxidation of Alcohols Using Ketoreductases

As we saw in Chapter 5, ketoreductase enzymes are typically used to catalyse the reduction of ketones and aldehydes to the corresponding alcohols. We mentioned in Chapter 5 that these enzymes are also capable of catalysing the reverse process, the oxidation of an alcohol to the ketone or aldehyde. Due to this reactivity, these enzymes are also known as alcohol dehydrogenases (ADHs). These are the same enzymes and the names are used somewhat interchangeably, so remember that when you see ADH, it is the same as Kred and *vice versa*. The use of ketoreductases for the oxidation of alcohols has been less studied than the reduction of ketones, in part due to the fact that the oxidation of secondary alcohols removes a chiral centre rather than creating one in the reduction direction. As the ADH name suggests, the oxidation of alcohols using Kreds is a dehydrogenation reaction with the nicotinamide co-factor this time acting as a hydride acceptor rather than a hydride donor. As shown in Figure 6.27, this is the reverse direction to the reaction we saw in Chapter 5.

One disadvantage to using these enzymes in the oxidation direction is that the redox potential of the nicotinamide co-factor couple $(NAD(P)H/NAD(P)^+)$ is more negative than a typical alcohol/ketone couple. This means that $NAD(P)^+$-driven oxidation is thermodynamically unfavourable and an additional driving force is required to shift the reaction equilibrium towards the oxidation product. One way to shift the equilibrium is to use a catalytic amount of the nicotinamide

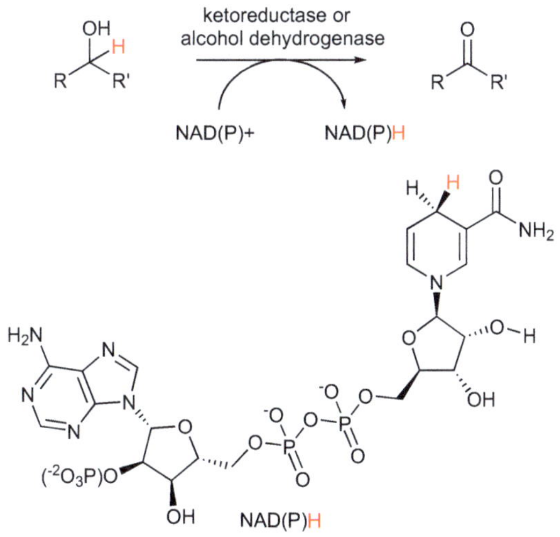

Figure 6.27 General transformation for the oxidation of alcohols catalysed by ketoreductases.

co-factor and a co-factor recycling system to drive the equilibrium. As we saw in Chapter 5, there are two main methods for co-factor recycling. The first is a substrate-coupled approach where we use a sacrificial substrate to recycle the nicotinamide co-factor. For oxidation reactions, chemists often use acetone as a co-substrate. As shown in Figure 6.28, each time the substrate is oxidised, a molecule of the reduced nicotinamide, NAD(P)H, is produced and this is recycled by reducing the co-substrate acetone to isopropyl alcohol. However, in order to drive this system in the direction we want, we have to use a large excess (>20 equivalents) of the sacrificial substrate, which may not be suitable for the enzyme. Other substrates used in the same way include acetaldehyde and 1-chloro-acetone.

The second option is to use a second enzyme (and substrate) to recycle the nicotinamide co-factor. In the oxidation direction, this approach typically uses an NAD(P)H oxidase enzyme in order to recycle the nicotinamide co-factor. This is called an enzyme-coupled recycling system because we use a separate enzyme to recycle the co-factor, and in the case of Figure 6.29 the terminal oxidant is molecular oxygen. The advantage of this system is that by using a second enzyme, we have an irreversible recycling reaction and so don't need to

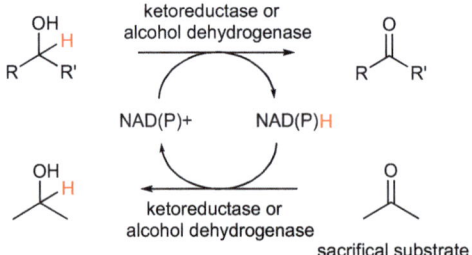

Figure 6.28   A substrate-coupled recycling system for the ketoreductase-catalysed oxidation of alcohols.

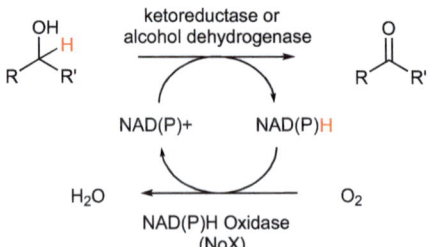

Figure 6.29   An enzyme-coupled recycling system for the ketoreductase-catalysed oxidation of alcohols.

worry about the equilibrium. The disadvantage of a two-enzyme system is that we need to find reaction conditions that are suitable for both enzymes and we need to make sure that the substrate and product from one enzyme reaction don't inhibit the other enzyme.

A range of primary and secondary alcohols can be oxidised to the corresponding aldehydes and ketones using ketoreductases. As shown in Figure 6.30, aliphatic and benzylic alcohols are oxidised and alkenes, amines, thiols and heteroaromatic substituents are tolerated. In each of the examples in Figure 6.30, the product aldehyde or ketone is shown.

Chemists have taken advantage of the high stereoselectivity of ketoreductases to carry out kinetic resolutions of secondary alcohol substrates. As shown in Figure 6.31, the enzyme selectively oxidises one enantiomer of the alcohol, leaving behind the other enantiomer in high enantiomeric excess. For each example in Figure 6.31, the product enantiomer is shown, whilst the other enantiomer is oxidised to the corresponding ketone. It is often possible to access either

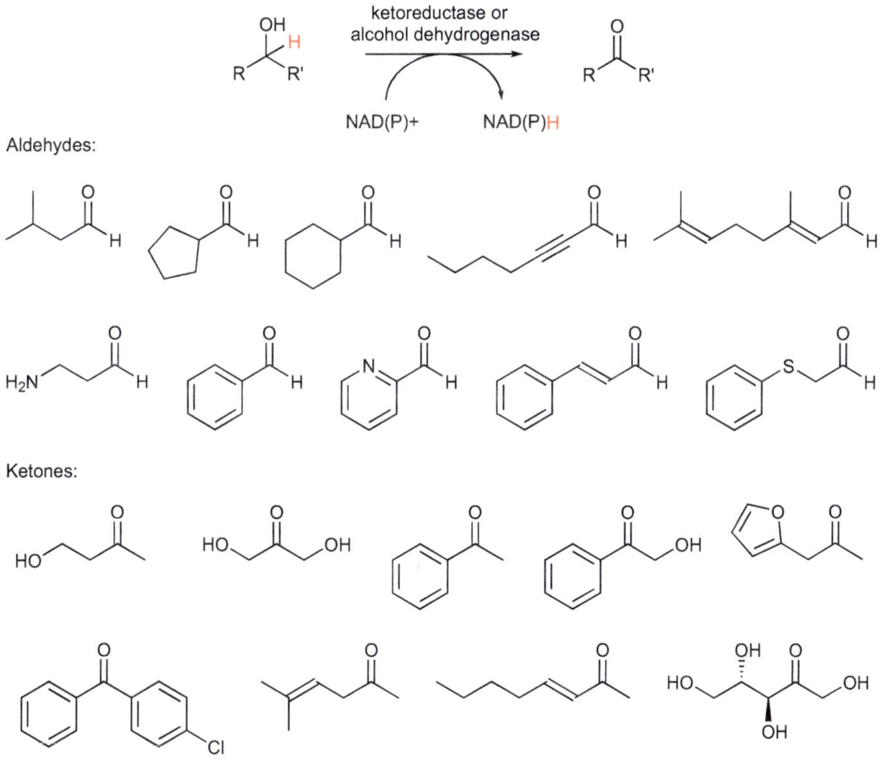

Figure 6.30 The product scope of ketoreductase-catalysed oxidation of alcohols.

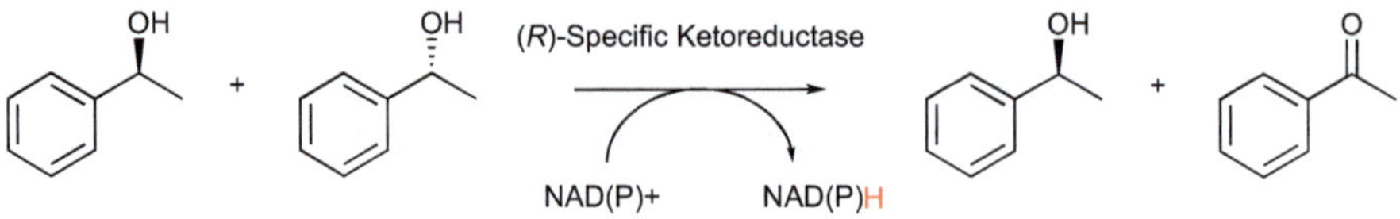

Figure 6.31 The resolution of alcohols catalysed by ketoreductases.

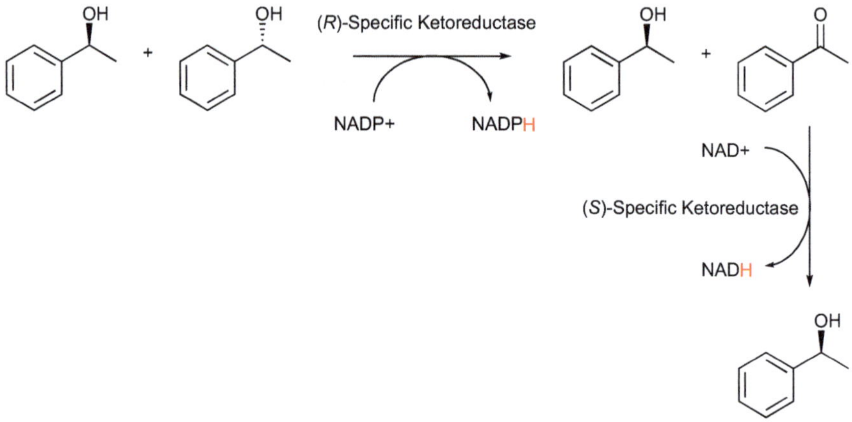

Figure 6.32 The deracemisation of alcohols catalysed by ketoreductases.

enantiomer of a substrate by choosing the appropriate enzyme to carry out the resolution.

The downside to a kinetic resolution method is that half of the material is converted into an undesired by-product. To get around this, chemists have used two ketoreductase enzymes to carry out the deracemisation of secondary alcohols. As shown in Figure 6.32, the first enzyme carries out a kinetic resolution reaction in an identical fashion to that shown in Figure 6.31. However, in the deracemisation process, a second enzyme is present, which then reduces the ketone back into the desired enantiomer of the alcohol. This enables a yield of greater than 50% to be obtained. The process works because the two enzymes are actually selective for different enantiomers of the alcohol, meaning that the desired enantiomer of the alcohol is produced. To avoid issues with co-factor recycling systems, the enzymes are also chosen so that they rely on different nicotinamide co-factors. For each of the examples below, the product enantiomer is shown. Again, access to either enantiomer of a substrate should be possible through the correct choice of enzymes for the reaction.

For both of these processes, the chiral environment of the enzyme active site determines the face of the alcohol that the hydride is abstracted from.

6.4.2 Oxidation of Alcohols Using Alcohol Oxidases

Alcohol oxidases are quite widespread in Nature, being found in bacteria, yeast, fungi, plants and insects. They are capable of catalysing the oxidation of alcohols to aldehydes and ketones and occasionally the oxidation of aldehydes to carboxylic acids. These enzymes differ from ketoreductases/alcohol dehydrogenases in that they do not use a nicotinamide co-factor. Instead, they use molecular oxygen as the terminal oxidant and require a flavin co-factor, which is bound in or near the active site. The flavin acts as a redox co-factor, shuttling electrons to molecular oxygen. As shown in Figure 6.33, this results in the formation of the oxidised product with hydrogen peroxide produced as a by-product.

Due to their good levels of regioselectivity, alcohol oxidases make ideal candidates for the selective oxidation of polyols for the resolution of secondary alcohols. Many of the flavin-dependent alcohol oxidases belong to one of two families with structurally similar members. Each family shares a similar flavin-binding domain, but variations in the substrate-binding domain are responsible for the substrate scope of each protein. A range of substrates oxidised by alcohol oxidases are shown in Figure 6.34. In general, most enzymes have a somewhat narrow substrate scope, but a large number of alcohol oxidases exist, so if the chosen enzyme doesn't work, then others can be considered. In each case the oxidation product is shown.

A number of non-flavin-dependent alcohol oxidases also exist, the most interesting of which is an enzyme called galactose oxidase. This enzyme naturally catalyses the oxidation of a wide range of carbohydrates, but it also catalyses the oxidation of a wide range of alcohols as well. The enzyme uses a single copper atom to carry out the oxidation, with molecular oxygen acting as the terminal oxidant as before. Galactose oxidase is very regio- and stereoselective. As shown in Figure 6.35, a range of substrates undergo oxidation. In each case, the product of the oxidation is shown.

As we saw previously, one potential drawback to enzymes that use molecular oxygen as the terminal oxidant is that hydrogen peroxide is

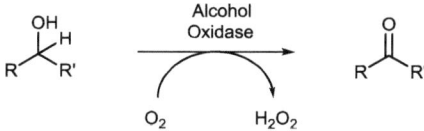

Figure 6.33 General transformation of alcohol oxidation catalysed by alcohol oxidases.

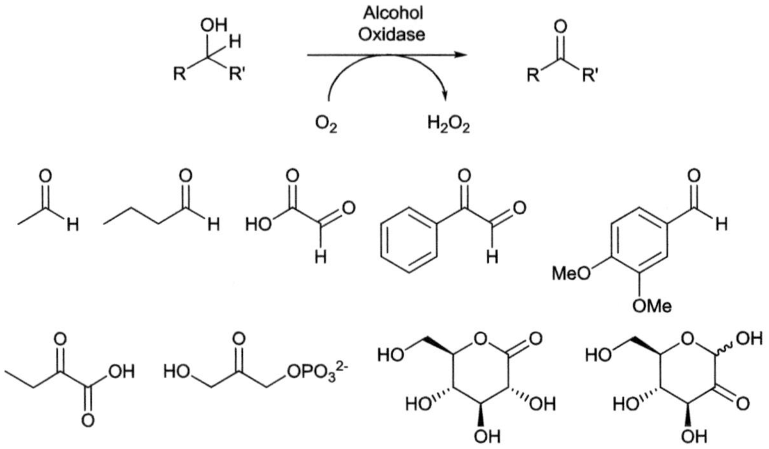

Figure 6.34 Selected products produced by the alcohol oxidase-catalysed oxidation of alcohols.

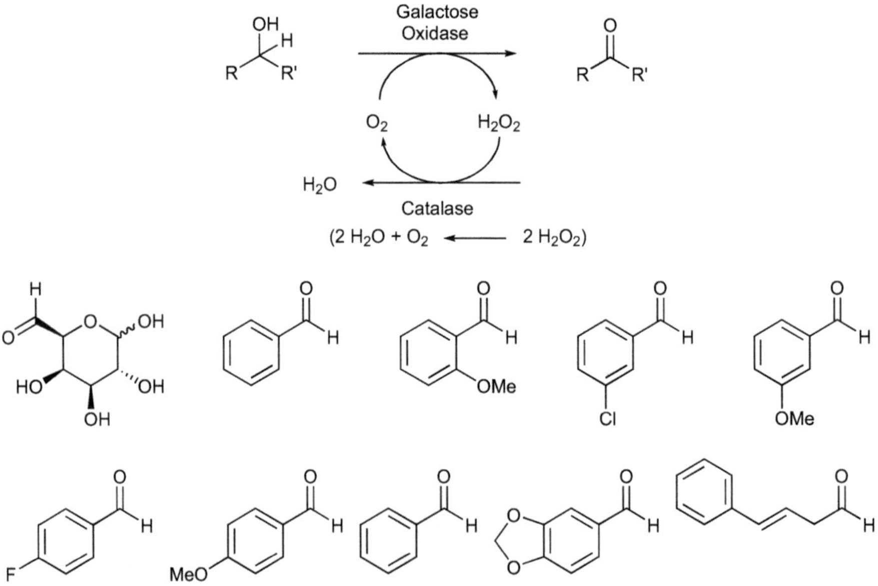

Figure 6.35 The scope of aldehydes and ketones produced by the galactose oxidase-catalysed oxidation of alcohols.

produced during the reaction. If this accumulates to sufficient levels it may prevent the enzyme from working or form unwanted side products. The most practical method of dealing with this problem is to add a second enzyme to the reaction called catalase. As shown in Figure 6.35, this enzyme decomposes hydrogen peroxide into water and oxygen, two much greener waste products.

6.4.3 Oxidation Using Laccases

Laccases are copper-containing enzymes found in a variety of micro-organisms, fungi and plants. Like the peroxygenases we saw earlier, they can be secreted outside of the cell by fungi to help with the decomposition of organic matter. They typically catalyse the oxidation of a wide range of organic and inorganic compounds through a single electron transfer process. The substrates for this process are electron rich phenols, which are converted to phenolic radicals that subsequently dimerise or oligomerise. As a result, they are used in the breakdown of lignocellulose, in the cross-linking of polysaccharides and in a variety of industries such as the bioremediation, food, textile and paper industries. Laccases can also be used to oxidise alcohols, although many substrates are unable to enter the active site or have too high a redox potential. Therefore, laccases are most often used with a chemical "mediator" that carries out the oxidation of the substrate before being reoxidised by the laccase, as we will see below. Different types of mediator exist, each with different mechanisms, but we will concentrate on the use of oxoammonium cations such 2,2,6,6-tetramethylpiperidin-1-oxyl (TEMPO). The general reaction is shown in Figure 6.36.

Like the other enzymatic oxidations in this section, these reactions occur under mild ambient conditions with aerial oxygen as the terminal oxidant. Laccase mediator systems work on a range of benzylic, allylic and aliphatic alcohols, as shown in Figure 6.37. In each case the product of the oxidation reaction is shown.

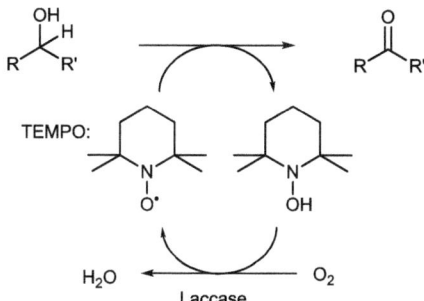

Figure 6.36 General transformation of alcohol oxidation catalysed by laccases.

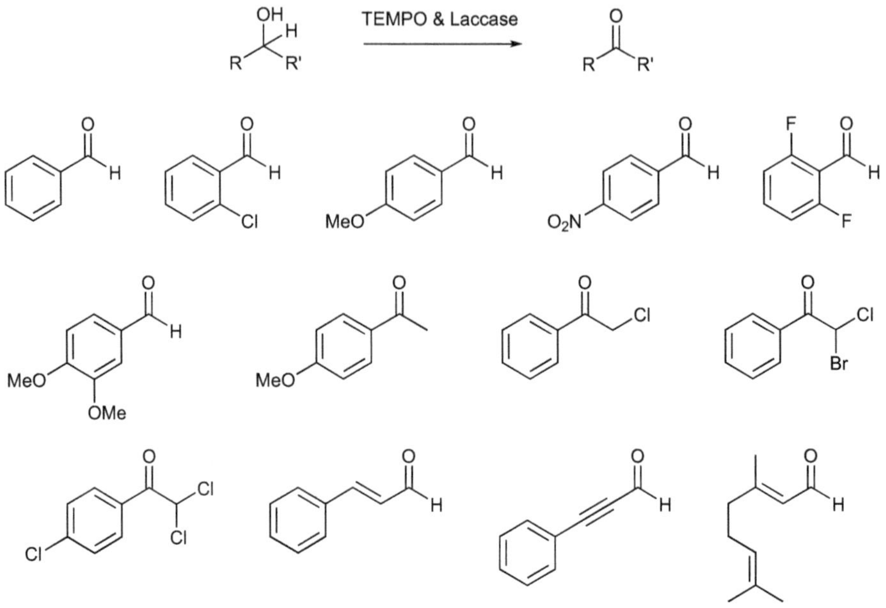

Figure 6.37   The scope of aldehydes and ketones produced by the laccase-catalysed oxidation of alcohols.

As mentioned previously with some traditional chemistry methods, over-oxidation of the initial product can be an issue with laccases. Chemists have taken advantage of this, however, to make lactones from 1,4- and 1,5-aliphatic alcohols, as shown in Figure 6.38. Although no stereoselectivity was observed, the lactones were formed with complete regioselectivity due to the selective oxidation of the primary alcohol.

You do not need to know the mechanism of laccase oxidations using a chemical mediator, but we have included it below for those readers who would like a little more detail. As shown in Figure 6.39, laccases typically contain four copper atoms. The catalytic cycle starts with the transfer of a single electron in the enzyme active site to oxidise the TEMPO (or similar derivative) mediator to form an oxo-ammonium cation. The mediator may then leave the active site and is attacked by the substrate alcohol to give an alternative oxoammonium cation. The alcohol is then oxidised, reducing the oxoammonium cation to the corresponding hydroxylamine with a proton generated as a by-product. The hydroxylamine then loses a proton to regenerate the TEMPO radical ready for another catalytic cycle. Finally, the laccase is oxidised back to its original oxidation state by molecular oxygen, which is converted into water using the two protons produced in the catalytic cycle.

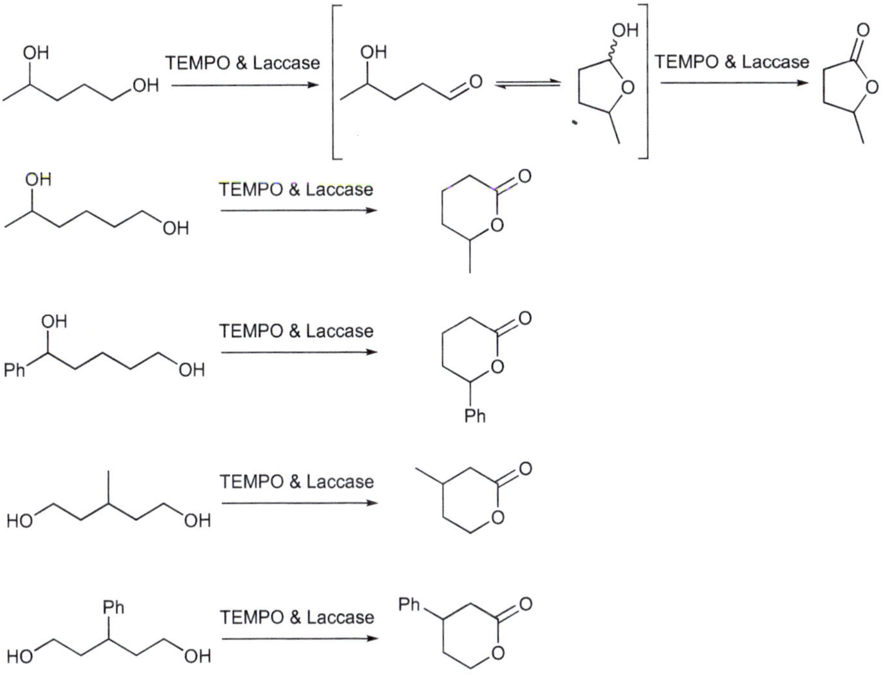

Figure 6.38 The formation of lactones from the laccase-catalysed oxidation of alcohols.

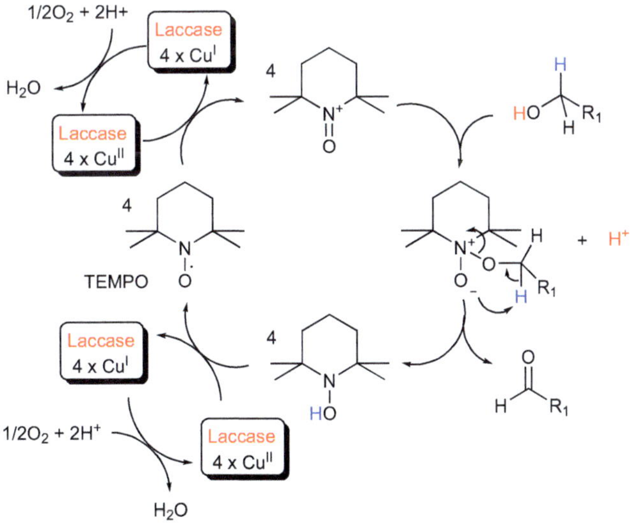

Figure 6.39 The mechanism of laccase-catalysed oxidation of alcohols.

6.5   Oxidation of Aldehydes and Ketones

We have already covered the importance of ketones and aldehydes earlier in this chapter and in Chapter 5. They are widespread in Nature and found in many industrial compounds. The oxygen atom of the carbonyl group in ketones and aldehydes is capable of forming hydrogen bond interactions with other functional groups and the carbonyl itself is a useful handle that can be interconverted into a range of other functional groups. One of those conversions is further oxidation to carboxylic acids and esters. Esters are very important compounds in both Nature and man-made applications. A large number of esters are naturally occurring, for example in lipids, essential oils, fruits and seeds. As a result, both natural and man-made esters are used in the perfume, flavour and cosmetics industries. They are also found in surfactants used in washing powders and soaps. Both esters and carboxylic acids are important functional groups in pharmaceutical and agrochemical compounds, as adding either of these functional groups changes the electronic properties of the molecule and may improve binding to a target receptor. A large number of chemical methods exist for the oxidation of aldehydes and ketones into carboxylic acids and esters. Many of these oxidative processes utilise metal catalysts, some of which may be too harsh for more sensitive functional groups and will require removal and disposal at the end of the reaction. A range of chemical methods for the oxidation of aldehydes and ketones are shown in Figure 6.40. Some of these methods proceed *via* the hydrate of the aldehyde, so adding water to these reactions can be required to facilitate the reaction.

   In contrast to the chemical methods detailed above, enzymatic methods offer the potential to carry out the same oxidations under mild ambient conditions without the need for environmentally unfriendly solvents or catalysts. The sections below detail enzymatic alternatives to the chemical oxidation of aldehyde and ketones.

6.5.1   Oxidation of Aldehydes with Aldehyde Dehydrogenases

As we previously discussed, the oxidation of aldehydes is an important method of carboxylic acid synthesis. The carboxylic acid products produced are useful compounds for the production of polymers, pharmaceuticals, solvents and food additives. Carboxylic acids are also prevalent in Nature. For example, they are present in amino

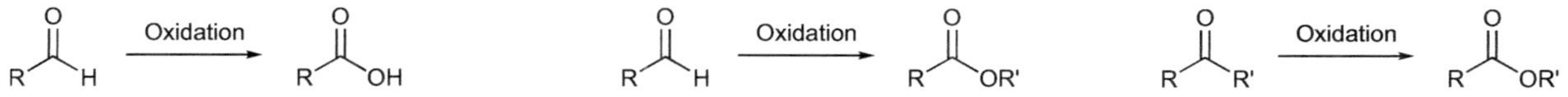

Figure 6.40 Selected examples of the chemical oxidation of aldehydes and ketones.

Chemical Reagents:

1) H_2SO_4, CrO_3 (Jones Oxidation)

2) Pyridinium Dichromate (Aqueous Conditions)

3) $KMnO_4$

4) H_2O_2 + Transition metal eg VO(acac) or $MeReO_3$

5) $KHSO_5$ (Oxone)

6) $NaBO_3$ + Acetic acid.

Chemical Reagents:

1) $Al(OR'')_3$ (Tishchenko Reaction)

2) MnO_2 + R'OH + N-Heterocyclic Carbene catalyst

3) R'OH + Pt cathode (Electrochemical)

4) H_2O_2 + R'OH + Transition metal eg VO(acac)

5) $KHSO_5$ + R'OH (Oxone)

6) $NaBO_3$ + Acetic acid.

Chemical Reagents:

1) Peracid eg mCPBA (Baeyer-Villiger Reaction)

2) H_2O_2, BF_3

acids, central metabolism (citric acid) and many natural products. The oxidation of aldehydes is catalysed by a family of enzymes known as aldehyde dehydrogenases. They are found in bacteria and animals and are similar to ketoreductases (or alcohol dehydrogenases), requiring a nicotinamide co-factor in order to carry out the oxidation reaction, as shown in Figure 6.41. Human aldehyde dehydrogenase is found throughout cells and plays a role in protecting stem cells. Recent research has focused on using aldehyde dehydrogenase as a marker for cancer stem cells in order to help treatment.

One drawback of carrying out the oxidation of primary alcohols in whole cells can be that after oxidation of the alcohol to the aldehyde, it may be oxidised further by aldehyde dehydrogenases naturally present in the cells. Aldehyde dehydrogenases catalyse the oxidation of a range of aldehydes as shown in Figure 6.42. In each case, the substrate aldehyde is shown.

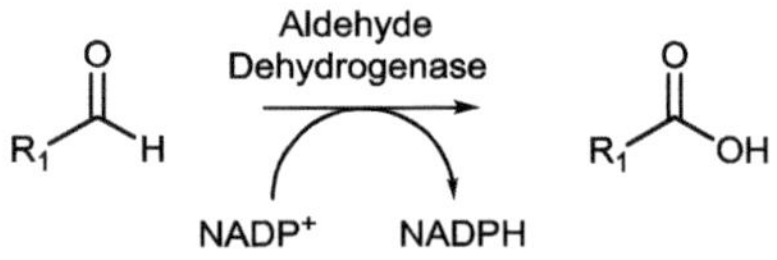

Figure 6.41   General transformation of aldehyde oxidation catalysed by aldehyde dehydrogenases.

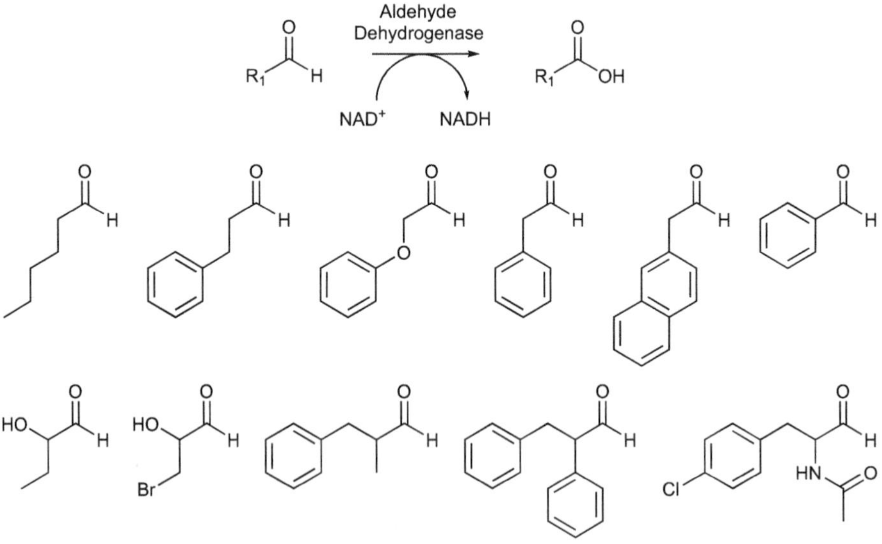

Figure 6.42   The substrate scope of aldehyde oxidation catalysed by aldehyde dehydrogenases.

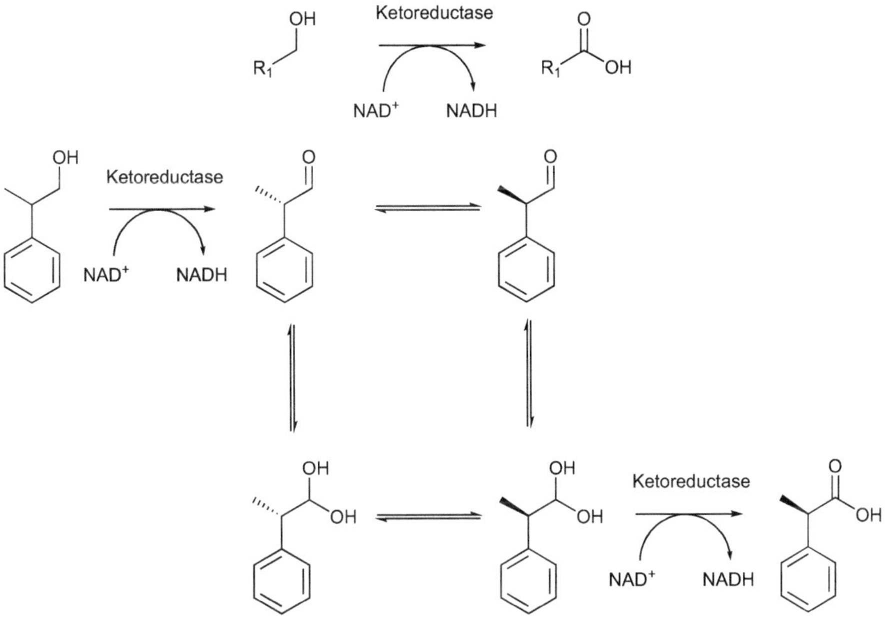

Figure 6.43 Dynamic kinetic resolution during alcohol oxidation catalysed by a ketoreductase.

In some very specific cases, the oxidation of aldehydes can also be catalysed by ketoreductase enzymes. Like many chemical oxidation methods, this occurs through an intermediate hydrate of the aldehyde. As shown in Figure 6.43, chemists can take advantage of the enzymes' high selectivity to oxidise only one enantiomer of the substrate, setting up a dynamic kinetic resolution process.

6.5.2 Oxidation of Aldehydes and Ketones with Baeyer–Villiger Monooxygenases

The Baeyer–Villiger oxidation was first described over one hundred years ago by the two scientists after whom it is named. It is used frequently in organic synthesis and is the oxidation of a ketone with an organic peracid to give an ester. The reaction is also catalysed by a family of enzymes called Baeyer–Villiger monooxygenases (BVMOs). These flavin-dependent enzymes catalyse exactly the same transformation as shown in Figure 6.44. Like many other flavin-containing enzymes, BVMOs also need a nicotinamide co-factor to reduce the flavin at the end of the reaction in order to regenerate the active form of the flavin for the next molecule of substrate.

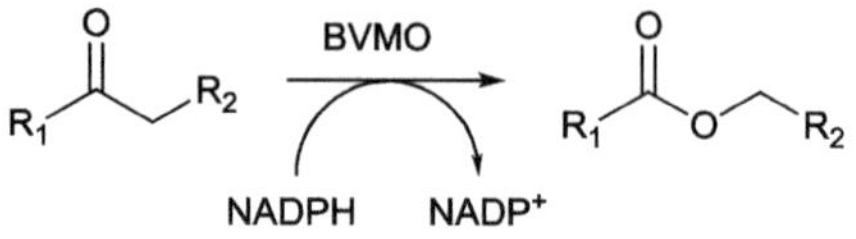

Figure 6.44 General transformation of ketone oxidation catalysed by Baeyer–Villiger monooxygenases (BVMOs).

The chemical Baeyer–Villiger reaction works on a wide range of substrates with reasonable functional group tolerance and the regiochemistry of the transformation is usually predictable. However, the use of organic peracids, which may be costly to produce and hazardous to handle, is a drawback, especially for running these reactions on a large scale. Peracids are strong oxidants and more sensitive functional groups may need to be protected, adding additional steps to a synthesis. In contrast, Baeyer–Villiger monooxygenases work under ambient aqueous conditions using molecular oxygen as the terminal oxidant. As shown in Figure 6.45, Baeyer–Villiger monooxygenases react with a wide range of aliphatic, cyclic and bicyclic ketones to selectively produce esters in good yields. A range of different functional groups are tolerated and in each case the product ester is shown.

Due to their high regio- and stereoselectivity, chemists have used Baeyer–Villiger monooxygenases to carry out the kinetic resolution of racemic substrates. As shown in Figure 6.46, a range of ketones undergo selective reaction to give both ketones and esters in high optical purity. The regiochemistry observed is consistent with a chemical Baeyer–Villiger reaction. Often, selective enzymes exist that allow access to either enantiomer of the desired ester (or ketone).

As with all kinetic resolution processes, the major drawback of these reactions is that we are only able to obtain a 50% maximum yield of the product. Instead, chemists have also used Baeyer–Villiger monooxygenases to carry out desymmetrisation reactions of prochiral ketone substrates. As shown in Figure 6.47, a range of prochiral ketones can be used as substrates and high yields and stereoselectivities can be achieved. It is often possible to make the opposite enantiomer to that shown through the choice of an alternative enzyme.

It is also possible to achieve a greater than 50% yield for some substrates by using a dynamic kinetic resolution process. In particular, α-alkylated ketones undergo dynamic kinetic resolutions with Baeyer–Villiger monooxygenases in high yields and stereoselectivities, as shown in Figure 6.48. The racemisation of the substrate was achieved by running the reaction at basic pH and using a weak anion exchange resin in the reaction mixture.

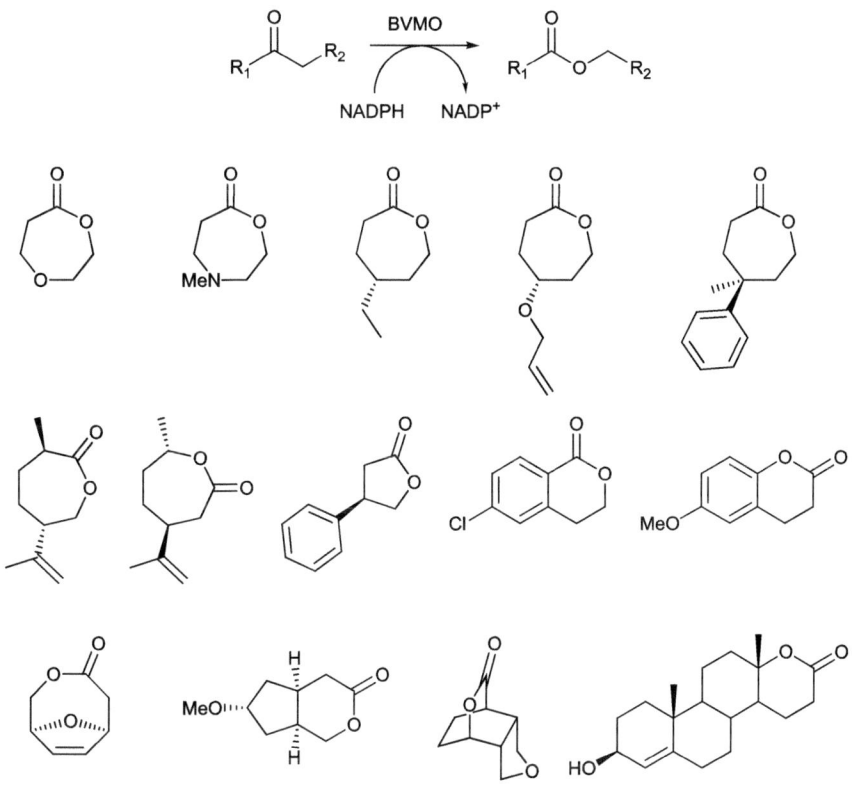

Figure 6.45 The scope of lactones produced by Baeyer–Villiger-catalysed oxidation of ketones.

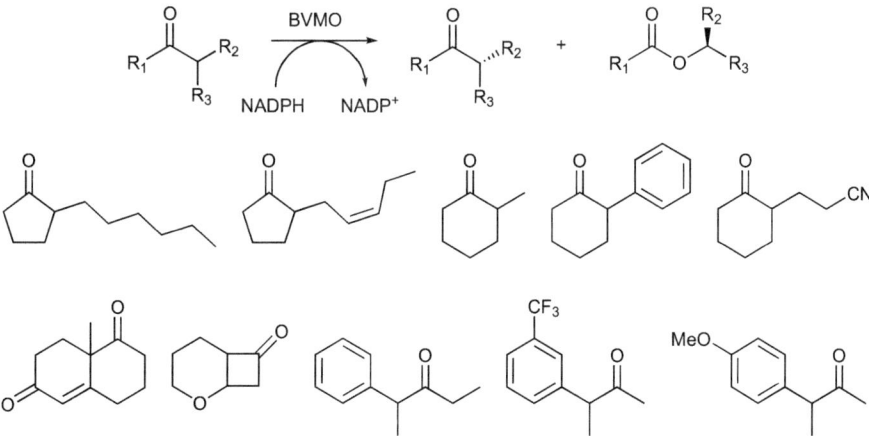

Figure 6.46 The scope of ketone substrates that undergo kinetic resolution catalysed by Baeyer–Villiger monooxygenases.

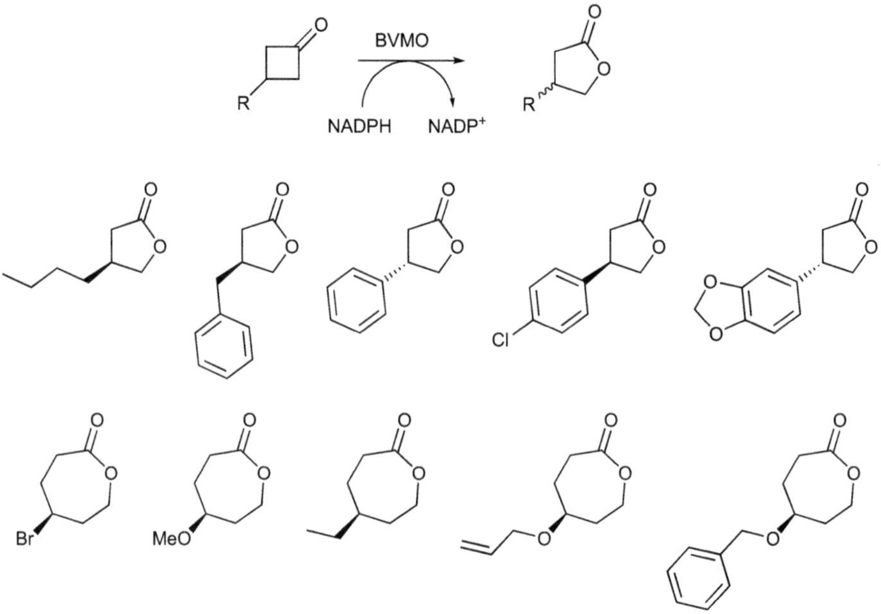

Figure 6.47 The scope of chiral lactone products from the desymmetrisation of ketones catalysed by Baeyer–Villiger monooxygenases.

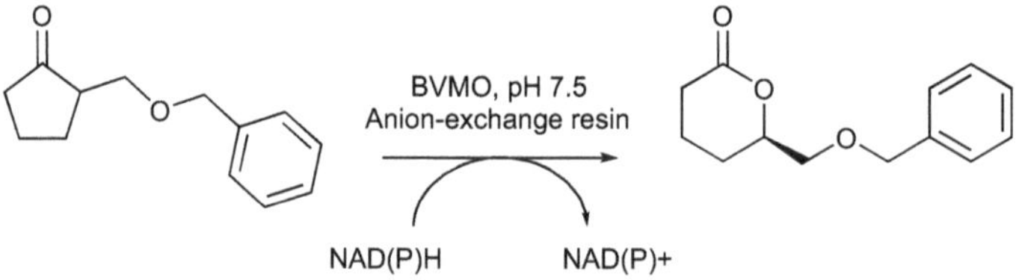

Figure 6.48 The dynamic kinetic resolution of a cyclic ketone catalysed by a Baeyer–Villiger monooxygenase.

As we mentioned previously, Baeyer–Villiger monooxygenases need a nicotinamide co-factor in order to reduce the flavin molecule present in the active site. We have seen previously that nicotinamide co-factors can be recycled by a variety of methods. These are equally applicable to Baeyer–Villiger-catalysed transformations. As shown in Figure 6.49, the dynamic kinetic resolution of β-ketoesters can be carried out in the presence of glucose 6-phosphate dehydrogenase, which efficiently recycles the NADPH co-factor.

The mechanism of Baeyer–Villiger monooxygenases is illustrated in Figure 6.50. Like the other mechanisms in this section, you do not need to know this but we have included it here for those who would

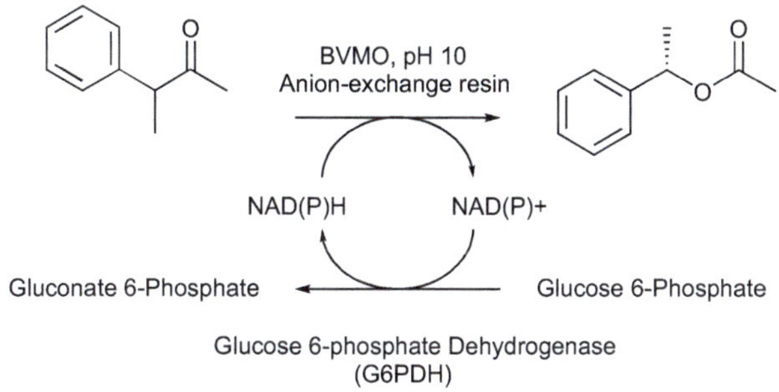

Figure 6.49 An enzyme-coupled co-factor recycling system for Baeyer–Villiger monooxygenases.

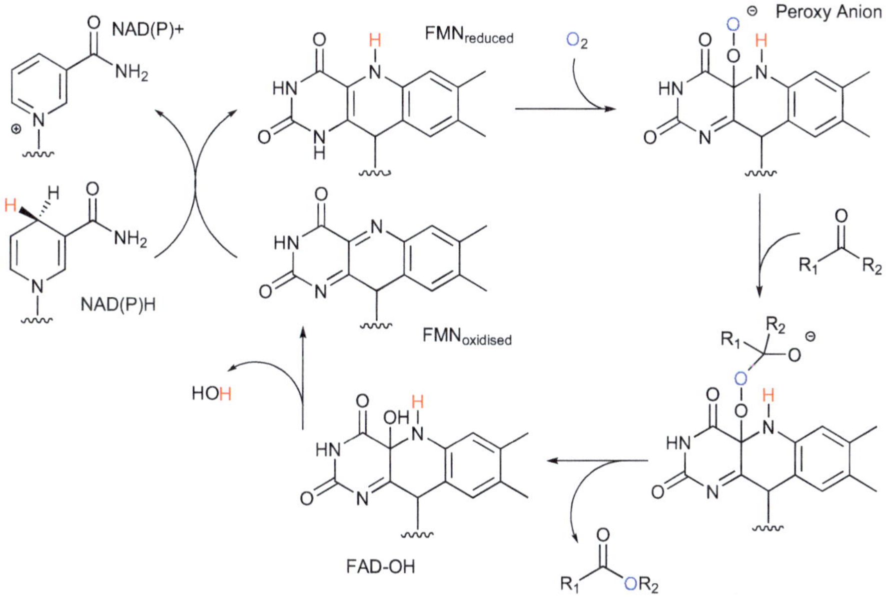

Figure 6.50 General mechanism of ketone oxidation catalysed by Baeyer–Villiger monooxygenases.

like to develop a greater understanding. As with many other flavin processes, the cycle starts with reduction of the flavin by the nicotinamide co-factor NADPH. The reduced flavin molecule is then able to undergo reaction with molecular oxygen, which forms an enzyme peroxy anion intermediate. This reaction now closely follows the classic Baeyer–Villiger transformation, with this intermediate attacking the ketone substrate. The tetrahedral intermediate formed then collapses, and migration occurs with a hydroxy flavin acting as the leaving group. This gives the product ketone and the hydroxy flavin, which loses water to give the oxidized form of the flavin. This must be reduced by NAPDH to begin the catalytic cycle once more.

6.6 Oxidation of Amines

Amines are ubiquitous, being found in natural products, pharmaceuticals and agrochemicals. The amine group is an important pharmacophore due to its ability to form hydrogen bonds with receptors and other biological molecules. Chiral amines are therefore valuable building blocks in the synthesis of medicinal compounds, especially if they can be produced by simple methods and with high regio- and stereoselectivity. Amines can also be converted into a number of other functional groups, for example imines, which are the nitrogen equivalent of ketones. Like their oxygen equivalents, the nitrogen atom of an imine is able to form hydrogen bond interactions with other functional groups and the imine itself is a useful handle that can be interconverted into a range of other functional groups. Imines are one oxidation level above amines without containing the acidic protons of the amine. Therefore, one of the more obvious methods to produce imines is to oxidise amines. A variety of chemical methods exist for the synthesis of imines, as shown in Figure 6.51.

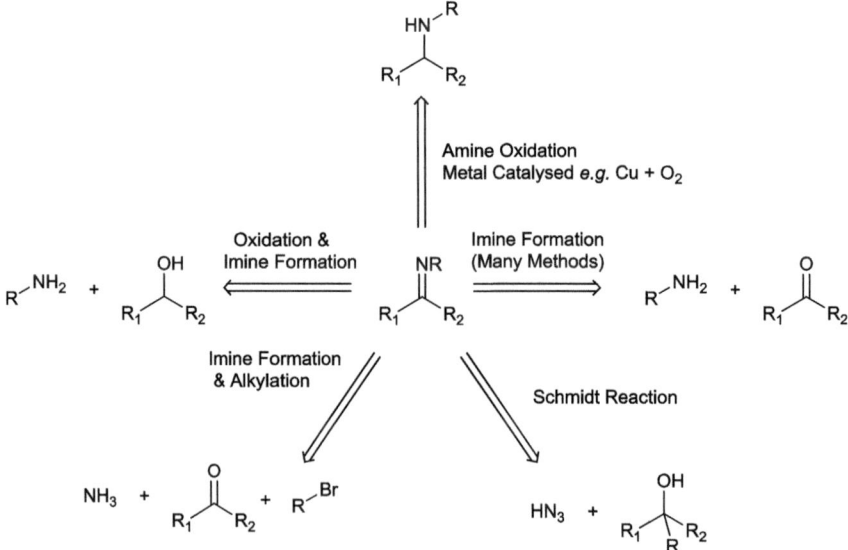

Figure 6.51 Selected chemical methods of imine synthesis.

Most methods of imine synthesis start from an alcohol or ketone and carry out formation of the carbon–nitrogen bond in the course of the reaction. In comparison, there are fewer methods for oxidation of amines into imines.

Chemical methods of imine formation, whether *via* oxidation of the corresponding amine or *via* condensation of an amine with carbonyl compounds, have some drawbacks. The processes often use metal catalysts and ligands, which may be expensive, and they may also require harsh reaction conditions and environmentally unfriendly organic solvents. Enzymatic oxidation methods offer a complementary alternative to chemical ones, as they are carried out under mild aqueous reaction conditions without the need for metal catalysts and ligands. Each of the sections below describes in more detail one of the families of enzymes capable of carrying out this transformation.

6.6.1 Oxidation of Amines Using Amine Oxidases

Amine oxidation is widespread throughout Nature, partly because amines are involved in cell growth and differentiation and as a growth source for a number of species of bacteria. One of the classes of enzymes capable of oxidising amines to imines is the amine oxidases. Amine oxidases can be further subdivided into two families. One uses copper atoms, which we will not cover, whilst the other family are flavin-dependent enzymes, which are described in more detail below. The flavin-dependent enzymes are known as monoamine oxidases and they are found in eukaryotes such as yeast, fungi and humans. They catalyse the oxidation of amines as shown in Figure 6.52.

Monoamine oxidases (MAOs) have been reported to oxidise a range of amine substrates including aliphatic and cyclic amines and interestingly a small range of *O*-methyl hydroxylamines, as shown in Figure 6.53. In general, tertiary amines are oxidised faster than secondary amines, which in turn are oxidised faster than primary amine substrates. In each case the substrate amine is shown.

Due to their stereoselectivity, monoamine oxidases were initially used to carry out the kinetic resolution of racemic substrates. This process was subsequently improved by adding a nonselective

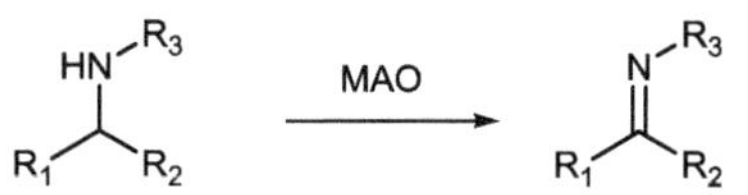

Figure 6.52 General transformation catalysed by monoamine oxidases.

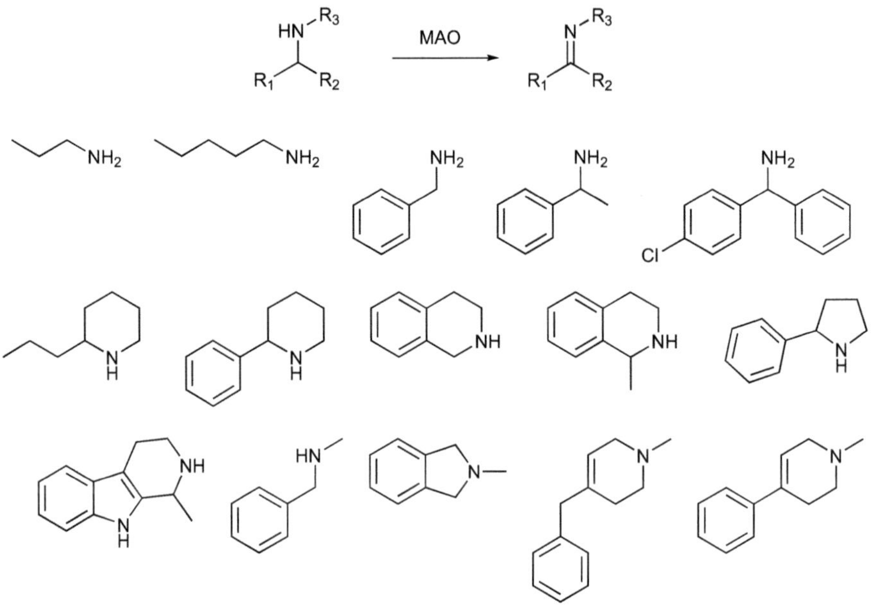

Figure 6.53 The substrate scope of amine oxidation catalysed by monoamine oxidases.

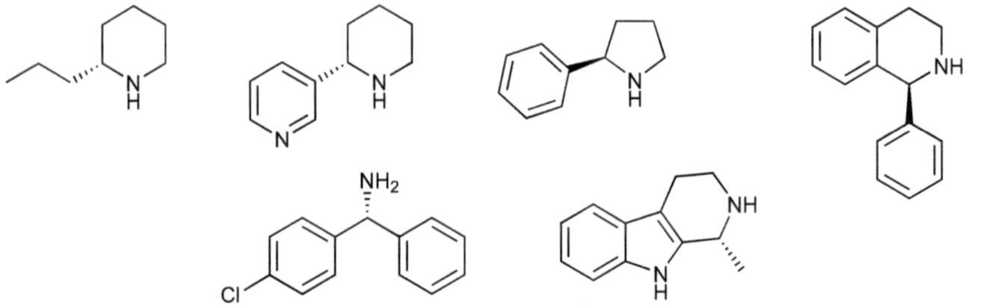

Figure 6.54 The scope of amine products produced by monoamine
oxidase-catalysed deracemisations.

reducing reagent to the reaction. Because the monoamine oxidase
only oxidises one enantiomer, the racemic amine is eventually deracemised to give a single enantiomer, as shown in Figure 6.54. This
approach has been demonstrated for a range of amines, which were

deracemised in good yields and high stereoselectivities. In each case the product amine is shown.

Monoamine oxidases have also been used to desymmetrise a range of amine substrates. In addition to selectively reacting with a single enantiomer of the starting material, the intermediate imine can be reacted *in situ* with a nucleophile to create an additional bond. Chemists used this approach towards the synthesis of boceprevir, a medicine used to treat hepatitis C infections. As shown in Figure 6.55, the chemists used a monoamine oxidase to desymmetrise a pyrrolidine derivative. The imine formed was reacted with sodium bisulfite to form the sulfonate intermediate, which was further reacted to produce boceprevir.

As mentioned previously, monoamine oxidases are flavoproteins that require a flavin adenosine dinucleotide (FAD) co-factor. However, unlike Baeyer–Villiger monooxygenases, they do not use a nicotinamide co-factor to recycle the flavin and instead use molecular oxygen. This results in hydrogen peroxide being produced as a by-product of the oxidation reaction. If the hydrogen peroxide accumulates to sufficient levels, it may prevent the enzyme from working or form unwanted side products. The most practical method of dealing with this problem is to add a second enzyme to the reaction called catalase. As shown in Figure 6.56, this enzyme decomposes hydrogen peroxide into water and oxygen, two much greener waste products.

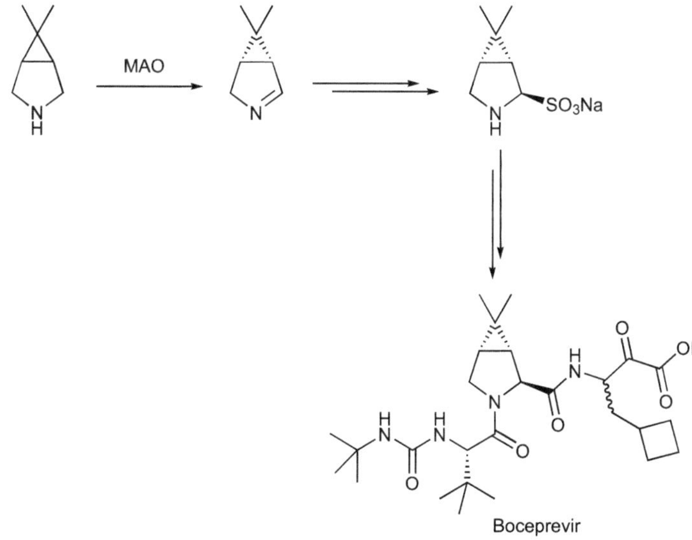

Figure 6.55 The synthesis of boceprevir utilising a monoamine oxidase-catalysed desymmetrisation.

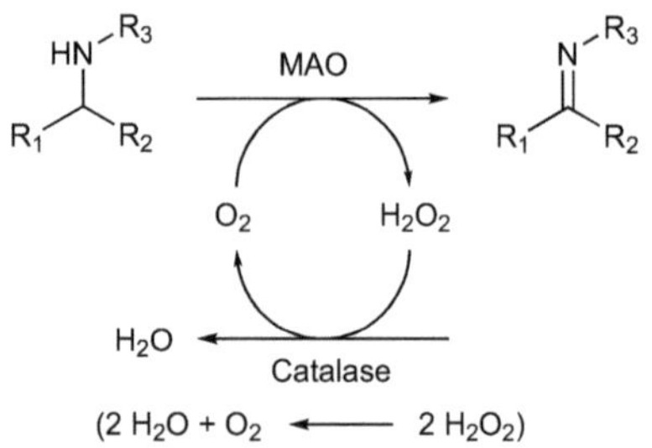

Figure 6.56   The use of catalase to remove toxic hydrogen peroxide from monoamine oxidase reactions.

Despite being researched in great detail, the mechanism of monoamine oxidase-catalysed oxidation of amines is still not entirely clear. Two different mechanisms have been proposed, one of which is a radical mechanism, whilst the other is a nucleophilic mechanism. As with the other mechanisms in this section, you do not need to know this but we have included extra detail here for those who are interested. In Figure 6.57, we have included the nucleophilic mechanism, as it is the easier of the two mechanisms to understand. Advanced readers are encouraged to seek out the radical mechanism, examine both and then draw their own conclusions. The nucleophilic mechanism starts with the attack of the amine substrate on the flavin co-factor. The flavin anion then extracts a proton from the substrate α to the amine forming the imine and further reducing the flavin. Finally, the imine is released and the reduced form of the flavin co-factor is reoxidised by molecular oxygen to give the oxidised flavin (ready to react with another molecule of substrate) and hydrogen peroxide as a by-product.

6.6.2   Oxidation of Amines Using Amino Acid Dehydrogenases/Amino Acid Oxidases

Amino acids are vital biological molecules, whether we consider the 23 proteinogenic amino acids that make up proteins or the hundreds of non-proteinogenic amino acids used by many different organisms in a huge variety of biological processes. Amino acids can be oxidised by a family of flavin-dependent proteins called amino acid oxidases. These enzymes can be divided into two distinct groups according to which amino acid enantiomer they prefer to oxidise. Those that prefer to oxidise the "natural" enantiomer substrates are referred to

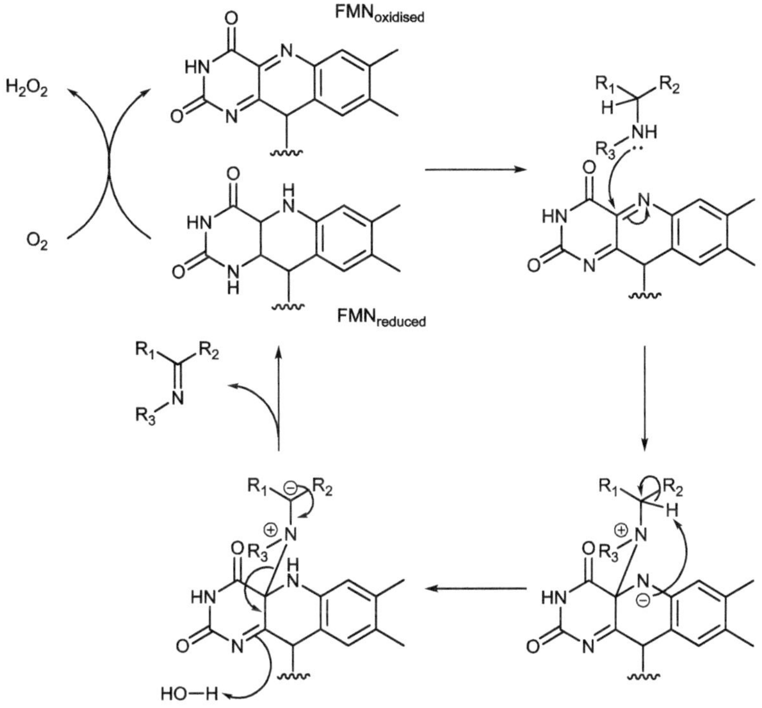

Figure 6.57 One of two suggested possible mechanisms for the mono-amine oxidase-catalysed oxidation of amines.

as L-amino acid oxidases (L-AAOs) and are found in a large range of both prokaryotes and eukaryotes from bacteria to snakes. In contrast to L-AAOs, D-amino acid oxidases are found very widely in eukaryotes such as fungi, yeast, insects, birds and mammals, but interestingly are not found in prokaryotes despite bacteria producing a larger number of D-amino acids. D-Amino acid oxidases are thought to be involved in making D-amino acids available for central metabolism or to detoxify D-amino acids that are ingested from the environment or formed within the organism. Both D- and L-amino acid oxidases catalyse the oxidation of the amine group of amino acids into the corresponding imine. Under physiological conditions, this is typically hydrolysed to produce the ketoacid as a product, as shown in Figure 6.58. As with other amine oxidases, these enzymes use molecular oxygen as the terminal oxidant and produce hydrogen peroxide as a by-product.

Both L- and D-amino acid oxidases tend to display a narrow substrate scope. However, as quite a few enzymes are known overall, a range of amino acids can be oxidised as shown in Figure 6.59. In each

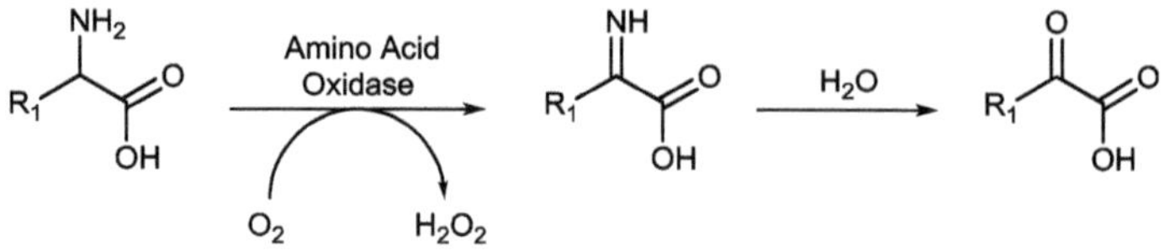

Figure 6.58 General transformation catalysed by amino acid oxidases.

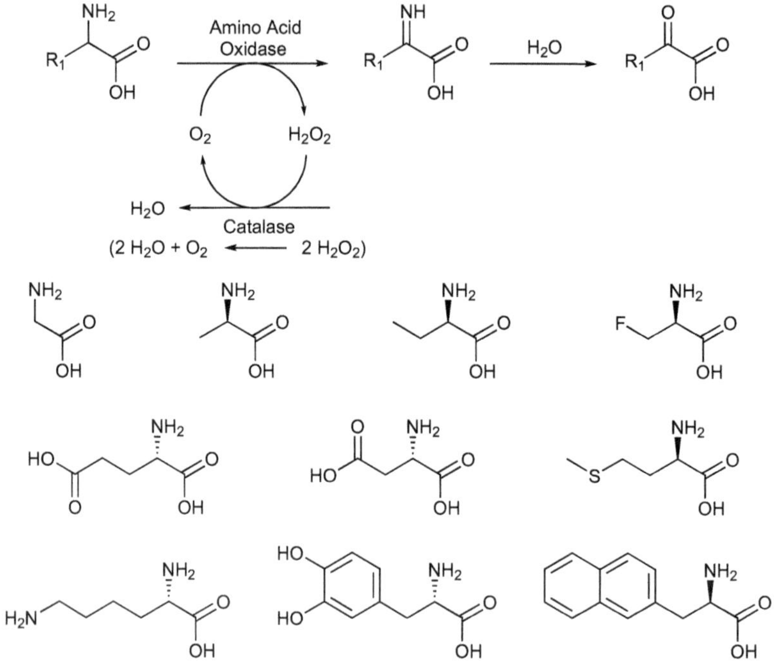

Figure 6.59 The scope of amino acids oxidised by amino acid oxidases.

case the amino acid starting material enantiomer is shown and often the other enantiomer of the amino acid can also be oxidised by choosing the amino acid oxidase with the opposite selectivity. As we have seen with other families of enzymes in this chapter, the production of hydrogen peroxide may prevent the enzyme from working or form unwanted side products if it reaches sufficient levels. The most practical method of dealing with this problem, as we saw with monoamine oxidases, is to add catalase to the reaction. This enzyme decomposes the hydrogen peroxide formed into water and oxygen, two much greener waste products.

Amino acid oxidases can be used to carry out kinetic resolutions of racemic amino acids. As shown in Figure 6.60, one enantiomer is

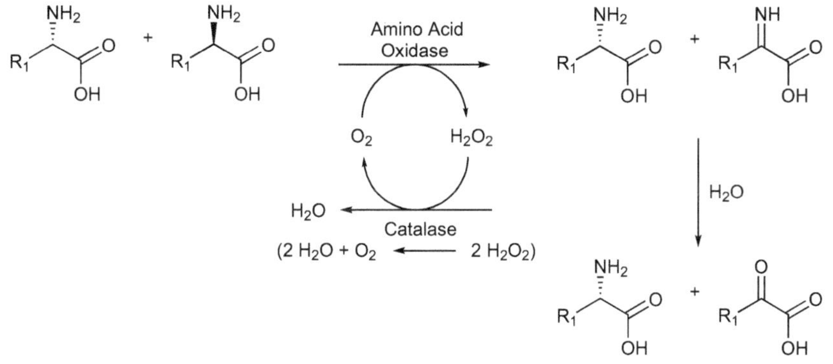

Figure 6.60 The kinetic resolution of amino acids catalysed by amino acid oxidases.

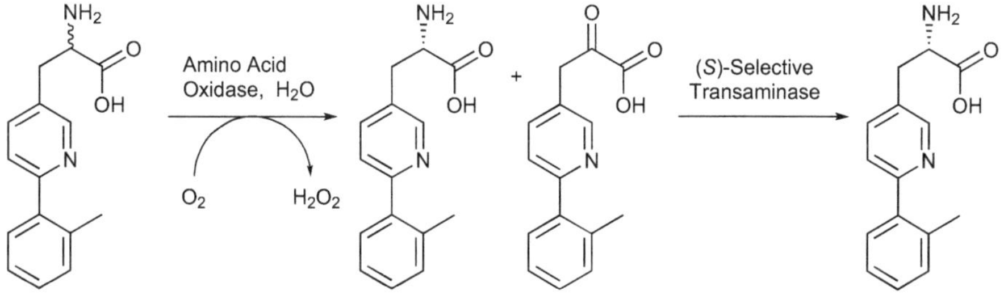

Figure 6.61 The deracemisation of amino acids utilising an amino acid oxidase and a transaminase.

converted to the keto acid, leaving the amine product in high stereoselectivity.

In order to improve the maximum 50% yield obtainable *via* a kinetic resolution process, chemists have designed a deracemisation method for amino acids. This method uses a transaminase enzyme, which we will see in Chapter 7, in addition to an amino acid oxidase. As shown in Figure 6.61, the amino acid oxidase carries out the oxidation reaction on one enantiomer of the substrate to produce the keto acid. The transaminase then converts the keto acid back into the amino acid, but importantly the transaminase selectively produces the same enantiomer, which is not oxidised by the amino acid oxidase. This method is therefore able to produce the desired amino acid enantiomer in both high yield and stereoselectivity.

Chemists in industry have used amino acid oxidases to produce semi-synthetic compounds, in particular antibiotics based on cephalosporins, a class of β-lactams. Cephalosporin C is produced

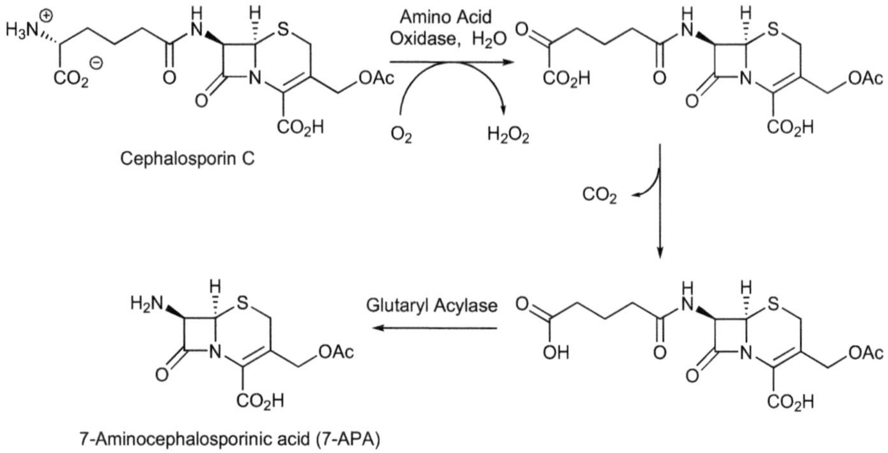

Cephalosporin C

7-Aminocephalosporinic acid (7-APA)

Figure 6.62 The synthesis of 7-aminocephalosporinic acid (7-APA) using an amino acid oxidase.

naturally by a fermentation process and then degraded to the building block 7-aminocephalosporinic acid (7-APA) by the use of toxic reagents and organic solvents. As shown in Figure 6.62, chemists have reported an enzymatic method of producing 7-APA from cephalosporin C, in which the first step is catalysed by a D-amino acid oxidase. The 7-APA produced can then be reacted with other chemicals to produce new semi-synthetic antibiotics.

6.7 Oxidation of Imines

Amides can be formed in a number of ways, one of which is the oxidation of imines. Amides are the same oxidation level as carboxylic acids and esters and like both of those functional groups, they are found widely in both natural and synthetic compounds. Amides are present in peptides, proteins, alkaloids and other natural products. They are also found in man-made materials, such as nylon and both pharmaceutical drugs and agrochemicals. One enzyme that is capable of oxidising imines is xanthine oxidase. This enzyme catalyses the oxidation of hypoxanthine to xanthine and subsequently to uric acid, as shown in Figure 6.63.

A large number of chemical methods exist for the synthesis of amides with most starting from the corresponding carboxylic acid, which is then activated and reacted with an amine nucleophile. However, as shown in Figure 6.64, a number of methods also exist to make amides by oxidising amines or imines. These methods typically

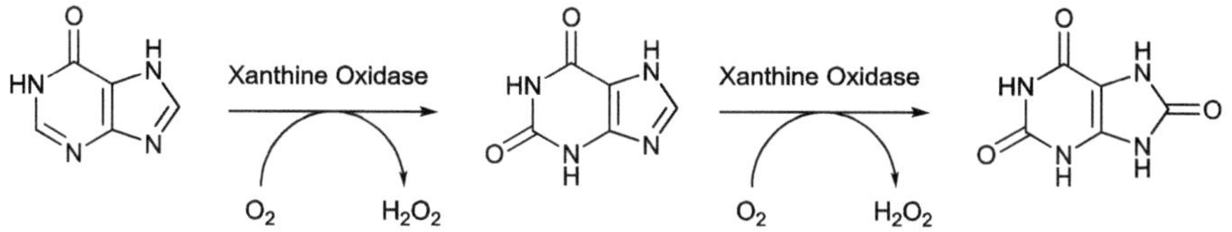

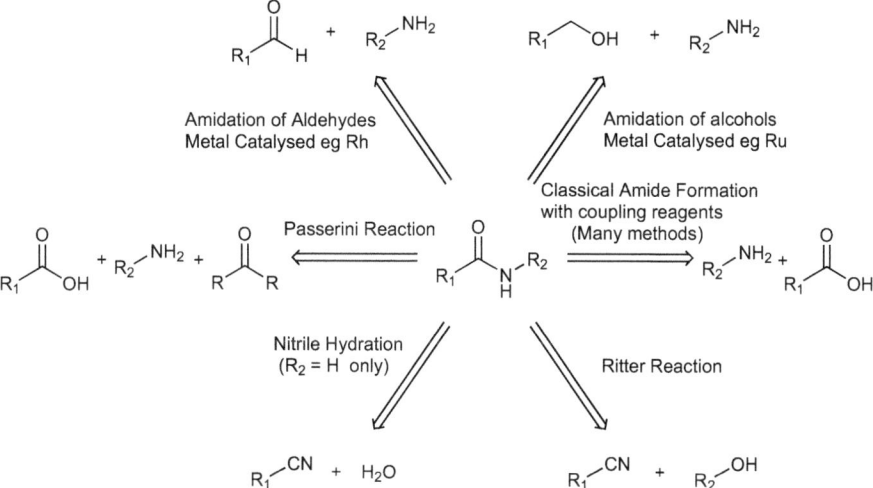

Figure 6.63 General transformation catalysed by xanthine oxidase.

Figure 6.64 Selected chemical methods of amide synthesis.

use transition metal catalysts, which need to be disposed of at the end of the reaction, and ligands that may take several steps to synthesise. They may also use organic solvents and harsh reaction conditions, which are not suitable for sensitive functional groups.

Enzymes offer an alternative to chemical methods of amide synthesis as they catalyse the formation of amides under mild, aqueous conditions that are compatible with many functional groups. In addition, they don't need metal catalysts or chiral ligands, making them a complementary technology. Of the enzymes that are capable of performing this reaction, xanthine oxidase has been reported to catalyse the oxidation of a small number of imines to the corresponding iminols. These then tautomerise to give the product amides, as shown in Figure 6.65.

Xanthine oxidase, like galactose oxidase which we saw earlier, requires the presence of a metal ion in the active site in order to catalyse the reaction. Galactose oxidase uses a copper atom, whilst xanthine oxidase uses a molybdenum atom to carry out the oxidation reaction.

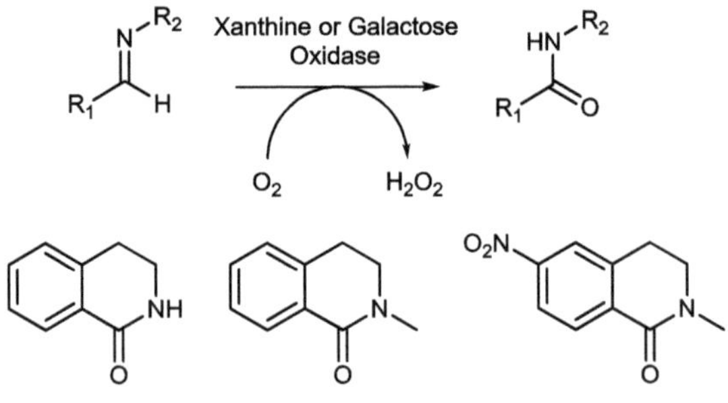

Figure 6.65 Selected examples of amides produced by xanthine or galactose oxidase.

6.8 Halogenation

Halogen atoms are useful for the synthetic chemist for a number of
reasons and a carbon with a single halogen substituent is at the alcohol oxidation level. The substitution of a hydrogen for a halogen
atom can change a molecule's properties on account of the electro-
negativity of the halogen. This may change the molecule's electronic
properties and its lipophilicity, which is an important factor in the
pharmaceutical or agrochemical industries where it may affect a
molecule's drug metabolism and pharmacokinetic profile. This may
lead to a change in the binding of the molecule to the target receptor
and/or other receptors, its bioavailability and its half-life if the halogen is placed at a site of metabolism. Halogen atoms are also good
leaving groups on account of their atomic properties, making them
very useful in synthetic bond forming reactions. They can be transformed into a number of other groups through substitution or elimination reactions. An increasing number of natural products are
found to contain halogen atoms, particularly those from marine environments where a higher concentration of halide ions may occur.
A number of the halogens are installed in a very selective manner,
showing the potential that halogenating enzymes may have for use in
organic synthesis. Chemical methods of halogenation can usually be
categorised into electrophilic, nucleophilic or radical halogenation,
with a range of methods shown in Figure 6.66.

Although many useful methods of chemical halogenation have
been developed, there still remain some drawbacks with this
methodology. Classical halogenation reactions use harsh conditions, which may not be suitable for complex substrates, whilst
more modern methods require the use of transition metal catalysts,

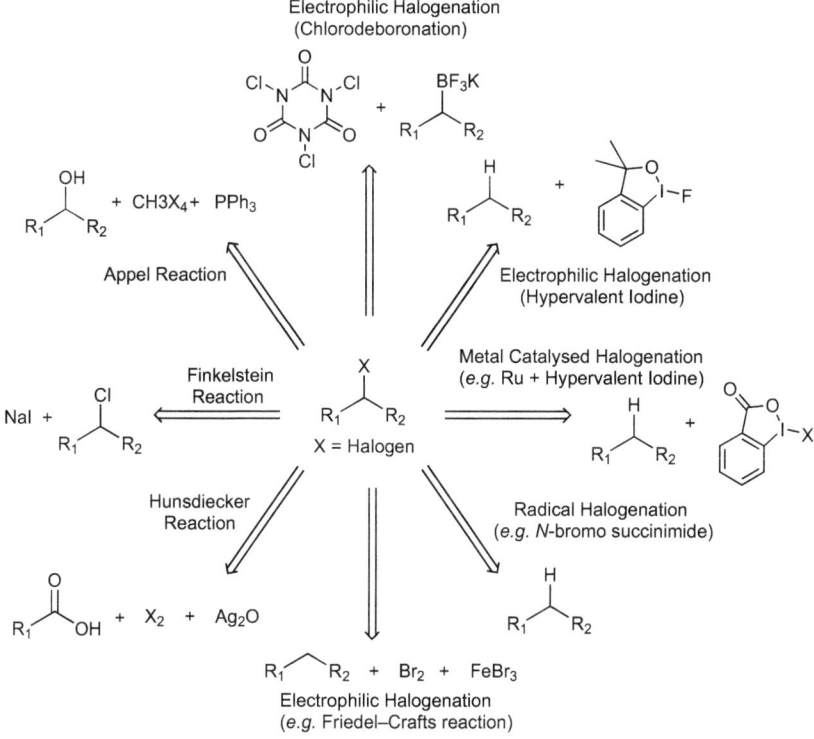

Figure 6.66 Selected chemical methods of halogenation.

which may be expensive or difficult to dispose of after the reaction. Some of the methods, such as radical processes, may also use environmentally toxic solvents, making them not very green or sustainable. In contrast, enzymatic methods of halogenation need only mild ambient conditions in aqueous reaction media. In addition, the reactions are often highly regio-, chemo- and sometimes stereoselective, making them complementary to chemical methods. Enzymatic halogenation is predominately carried out by three different types of enzymes, each of which is discussed in more detail below. Although other enzymes exist that are capable of carrying out the transformation, they are not discussed here.

6.8.1 Halogenation Using Halogenases

Although all of the enzymes in this section may be called halogenases, this name is usually used to refer to a subset of flavin-dependent enzymes. As the name indicates, flavin-dependent halogenases require a

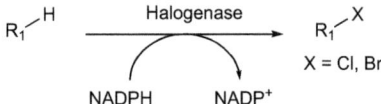

Figure 6.67 General transformation catalysed by halogenases.

flavin co-factor in order to catalyse halogenation reactions. They are predominately found in prokaryotes, such as bacteria, and of the eukaryotic halogenases known, almost all are found in fungi. Most members of this enzyme class have been found to be involved in the synthesis of secondary metabolites, where they are responsible for the introduction of a halogen. As shown in Figure 6.67, these enzymes work by using oxygen and a halide source to generate hypohalous acid, an electrophilic halogenating reagent. This is then used to very selectively halogenate the substrate. In general, halogenases prefer to perform chlorination rather than bromination reactions, in part because high bromine concentrations have a negative effect on cell growth and gene expression.

As flavin-dependent halogenases generate an electrophilic halogen source, it is unsurprising that they primarily react with electron-rich aromatic systems and activated aliphatic substrates. Depending on the particular enzyme, halogenases may act upon the free substrate or as part of a biosynthetic pathway where the substrate is bound to a carrier protein. This is worth taking into consideration if we want to carry out a reaction *in vitro* with a specific protein. Because halogenases are typically very selective, they may often display narrow substrate selectivity. However, chemists are busy trying to improve the substrate scope of these enzymes and a range of substrates are shown in Figure 6.68. In each case, the halogenated product is shown.

As can be seen in Figure 6.68, some substrates, in particular tryptophan, can be halogenated with excellent regioselectivity, making this a useful method of installing a halogen prior to carrying out a palladium cross-coupling reaction. The mechanism of a tryptophan halogenase is shown in Figure 6.69. As with the other mechanisms, you don't need to know this level of detail, but it is included here for those who would like further understanding. As shown in Figure 6.69, the mechanism begins with the reduced flavin (in this case flavin adenine dinucleotide). The flavin is reduced by an associated flavin reductase, which in turn uses a nicotinamide co-factor as the terminal reductant in a similar manner to Ereds, which we have seen before in the Reduction chapter. The reduced flavin then reacts with oxygen to form a hydroperoxide flavin intermediate. This is analogous to the peroxy anion intermediate we saw previously in the Baeyer–Villiger monooxygenase

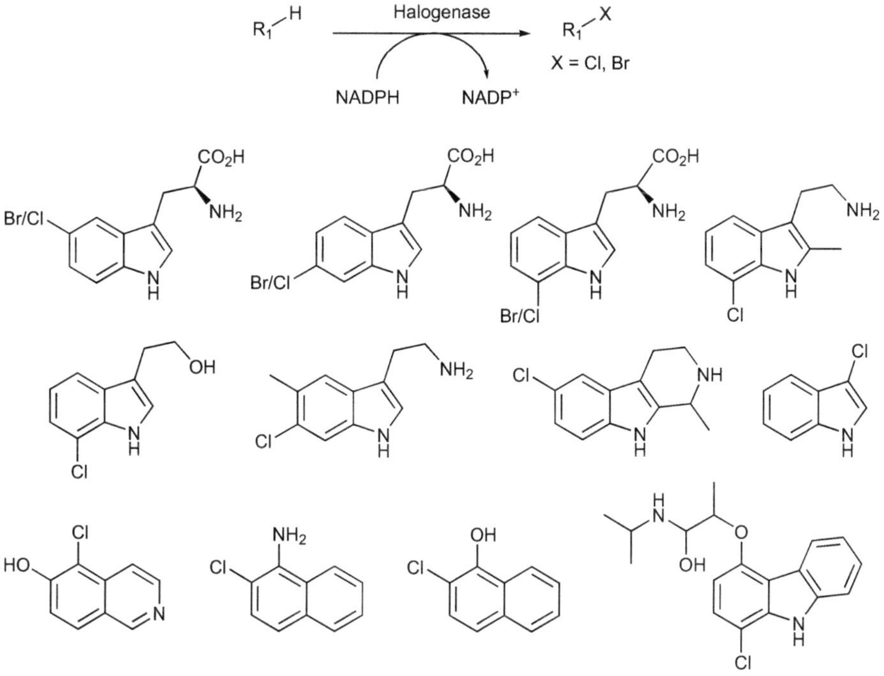

Figure 6.68 The scope of products produced by halogenation catalysed by halogenases.

mechanism in Figure 6.50. In the halogenase reaction, the hydroperoxide intermediate is then attacked by the halide source, in this case chloride, to form hypochlorous acid and a hydroxy flavin. The hydroxy flavin then loses water to regenerate the reduced flavin for the next catalytic cycle. The hypochlorous acid is then transferred *via* a tunnel within the enzyme to the substrate binding site. A lysine residue positioned at the end of the tunnel is thought to interact with the hypochlorous acid, placing it in the correct position to chlorinate the substrate.

6.8.2 Halogenation Using Haloperoxidases

The second group of enzymes capable of carrying out halogenation reactions is the haloperoxidases. This family of enzymes can be further subdivided into two groups based on how the enzymes work, and we will mention each in further detail. Like all peroxidase enzymes, they use hydrogen peroxide; however, in the case of haloperoxidases as the name suggests, the hydrogen peroxide is used to convert a halide ion into hypohalous acid. This is the same overall process as that for the flavin-dependent halogenases above, but the

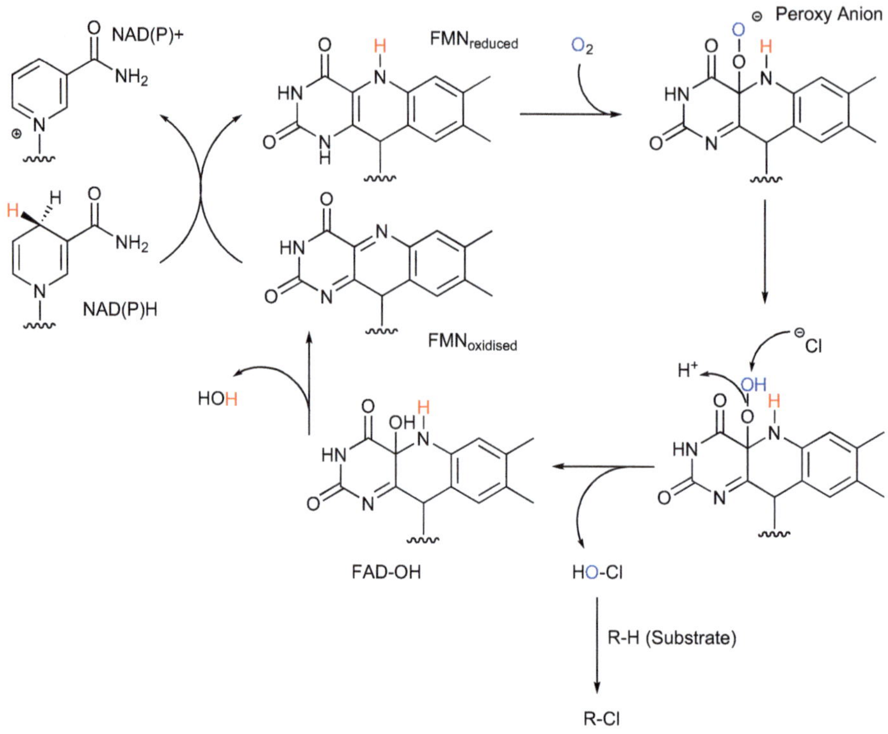

Figure 6.69 The mechanism of halogenation catalysed by halogenases.

mechanism to make the hypohalous acid is different. The first of the subgroups is the heme iron-containing haloperoxidases. These enzymes are found in bacteria, fungi and plants and require an iron atom stabilised by a heme ligand to carry out the reaction. Although we won't discuss the mechanism in detail here, it is very similar to the mechanism of P450 monooxygenases discussed previously in this chapter. The main difference is that hydrogen peroxide, rather than molecular oxygen, is bound to the iron. We still produce an active iron oxo complex, which in the case of haloperoxidases is attacked by a halide ion to produce the hypohalous acid. These enzymes react with electron-rich aromatic rings and with alkenes as shown in Figure 6.70. In each case the halogenated product is shown and where multiple halogens are present, the substituent added in the reaction is indicated in red.

Haloperoxidases have also been used in the synthesis of pharmaceutical drugs. Chemists from Merck investigated the use of a haloperoxidase in the synthesis of Crixivan, an HIV-1 protease inhibitor. As shown in Figure 6.71, starting from indene, reaction with a

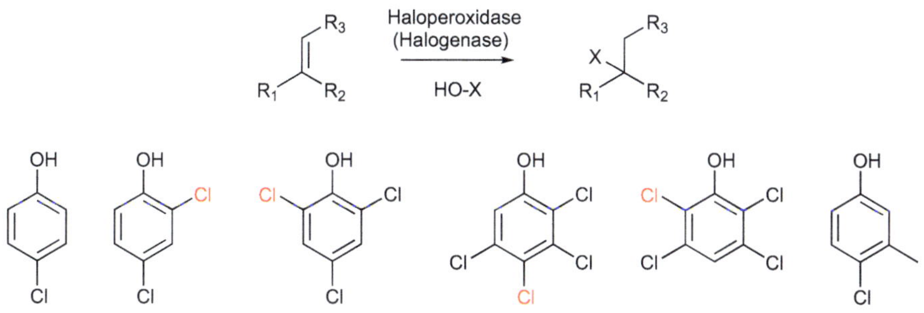

Figure 6.70 Selected examples of halogenated products produced by haloperoxidases.

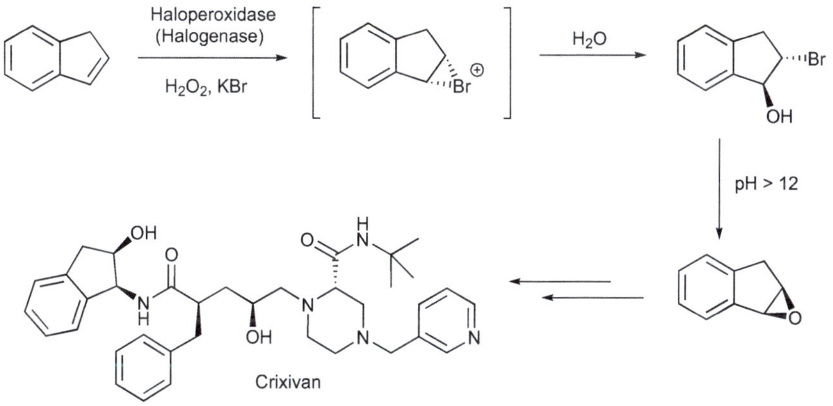

Figure 6.71 The synthesis of Crixivan using a halogenase.

haloperoxidase and excess potassium bromide gave the desired (*S,S*)-bromoindanol in good yield and excellent selectivity. In this case, the reaction with hypobromous acid is thought to form a bromonium ion, which is rapidly opened by water to give the bromohydrin product. The bromoindanol was then converted into the epoxide *in situ* by increasing the reaction pH to greater than 12. The epoxide produced was then reacted further to produce Crixivan.

The second group of haloperoxidases is the vanadium-dependent haloperoxidases. These enzymes are found mainly in marine environments, being present in algae, fungi and bacteria. These enzymes require vanadium to be present in order to catalyse halogenation reactions. Without vanadium present, they show alternative phosphatase reactivity, suggesting a common evolutionary link between these two enzymes. The mechanism of these enzymes is not completely clear, but it is known that the enzymes use an anionic vanadate complex, hydrogen peroxide and a halide ion, as shown in Figure 6.72.

It is not completely clear what the actual halogenating agent is in reactions catalysed by these enzymes, but for now we can certainly use hypohalous acid as a potential halogenating reagent given that it appears to be formed in the reaction. As with the other halogenating enzymes we have already discussed, haloperoxidases are capable of halogenating electron-rich aromatics and alkenes. Depending on the enzyme, these proteins can also show selectivity towards chloride, bromide or iodide, although fluoride is not transferred. The regioselectivity of haloperoxidases tends to mirror that of chemical electrophilic halogenating reagents such as *N*-halosuccinimides. A range of substrates that undergo halogenation by haloperoxidases are shown in Figure 6.73. In each case, the product (sometimes there may be more than one) of the halogenation reaction is shown.

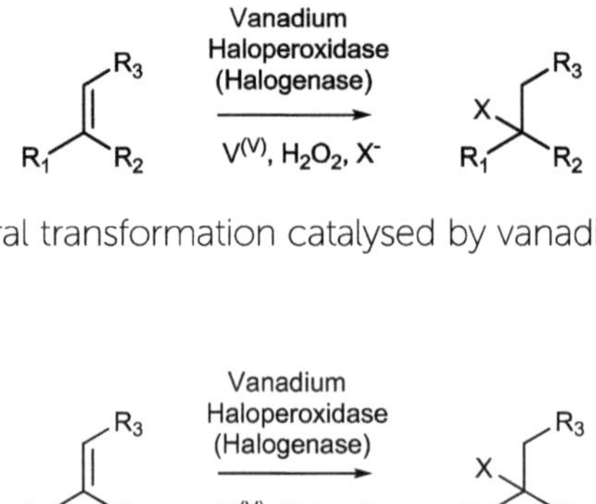

Figure 6.72 General transformation catalysed by vanadium haloperoxidases.

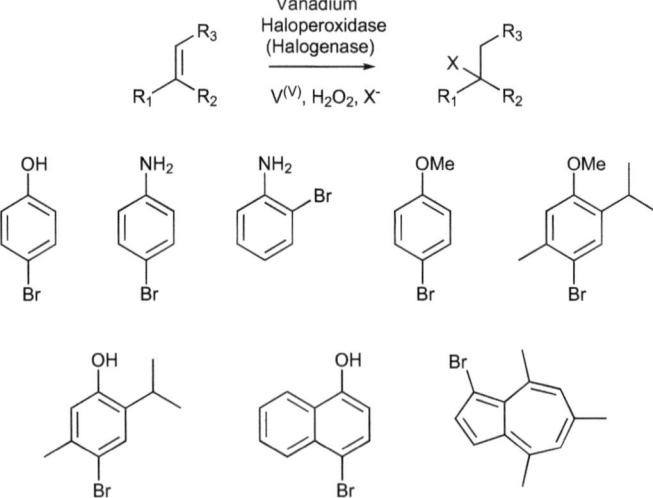

Figure 6.73 The scope of brominated compounds produced by vanadium haloperoxidases.

6.8.3 Halogenation Using *S*-Adenosylmethionine-dependent Halogenases

So far, we have seen enzymes that generate an electrophilic halide source in order to carry out halogenation reactions. We will now look at enzymes that are capable of catalysing reactions where the halide acts as a nucleophile. These enzymes use *S*-adenosylmethionine (SAM) as a co-substrate and have very limited substrate range. They are included here as the reaction products may be further reacted into more useful compounds. The first enzyme we will look at has been termed a fluorinase and comes from a type of bacteria called *Streptomyces*. As shown in Figure 6.74, the enzyme catalyses the conversion of SAM into 5′-fluoro-5′-deoxyadenosine, which is then further converted by the organism into fluoroacetic acid and L-fluorothreonine, two potentially useful materials for organic synthesis.

A closely related enzyme to the fluorinase is an enzyme isolated from the marine organism *Salinispora tropica*. This enzyme catalyses an identical reaction to make 5′-halo-5′-deoxyadenosine compounds, utilising either chloride, bromide or iodide. Interestingly, these reactions can be run in reverse in order to make non-natural analogues of *S*-adenosylmethionine. This in turn has the potential to widen the number of groups that can be transferred by methyltransferase enzymes, which we will see more of in later chapters. An example of this methodology is shown in Figure 6.75.

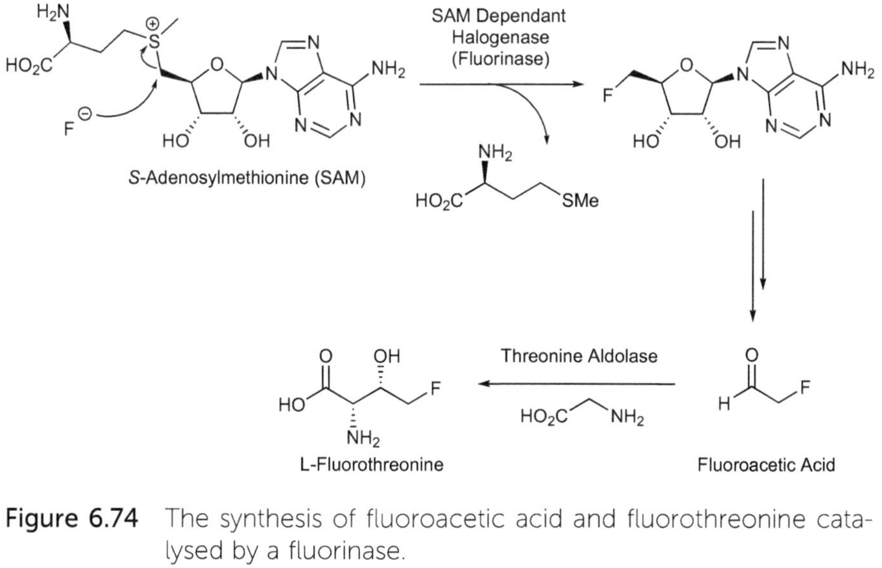

Figure 6.74 The synthesis of fluoroacetic acid and fluorothreonine catalysed by a fluorinase.

Figure 6.75 General transformation catalysed by *S*-adenosylmethionine (SAM)-dependent halogenases.

6.9 Oxidation of Sulfides to Sulfoxides and Sulfones

Sulfur is an abundant, naturally occurring substance that is found in its elemental form or in organo and metal sulfides. It is an essential element for all life, present in two proteinogenic amino acids and two vitamins. It is also found in many biological compounds, such as co-factors, including glutathione and thioredoxin, along with iron–sulfur proteins. Other sulfur functional groups also exist such as disulfides, which confer extra stability and strength to proteins. Elemental sulfur itself is also a product from and an oxidant for various species of bacteria. The more oxidised forms of sulfur (sulfoxides and sulfones) are found in a variety of natural products and man-made compounds, such as pharmaceuticals, agrochemicals, chiral auxiliaries and ligands. Sulfoxides, in particular, are of interest as the sulfur and oxygen atoms do not share a typical p-orbital π-bond like a carbonyl group. Instead, the sulfur assumes an sp^3 hybridisation with a lone pair of sulfur electrons occupying one of the orbitals. Sulfoxides are con-formationally stable at room temperature and can be separated into enantiomers. Importantly, this means that it is possible to selectively oxidise sulfides to produce a single sulfoxide enantiomer. Sulfones are present in a wide array of pharmaceutical drugs, in particular drugs used to treat leprosy and a number of anti-cancer drugs. A wide range of chemical methods exist for the oxidation of sulfides to sulfoxides and sulfones using a variety of different oxidising reagents, as shown in Figure 6.76. In contrast to the oxidation of alcohols, where the products of the initial oxidation (aldehydes and ketones) are more susceptible to further oxidation, the oxidation of sulfides can often be slower than the initial oxidation to the sulfoxide. Despite this, over-oxidation of the sulfoxide may still occur when using chemical methods.

Using the chemical methods in Figure 6.76, it is possible to oxidise sulfides to either sulfoxides or sulfones. However, many of these pro-cesses require the use of metal catalysts, which need to be disposed of at the end of the reaction. Some of the methods also use oxidising reagents such as hydrogen peroxide or hydrazine, which are difficult to handle on a large scale. Asymmetric oxidations also require chiral ligands or auxiliaries, which need to be synthesised. In contrast, en-zymatic oxidation methods tend to be carried out under mild con-ditions at ambient temperature in aqueous conditions, without the need for metal catalysts or chiral ligands. As a result, enzymatic methods offer a complementary alternative to traditional chemical

Oxidation

Chemical Reagents:

1) 1 Equivalent of Peracid *e.g.* mCPBA

2) NaOCl

3) H_2O_2 + Organocatalyst *e.g.* Chiral Phosphoric Acid

4) H_2O_2 + Transition metal *e.g.* VO(acac)

5) H_5IO_6 + $FeCl_3$

6) Hypervalent Iodine Reagents *e.g.* $NaIO_4$

Oxidation

Chemical Reagents:

1) Greater than 1 Equivalent of Peracid

2) Urea Hydrogen Peroxide adduct

3) $KMnO_4$ / MnO_2

4) H_2O_2 Transition metal *e.g.* Ammonium Tungstate

5) $KHSO_5$ - Oxone

6) Magnesium Monoperoxyphthalate (MMPP)

Figure 6.76   Selected chemical methods of sulfide oxidation.

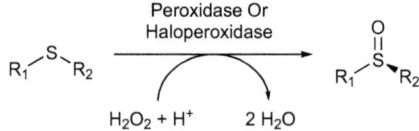

Figure 6.77   General transformation catalysed by peroxidases and haloperoxidases.

approaches. Several classes of enzymes are capable of oxidising sulfides and we will consider each separately in the sections that follow.

6.9.1   Oxidation of Sulfides Using Peroxidases and Haloperoxidases

One of the first classes of enzymes that was discovered to carry out the oxidation of sulfides was the peroxidases. We have already seen this class of enzymes in previous sections of this chapter, along with the related haloperoxidase enzymes. These enzyme classes are found in bacteria, fungi, plants and animals. Both of these classes of enzymes utilise hydrogen peroxide as the terminal oxidant and are able to oxidise sulfides to the corresponding sulfoxides as shown in Figure 6.77.

In the case of haloperoxidases, these reactions are run without any halide ions being present. This prevents the formation of hypohalous acid and subsequent halogenation reactions, which we covered in detail in the previous section. Both peroxidases and haloperoxidases are capable of oxidising a range of substrates with good yields and high levels of stereoselectivity. A range of substrates are shown in Figure 6.78 and in each case the product of the oxidation reaction is

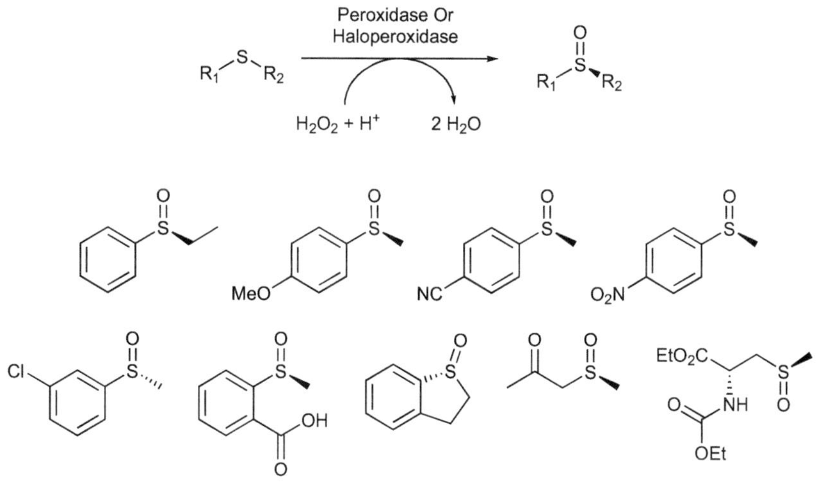

Figure 6.78 The scope of products produced by the (halo)peroxidase-catalysed oxidation of sulfides.

shown. It is often possible to access either enantiomer through the appropriate choice of the enzyme.

In general, these enzymes are less active with substrates containing electron-withdrawing substituents or bulkier substrates. It is also worth noting that slow addition of hydrogen peroxide to these reactions results in higher yields of the products, possibly due to the enzymes being deactivated when high levels of hydrogen peroxide are present.

6.9.2 Oxidation of Sulfides Using Baeyer–Villiger Monooxygenases

The second class of enzymes capable of oxidising sulfides is the Baeyer–Villiger monooxygenases. We have already met these flavo-proteins in this chapter, where they catalyse the oxidation reaction they are named after, which transforms ketones into esters. These enzymes are flavin-dependent and require a nicotinamide co-factor as the terminal oxidant. They are also capable of catalysing the oxidation of sulfides to sulfoxides, as shown in Figure 6.79.

As with peroxidases, they oxidise sulfides in good yields and with high selectivity. A range of sulfides can be oxidised, as shown in Figure 6.80. In each case, the product sulfoxide is shown.

As we have seen in many other reactions, the nicotinamide co-factor can be recycled by many different recycling systems. We have chosen glucose dehydrogenase to illustrate this in Figure 6.80.

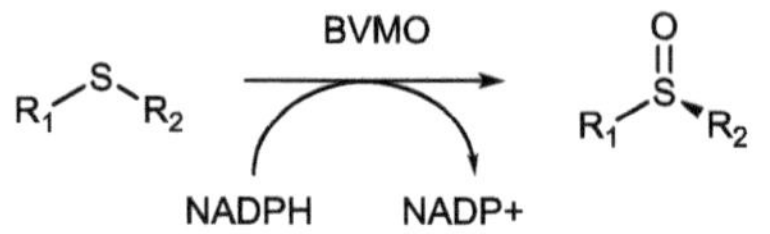

Figure 6.79 General transformation catalysed by Baeyer–Villiger monooxygenases.

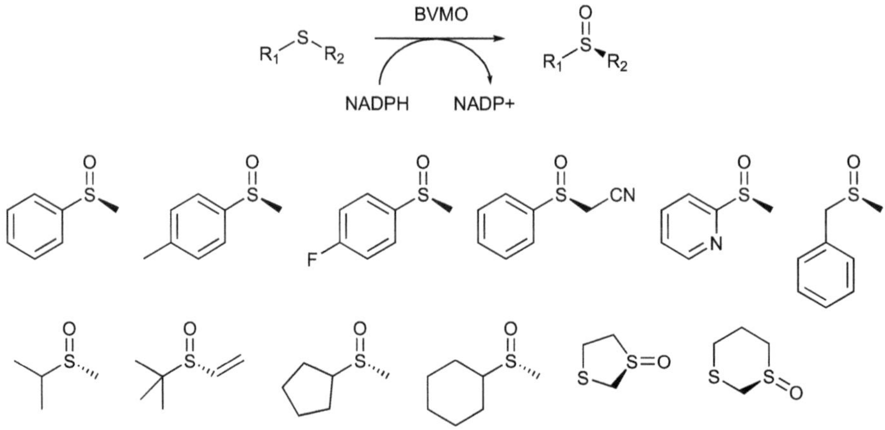

Figure 6.80 The scope of products produced by the Baeyer–Villiger monooxygenase-catalysed oxidation of sulfides.

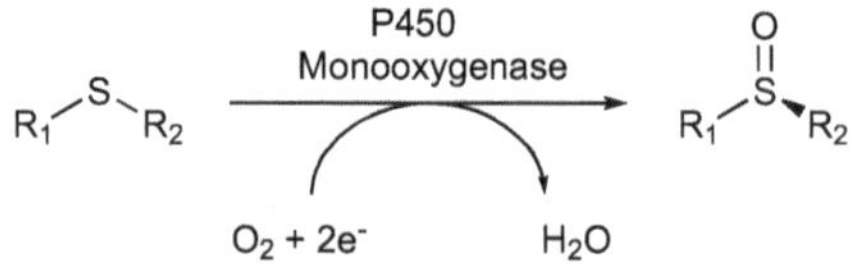

Figure 6.81 General transformation catalysed by P450 monooxygenases.

6.9.3 Oxidation of Sulfides Using P450 Monooxygenases

We have already seen in this chapter that P450 monooxygenases are capable of carrying out a range of different oxidation reactions. These heme-containing, iron-dependent enzymes are also able to carry out the oxidation of sulfides to sulfoxides, as shown in Figure 6.81.

There are fewer examples of P450-catalysed oxidation reactions described, which could in part be due to the complexity of these systems. As we previously mentioned, in addition to the P450 monooxygenase, we need to use the correct redox enzyme partners in order to transfer electrons from the nicotinamide co-factor, which is the terminal oxidant. As a result, it may be easier to use some of the other enzymes

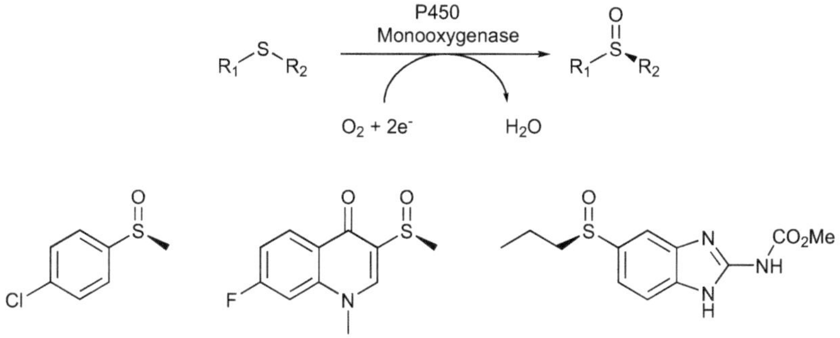

Figure 6.82 Selected examples of sulfoxides produced by the P450 monooxygenase-catalysed oxidation of sulfides.

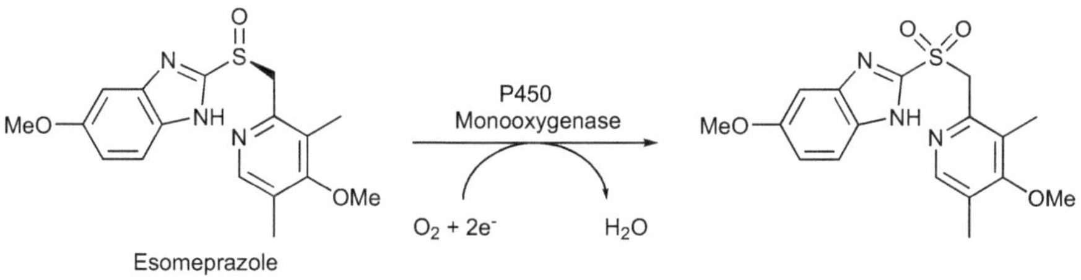

Figure 6.83 The oxidation of esomeprazole catalysed by a P450 monooxygenase.

described in this section. Despite this, P450 enzymes are particularly relevant as they oxidise sulfur-containing compounds within our bodies. The smaller range of substrates for the P450 monooxygenase-catalysed oxidation of sulfides is shown in Figure 6.82. In each case, the product of the oxidation reaction is shown.

Interestingly, P450 monooxygenases have also been shown to carry out the oxidation of sulfoxides to sulfones. Of particular interest is the oxidation of esomeprazole, a proton-pump inhibitor that lowers the amount of stomach acid produced in patients for the treatment of peptic ulcers and gastroesophageal reflux. As shown in Figure 6.83, the *S*-enantiomer of esomeprazole is oxidised by a human P450 monooxygenase to the sulfone around ten times faster than the corresponding *R*-enantiomer.

6.10 Oxidative Dealkylation of *N*-, *O*-, and *S*-ethers Using P450 Monooxygenases

As we will see in Chapters 7 and 8, a large family of enzymes called the methyltransferases is capable of catalysing the addition of a methyl group to a range of atoms, including oxygen, nitrogen and sulfur. The selective addition of a methyl group can have a large effect upon both natural products and pharmaceutical compounds. Regiospecific alkylation can alter a molecule's physicochemical properties and often results in a strongly increased affinity toward membranes and receptors. The methyl group can also be used biologically, for example in the epigenetic processes that determine gene expression, stem cell maturation, and DNA methylation. The methyl group can also be used as a protecting group for heteroatoms. It is added to the heteroatom in order to prevent undesired reactions occurring and then removed when we want the heteroatom to participate in a desired reaction. A large number of protecting groups exist, each with their own advantages and disadvantages. The methyl group is small and therefore atom efficient, but its removal can be problematic. Methyl groups are traditionally used in synthesis as a protecting group for phenols and one method of removing methyl groups from these compounds is to use a P450 monooxygenase enzyme, as shown in Figure 6.84.

A number of more classical chemistry methods also exist for the deprotection of phenylmethyl ethers. These typically involve the use of very strong Lewis acids and harsh reaction conditions, making them unsuitable for molecules that contain sensitive functional groups. They are also unselective, deprotecting all of the methyl ethers present and potentially also carrying out undesired deprotection of other functional groups. More modern methods use redox photochemistry which, whilst occurring under milder conditions, requires the use of transition metal catalysts and ligands. A range of chemical methods to carry out the deprotection of methyl ethers are shown in Figure 6.85.

In contrast to chemical methods, the demethylation of heteroatoms catalysed by P450 monooxygenases occurs in aqueous reaction media and mild temperatures. There is no need for transition metal catalysts and ligands or strong Lewis acids. In addition, P450 monooxygenase-catalysed demethylations are very selective, often removing only one methyl group in a substrate that contains several methyl ethers. Some examples of the substrate scope of these enzymes are shown in Figure 6.86. The majority of substrates

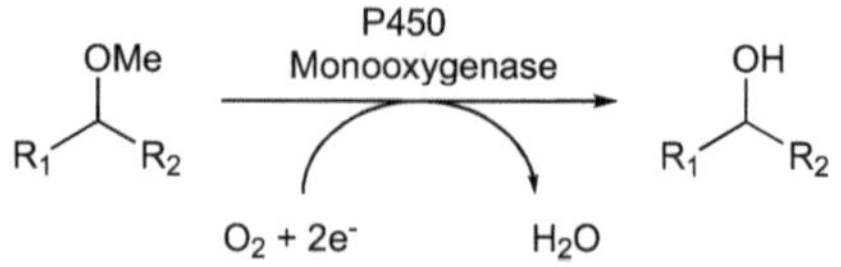

Figure 6.84 General mechanism catalysed by P450 monooxygenases.

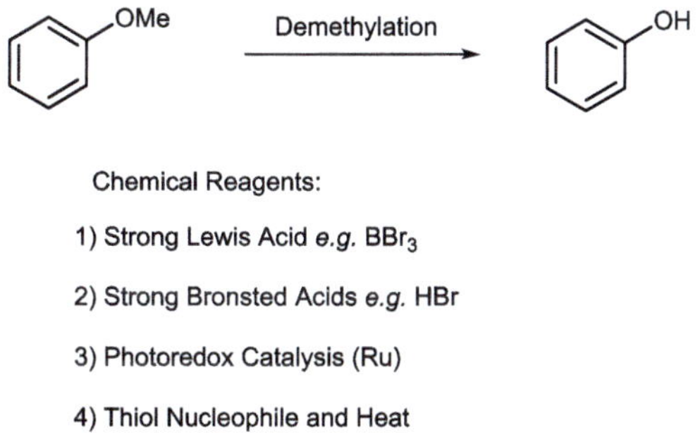

Figure 6.85 Selected chemical methods of demethylation.

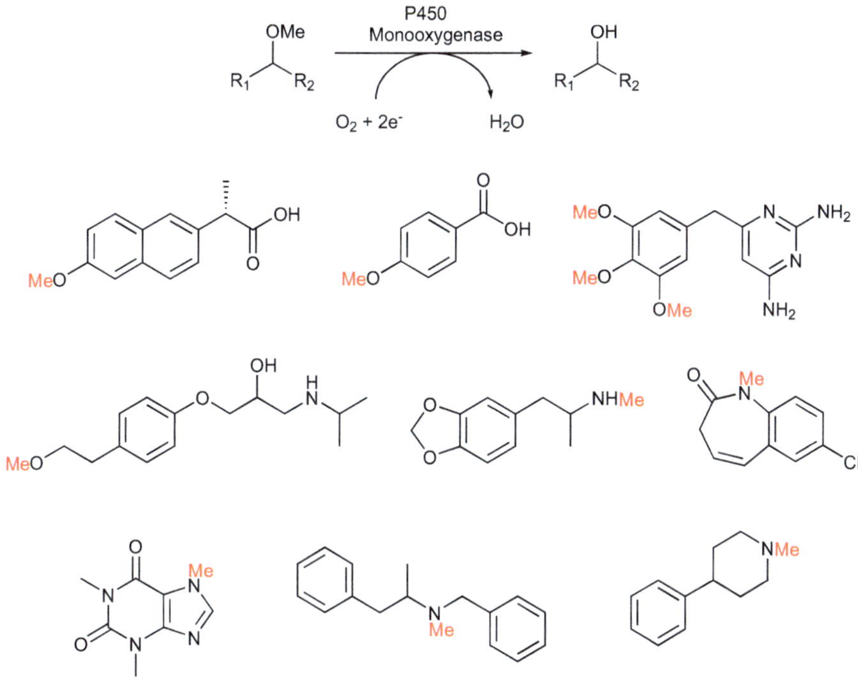

Figure 6.86 Selected substrates for demethylation reactions catalysed by P450 monooxygenases.

are phenylmethyl ethers or *N*-methyl heterocycles as discussed previously and in each case the methyl removed is coloured in red. Many of these substrates are drug molecules and the metabolism of these structures in mammalian species in a very active field of research.

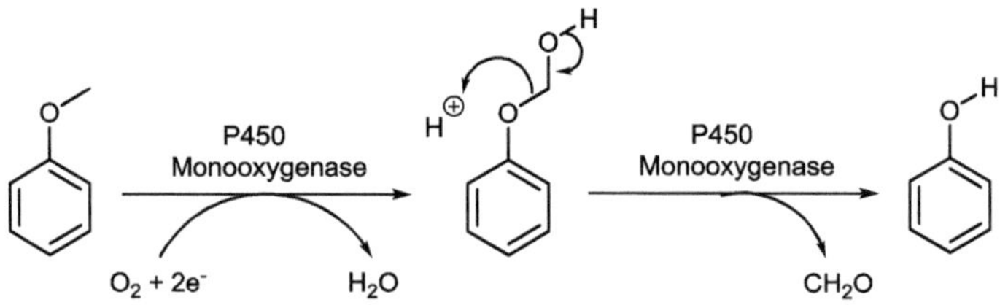

Figure 6.87 A condensed version of the mechanism of demethylation catalysed by P450 monooxygenases.

In general, O- and N-demethylation is much more common than S-demethylation, although this is in part because more research has been carried out on O- or N-methylated substrates. De-ethylation has also been reported, but the substrate scope is much narrower and it will not be covered here. A shortened version of the mechanism of demethylation is shown Figure 6.87. You do not need to know this, but it is included here for those readers who would like more detail. The majority of the mechanism proceeds exactly like the hydroxylation mechanism we have already discussed in this chapter in Figure 6.8. This results in the methyl group being hydroxylated as shown in Figure 6.87. Rather than finish here, as in the case of hydroxylation, this intermediate undergoes further reaction. As shown in Figure 6.87, it breaks down to release formaldehyde and the demethylated product.

O- and N-demethylation is also very common in plant secondary metabolism, although in those cases the reactions are not carried out by P450 enzymes, but by another enzyme class instead.

7 C–X Bond Formation

7.1 Introduction

This chapter will cover a number of enzymes that are capable of making carbon–heteroatom (C–X) bonds. These enzymes come from a range of different groups within the enzyme classification system. Some are from group number 2, whilst others are from group 4, and there is one enzyme from group 5. Group 2 enzymes are transferases defined as enzymes that transfer a functional group from one substance to another. In contrast, group 4 enzymes are lyases, defined as enzymes that catalyse the non-hydrolytic addition or removal of groups from substrates. Finally, enzymes in group 5 are isomerases, which as the name suggests catalyse the isomerisation of a molecule. Using these enzymes, we can make a range of C–X bonds, and in particular, we will cover C–N and C–O bond formation in more detail in this chapter. Nitrogen and oxygen are the most prevalent atoms found in chemical structures after carbon and hydrogen. They are present in a variety of functional groups (acidic or basic) that are able to interact with proteins or other biological molecules, often through hydrogen bonding. As a result, most natural products and nearly all pharmaceutical compounds contain a nitrogen or oxygen atom and usually more than one of each. These enzymes are also able to work in the reverse direction, although in most cases running the reaction in the "reverse" direction is not as synthetically useful. The transformations covered in more detail within this chapter are shown in Figure 7.1 and the synthesis reactions are discussed in more detail in separate sections for each type of bond formed.

Biocatalysis in Organic Synthesis: The Retrosynthesis Approach
By Nicholas J. Turner and Luke Humphreys
© Nicholas J. Turner and Luke Humphreys 2018
Published by the Royal Society of Chemistry, www.rsc.org

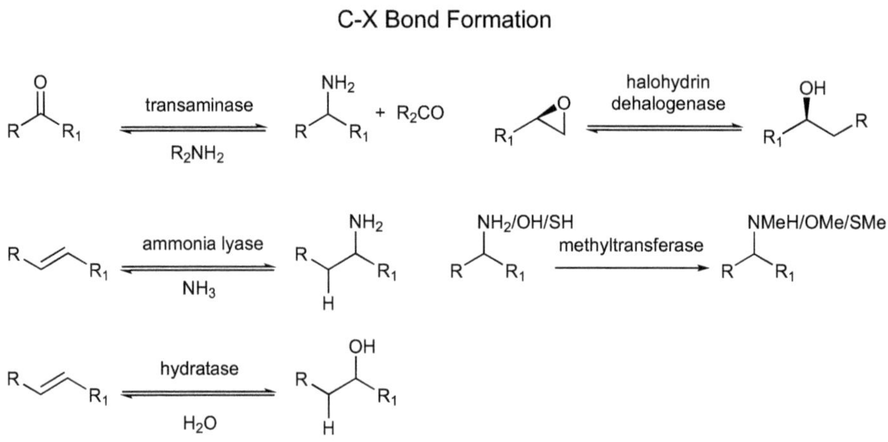

C-X Bond Formation

Figure 7.1 Overview of transformations presented in this chapter.

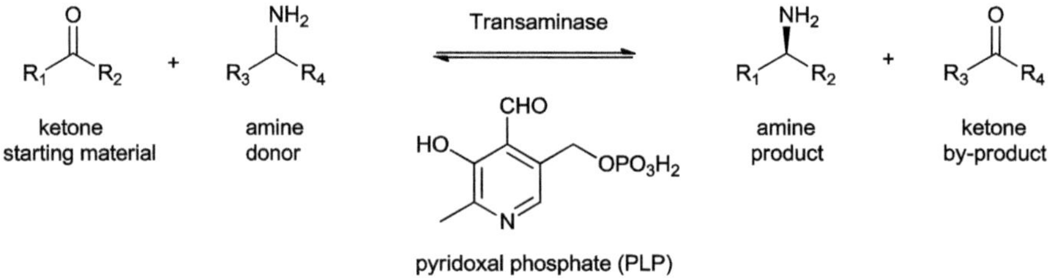

pyridoxal phosphate (PLP)

Figure 7.2 General transformation catalysed by transaminases.

7.2 Formation of C−N Bonds

7.2.1 Formation of C−N Bonds Using Transaminases

Transaminases are typically used by organisms to synthesise α-amino acids from an α-keto acid and an amine source (typically another amino acid). As this process is reversible, transaminases are also used to metabolise amino acids from proteins. As shown in Figure 7.2, the general transformation in the C–N bond forming direction is the conversion of a ketone to a chiral secondary amine. This transformation can also be carried out on aldehydes to produce the corresponding primary amines. As we can see in Figure 7.2, in addition to the starting material and product, we also require an amine donor, which is transformed into a ketone by-product. Transaminases also use a co-factor, pyridoxal phosphate (PLP), to catalyse the transformation.

Notice in Figure 7.2 that this is an equilibrium process and therefore the equilibrium position will depend upon the substrates

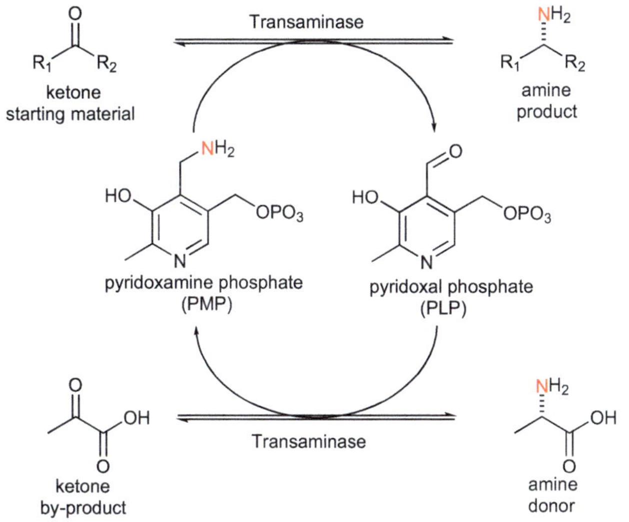

Figure 7.3 A substrate-coupled co-factor recycling system for transaminases.

and products. Unfortunately, for most transaminase-catalysed reactions the equilibrium tends to lie towards the side of the starting materials, but we will look at a number of methods chemists have used to overcome this. The formation of a chiral amine directly from a ketone is rather difficult to achieve chemically. The basic mechanism of a transaminase reaction is shown in Figure 7.3. The amine that is transferred during the reaction is coloured in red for clarity. The amine donor is initially transferred to the co-factor pyridoxal phosphate (PLP), which is converted into pyridoxamine phosphate (PMP). As a consequence, the amine donor is converted into a ketone by-product. The amine is then transferred from the co-factor PMP to the ketone starting material to produce the amine product. This reaction also converts the co-factor back to its original structure, PLP. Note that in this reaction, the transaminase transfers the chirality of the amine donor to the amine product.

Typically, if we wanted to carry out a reductive amination of a ketone chemically, we would use an amine to make the intermediate imine and then reduce this with a mild reducing agent such as sodium cyanoborohydride. However, if we try this process using ammonia, then the product amine can undergo reductive amination with another molecule of the starting material to give a secondary (or even tertiary) amine, as shown in Figure 7.4.

To prevent this unwanted over-reaction, chemists use a protecting group on the amine or an alternative functional group that can then

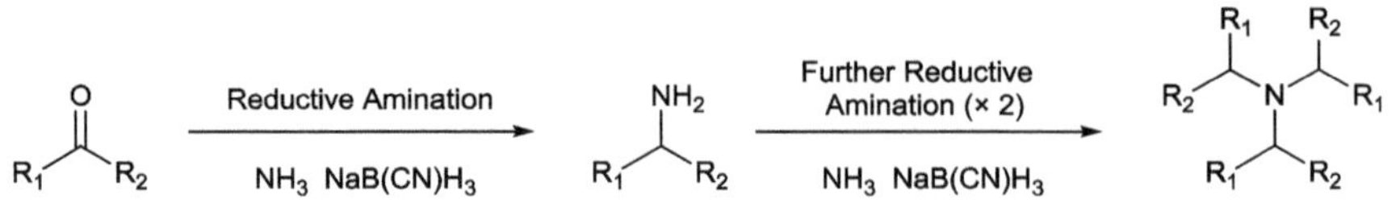

Figure 7.4 The issue of overreaction in reductive aminations using chemical reagents.

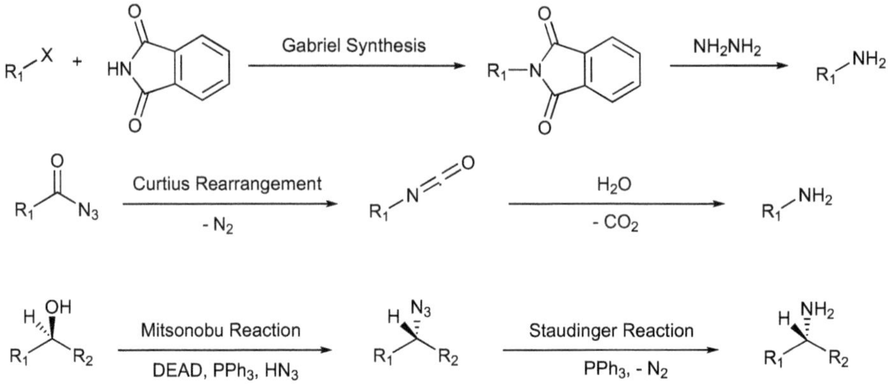

Figure 7.5 Protecting group strategies to avoid over-reaction in reductive aminations.

be transformed into the amine. However, this requires an extra step to attach the protecting group and/or remove it, or to convert the "amine equivalent" into the desired amine, as shown in Figure 7.5. The reaction conditions required for these steps can also be incompatible with more sensitive functional groups or chiral centres.

There are a wide range of chiral methods with which to prepare chiral amines, as shown in Figure 7.6. These approaches are capable of producing a huge range of amines with excellent stereoselectivity. However, they either require a chiral group attached to the molecule (often also doubling as the amine protecting group) or use a protecting group and a chiral metal ligand complex, which can often be expensive to use on a large scale. All of these approaches require the amine protecting group to be removed to access the primary amine.

In contrast to the methods above, transaminases produce the chiral primary amine under mild reaction conditions without the need for protecting groups, chiral auxiliaries or catalysts and chiral ligands. A wide variety of ketones undergo reaction with transaminases, as shown in Figure 7.7. In each case, the product amine is shown, although in many cases it is possible to produce the opposite enantiomer by choosing a transaminase with the opposite stereoselectivity.

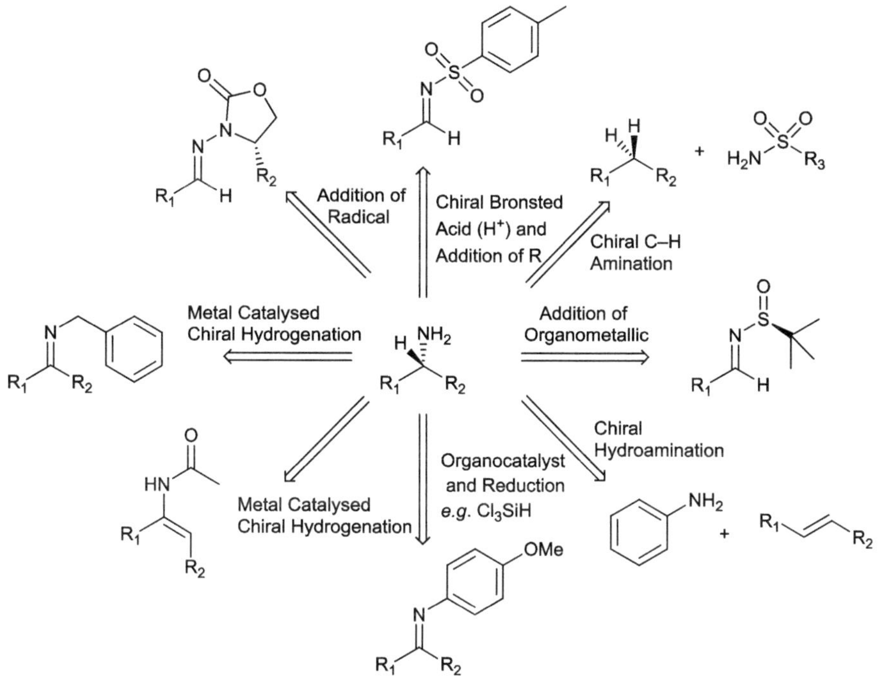

Figure 7.6 Selected chemical methods of preparing chiral amines.

The reaction of ketones with transaminases can be extended if the substrate molecule also contains an electrophilic group to enable a cyclisation reaction to take place. As shown in Figure 7.8 a variety of leaving groups can be used to make a range of cyclic compounds. In each case the starting material and product are shown. The last three examples in Figure 7.8 also illustrate that transaminases are highly regioselective. For a range of 1,5-diketones, the less hindered methyl ketone selectively undergoes transamination. The amine produced is then cyclised onto the second ketone to give the cyclic imine product shown. Although one enantiomer is shown in Figure 7.8, it is also possible to access the other enantiomer by choosing a transaminase with the opposite stereoselectivity. The product imine can then be reduced to either the *cis* or *trans* compound, giving access to all four possible diastereoisomers.

Transaminases can also be used for the amination of aldehydes to produce primary amines, with the aldehyde being a more reactive amine acceptor than the corresponding ketone. As shown in Figure 7.9, a range of primary amines can be produced by the amination of aldehydes.

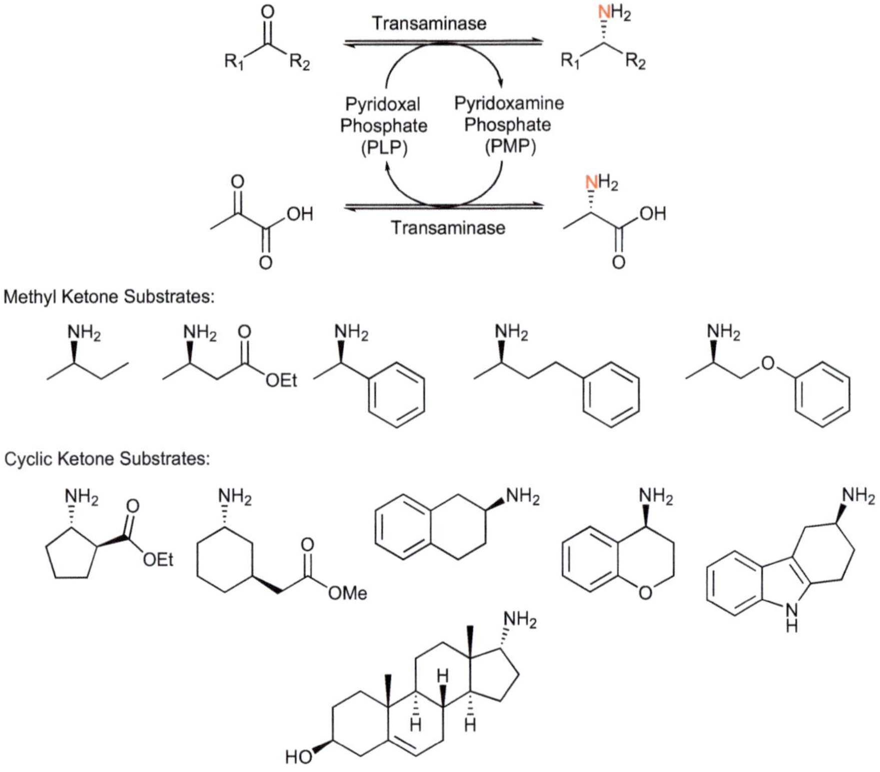

Figure 7.7 The scope of product amines produced using transaminases.

In addition to amination reactions, it is also possible to use transaminases to carry out the kinetic resolution of racemic amines. In this process, the transaminase works in the reverse direction, converting one enantiomer into the ketone, whilst leaving behind the other enantiomer of the amine. As shown in Figure 7.10, this approach can be used to prepare building blocks or fragments such as 3-amino pyrrolidine.

The disadvantage to carrying out kinetic resolution is that we lose half of our starting material. It is, however, possible to extend this methodology if we react the ketone formed from deamination with another transaminase to convert it back into the desired amine. If we pick a transaminase with the opposite selectivity to the one carrying out the deamination, then we are able to convert all of our racemate into a single enantiomer as shown in Figure 7.11. This process is now a deracemisation of the amine starting material with enantiocomplementary transaminases.

Transaminases have been used in the synthesis of both natural products and pharmaceutical intermediates. Chemists from Merck

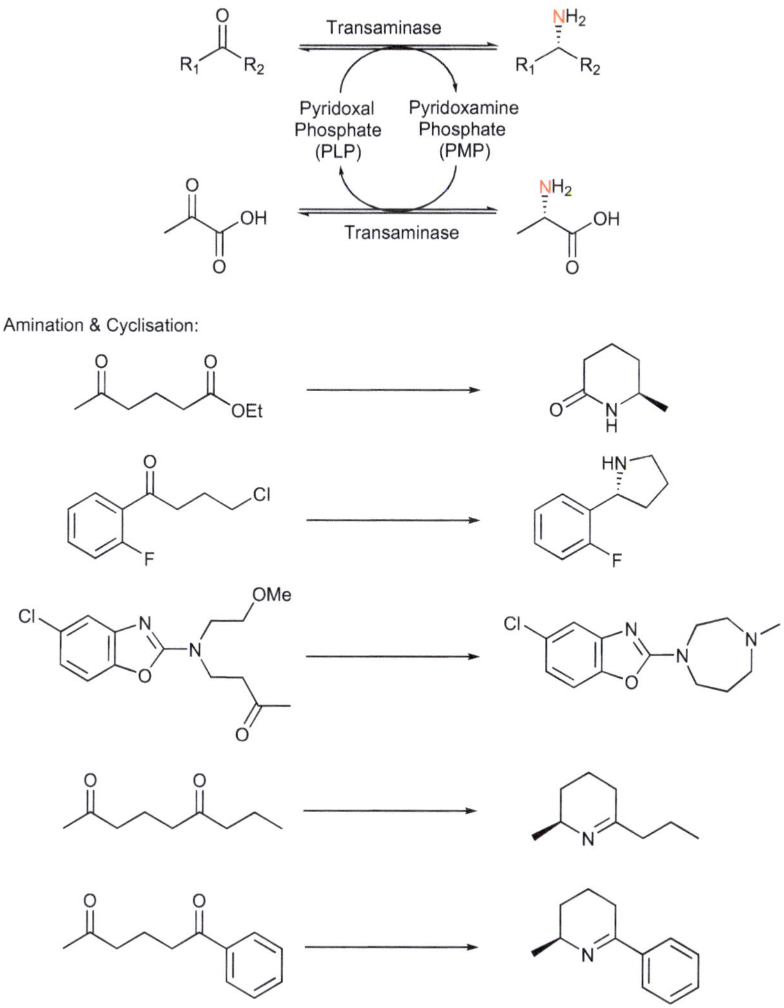

Figure 7.8 The synthesis of cyclic imines using transaminases.

used a transaminase in the synthesis of sitagliptin, a drug used to treat diabetes. As shown in Figure 7.12, isopropylamine is used as the amine donor in this reaction.

As we mentioned earlier, the equilibrium of transaminase reactions often lies in favour of the starting materials rather than the products, forcing chemists to devise methods to overcome this. If we use the amino acid alanine as the amine donor, then the ketone by-product formed is pyruvate. One method of driving the equilibrium of the transaminase reaction is therefore to remove the pyruvate from the reaction by using a ketoreductase to convert it into lactic acid. The ketoreductase required

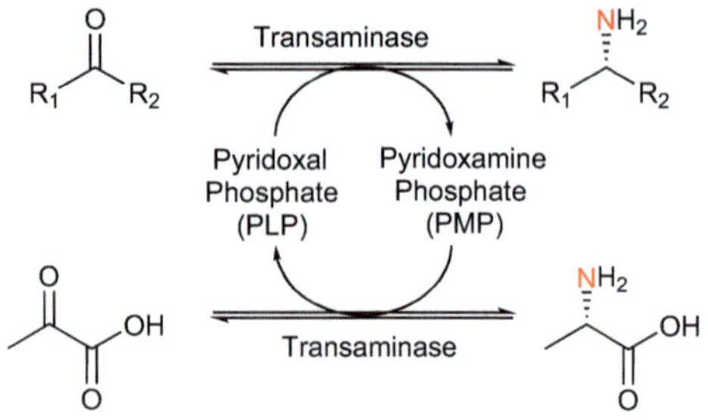

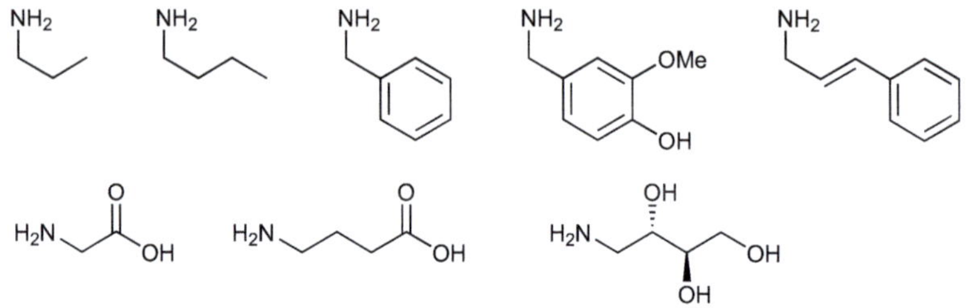

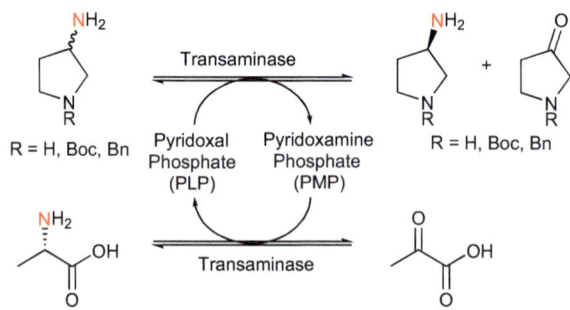

Figure 7.9 The substrate scope of primary amines produced from aldehydes using transaminases.

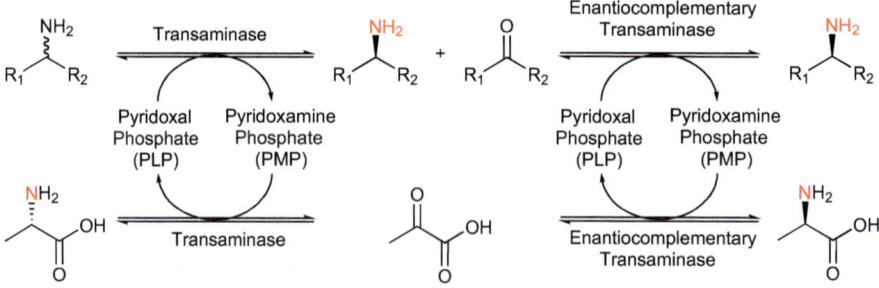

Figure 7.10 The kinetic resolution of amines using transaminases.

Figure 7.11 The deracemisation of amines using enantiocomplementary
transaminases.

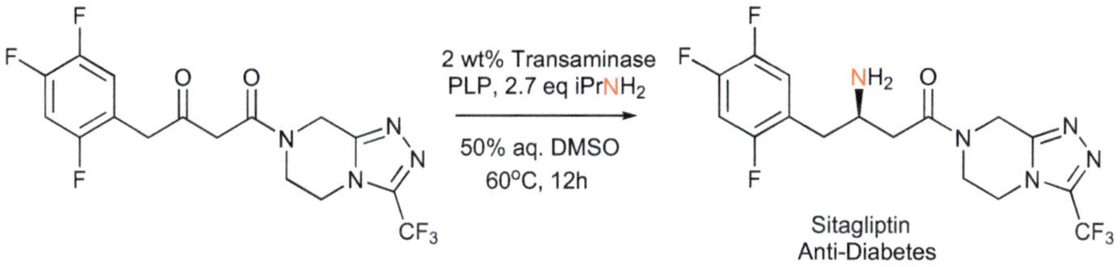

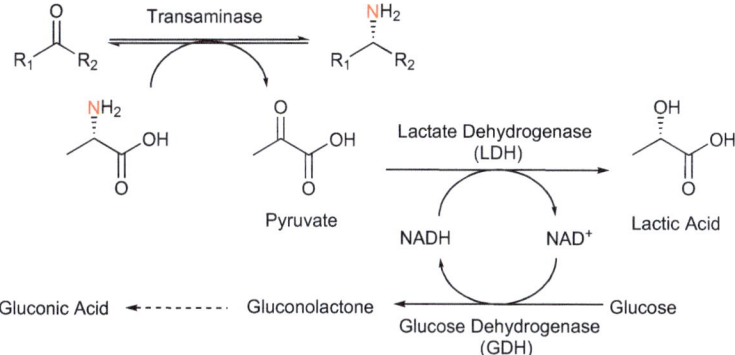

Figure 7.12 The synthesis of sitagliptin using a transaminase.

Figure 7.13 Using a ketoreductase to help drive the equilibrium of transaminase reactions.

(lactate dehydrogenase) uses a nicotinamide co-factor and therefore we also need a third enzyme to recycle the nicotinamide. As shown in Figure 7.13, glucose dehydrogenase can be used for this purpose.

As shown in Figure 7.14, this method has been demonstrated to work well in the synthesis of rivastigmine, a drug used to treat Alzheimer's or Parkinson's disease. We have already seen a synthesis of this compound using a ketoreductase in Chapter 5, but the conversion of the ketone directly into the amine by using a transaminase is a shorter, more efficient, route.

Although this method works well, we require a stoichiometric (or sometimes larger) amount of the amine donor. As an alternative, chemists have developed a system in which the pyruvate is removed (still driving the equilibrium) but is recycled back into the amine donor. As shown in Figure 7.15, by using an enzyme called alanine dehydrogenase and ammonia, pyruvate is recycled back into alanine. This is an amino acid dehydrogenase, which we saw used for the synthesis of α-amino acids in Chapter 5. As in the previous method, this enzyme uses a nicotinamide co-factor and therefore a third enzyme (glucose dehydrogenase again) is required to recycle the co-factor.

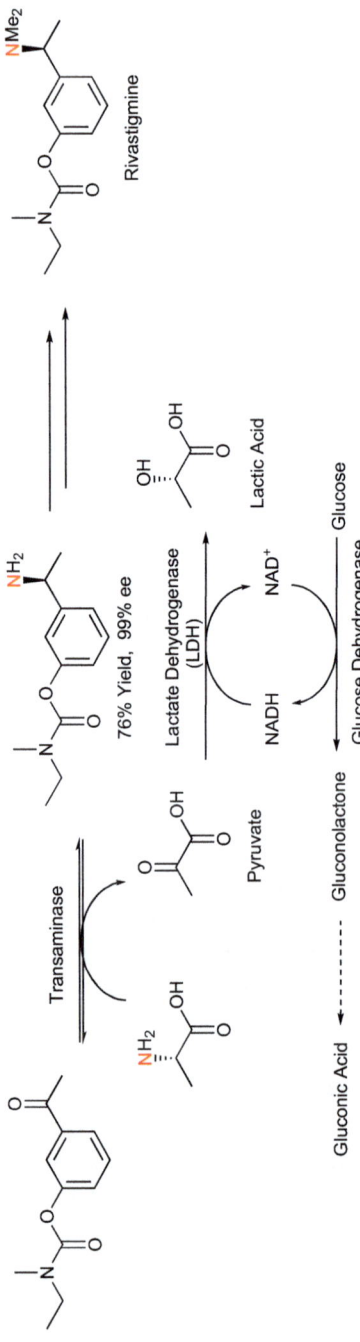

Figure 7.14   The synthesis of rivastigmine using a transaminase.

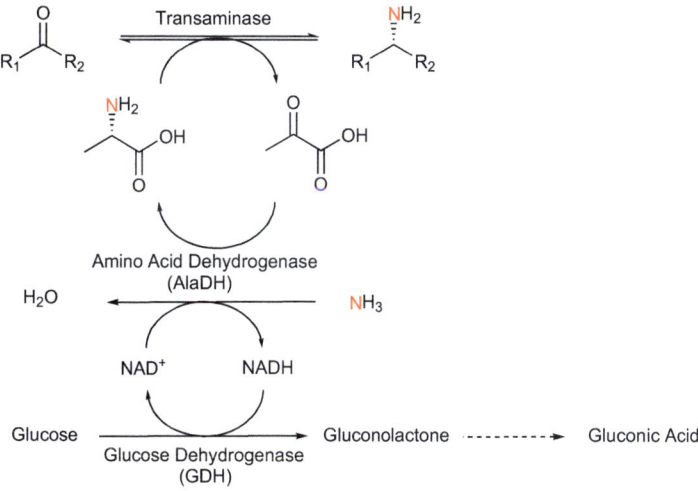

Figure 7.15 Using an amino acid dehydrogenase to recycle the amine donor alanine.

An additional but equally important consideration when using transaminases is the choice of amine donor. Important factors to consider in the choice of the amine donor are the rate of reaction with the transaminase of interest, cost, recyclability and ease of separation from the products. A range of donors are available and each has its advantages and disadvantages, as shown in Figure 7.16.

As we have already seen, transaminases are stereospecific. This means that if we use L-alanine as an amine donor, we produce the L-configured amine product. Therefore, for the opposite amine product we need to use the much more expensive D-alanine as the amine donor. Chemists have developed a method to overcome this limitation, as shown in Figure 7.17. By using a racemase enzyme, L-alanine can be converted *in situ* to D-alanine, which is then processed as the amine donor by the transaminase to give the product. The pyruvate produced as a by-product can then be removed using a ketoreductase or recycled using an amino acid dehydrogenase as we saw in Figures 7.13 and 7.15. For convenience, only the amino acid dehydrogenase system is shown in Figure 7.17.

The mechanism of transaminases has been investigated in great detail and is well understood. It is included here for more advanced readers or those with a greater interest. As we have seen, the amino group transfer is mediated by the co-factor pyridoxal phosphate (PLP), which is reversibly bound to the enzyme by linkage to a lysine residue to form an imine. This is the start of the catalytic cycle shown in

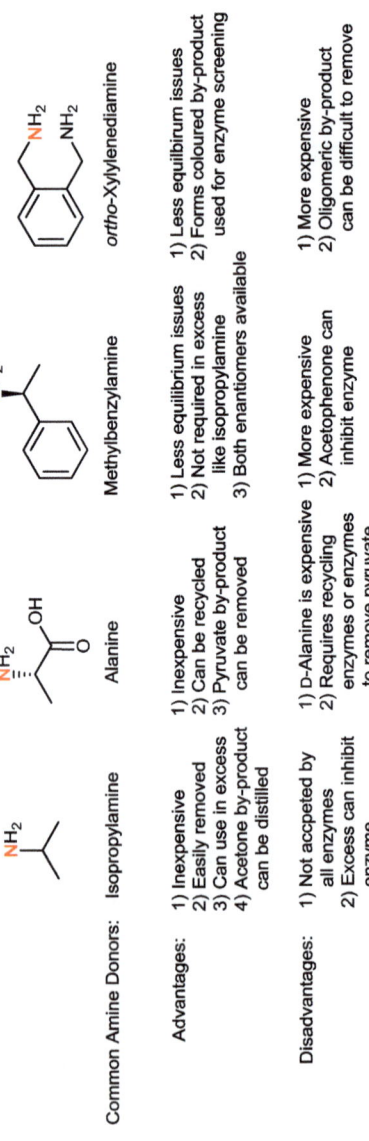

Common Amine Donors:	Isopropylamine	Alanine	Methylbenzylamine	*ortho*-Xylylenediamine
Advantages:	1) Inexpensive 2) Easily removed 3) Can use in excess 4) Acetone by-product can be distilled	1) Inexpensive 2) Can be recycled 3) Pyruvate by-product can be removed	1) Less equilibrium issues 2) Not required in excess like isopropylamine 3) Both enantiomers available	1) Less equilibirum issues 2) Forms coloured by-product used for enzyme screening
Disadvantages:	1) Not accpeted by all enzymes 2) Excess can inhibit enzyme	1) D-Alanine is expensive 2) Requires recycling enzymes or enzymes to remove pyruvate	1) More expensive 2) Acetophenone can inhibit enzyme	1) More expensive 2) Oligomeric by-product can be difficult to remove

Figure 7.16 Comparison of amine donors typically used in transaminase reactions.

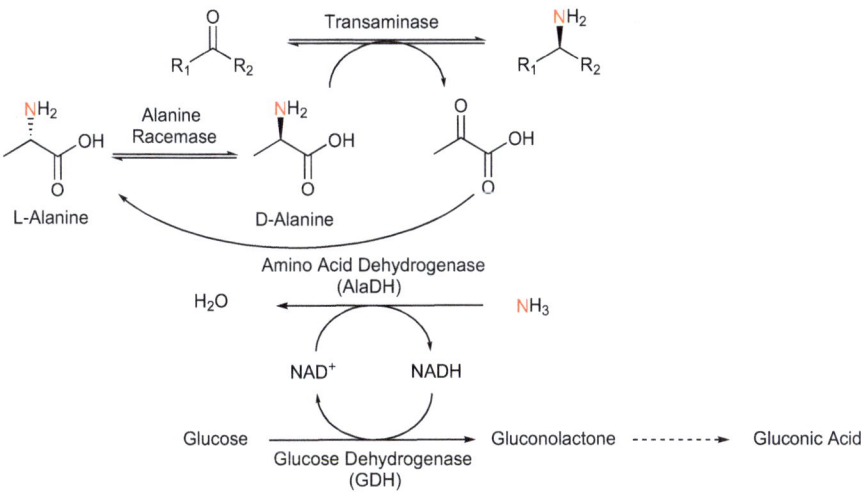

Figure 7.17 The use of alanine racemase to generate D-alanine as an amine donor for transaminase reactions.

Figure 7.18. The mechanism can be thought of in two discrete steps. Firstly, the compound from which the amine of the product originates (the amine donor, *e.g.* alanine) transfers its amine (in red) onto the pyridoxal phosphate to generate a pyridoxamine phosphate intermediate (PMP). This occurs *via* a series of imine formation reactions. The imine between the PLP and the lysine residue is cleaved and a second imine formed between the amine donor and the PLP. Tautomerisation of the imine formed from the amine donor and PLP then gives a second imine, which can be hydrolysed to give the PMP intermediate, transferring the nitrogen to the co-factor. In the second step, this process is reversed, transferring the amino group from the PMP co-factor to the starting material ketone. As with the first half of the cycle, we form an imine between the PMP and the ketone. This imine is then tautomerised to give a second imine, which can be hydrolysed to generate the product and reform the PLP to allow the catalytic cycle to continue. All of these steps are in equilibrium, which as we have discussed previously can lead to equilibrium issues. In addition, substrate and product inhibition can also be a problem.

7.2.2 Formation of C–N Bonds Using Ammonia Lyases

Ammonia lyases are found in a range of organisms and have diverse biological functions. One of their uses is the catabolism of amino acids by deaminating the amino acid to produce an unsaturated

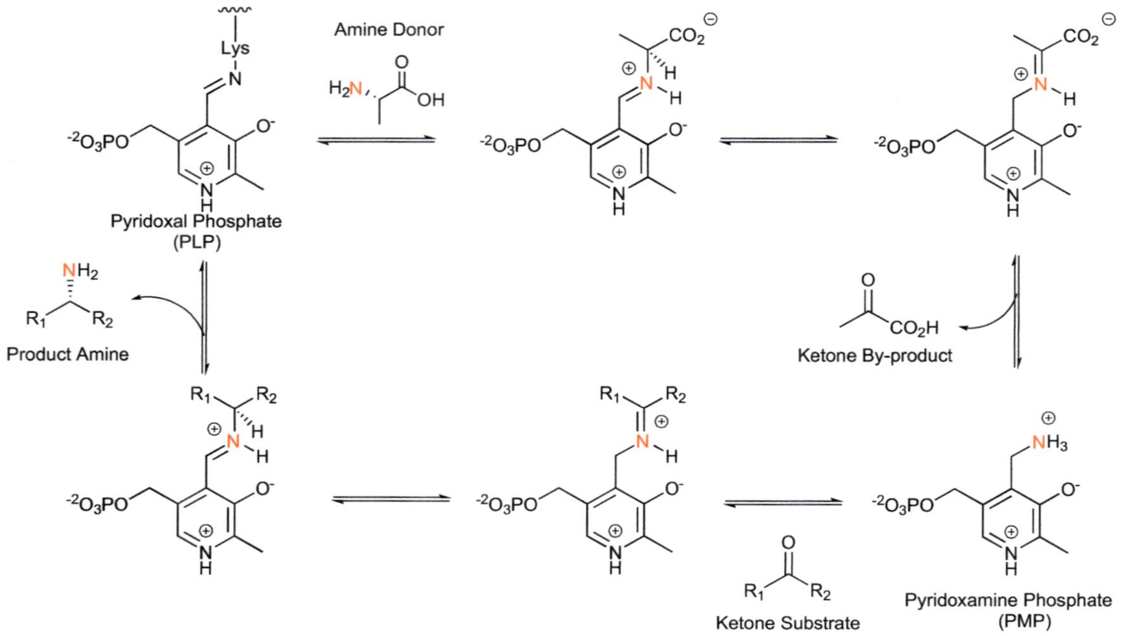

Figure 7.18 The mechanism of transaminase reactions.

carboxylic acid. These substrates are then used in a variety of functions, *e.g.* central metabolism (the citric acid cycle), lignin synthesis, the urea cycle and production of β-amino acids. They are found in biosynthetic pathways to make natural products and can be used by chemists to make a range of chiral amines. As shown in Figure 7.19, ammonia lyases catalyse the stereo- and regiospecific addition of ammonia to a carbon–carbon double bond.

The regio- and stereospecific addition of ammonia to an alkene is a challenging reaction to carry out using chemical methods. Direct addition of ammonia is only possible under high temperatures and pressures, which are unsuitable for industrial processes and not suitable for substrates with functional groups that are unstable to harsh conditions. As an alternative, chemists have carried out the same overall transformation using methods that involve an extra step. As shown in Figure 7.20, these include hydroboration (followed by oxidative work-up with a nitrogen source to give the amine) or aziridination (followed by ring opening). Whilst these methods work, they

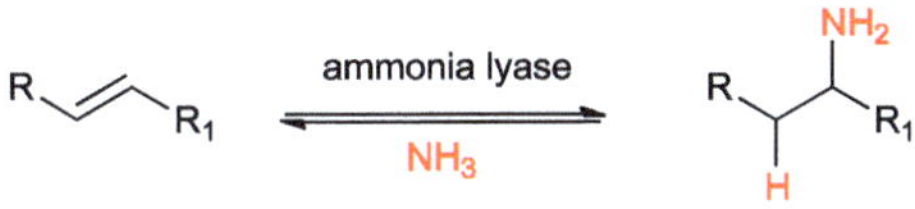

Figure 7.19 General transformation catalysed by ammonia lyases.

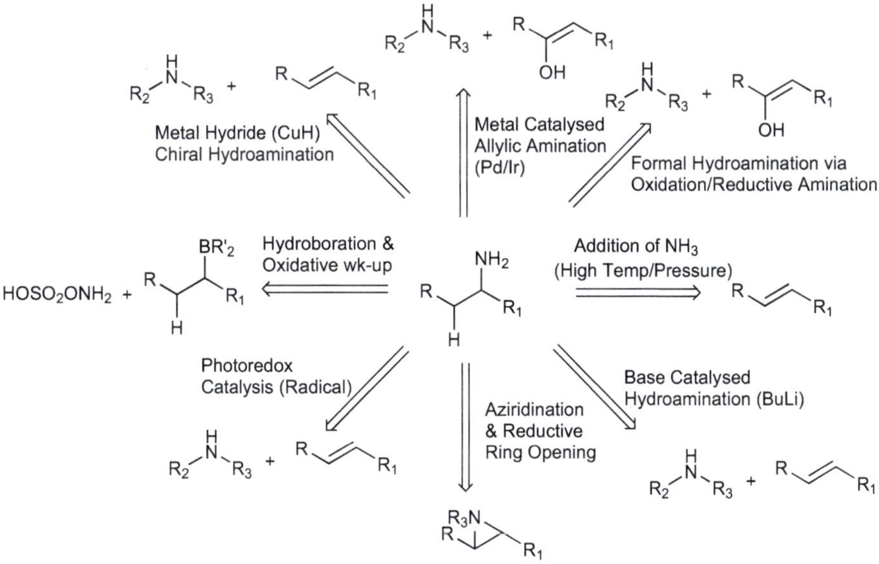

Figure 7.20 Selected chemical methods for the addition of ammonia to an alkene or equivalent.

can be several steps and may not produce products with the required selectivity. An alternative is base-catalysed hydroamination. This reaction produces the desired product in one step, but requires strong bases (*e.g.* BuLi) and a protected nitrogen source, and is unselective. In contrast, more modern methods carry out the hydroamination of carbon–carbon double bonds using transition metal catalysts. These systems can show good selectivity and work under mild conditions, but require the use of potentially expensive metals and chiral ligands. In addition, the nitrogen atom has to be protected so an extra step is required afterwards to remove the protecting groups. Alternative methods carry out a formal hydroamination *via* a multistep procedure of oxidation followed by reductive amination. These methods can also be selective, but suffer from being more than one step and the use of chiral ligands and metal catalysts. A variety of these chemical alternatives are shown in Figure 7.20.

In contrast to the methods shown in Figure 7.20, ammonia lyases work under mild conditions and display high selectivity. However, a drawback is their substrate range, which is more limited than some other enzyme classes. One important family of ammonia lyases is the aspartase family. These enzymes are responsible for the deamination of L-aspartate and 3-methylaspartate to give fumarate or mesaconate and ammonia. Chemists are more interested in using these enzymes in the reverse direction to add ammonia to a carbon–carbon double bond. As shown in Figure 7.21, if we use 3-methylaspartase, we are able to produce L-aspartic acid from fumaric acid with high enantioselectivity even though this is not the enzymes natural substrate. The addition of the ammonia (in red) occurs to opposite faces of the double bond, such that the amine and proton are *anti* to one another.

These enzymes have active site lysine or serine residues, which act as a general base in the deamination direction as shown in Figure 7.22. In the amination direction, these residues are the source of the proton after the ammonia carries out the initial Michael addition. Hopefully you recall the synthesis of the artificial sweetener aspartame from Chapter 4. In that synthesis, we used racemic amino acids to make the

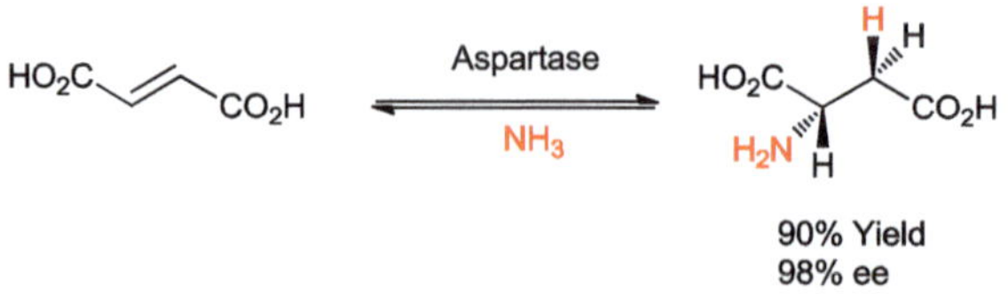

90% Yield
98% ee

Figure 7.21   The synthesis of L-aspartate using an aspartase.

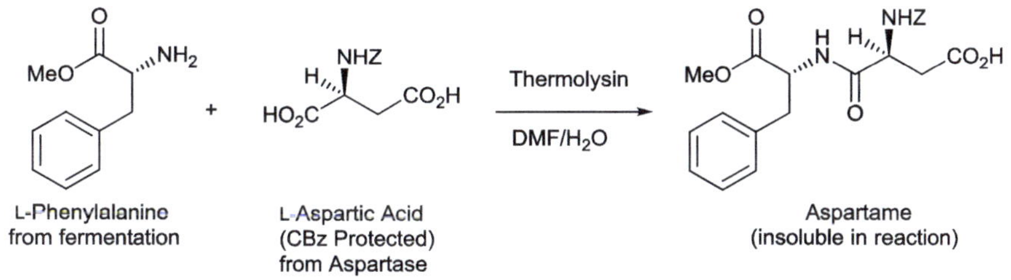

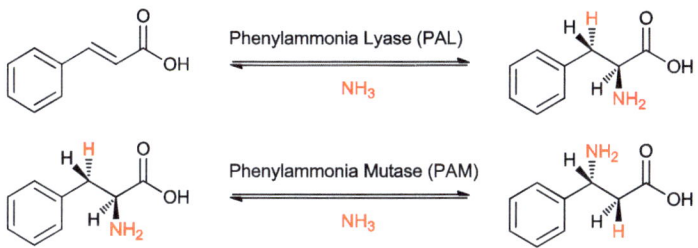

Figure 7.22 The synthesis of aspartame from an aspartic acid derivative prepared using aspartase.

Figure 7.23 General transformations catalysed by phenylammonia lyase (PAL) and phenylammonia mutase (PAM).

product. As a result, half of the input material was wasted as it was the wrong enantiomer. However, using 3-methylaspartase, we can produce L-aspartic acid selectively and we can also produce L-phenylalanine selectively from fermentation. As shown in Figure 7.22, we can now directly couple the desired enantiomer of the two compounds and not have to worry about wasting the undesired enantiomer.

A second important group of ammonia lyase enzymes is the aromatic amino acid lyases. These enzymes catalyse the reversible deamination of the amino acids histidine, tyrosine and phenylalanine. As with the aspartase family of enzymes, these proteins work to catabolise the respective amino acids. However, as chemists, we are more interested in using them to produce amino acids enantioselectively. Also of interest are two enzymes that are very structurally similar to the aromatic amino acid lyases. These are amino acid mutases, which catalyse the exchange of the amino group and an adjacent hydrogen allowing an α-amino acid to be converted into a β-amino acid. These enzymes are classified as isomerases and given the enzyme classification number 5. The transformations catalysed by these enzymes are shown in Figure 7.23.

Although amination is the opposite of the reaction that these enzymes naturally catalyse, if we use an excess of ammonia we can get the reaction to go in the reverse direction. This reaction has been studied in most detail for the enzyme phenylammonia lyase (PAL) and a range of substituted aryl acrylates undergo reaction, some of which are shown in Figure 7.24. In each case, the product enantiomer is shown (although the opposite enantiomer can also be made using a PAL enzyme and an L-amino acid oxidase). The enantioselectivities are typically very good, whilst the yields can be slightly more variable.

Phenylammonia lyase (PAL) has also been used in the synthesis of pharmaceutical intermediates, as shown in Figure 7.25. The drug trandolapril is used for the treatment of high blood pressure. In a route to make trandolapril, phenylammonia lyase was used to synthesise 2,3-dihydroindole-2-carboxylic acid, an intermediate compound.

As we have already seen, phenylammonia mutase (PAM) is an isomerase enzyme that reversibly converts α-amino acids into β-amino acids. Most of the compounds that are substrates for phenylammonia lyase (PAL) are also substrates for PAM. As with PAL the enantioselectivities for reactions catalysed by PAM are generally very good, although the yields can be variable. Reactions with PAM can either use the α-amino acid as the starting material, or potentially more

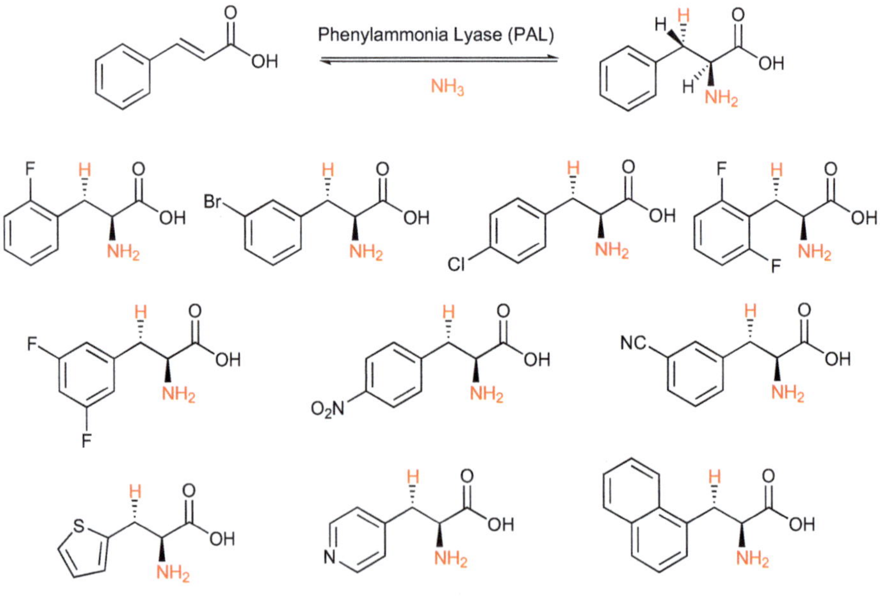

Figure 7.24 The scope of product amines produced by phenylammonia lyase (PAL).

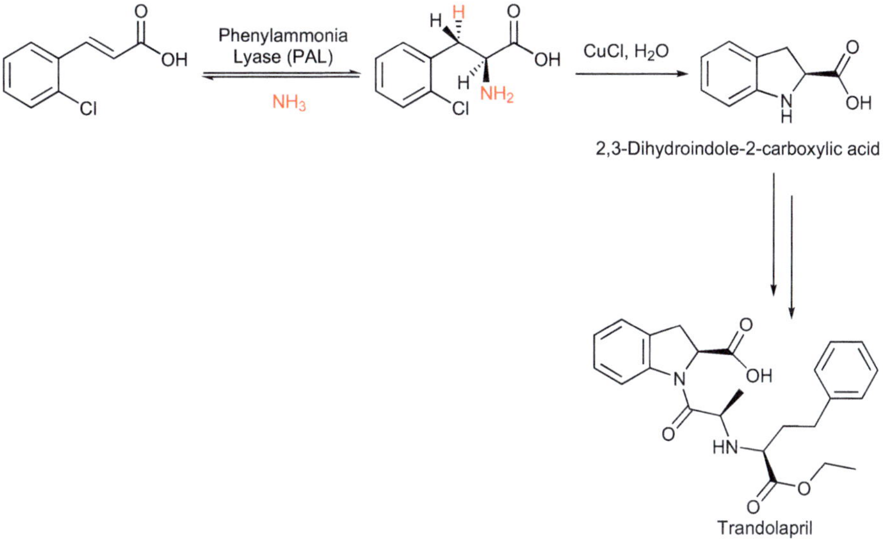

Figure 7.25   The synthesis of trandolapril using a phenylammonia lyase (PAL).

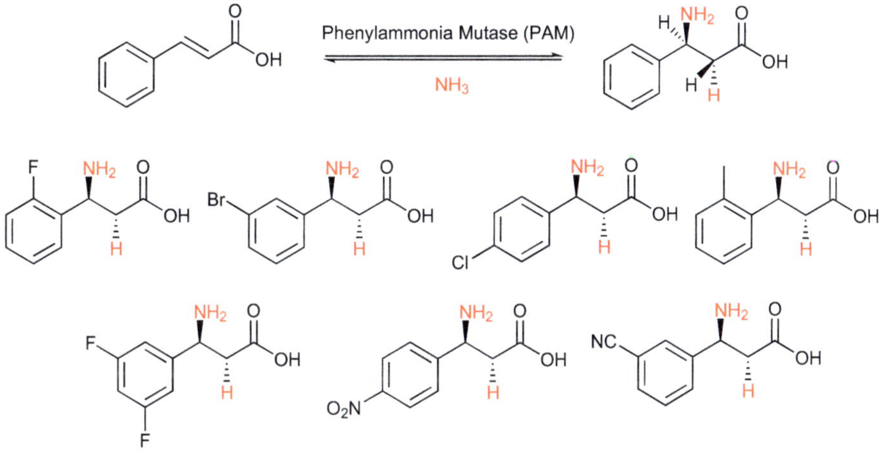

Figure 7.26   The scope of product amines produced by phenylammonia mutase (PAM).

useful from a synthetic point of view, we can start with the carboxylic acid, as shown in Figure 7.26.

The two enzymes PAL and PAM can be used together to carry out kinetic resolution of β-amino acids. As shown in Figure 7.27, by using a PAM enzyme, one enantiomer of the β-amino acid is converted into its α-amino acid regioisomer. The α-amino acid then undergoes

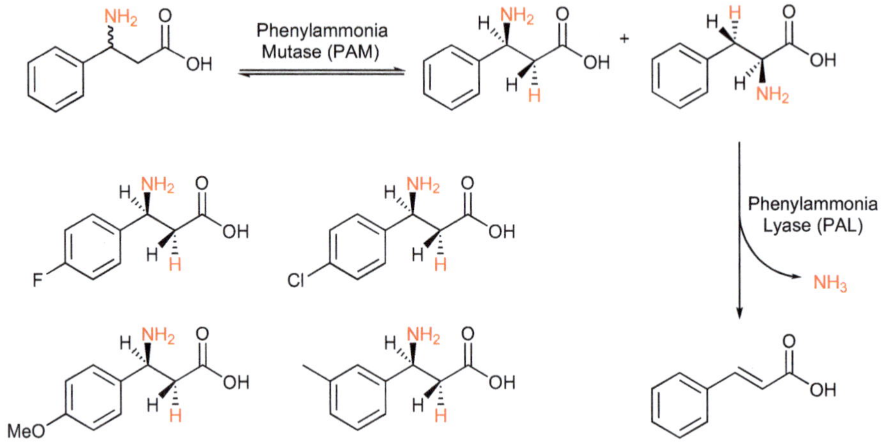

Figure 7.27   The kinetic resolution of β-amino acids using both phenylammonia mutase (PAM) and phenylammonia lyase (PAL).

deamination catalysed by the PAL enzyme to give a cinnamic acid and
ammonia as by-products, along with the desired enantiomer of the
β-amino acid. In each case, the product enantiomer is shown.

The detailed mechanism of the PAL and PAM enzymes is shown in
Figure 7.28. As with the other more detailed mechanisms in this
book, we have included it here for completeness and for readers who
would like to develop a deeper understanding. Both PAL and PAM
enzymes do not use a co-factor, using instead a prosthetic group (part
of the enzyme structure) known as MIO (5-methylene-3,5-dihydro-4*H*-
imidazol-4-one). This group, which plays a key role in the mechanism,
is produced from a sequence motif present in all PAL and PAM
enzymes consisting of threonine or alanine, serine and glycine. For
the enzyme to be active, this motif must first cyclise, releasing water
to form the MIO structure. The other important residue for this
mechanism is a tyrosine, which acts as a proton source.

If we start from the top of Figure 7.28, we have the MIO group
already made in the active site and the cinnamic acid substrate
bound. The reaction starts by Michael addition of ammonia to the
MIO prosthetic group to covalently attach the nitrogen, which will be
transferred to the substrate. With the amine now in the correct
orientation, attack on the alkene occurs with the tyrosine acting as a
general acid to provide the proton on the opposite face to the attack of
the amine. The MIO group is also regenerated in this step ready for
the next catalytic cycle. The regioselectivity of the attack on the alkene
is determined by the active site orientation of the PAL or PAM

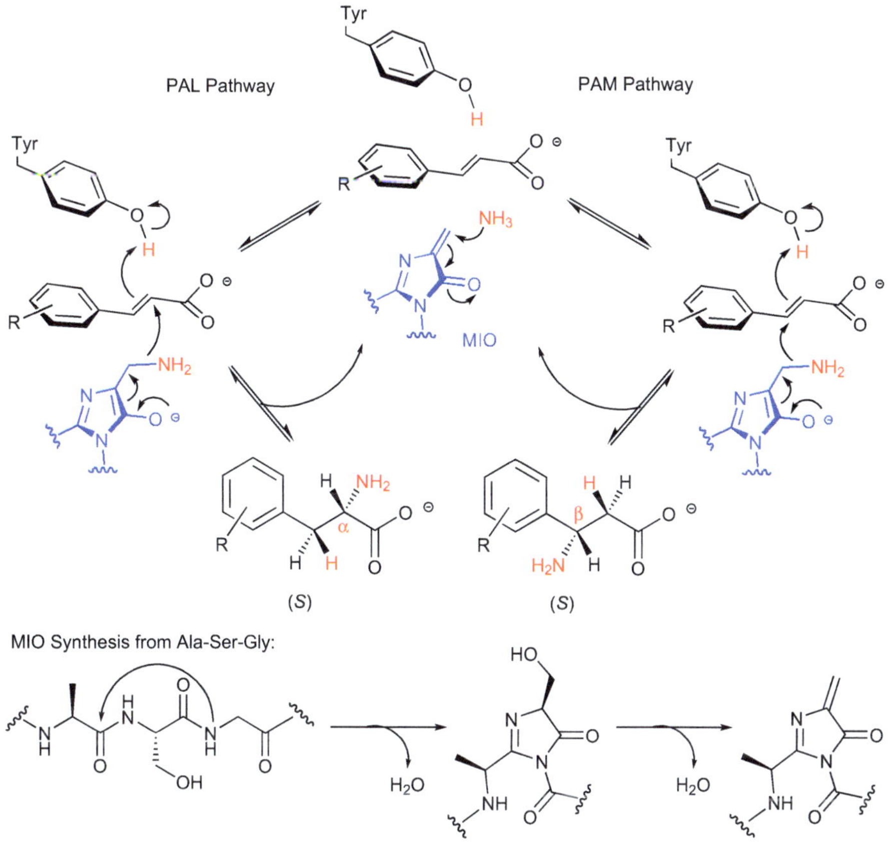

Figure 7.28 The mechanism of both phenylammonia mutase (PAM) and phenylammonia lyase (PAL) and the synthesis of the prosthetic group MIO.

enzymes respectively. As we have seen, the addition of ammonia to the alkene gives the *anti*-product with the proton and amine on opposite faces of the double bond. The left-hand pathway in Figure 7.28 illustrates the mechanism for PAL, whilst the right-hand pathway shows the mechanism for PAM.

7.3 Formation of C–O Bonds

7.3.1 Formation of C–O Bonds Using Hydratases

Like carbon–nitrogen bonds, carbon–oxygen bonds are important bonds within molecules due to the oxygen atom's ability to form hydrogen bonds in a similar manner to nitrogen. As a result, carbon–oxygen bonds are very prevalent in natural compounds and synthetic drugs, and therefore methods to make carbon–oxygen bonds are useful to chemists. We have already seen how we can make carbon–oxygen bonds using oxidation processes. However, we can also make carbon–oxygen bonds without adjusting the oxidation state of the molecule by hydration (addition of water) of alkenes. This is the analogous transformation to that catalysed by ammonia lyases, except that instead of ammonia this time we add water across the double bond. As shown in Figure 7.29, this transformation is catalysed by hydratase enzymes. These enzymes are found widely in nature as the addition of water is an essential reaction in biosynthesis. Hydratases are involved in central metabolism (the citric acid cycle) and fatty acid metabolism (degradation of fatty acids), and are used in the biosynthesis of many natural products such as terpenes.

The direct addition of water to an alkene using chemical methods can be difficult to achieve. This is in part because water itself is a poor nucleophile, as being uncharged it has poor electron density. As chemists, we are much more likely to use hydroxide (OH^-) instead. The chemical addition of water to an alkene asymmetrically to give one enantiomer is also challenging. For example, if we use acidic conditions in order to protonate the alkene then the carbocation formed is planar and therefore will be quenched from either side resulting in a racemic mixture of alcohols. Traditional chemical methods to carry out this transformation often require a strong acid or the use of mercury acetate, which is highly toxic. More modern methods tend to carry out the transformation in two steps, *e.g.* hydroboration and oxidative work-up or epoxide formation and opening with a hydride reagent. Several methods of chemical alkene hydration are shown in Figure 7.30.

In contrast to the chemical methods above, hydratases are an elegant solution to the problem of direct alkene hydration using water as both a solvent and the nucleophile with the enzyme activating the nucleophile and the substrate. This allows the

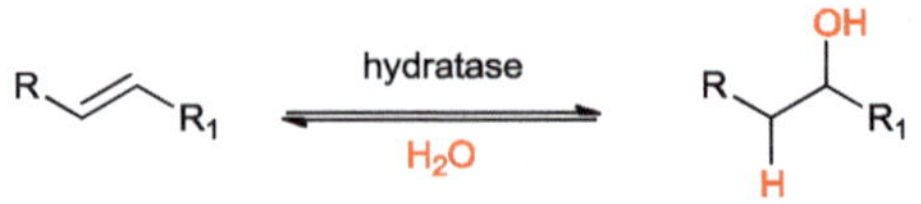

Figure 7.29 The general transformation catalysed by hydratases.

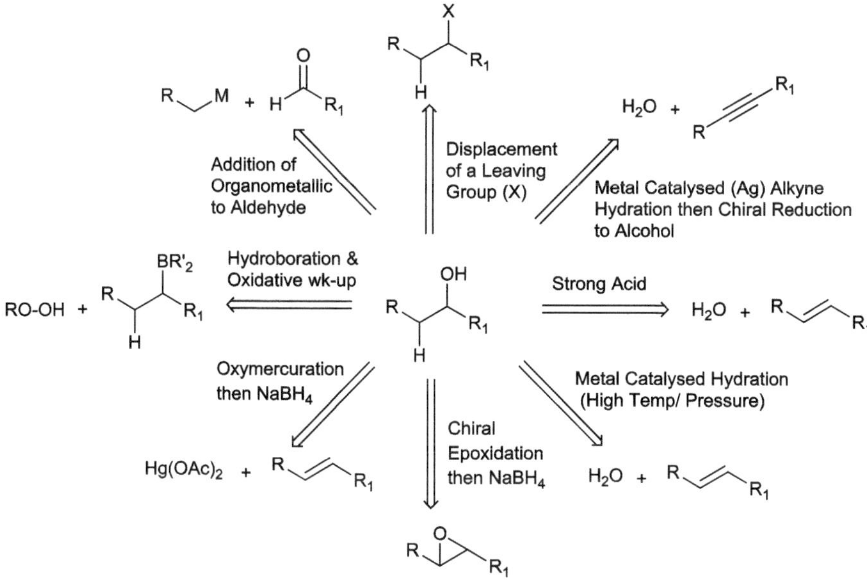

Figure 7.30 Selected chemical methods for the addition of water to an alkene or equivalent.

reaction to occur under very mild conditions, whilst being highly regio- and stereoselective, making hydratases a good alternative to chemical methods. The addition to or elimination of water from an alkene can occur with two different stereochemical outcomes. As shown above, chemical addition or elimination occurs in an *anti*-fashion (if going *via* a bimolecular mechanism such as S_N2 or E2) or with no facial selectivity at all (if going *via* an S_N1 or E1 mechanism). In contrast, depending upon the enzyme we choose, hydratases can catalyse addition or elimination in either an *anti* or *syn* manner. As shown in Figure 7.31, water (in red) can be added to the same face or opposite faces of the alkene. We will discuss each of these enzymes in more detail below.

Hydratase enzymes can be grouped according to the types of carbon–carbon bond they use as substrates. We look first at enzymes that catalyse the addition of water to electron rich alkenes. There are several families of enzymes that catalyse the addition of water to electron-rich substrates and as an example we will cover enzymes known as oleate hydratases. These enzymes have successfully been applied in synthesis and are named after oleate hydratase, the enzyme that catalyses the reversible addition of water to

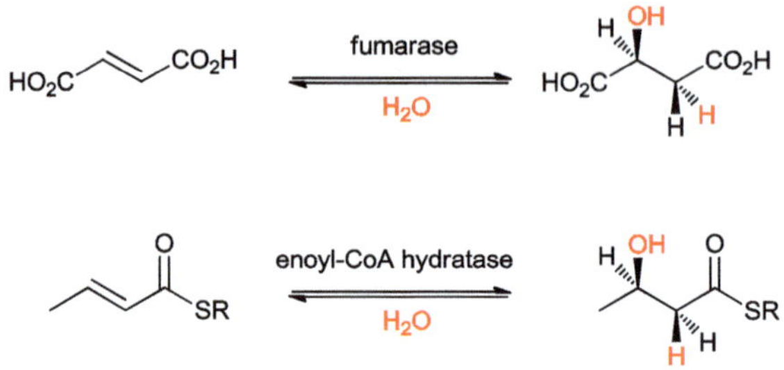

Figure 7.31   The two classes of hydratases that catalyse addition or elimination in either an *anti* or *syn* manner.

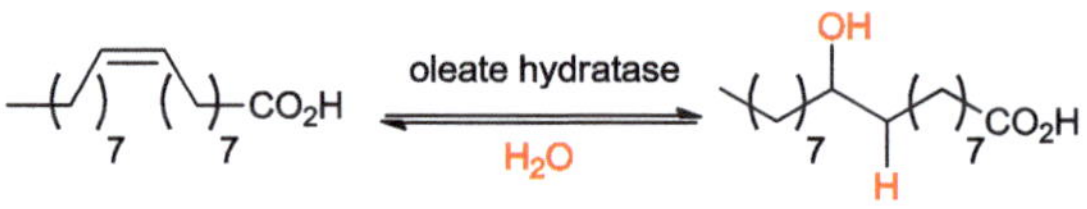

Figure 7.32   The general transformation catalysed by oleate hydratase.

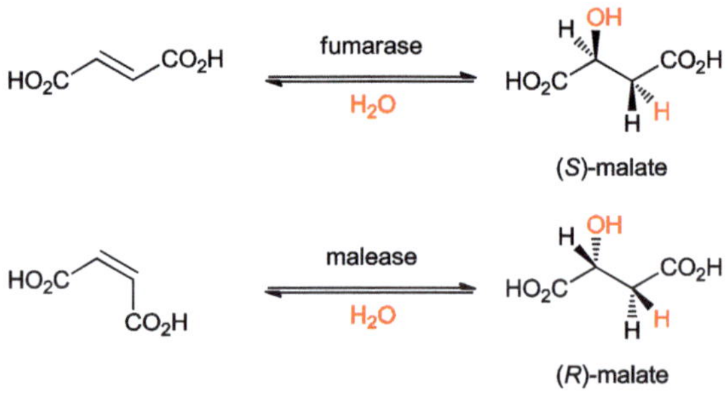

Figure 7.33   The production of either enantiomer of malate by using either
fumarase or malease.

oleic acid as shown in Figure 7.32. This particular enzyme is only
active on the Z-isomer of the double bond, although other oleate
hydratases are active on other substrates and double bond
configurations.

   In comparison to oleate hydratases, other families of enzymes
catalyse the addition of water to electron-deficient alkenes. One of
the most studied enzymes in this class is fumarase, the enzyme
that catalyses the reversible addition of water to fumaric acid to form
(S)-malate, as shown in Figure 7.33. This enzyme is a vital part of
central metabolism as it is involved in the citric acid cycle. Notice

that these enzymes catalyse an *anti*-addition of water across the alkene in the same way as aspartase added ammonia across an alkene. This enzyme only accepts the *trans*-alkene as a substrate. As shown in Figure 7.33, there is another enzyme that accepts the *cis*-alkene (malic acid) to form (*R*)-malate, which is called malease. It should be noted that these enzymes both have a very narrow substrate scope.

There is another family of enzymes that is capable of hydrating electron-deficient double bonds and these are the enoyl-CoA hydratases. These enzymes are involved in the degradation of fatty acids and catalyse the addition of water to the alkene in a similar manner to a Michael reaction. The enzymes only work on substrates that contain thioesters with a specific group attached called coenzyme A or CoA for short. They work on substrates with a variety of chain lengths and depending on which enzyme is used, both the *R*- and *S*-enantiomers can be obtained, as shown in Figure 7.34. Remember, in contrast to fumarase above, these enzymes catalyse a *syn*-addition of water across the alkene.

Enoyl-CoA hydratases have been used industrially for the synthesis of vanillin, an important compound for the food and flavour

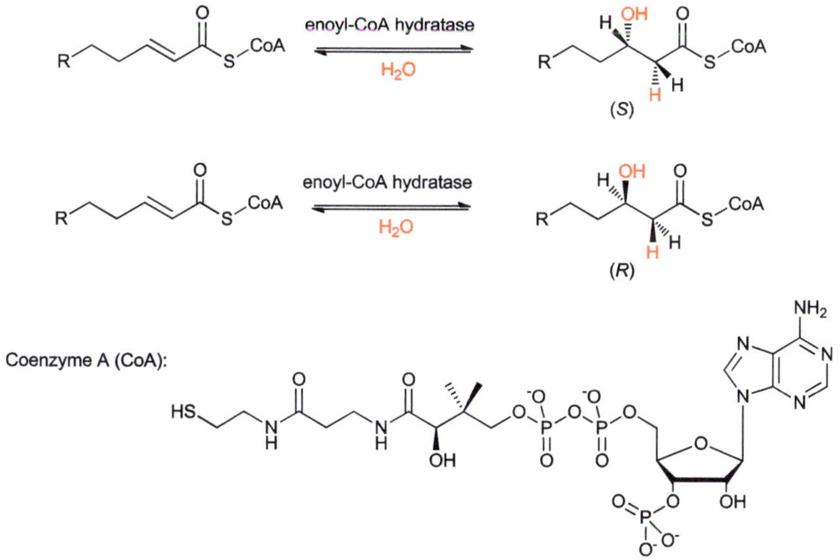

Figure 7.34 The general transformation catalysed by enoyl-CoA hydratases.

industries. We previously saw in Chapter 5 how this compound can also be prepared using a carboxylic acid reductase enzyme. The starting material for the hydratase enzyme-catalysed step is ferulic acid, a chemical present in plant cell walls and therefore available from agricultural waste. As shown in Figure 7.35, the enoyl-CoA hydratase first catalyses the hydration of the ferulic acid alkene, before then also catalysing a retro-Claisen reaction to produce vanillin and acetyl-CoA as a by-product.

Hydratase activity has also been reported for an enzyme that wouldn't normally be expected to catalyse this transformation. Chemists have found that for certain substrates, phenolic acid decarboxylase can also show hydratase activity. As shown in Figure 7.36, this useful transformation is enantioselective for the (*S*)-enantiomer, but is limited to a small range of substrates. For reaction to occur, the phenol must first tautomerise to the keto form in order to generate the alkene, which is hydrated.

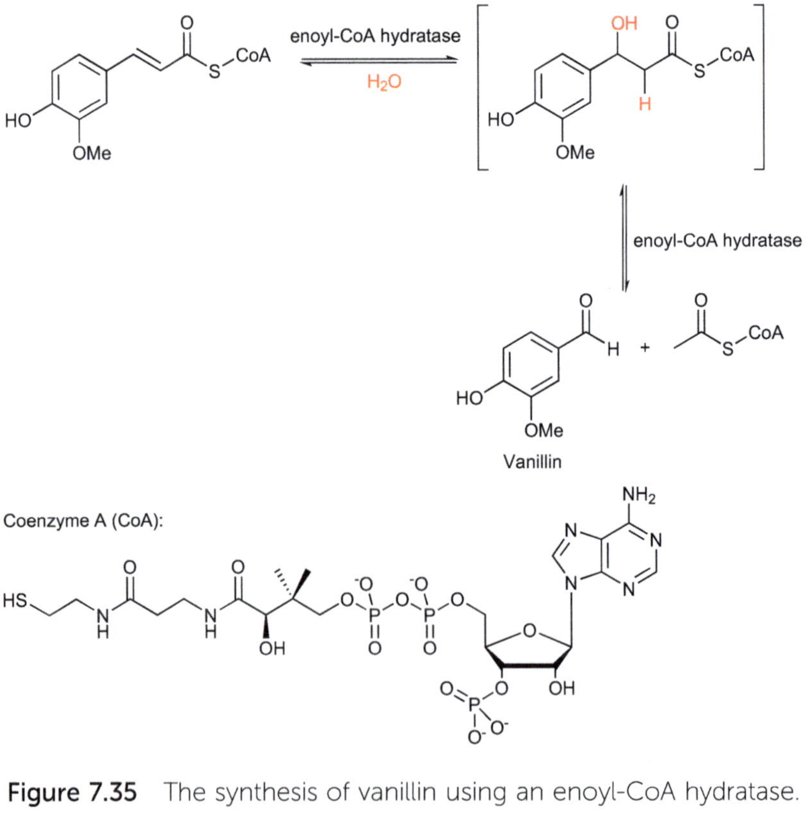

Figure 7.35   The synthesis of vanillin using an enoyl-CoA hydratase.

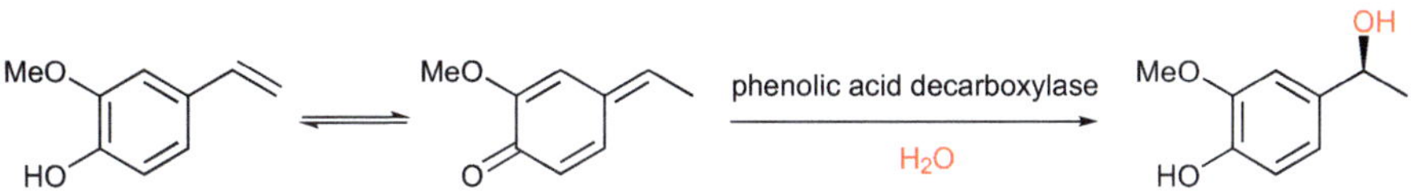

Figure 7.36 The unexpected hydratase activity of phenolic acid decarboxylase.

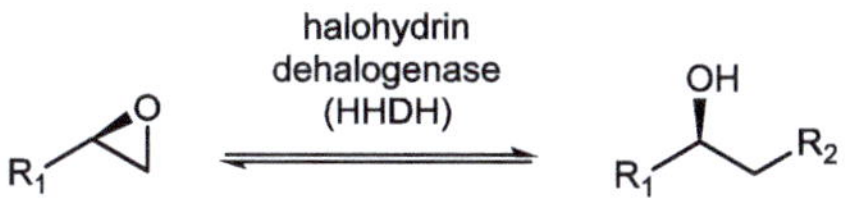

Figure 7.37 The general transformation catalysed by halohydrin dehalogenases.

7.4 Formation of C–X Bonds Using Halohydrin Dehalogenases

Halohydrin dehalogenases are enzymes that are capable of forming a range of carbon heteroatom bonds (and carbon–carbon ones as well). Halohydrin dehalogenases are lyases and therefore belong to enzyme class 4 in the enzyme classification system. The natural activity of these enzymes is to catalyse the formation of epoxides from an α-halo alcohol, as shown in Figure 7.37. They are found in bacteria that metabolise halogenated compounds, and in a similar role to haloalkane dehalogenases, halohydrin dehalogenases remove the halogen atom from compounds, which are then further metabolised by the organism. However, the reaction is reversible, such that it is also possible to use these enzymes to catalyse the opening of epoxides with a nucleophile. Although fewer halohydrin dehalogenases have been reported than other enzyme classes, they are useful industrial biocatalysts because epoxides are valuable synthetic intermediates.

We previously saw a range of chemical methods for the formation of epoxides in Chapter 6. These typically involve the use of peracids or hydrogen peroxide, metal catalysts and chiral ligands or a chiral organocatalyst. The other potential products of a halohydrin dehalogenase-catalysed reaction, such as amino alcohols or halo alcohols, are often produced *via* the opening of an epoxide or *via* addition reactions directly across an alkene. Some of these methods are shown in Figure 7.38.

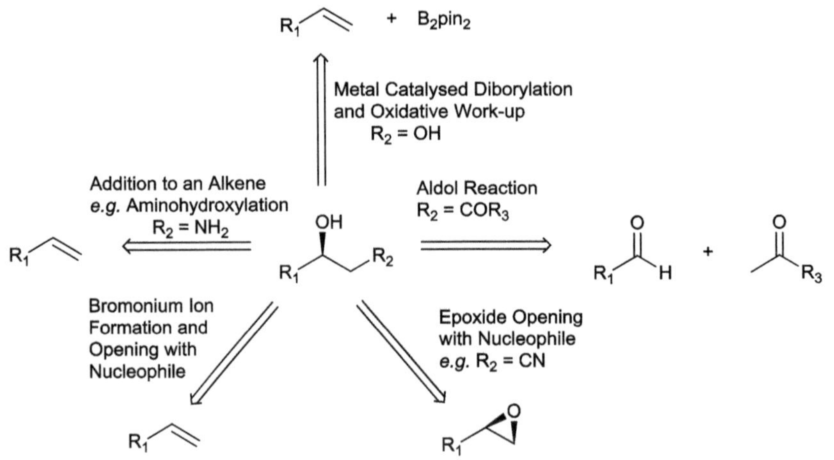

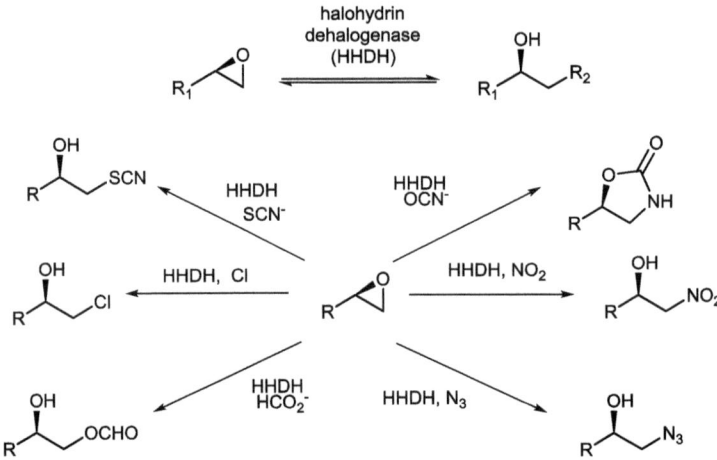

Figure 7.38 Selected chemical methods of epoxide opening or equivalents.

Figure 7.39 The reaction scope of halohydrin dehalogenases.

In comparison, halohydrin dehalogenases work under mild reaction conditions without the need for metal catalysts or chiral ligands. They are co-factor-independent enzymes, which makes them easy to use as no extra co-factors are required. They are highly stereoselective and in the ring opening direction, are able to catalyse the reaction with a range of nucleophiles as shown in Figure 7.39.

These reactions are also very regioselective with attack of the nucleophile at the least hindered end of the epoxide being highly favoured.

For each of the ring opening reactions shown in Figure 7.38, the enzymes have a fairly narrow substrate range. Examples of substrate epoxides that undergo ring opening reactions are shown in Figure 7.40. In each case, the substrate enantiomer that reacts fastest is shown.

Halohydrin dehalogenases are also used to catalyse the reaction in the ring closing direction in order to synthesise epoxides. This is a less frequently used reaction and as a result the substrate range reported is rather limited, as shown in Figure 7.41.

Halohydrin dehalogenases have been used for the synthesis of pharmaceutical drug intermediates. We have already seen a number of enzyme-catalysed reactions used in the synthesis of the statin drug, Lipitor. The statin family of drugs is used to treat high cholesterol levels and prevent heart disease. The side-chain of Lipitor contains two stereocentres with only the *R,R*-diastereoisomer used in the active compound. As we have already seen in Chapter 5, chemists have developed a three-enzyme route to make the Lipitor side-chain, in

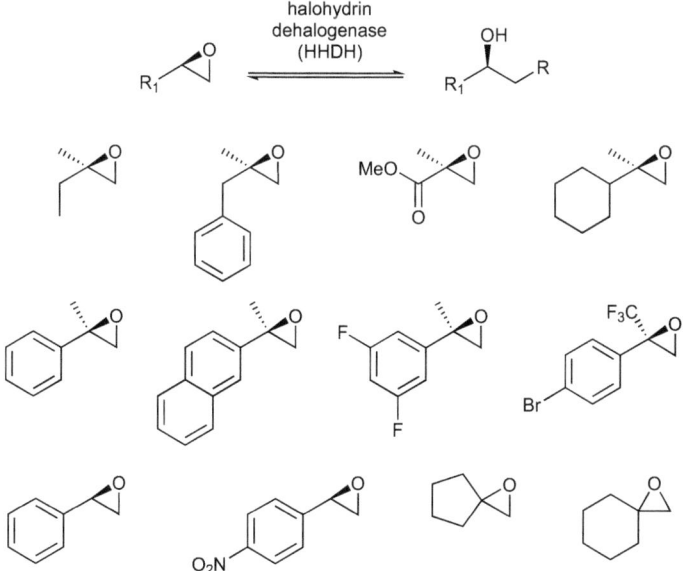

Figure 7.40 The substrate scope of halohydrin dehalogenases.

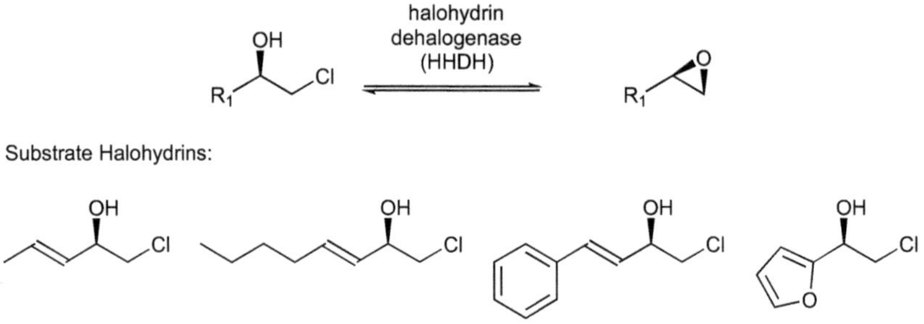

Figure 7.41   The halohydrin substrate scope of halohydrin dehalogenases.

which two of the steps are catalysed by ketoreductases. The third enzyme, as shown in Figure 7.42, is a halohydrin dehalogenase that catalyses the conversion of the chloro-alcohol into an epoxide, which is then opened by cyanide (again catalysed by the halohydrin dehalogenase). Notice that the ketoreductase produces the enantiomer that is accepted by the halohydrin dehalogenase.

Halohydrin dehalogenases have also been used to carry out dynamic kinetic resolutions. In Figure 7.43, epibromohydrin is racemised by the halohydrin dehalogenase enzyme. However, the same enzyme also catalyses the ring opening reaction of one enantiomer of the epoxide with azide. This enables dynamic kinetic resolution to take place giving the desired compound in high yield and stereoselectivity.

A detailed mechanism for these enzymes is shown in Figure 7.44. As with the other mechanisms in this section, it is included for those readers who would like more detail, but it is not essential that you know it. Halohydrin dehalogenases use a catalytic triad of serine, tyrosine and arginine to catalyse both ring closing and ring opening reactions. The ring opening reaction is shown in Figure 7.44. As we can see, the serine and tyrosine residues activate the epoxide through hydrogen bonds to the epoxide oxygen atom. This hydrogen bonding network is further stabilised by an interaction between the tyrosine and arginine residues. Once the substrate epoxide is bound and activated, the nucleophile (in this case cyanide) can attack the epoxide. Attack from the least hindered end is favoured and the epoxide is opened to give the cyano alcohol shown in the scheme. The alcohol proton comes from the tyrosine residue, which in turn is protonated by the arginine. Once the product is released from the active site, the enzyme is ready to bind the next molecule of epoxide and the catalytic cycle can begin again.

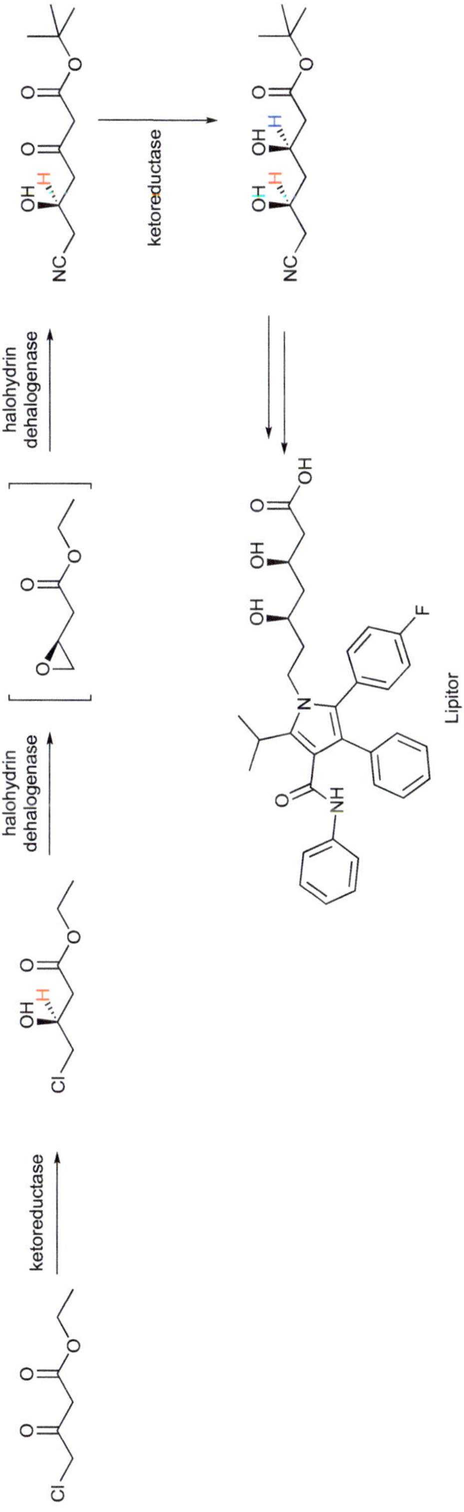

Figure 7.42 The synthesis of Lipitor using a halohydrin dehalogenase.

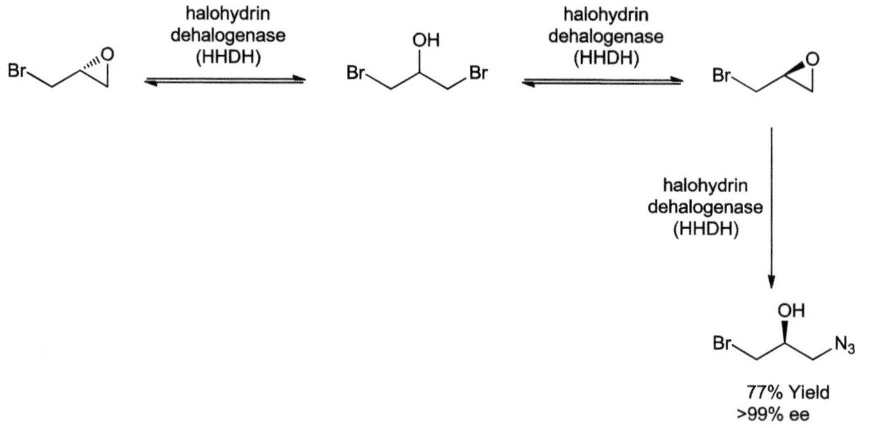

Figure 7.43 The reaction of epibromohydrin with a halohydrin dehalogenase.

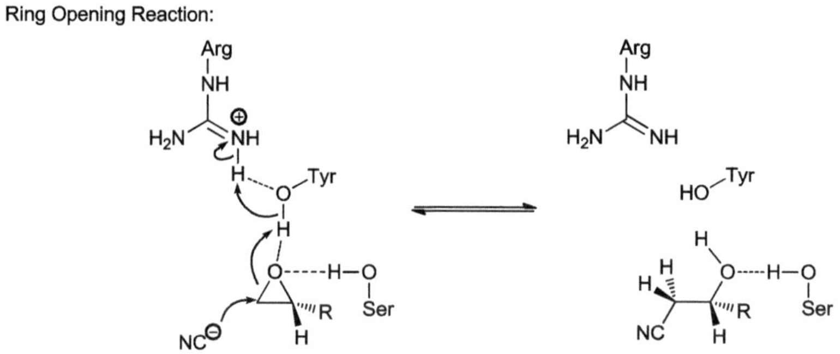

Figure 7.44 The general mechanism of halohydrin dehalogenases.

7.5 Formation of C–X Bonds Using Methyltransferases

Methyltransferases are found in a wide variety of organisms from bacteria, fungi, plants and animals. They are capable of forming a range of carbon–heteroatom bonds, and as we will see in Chapter 8, they are also able of catalysing carbon–carbon bond formation. They are involved in a huge range of biological processes, such as natural product synthesis, and they are especially abundant in plant metabolic pathways leading to lignins, flavonoids, stilbenes and alkaloids. Methyltransferases are also involved in a wide range of metabolic recognition processes. For example, histones are proteins that are responsible for packaging the DNA in cells into tightly wound chromatin. In order to access the DNA, the histone proteins must be methylated and this epigenetic process determines gene expression, genomic stability, stem cell maturation, DNA methylation, and cell mitosis. In order to selectively carry out these processes, the methylation reaction must be very specific. The majority of methyl-transferases catalyse the methylation of a heteroatom using a co-factor, *S*-adenosyl-L-methionine (SAM), as shown in Figure 7.45. Depending on the atom they methylate, these enzymes are classified as *O*-methyltransferases, *N*-methyltransferases or *S*-methyltransferases.

The selective addition of a methyl group can have a large effect on both natural products and pharmaceutical compounds. Regiospecific alkylation can alter a molecule's physicochemical properties and often results in a strongly increased affinity towards membranes and receptors. The methyl group can also be used biologically to mediate 'hydrophobic masking' of a certain sub-structure or functional group of compounds or larger biological structures. A wide variety of chemical methylation methods exist and can be split into three groups. Electrophilic methylating reagents, such as alkylhalides or sulfates, are often used, but these reagents can be toxic and react with little selectivity. On the other hand, nucleophilic methylating reagents, such as organolithiums or Grignard reagents, are highly reactive and therefore may decompose the substrate material. In addition, because heteroatoms are typically electronegative, they must be activated by the addition of a leaving group to the heteroatom adding extra steps to a synthesis. The third option is the use of radical chemistry. Several methods have been demonstrated to transfer methyl groups to carbon and heteroatoms. However, some of these approaches require several steps to make the methylating reagent. Metal-catalysed couplings are not often successful as the heteroatoms tend to bind strongly to the metal centres as well. The other option instead of alkylation is reductive amination, which we have covered already in this chapter. Figure 7.46 shows a number of alkylation methods.

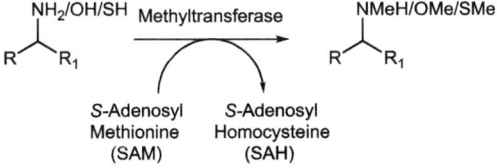

Figure 7.45 The general transformation catalysed by methyltransferases.

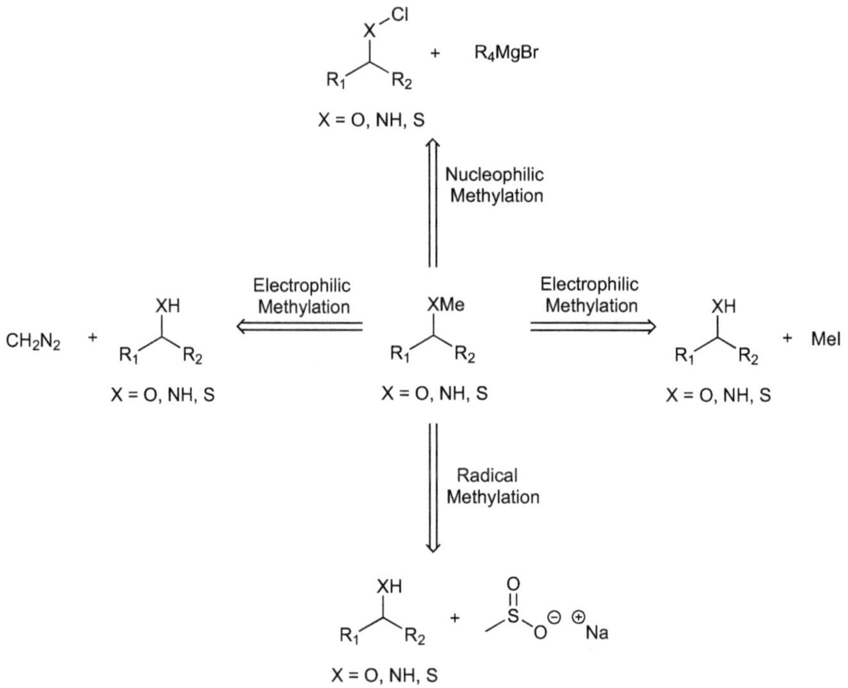

Figure 7.46 Selected chemical methods of heteroatom methylation.

In comparison to chemical methods, methyltransferases catalyse the addition of a methyl group under mild conditions at ambient temperature and using aqueous solvents. They are also very regio- and chemoselective, often exclusively alkylating a single heteroatom in the presence of other heteroatoms. For synthetic purposes, however, this selectivity can sometimes be a drawback. Because they are very selective, methyltransferases often display a narrow substrate scope. Fortunately, many methyltransferases have been reported so it pays to do some research on a particular substrate before performing reactions in the laboratory and to screen a variety of enzymes in order to find one that catalyses the methylation of the desired substrate.

We will look at *O*-methyltransferases first. As can be seen in Figure 7.47, as a whole they catalyse the addition of methyl groups to a wide range of substrates. In each case, the product of the methylation reaction is shown to give you an idea of how selective these enzymes can be.

The second family are *N*-methyltransferases, which also display a wide substrate scope as a family. A range of substrates are shown in Figure 7.48, with the product of methylation shown in each case.

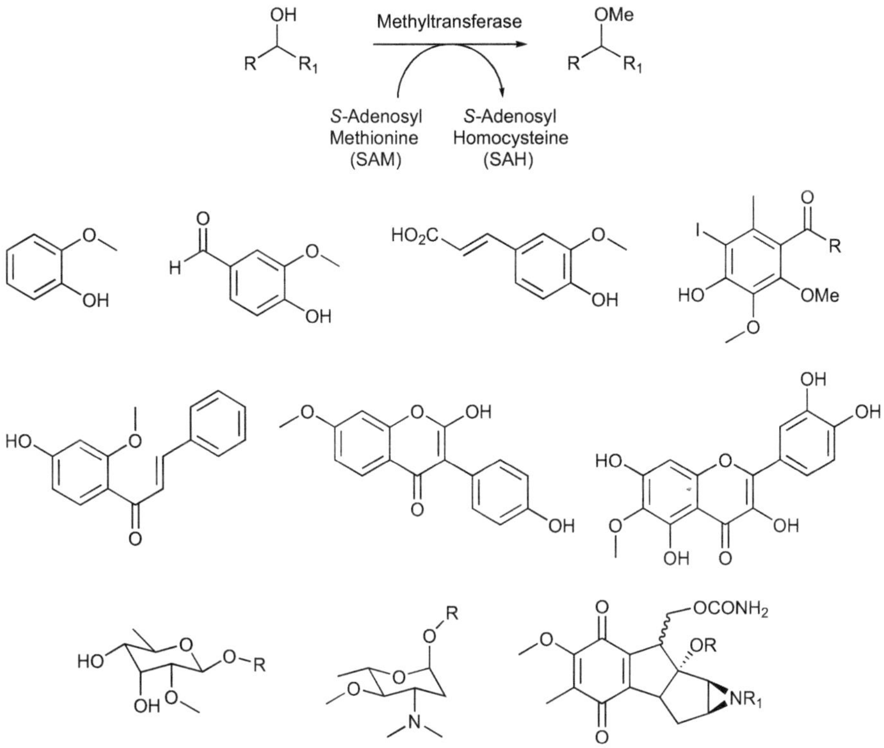

Figure 7.47 Selected examples of the scope of products produced by *O*-methyltransferases.

Methyltransferases are also able to react with sulfur atoms. Fewer enzymes have been reported to catalyse this reaction and therefore the substrate scope is more limited than the *O*- and *N*-methyltransferases detailed above. The substrate scope is shown in Figure 7.49, with the product of the methylation reaction drawn.

The mechanism of methyltransferases is the same irrespective of which heteroatom is being alkylated. You do not need to know the mechanism, but we have included it below for those who would like more detail. Methyltransferase enzymes, which are *S*-adenosyl-L-methionine (SAM)-dependent, all share a highly conserved SAM binding site within the protein. The substrate binding site differs widely between enzymes and it is the orientation of the substrate within the enzyme active site that is responsible for the observed selectivity. In Figure 7.50, we have chosen catechol-*O*-methyltransferase (COMT) as a "typical" protein. The mechanism starts with SAM and the substrate being bound to the enzyme. In the case of COMT, a

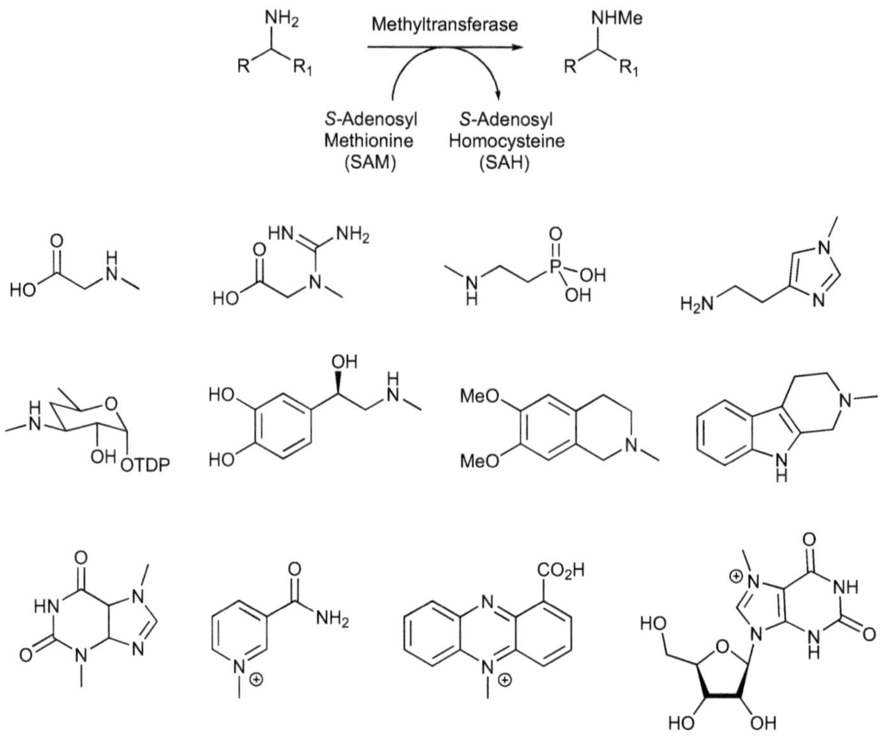

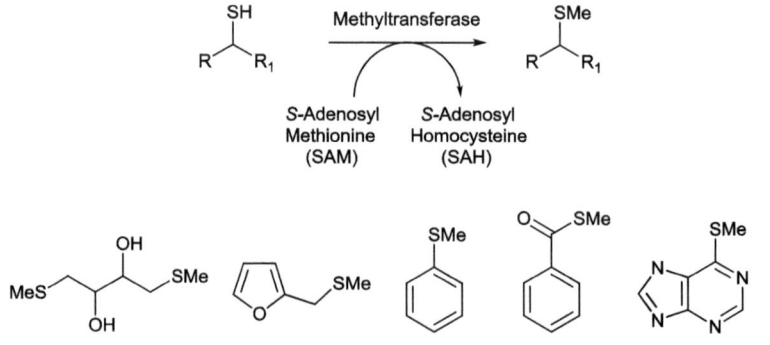

Figure 7.48   Selected examples of the scope of products produced by *N*-methyltransferases.

Figure 7.49   The scope of methylated products produced by *S*-methyltransferases.

magnesium atom is important in activating the substrate along with a lysine residue, which deprotonates the substrate. In addition, a glutamic acid also helps activate the substrate. Once the substrate and

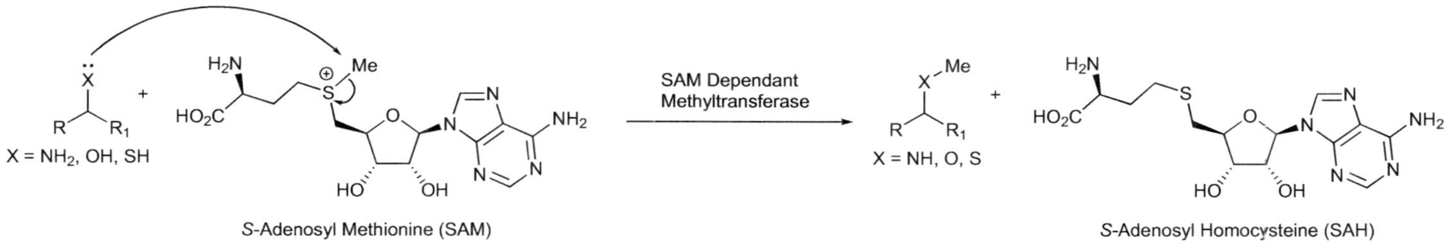

Figure 7.50 The general mechanism of methyltransferases.

SAM co-factor are brought into close proximity, the heteroatom (in this case oxygen) attacks the methyl group attached to the SAM co-factor. This can be thought of as an S_N2 reaction with *S*-adenosyl-L-homocysteine (SAH) as the leaving group. The methylated product and SAH by-products are then released so that the next substrate and co-factor molecules can be bound for the next reaction.

8 C–C Bond Formation

8.1 Introduction

This chapter will cover a number of enzymes that are capable of making carbon–carbon (C–C) bonds. These enzymes come from two different groups within the enzyme classification system. Some are from group number 2, whilst others are from group number 4. Group 2 enzymes are transferases, defined as enzymes that transfer a functional group from one substance to another. In contrast, group 4 enzymes are lyases, defined as enzymes that catalyse the non-hydrolytic addition or removal of groups from substrates. Using both of these classes of enzymes, we can make a range of C–C bonds and we will cover each class of enzyme in more detail in this chapter. The formation of carbon–carbon bonds is the basis for organic chemistry, with C–C bonds being fundamental to biological structures and life itself. This makes methods of carbon–carbon bond formation, whether chemical or enzymatic, particularly important. From a synthesis perspective, these methods allow us to make a much wider range of disconnections when carrying out a retrosynthetic analysis of a target molecule, as there are typically many C–C bonds that could be disconnected. This in turn gives us more options than disconnecting only the carbon–heteroatom (C-X) bonds present. The transformations covered within this chapter are shown in Figure 8.1. Each transformation and its synthetic use are discussed in more detail in a separate section within this chapter.

Biocatalysis in Organic Synthesis: The Retrosynthesis Approach
By Nicholas J. Turner and Luke Humphreys
© Nicholas J. Turner and Luke Humphreys 2018
Published by the Royal Society of Chemistry, www.rsc.org

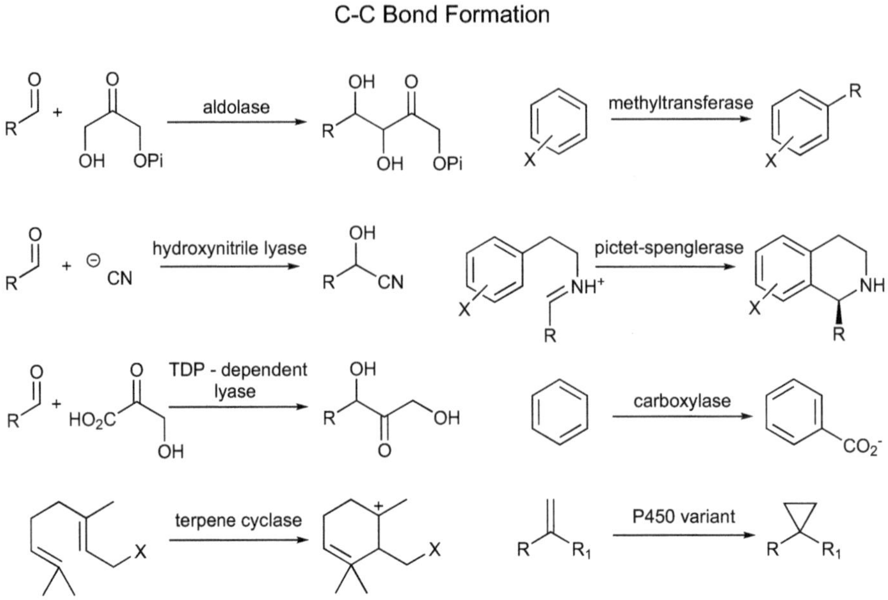

C-C Bond Formation

Figure 8.1 Overview of transformations presented in this chapter.

8.2 Carbon–Carbon Bond Formation Using Aldolases

The aldol reaction is one of the cornerstones of organic synthesis. The reaction of the enolate of an aldehyde or a ketone at the α-carbon with the carbonyl of another molecule is one of the most versatile, effective and general methods for the formation of C–C bonds in modern organic synthesis. The β-hydroxy aldehyde or ketone product is otherwise known as an aldol (an abbreviation of aldehyde and alcohol) and can be formed under basic or acidic conditions. The products of the aldol reaction are valuable synthetic building blocks, which is why this reaction has frequently been used as a tool to produce the polyol structures present in many natural products, particularly polyketides. Aldol reactions also occur in Nature and are often catalysed by enzymes known as aldolases. These enzymes are lyases, belonging to the enzyme classification group 4. They are found in a variety of sources and are typically involved in the biosynthesis of sugars. Several different classes of aldolases exist and although there are mechanistic differences, all of them catalyse the general aldol reaction shown in Figure 8.2.

The enzyme-catalysed reaction occurs between a "donor" substrate (the nucleophile in the reaction) and an "acceptor" aldehyde (the

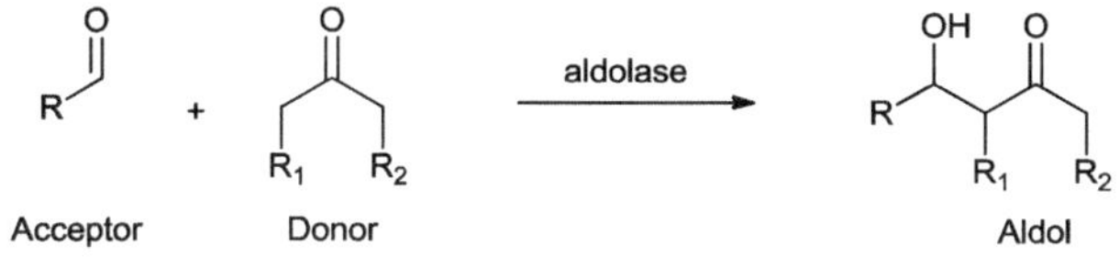

Figure 8.2 General transformation catalysed by aldolases.

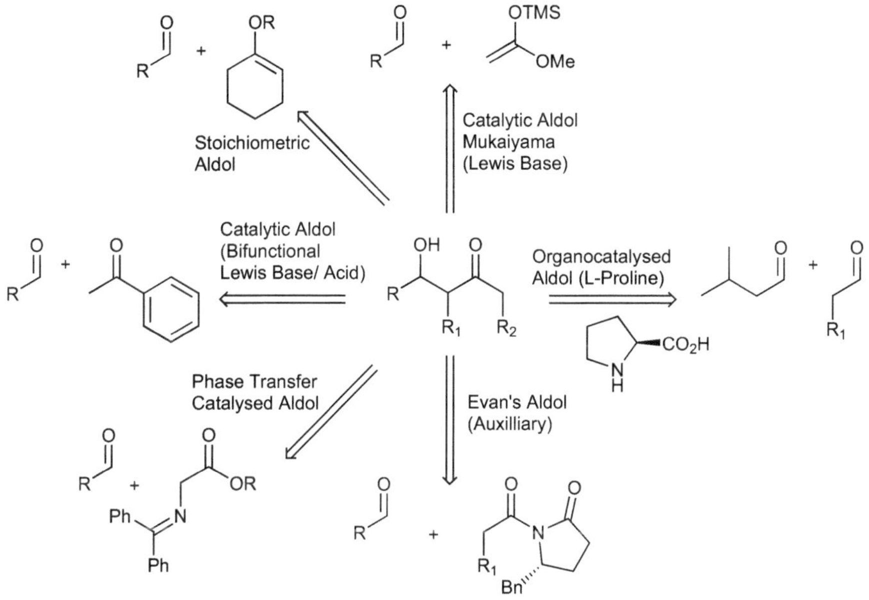

Figure 8.3 Selected chemical methods for aldol reactions.

electrophile). As mentioned above, the aldol reaction is one of organic chemistry's "classic" reactions and a wide variety of methods have been reported in order to selectively produce one product over another. Figure 8.3 shows a number of the methods that have been reported, including the use of catalytic methods as well as chiral auxiliaries.

Catalytic methods tend to be preferred over the use of chiral auxiliaries as the auxiliary has to be synthesised and then removed at the end of the reaction, adding extra steps to the synthesis. In addition, low temperatures, strong bases and Lewis acids are used in order to control the geometry of the enolate formed at the beginning of the reaction. Catalytic aldol reactions have the advantage of not requiring a stoichiometric auxiliary, but may still use Lewis acids and chiral ligands or chiral bases. Modern organocatalytic methods are a good alternative to more traditional approaches, but may require the synthesis of complex chiral ligands and need long reaction times.

In contrast to the chemical methods above, aldolases are able to catalyse the aldol reaction under mild, aqueous conditions and ambient temperatures without the need for organic solvents or cryogenic temperatures. There is also no need for auxiliaries, chiral ligands, or metal catalysts for stereoinduction. Due to the mild conditions of aldolase-catalysed reactions, the need to protect sensitive or reactive functional groups is also removed. This increases the atom efficiency of a synthesis and eliminates protection and deprotection steps. In order to achieve high diastereoselectivity, aldolases often tolerate little variation in the donor substrate they accept. Fortunately, the range of aldehyde acceptors that can be reacted with a particular donor substrate is typically much wider, such that these enzymes can be used in synthesis. In the figures below, each different donor substrate and a range of electrophiles with which it reacts are shown. One of the most well-studied donor substrates is 1,3-dihydroxyacetone phosphate. As shown in Figure 8.4, a wide range of aldehydes have been reported to react with this substrate. In each case, the substrate

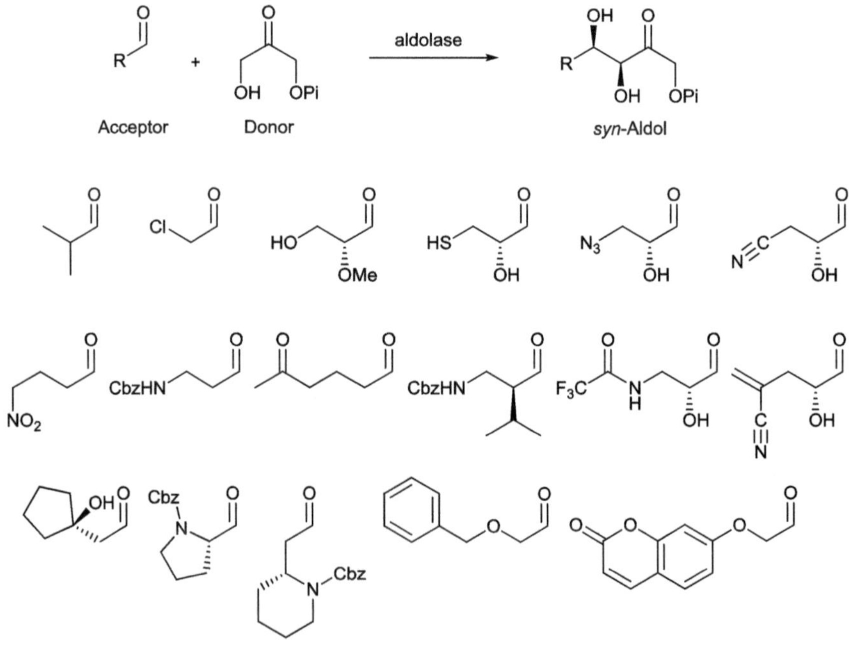

Figure 8.4 The scope of aldehyde substrates that react with 1,3-dihydroxy-acetone phosphate in aldolase-catalysed reactions.

aldehyde is shown and typically the *syn*-aldol product is produced. In some cases, the *anti*-aldol product can be obtained through the choice of an alternative enzyme.

Several aldolases involved in the metabolism of sugars have been shown to use pyruvate rather than 1,3-dihydroxyacetone phosphate as a donor. As shown in Figure 8.5, these enzymes typically use sugar derivatives as electrophiles. In each case, the aldehyde substrate is shown and the reactions proceed with very high enantioselectivity, typically producing the *S*-enantiomer due to attack of the nucleophile from the *Si* face. For particular substrates, however, enantiocomplementary aldolases exist allowing access to both *syn*- and *anti*-aldol products.

A further set of aldolases use glycine as the donor substrate. As you might have guessed, these aldolases are involved in the synthesis of certain amino acids, such as threonine. A range of aldehyde substrates can be used as shown on Figure 8.6. Enantiocomplementary aldolases give access to either L- or D-amino acids and the *syn*- or *anti*-products can be produced in high selectivity, depending on the choice of enzyme.

Acetaldehyde can also be used as the donor substrate (nucleophile) by the enzyme 2-deoxy-D-ribose 5-phosphate aldolase (DERA). This enzyme catalyses the self-condensation or cross-aldol condensation

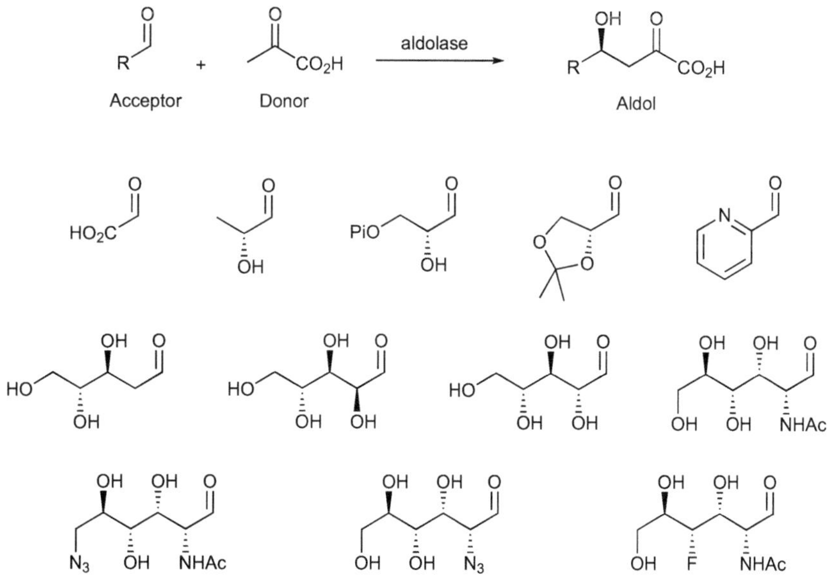

Figure 8.5 The scope of aldehyde substrates that react with pyruvate in aldolase-catalysed reactions.

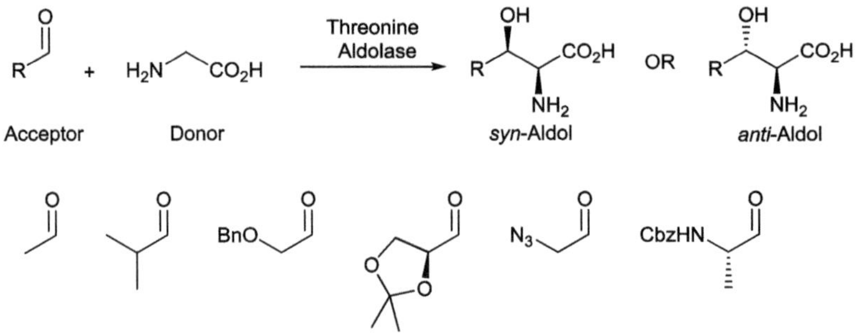

Figure 8.6 The scope of aldehyde substrates that react with glycine in reactions catalysed by threonine aldolase.

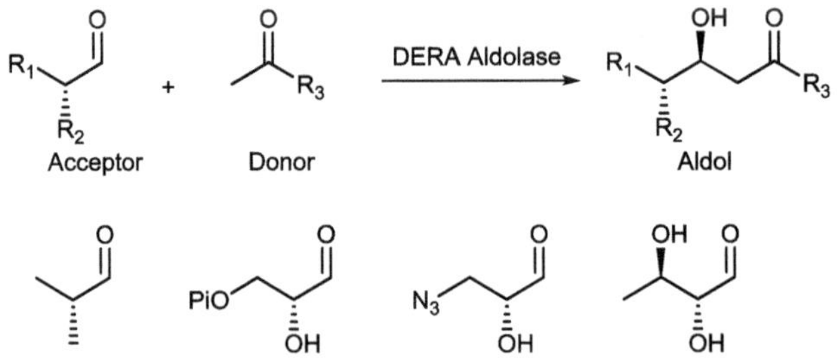

Figure 8.7 The scope of aldehyde substrates that react in reactions catalysed by 2-deoxy-D-ribose 5-phosphate aldolase (DERA).

of acetaldehyde, something which is challenging to carry out using chemical methods. As shown in Figure 8.7, DERA catalyses the reaction with a small range of acceptor substrates.

It is also possible to use this enzyme to carry out sequential two-step aldol reactions. In these processes, two molecules of acetaldehyde can be added one after another to an aldehyde starting material. This method was used to make the side-chain of Lipitor, the statin drug used to treat high cholesterol levels and prevent heart disease. We have seen other syntheses of this drug previously in Chapters 5 and 7. As shown in Figure 8.8, in this approach, chloroacetaldehyde undergoes two sequential aldol reactions with acetaldehyde catalysed by DERA. The resulting aldehyde is oxidised and cyclises *in situ* to give the chlor-olactone. This intermediate is then transformed into Lipitor.

Aldolases have also been used for the synthesis of aza-sugars. These are sugar analogues in which the oxygen atom in the pyranose ring is replaced with a nitrogen atom to give a piperidine instead of a tet-rahydropyran. Aza-sugars are found in Nature and have been widely

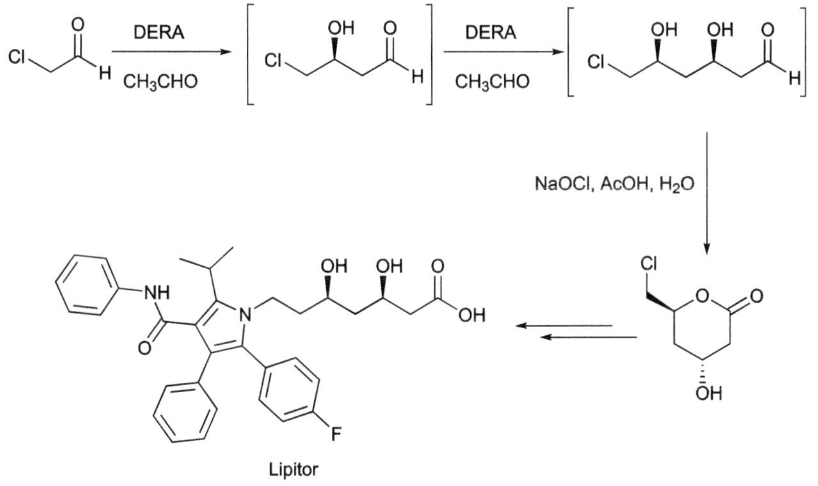

Figure 8.8 The synthesis of Lipitor using 2-deoxy-D-ribose 5-phosphate aldolase (DERA).

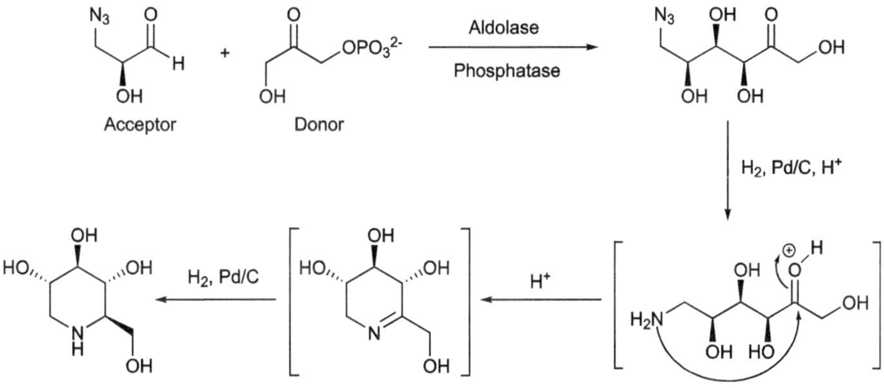

Figure 8.9 The synthesis of deoxynojirimycin using an aldolase.

studied as potential antiviral and anticancer agents. A range of aza-sugars have been synthesised using aldolases, with the natural product deoxynojirimycin shown in Figure 8.9 as an example. An aldolase is used to form a carbon–carbon bond between the two fragments. The phosphate group in the donor molecule is required for recognition by the aldolase. It is then removed from the product by the use of a second enzyme, a phosphatase. The azide group is then reduced using palladium on carbon under acidic conditions and the resultant amine cyclises to form an imine. This is reduced further to give the aza-sugar product.

8.3   Carbon–Carbon Bond Formation Using Hydroxynitrile Lyases

Cyanohydrins are versatile intermediates in organic synthesis, particularly when prepared with high stereoselectivity. Both the alcohol and nitrile group can undergo further manipulation making them useful as intermediates towards compounds of interest for the pharmaceutical and agrochemical industries. The formation of cyanohydrins is catalysed by a family of enzymes known by a number of names: hydroxynitrile lyase, oxynitrilase and hydroxynitrilase. However, as they belong to lyases under the enzyme classification system, we will refer to them as hydroxynitrile lyases. These enzymes are commonly found in plants, having been first discovered in almonds, but are also present in bacteria and fungi. They are thought to be used as a defence mechanism against herbivores and other microbes, as the release of cyanide from cyanohydrins renders the plant toxic. The formation of cyanohydrins using these enzymes is shown in Figure 8.10.

Many chemical methods of cyanohydrin formation have been previously reported. As shown in Figure 8.11, these typically involve the use of metals and chiral ligands, N-heterocyclic carbenes or N-oxides as catalysts. Chemical procedures often result in the alcohol of the cyanohydrin being derivatised *in situ* (as an ester or silyl ether), which then requires further steps to be converted to the desired alcohol. This is often carried out in order to avoid using hydrogen cyanide, which is highly toxic and explosive.

In contrast, cyanohydrin formation catalysed by hydroxynitrile lyases occurs under mild reaction conditions and with high stereoselectivity. Despite the drawbacks mentioned above, hydroxynitrile lyase-catalysed reactions use hydrogen cyanide under aqueous conditions and work on a wide range of substrates as shown in Figure 8.12. In each case, the product cyanohydrin enantiomer is shown but it is often possible to access the other enantiomer through the choice of an alternative enzyme.

Hydroxynitrile lyase reactions have also been reported using alternative cyanide sources. This avoids the need for hydrogen cyanide, which is highly toxic and explosive. Instead, acetone cyanohydrin has

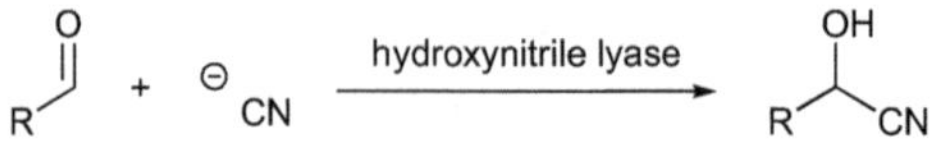

Figure 8.10   General transformation catalysed by hydroxynitrile lyases.

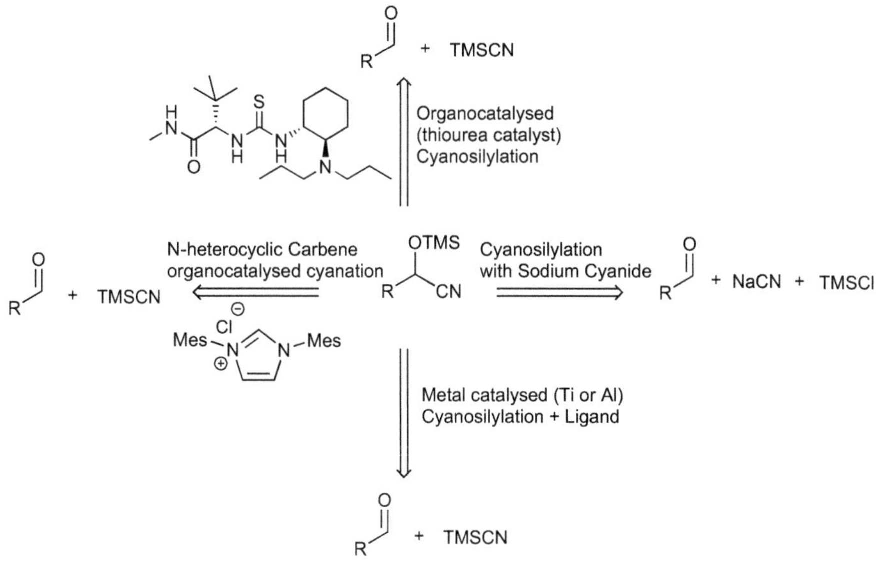

Figure 8.11 Selected chemical methods of producing cyanohydrins.

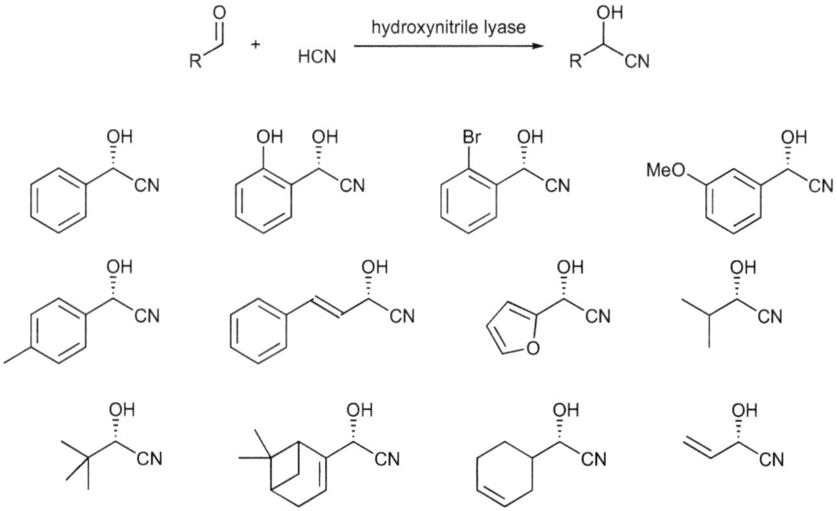

Figure 8.12 The scope of products produced by hydroxynitrile lyases.

been used as the cyanide source in what is a *trans* hydrocyanation reaction, as shown in Figure 8.13.

Although the cyanohydrin synthesis direction is more useful synthetically, hydroxynitrile lyases have also been reported for the

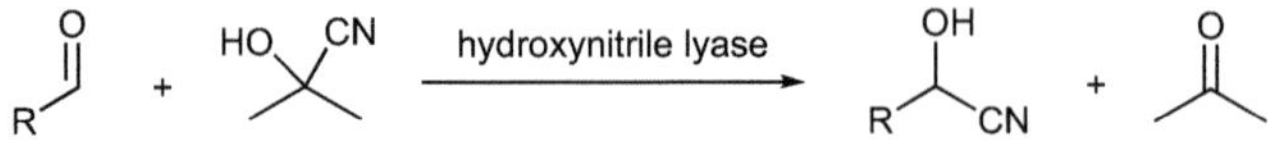

Figure 8.13 The synthesis of cyanohydrins using acetone cyanohydrin in place of hydrogen cyanide.

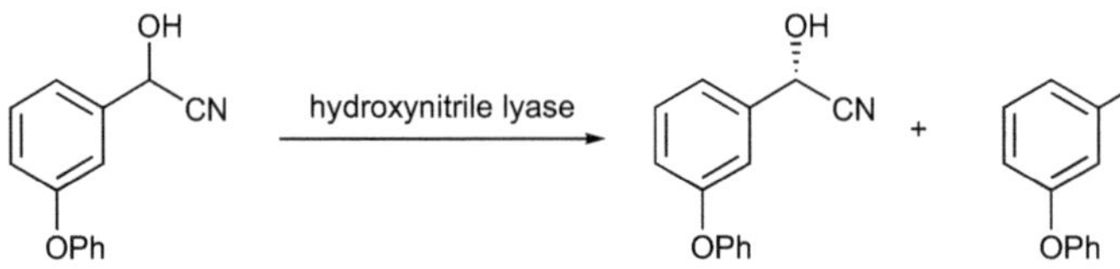

Figure 8.14 The resolution of cyanohydrins using hydroxynitrile lyases.

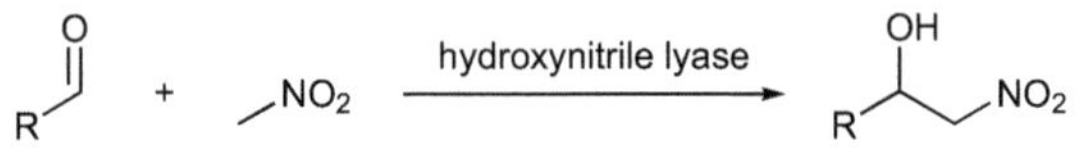

Figure 8.15 General transformation of nitroalkane addition reactions catalysed by hydroxynitrile lyases.

resolution of racemic cyanohydrins. As shown in Figure 8.14, in this process one enantiomer is converted into the corresponding aldehyde leaving behind a single enantiomer of the cyanohydrin.

Hydroxynitrile lyases have also been reported to catalyse the addition of nitroalkanes to aldehydes, as shown in Figure 8.15.

This is the biocatalytic equivalent of the organic reaction between the same compounds known as the nitroaldol or Henry reaction (named after its discoverer). The chemocatalysed Henry reaction typically requires the use of a metal catalyst, chiral ligands (to achieve high stereoselectivity) and organic solvents. The reaction times are often quite long as the reaction is slow. This reaction is of synthetic use because the product nitro alcohols can be further transformed into other functional groups, such as 1,2 amino alcohols, nitro ketones or hydroxy carboxylic acids. In contrast, hydroxynitrile lyase-catalysed Henry reactions occur under aqueous conditions without the need for metal catalysts and ligands. The reactions are highly stereoselective and work for a range of aryl aldehydes and a handful of nitroalkanes, as shown in Figure 8.16. In each case, the product enantiomer is shown; although, as with the synthesis of cyanohydrins, access to the other enantiomer is often possible through the choice of an alternative enzyme.

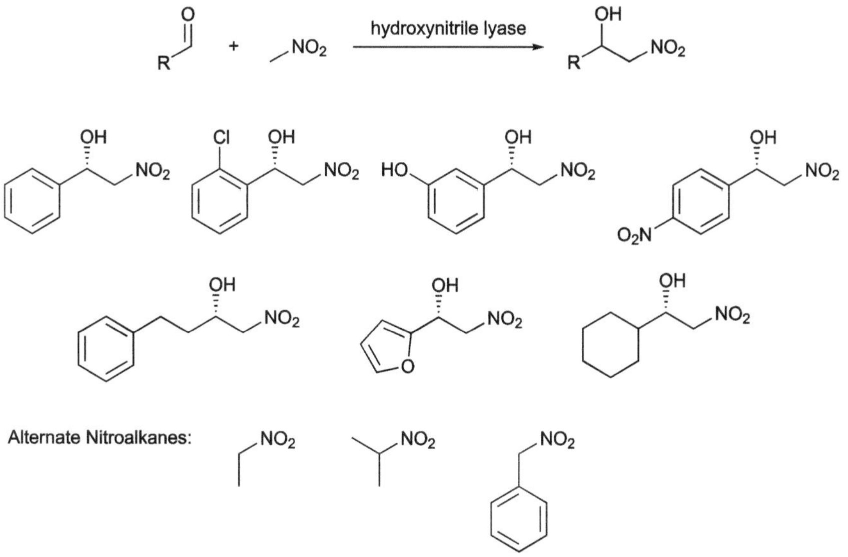

Figure 8.16 The scope of biocatalytic nitroaldol reactions catalysed by hydroxynitrile lyases.

Several different types of hydroxynitrile lyase have been discovered, each with differences in their mechanism. In Figure 8.17, we will concentrate on a more detailed mechanism for flavin-independent enzymes. You do not need to know this mechanism, but it is included here for those who would like to have a more detailed understanding. Hydroxynitrile lyases have the same structural fold as the hydrolases we saw in Chapters 3 and 4. It may come as no surprise then to discover that they use a catalytic triad of serine, aspartic acid and histidine and an oxyanion hole (the same as serine proteases). The residues of the oxyanion hole are typically cysteine, threonine and lysine, all of which are capable of hydrogen bonding in order to stabilise an anion. Figure 8.17 details the mechanism for the synthesis of cyanohydrins catalysed by this type of hydroxynitrile lyase. The carbonyl of the aldehyde substrate is bound in the active site and activated through hydrogen bonding to the serine residue. The hydrogen cyanide nucleophile enters the active site and is deprotonated by the histidine residue to give the cyanide anion. This subsequently attacks the aldehyde carbonyl to give an oxyanion which is stabilised by the residues of the oxyanion hole. Protonation of the oxyanion gives the cyanohydrin, which is then released ready for the next catalytic cycle.

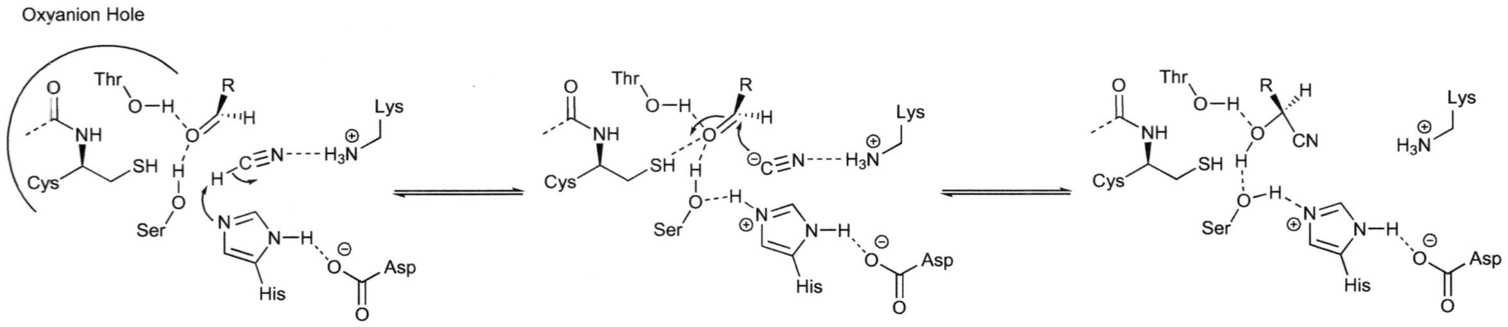

Figure 8.17 The general mechanism of flavin-independent hydroxynitrile lyases.

8.4 Carbon–Carbon Bond Formation Using Thiamine-dependent Lyases

The benzoin reaction, the name given to the coupling of two molecules of benzaldehyde in the presence of cyanide, is one of the oldest methods of making α-hydroxy ketones. Sometime later, the coupling of two different aldehydes was reported, the "cross-benzoin reaction". The α-hydroxy ketones or acyloins produced by these reactions are useful building blocks for the synthesis of larger molecules and are also found in a number of pharmaceutical drugs ranging from antidepressants to antibiotics. These compounds can also be produced by a family of enzymes known as thiamine phosphate-dependent lyases (ThDP). These reactions are interesting as a carbon–carbon bond is made between what appear to be two electrophilic groups. However, in the course of the reaction, the polarity of one of the reactive groups is reversed making it nucleophilic. For this reason, this type of reaction is termed "umpolung" from the German word meaning reversed polarity. We will see in more detail how this occurs at the end of this section. As shown in Figure 8.18, the aldehyde that reacts in an umpolung fashion is termed the donor and the other aldehyde the acceptor. As their name suggests, thiamine-dependent lyases require the co-factor thiamine diphosphate in order to be active.

Chemical methods of umpolung chemistry have been known for some time. In addition to the early reactions, which were catalysed with cyanide, subsequent methods of α-hydroxy ketone synthesis involved the use of dithiane acetals and a strong base. Modern methods

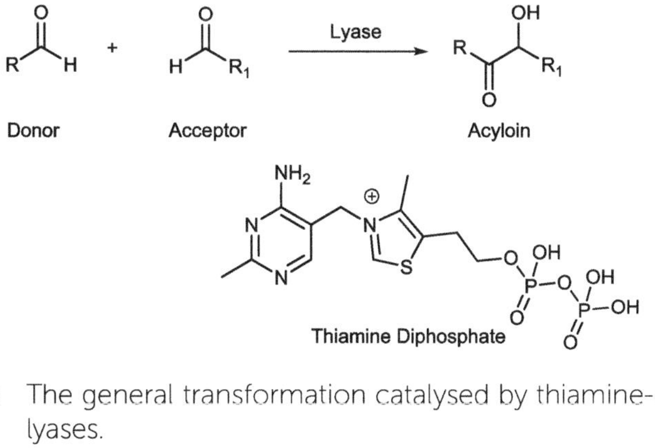

Figure 8.18 The general transformation catalysed by thiamine-dependent lyases.

use organocatalysts, such as N-heterocyclic carbenes, to catalyse cross-benzoin reactions with high stereoselectivities. The other major method of α-hydroxy ketone synthesis is hydroxylation of a carbonyl compound. A variety of methods for this transformation have been reported, although they typically require stoichiometric oxidants and may use high temperatures in order to achieve reactivity. Some of the oxidants used can be hazardous or explosive, such as hypervalent iodine reagents. A range of methods for producing α-hydroxy ketones are shown in Figure 8.19.

However, these chemical methods require the use of harsh reaction conditions or the synthesis of an organocatalyst and may only work on a limited range of substrates. In comparison, thiamine-dependent lyases work under mild conditions and display high stereo- and regioselectivity. When carrying out a cross-benzoin reaction with two different aldehydes, two products may be formed depending on which aldehyde is the donor (nucleophile). Often the selectivity of this process can be controlled though the correct choice of donor and acceptor molecules in combination with the appropriate enzyme. An example of this is shown in Figure 8.20 for the synthesis of phenylacetylcarbinol (PAC) derivatives. In each case, one of four cross-coupled products is produced in high selectivity showing the power of these enzymes.

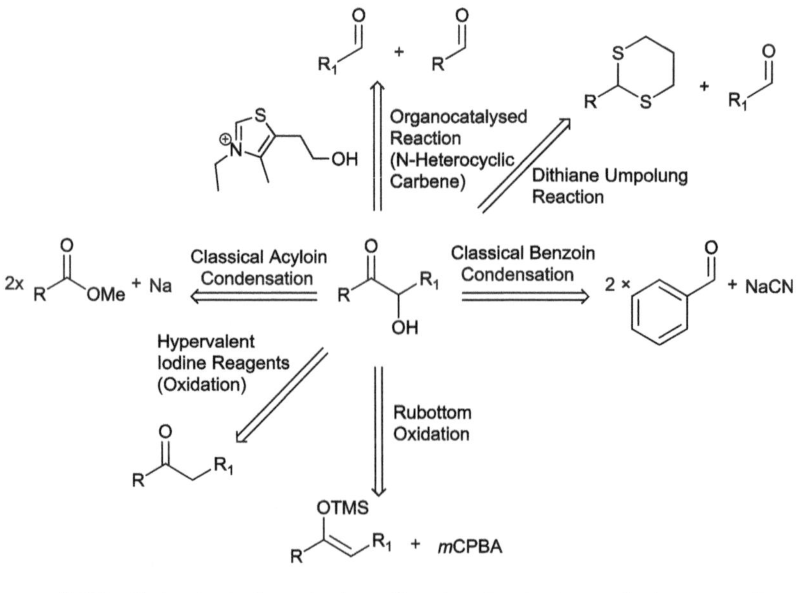

Figure 8.19   Selected chemical methods of α-hydroxy ketone synthesis.

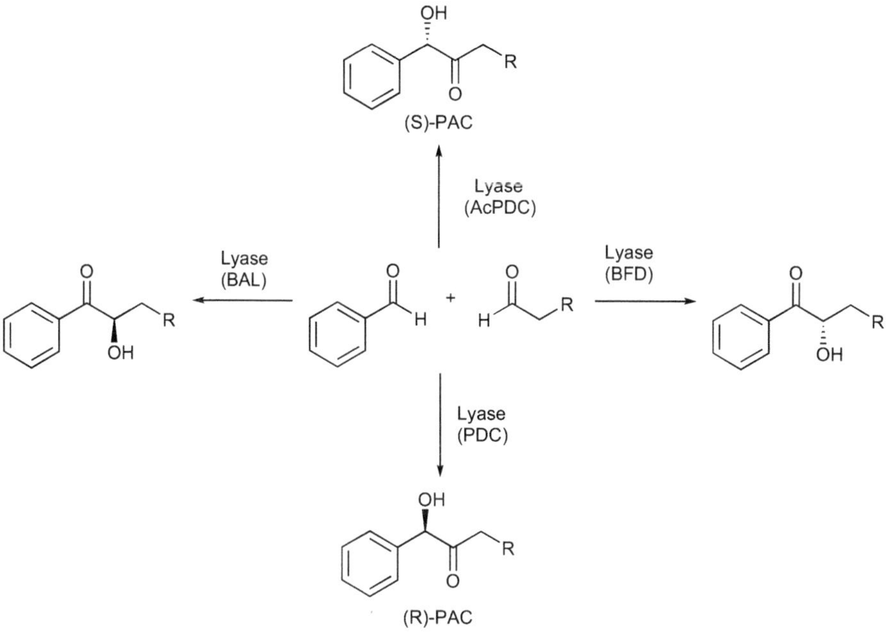

Figure 8.20 The divergent synthesis of phenylacetylcarbinol (PAC) derivatives using thiamine-dependent lyases.

As shown above, high regio- and stereoselectivities can be achieved through the correct choice of reactants and enzyme. Due the large range of possible combinations, only a sample of potential donors and acceptors are detailed in the next section. A range of donors that react with benzaldehyde substrates are shown in Figure 8.21.

It is also possible to use the benzaldehyde component as the umpolung (nucleophilic) partner in the reaction. As shown in Figure 8.22, benzaldehydes can be reacted with a range of aldehydes to give α-hydroxy ketone products, which are regioisomeric to those shown in Figure 8.21.

In addition to aromatic aldehydes, aliphatic aldehydes can also be used in these reactions. Figure 8.23 shows a range of aliphatic donors and acceptors that are capable of undergoing reactions catalysed by thiamine-dependent lyases. In general, both the yield and stereoselectivity of these reactions are high.

If a ketone is used as the acceptor rather than an aldehyde, then tertiary alcohols can be formed. These reactions have been studied far less by researchers and as a result, the yield and stereoselectivity of these reactions is not always as high. A small range of substrates undergo this reaction, as shown in Figure 8.24.

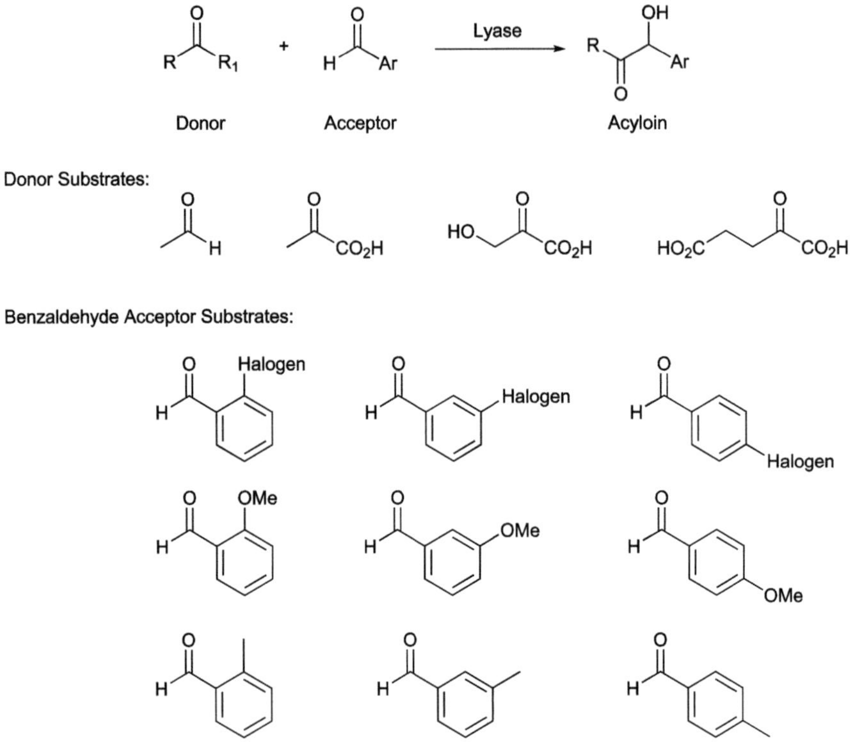

Figure 8.21 The scope of donor substrates in thiamine-dependent lyase-catalysed reactions with benzaldehydes.

The α-hydroxy ketone products formed in these reactions can be further functionalised into more valuable building blocks. An example of this is the synthesis of ephedrine, a medication that is used to prevent low blood pressure during anaesthesia and as a decongestant. As shown in Figure 8.25, the α-hydroxy ketone formed by the thiamine-dependent lyase then undergoes a reductive amination with methylamine to give the product.

The mechanism of these thiamine-dependent lyase-catalysed reactions is thought to be very similar to analogous organocatalytic reactions using N-heterocyclic carbenes. You don't need to know this mechanism, but like others in the book it is included in Figure 8.26 for those who would like further detail. The mechanism starts with the co-factor thiamine diphosphate being deprotonated, typically by an active site glutamic or aspartic acid residue. At this point, we can draw the structure in a resonance form that is a carbene, hence the similarity in mechanism to

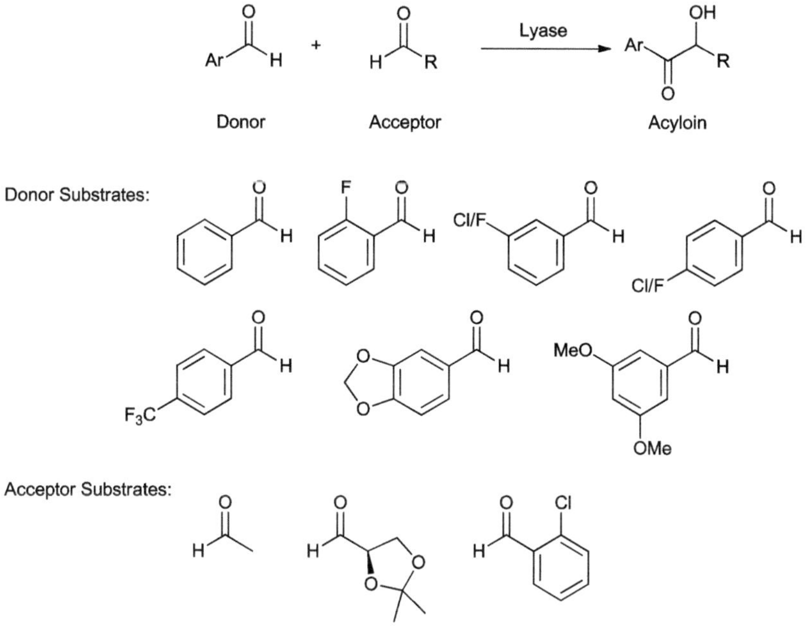

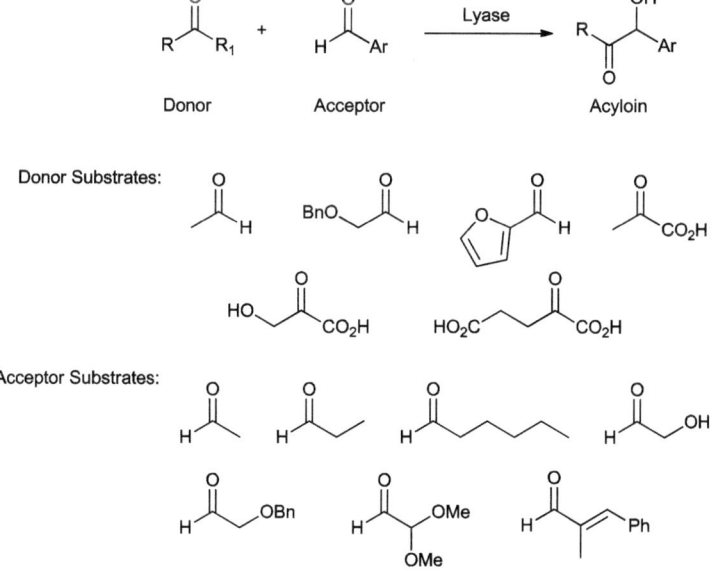

Figure 8.22 Reactions catalysed by thiamine-dependent lyases using benzaldehydes as donor substrates.

Figure 8.23 Reactions catalysed by thiamine-dependent lyases using aliphatic substrates.

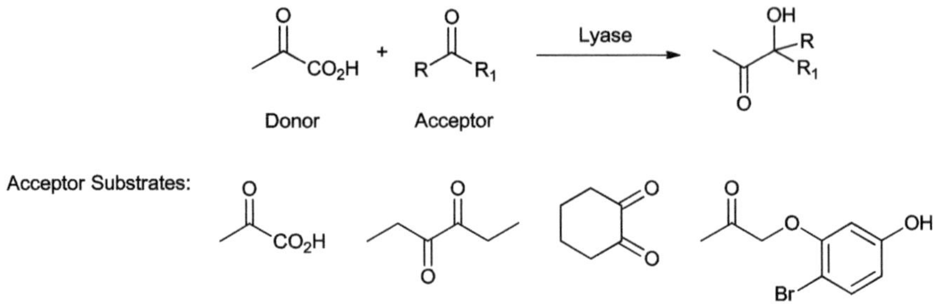

Figure 8.24 Reactions catalysed by thiamine-dependent lyases using ketone substrates.

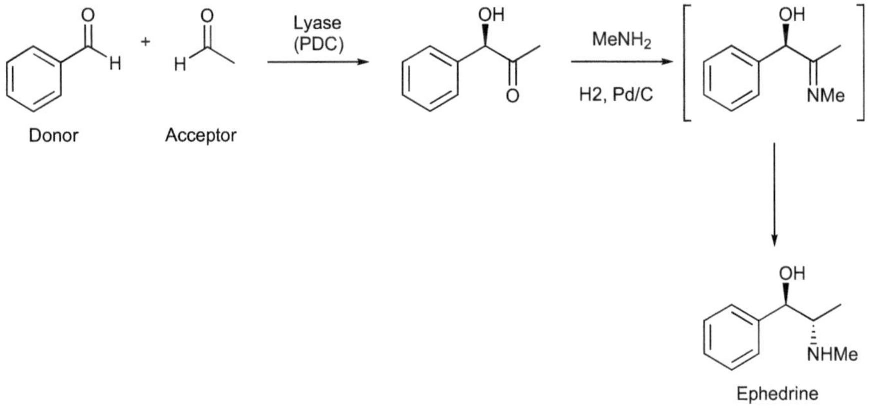

Figure 8.25 The synthesis of ephedrine using a thiamine-dependent lyase.

organocatalysed processes. The resulting anion then attacks the carbonyl carbon of the donor substrate to give a tetrahedral intermediate. Rather than collapsing back to give a carbonyl once more, the tetrahedral intermediate then loses a proton or carbon dioxide depending on which donor is being used in the reaction. Loss of either substituent generates an anion, which can be de-localised onto the thiamine molecule, stabilising the positively charged nitrogen atom (iminium ion). Drawn in the other reson-ance form, however, this gives an anion at the carbonyl carbon, a reverse in polarity as this carbon was originally electrophilic (being the carbon atom of a carbon–oxygen double bond). This carbon now displays umpolung reactivity with the anion carrying out nu-cleophilic attack on the carbonyl carbon of the acceptor substrate to give a diol. At this stage, we still have the tetrahedral

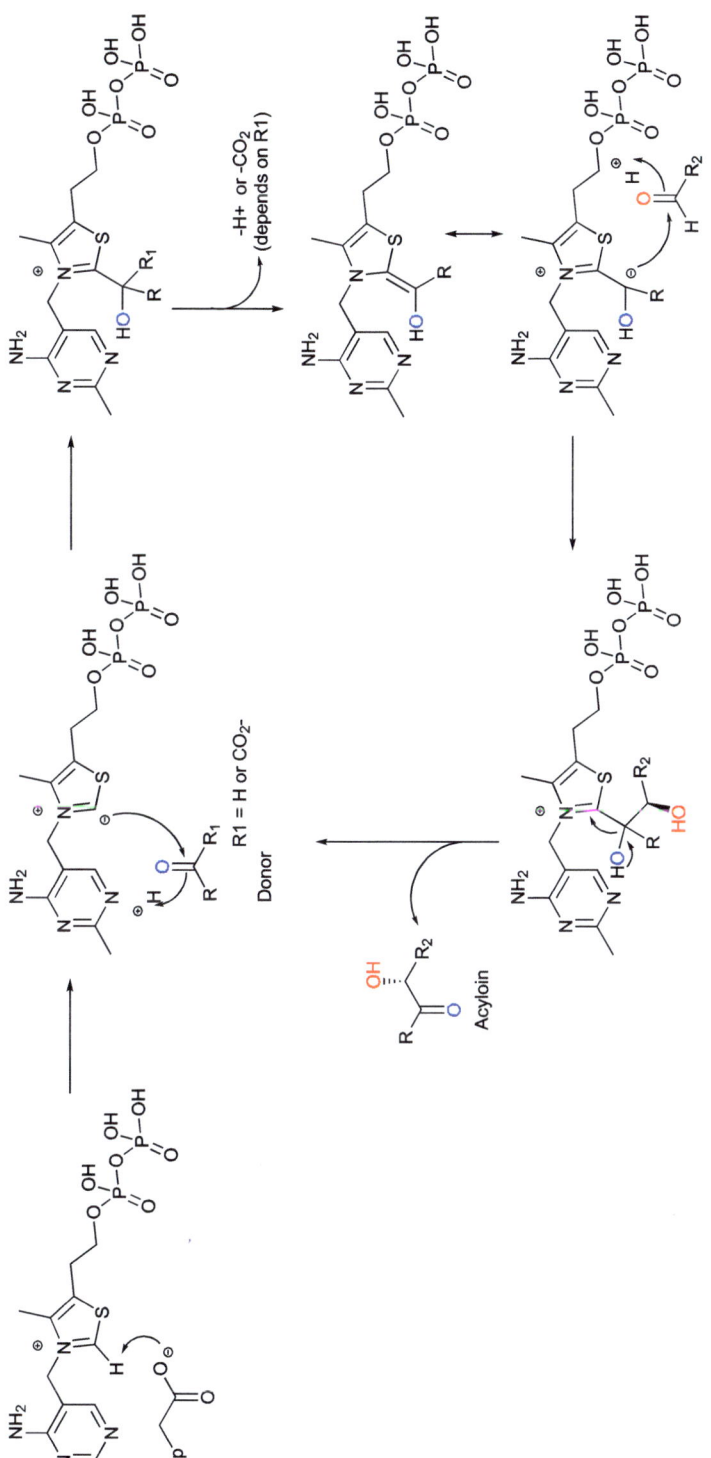

Figure 8.26 The general mechanism of thiamine-dependent lyases.

intermediate present from the initial attack of the thiamine di-phosphate on the donor carbonyl. Rather than deprotonation, the tetrahedral intermediate now undergoes more typical reactivity, collapsing back to form the carbonyl carbon–oxygen double bond and kicking out thiamine as a leaving group. This allows the thiamine to begin the catalytic cycle once more by attacking an-other donor substrate carbonyl.

We have seen above how thiamine-dependent lyases catalyse the umpolung addition to the carbonyl group of carbonyl compounds (1,2-addition). However, thiamine-dependent lyases also catalyse the umpolung reaction of an aldehyde to an α,β-unsaturated carbonyl compound (1,4-addition). This is the biocatalytic equivalent of the traditional "Stetter" reaction and is shown in Figure 8.27.

Modern chemistry uses organocatalysts, such as N-heterocyclic carbenes, to catalyse this reaction, although some transformations require further research, such as intermolecular asymmetric Stetter reactions, which can be challenging. Thiamine-dependent lyase-catalysed biotransformations may offer an alternative to chemical methods in some cases giving high stereoselectivities. Pyruvate is typically the donor in these reactions and a range of α,β-unsaturated carbonyl compounds undergo reaction, as shown in Figure 8.28.

The mechanism of these reactions is analogous to that which we saw above for 1,2-addition except that in this case the umpolung nucleophile carries out a 1,4-addition to the acceptor (electrophile) substrate. For those who are interested, drawing this out makes a good exercise!

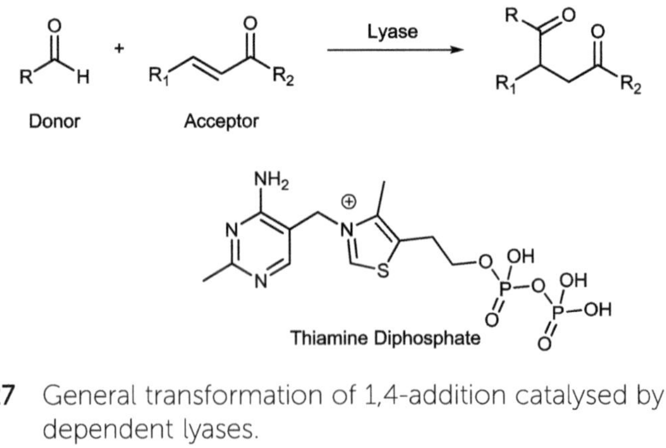

Figure 8.27 General transformation of 1,4-addition catalysed by thiamine-dependent lyases.

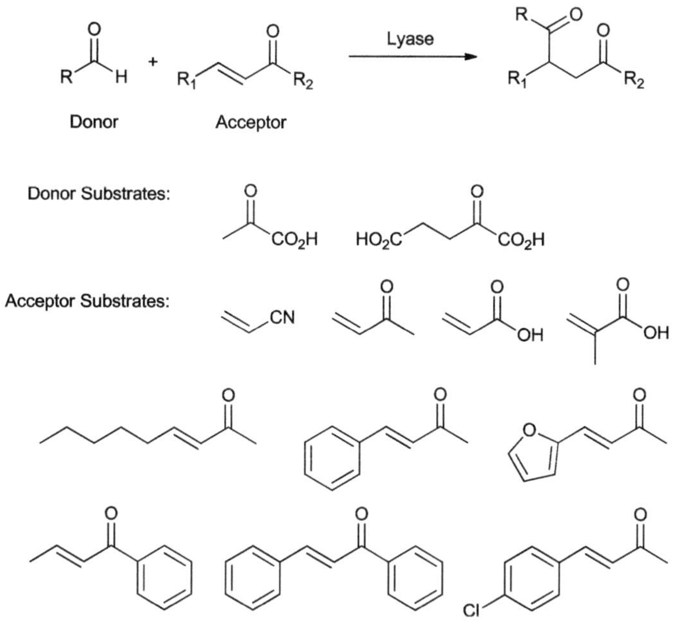

Figure 8.28 The scope of substrates that undergo 1,4-addition catalysed by thiamine-dependent lyases.

8.5 Carbon–Carbon Bond Formation Using P450 Monooxygenase Variants

We saw in Chapter 6 that P450 monooxygenase enzymes are capable of carrying out a range of different oxidation reactions. These enzymes are found in a large number of organisms from bacteria to plants and humans. These enzymes require an iron atom ligated by a heme ligand and a cysteine residue for activity and use molecular oxygen as the oxidant. As we saw in chapter 6, they are complex proteins that require additional proteins known as redox partners to facilitate the reaction. More recently, P450 monooxygenases have been mutated to allow them to catalyse non-natural transformations. As shown in Figure 8.29, these new P450 monooxygenase variants are able to catalyse cyclopropanation reactions in the absence of oxygen.

Several methods of cyclopropanation have been reported using chemical reagents. Both the Simmons–Smith and Corey–Chaykovsky reactions are classic organic chemistry cyclopropanation reactions.

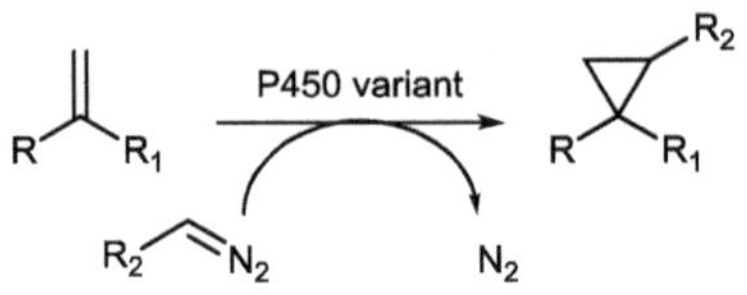

Figure 8.29 General transformation catalysed by P450 monooxygenase variants.

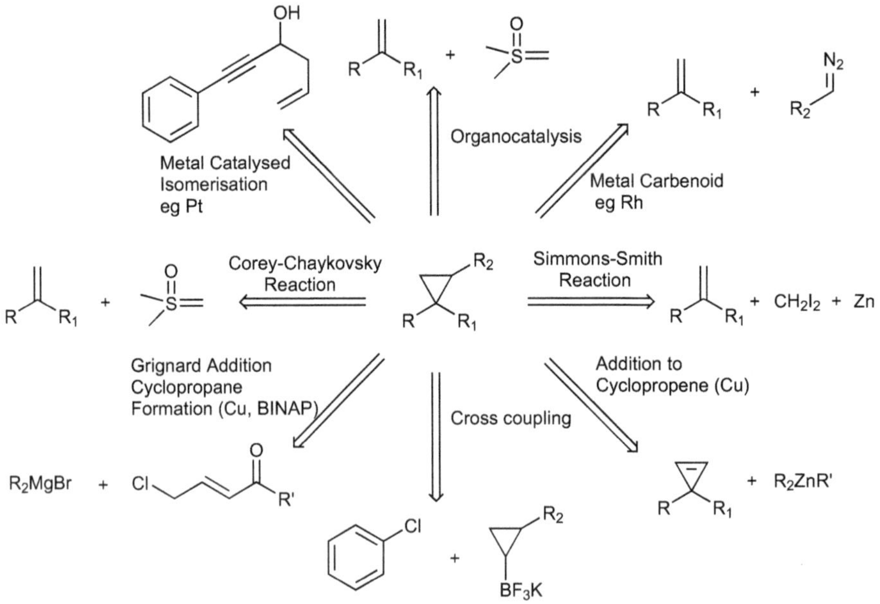

Figure 8.30 Selected chemical methods of cyclopropanation or equivalent.

More modern methods use a variety of metal carbenoids and chiral ligands for stereoinduction or organocatalysed reactions. The chiral ligands of catalysts used in these processes can often be expensive or require several steps to synthesise. The carbene precursors may also be toxic or difficult to handle, such as diazo compounds. A number of chemical cyclopropanation methods are shown in Figure 8.30.

Enzymatic cyclopropanation offers a complementary alternative to traditional chemical methods. The reactions do not require metal catalysts and chiral ligands and the reactions occur under mild aqueous conditions at ambient temperature. Enzymatic cyclopropanations do use diazoacetate carbene precursors in the same manner as chemical methods, although surprisingly they work in water. So far, only a small range of substrates undergo cyclopropanation reactions with ethyldiazoacetate, but the utility of this reaction was demonstrated in the synthesis of levomilnacipran, a selective serotonin reuptake inhibitor for the treatment of depression. As shown in Figure 8.31, the cyclopropanation proceeded in high yield and stereoselectivity, and the product could be carried on to the desired target.

As mentioned above, chemists have designed changes to the amino acid sequence of P450 monooxygenases in order accomplish this new reactivity. One of the most important changes made to these proteins was to replace the iron co-ordinating cysteine residue with a serine. The result of this substitution is shown in Figure 8.32. You do not

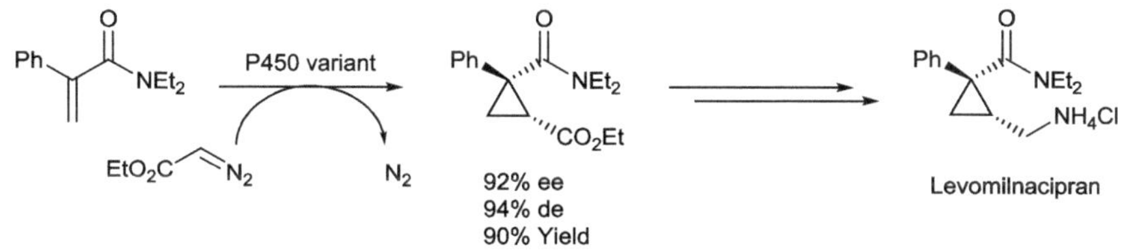

Figure 8.31 The synthesis of levomilnacipran using a P450 monooxygenase variant.

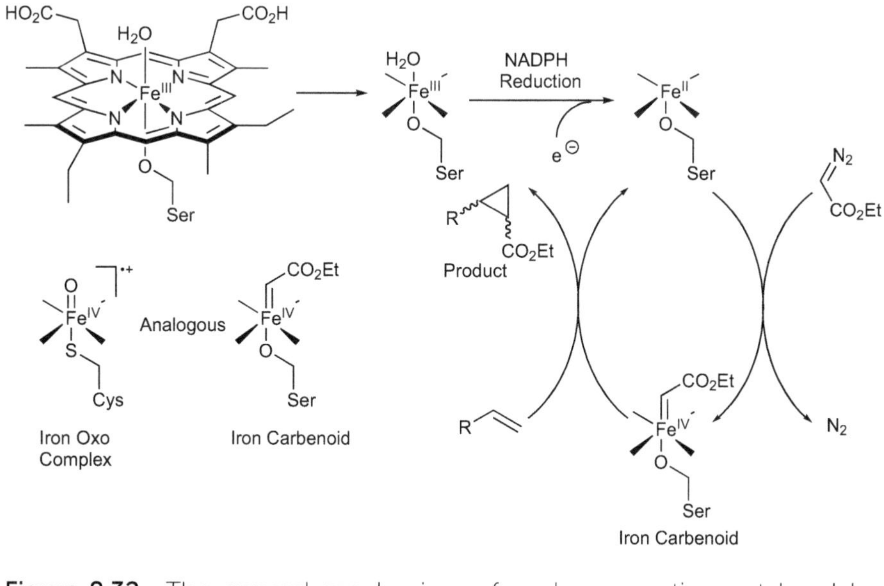

Figure 8.32 The general mechanism of cyclopropanation catalysed by P450 monooxygenase variants.

need to know the details of why these modified enzymes show different reactivity, but we have included it here for more advanced readers. By changing the iron ligating residue to a serine, the oxidation potential of iron(III) to iron(II) is lowered, such that it can now be reduced by NAD(P)H in the absence of oxygen. The iron(II) intermediate can now react with the ethyldiazoacetate to form an iron(IV) carbenoid, which is analogous to the iron(IV) oxo complex in the typical P450 monooxygenase hydroxylation reaction. The iron(IV) carbenoid then reacts with the substrate alkene to carry out the cyclopropanation reaction, regenerating the iron(II) catalyst.

Other transformations have also been reported with these P450 monooxygenase variants and the area of using mutated enzymes to catalyse non-natural reactivity is a rapidly developing area of research.

8.6 Carbon–Carbon Bond Formation Using Pictet–Spenglerases

The Pictet–Spengler reaction is a classical method of chemically synthesising tetrahydroisoquinolines. Named after its two discoverers, the reaction can be thought of as proceeding in two steps. Firstly, a β-aryl amine condenses with an aldehyde to form an intermediate iminium ion. This intermediate then undergoes an intramolecular Mannich-like cyclisation reaction, forming an asymmetric centre in the process. An analogous cyclisation reaction is catalysed by a class of enzymes called Pictet–Spenglerases, as shown in Figure 8.33. These enzymes were first identified from plants and were found to be involved in secondary metabolism, such as alkaloid synthesis. Like many of the other enzymes in this chapter, Pictet–Spenglerases belong to the lyase family in the enzyme classification system.

Chemical methods for asymmetric Pictet–Spengler reactions are shown in Figure 8.34. These typically involve the use of chiral catalysts, which must be prepared through a number of chemistry steps. These reactions may also require the use of protecting groups on the tetrahydroisoquinoline nitrogen. Methods that have been successfully

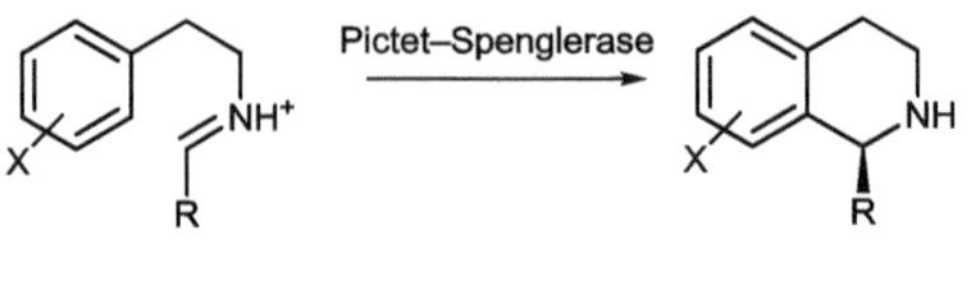

Figure 8.33 General transformation catalysed by Pictet–Spenglerases.

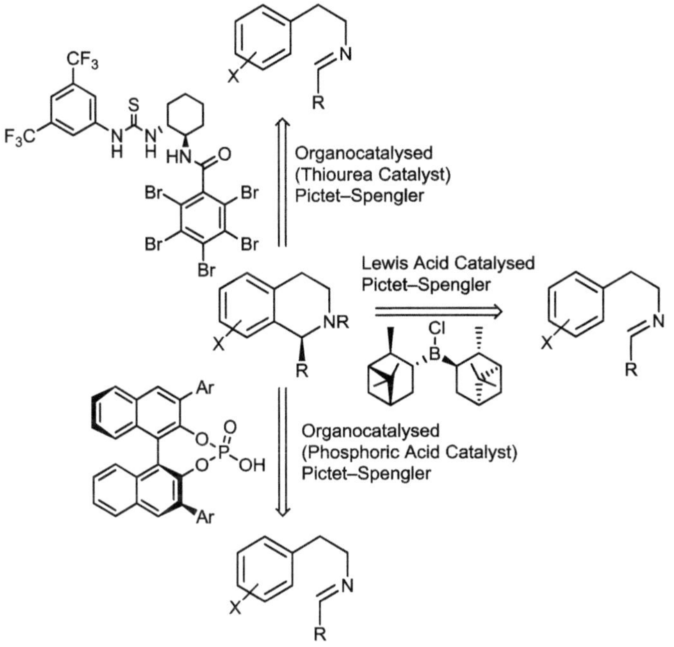

Figure 8.34 Selected chemical methods of asymmetric Pictet–Spengler reaction.

employed include the use of chiral phosphoric acids, chiral thioureas and chiral boron catalysts.

In comparison to chemical methods, Pictet–Spenglerase enzymes are able to catalyse the formation of tetrahydroisoquinolines in high stereoselectivity under mild reaction conditions, making them a useful alternative. The range of substrates for enzyme-catalysed Pictet–Spengler reactions is quite wide with respect to the aldehyde, but very narrow with respect to the amine component. Two amines have been shown to undergo reactions with a range of aldehyde partners and these are detailed in Figure 8.35.

One of the biosynthetic pathways in which Pictet–Spenglerases have been found is in the synthesis of benzylisoquinoline alkaloids. These compounds are believed to be made as secondary metabolites as part of a plant defence system against herbivores. One of the precursors in this pathway is the compound norcoclaurine. Chemists have shown that they can make (*S*)-norcoclaurine in one pot from tyrosine and dopamine using a Pictet–Spenglerase-catalysed cyclisation, as shown in Figure 8.36.

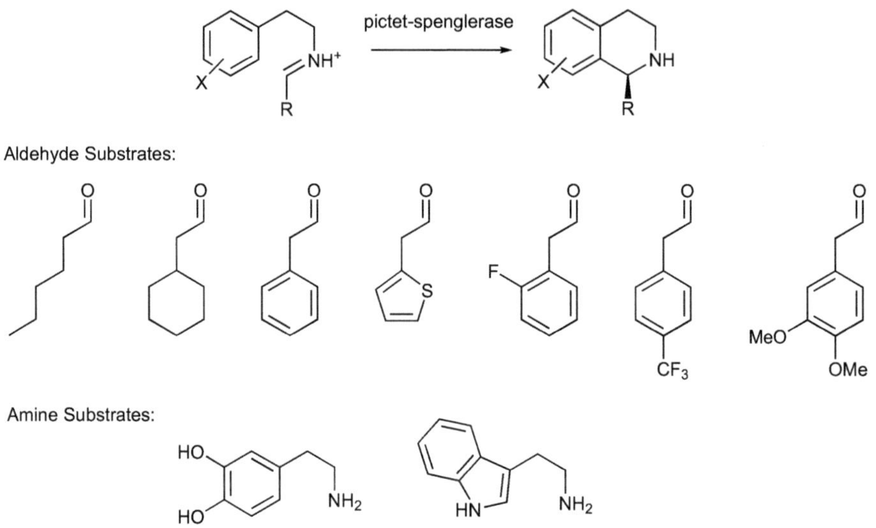

Figure 8.35   The substrate scope of reactions catalysed by Pictet–Spenglerases.

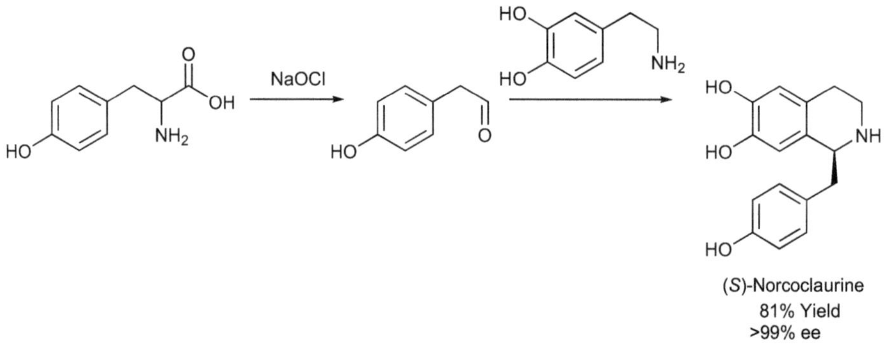

Figure 8.36   The synthesis of norcoclaurine using a Pictet–Spenglerase.

8.7   Aromatic Carbon–Carbon Bond Formation Using Transferase Enzymes

The direct functionalisation of aromatic rings to make a carbon–carbon bond typically has a high energy of activation due to the fact that the aromaticity of the ring must be initially broken during the reaction. As a result, most biological compounds containing substituted aromatic rings are formed as non-aromatic precursors before being aromatised. However, Nature has also evolved methods to alkylate benzene rings using carbon electrophiles. These processes play an important role in plant secondary metabolism, helping to produce a wide range of natural products that are involved in crucial biological processes. Examples include the production of vitamin E or the methylation of benzoates as part of ubiquinone synthesis. Enzymes catalyse the addition of several different functional groups to aromatic ring systems and each is described in more detail in a separate section below.

8.7.1 Carbon–Carbon Bond Formation Using Methyltransferases

Methylation of organic compounds occurs widely in Nature, with DNA, RNA and proteins all undergoing methylation reactions. These are often highly important as the methylation of biological molecules such as RNA and DNA is involved in signalling pathways. Methylation can not only change the sterics of the molecule, but also its lipophilicity increasing its membrane permeability. The methylation of aromatic substrates is carried out by a class of enzymes called methyltransferases, which we met previously in Chapter 7. These enzymes use a co-factor called *S*-adenosyl-L-methionine (SAM) as the source of the methyl group. They are a large family of enzymes with other members capable of methylating heteroatoms such as oxygen, nitrogen or sulfur, as we saw in Chapter 7. The general scheme for methyltransferase-catalysed methylation is shown in Figure 8.37.

The chemical alkylation (and acylation) of aromatic rings is traditionally carried out using the Friedel–Crafts reaction. Named after its two discoverers, the reaction uses an alkyl halide and a Lewis acid to

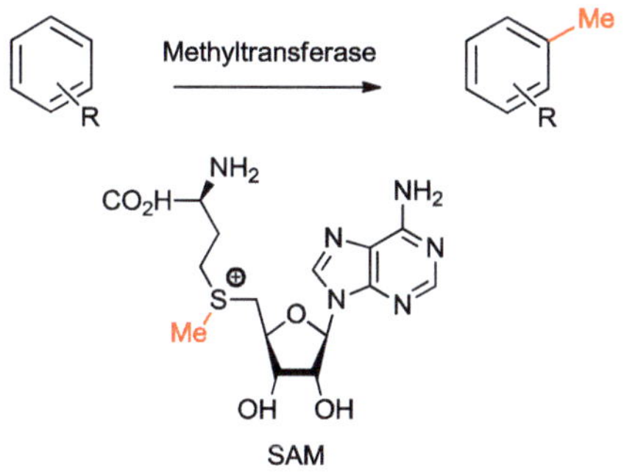

Figure 8.37 General transformation catalysed by carbon methyltransferases.

catalyse the reaction. These processes often require forcing conditions, which may not be suitable for sensitive functional groups, and the use of Lewis acids, generating metal waste that must be disposed of. There may also be issues of selectivity or rearrangement products during Friedel–Crafts reactions. Other alternatives include the alkylation of an aromatic anion or the generation of an aryl radical, which in turn may require the use of strong organometallic bases or harsh conditions. More modern synthetic methods use transition metal catalysts to couple an aryl halide with an alkyl equivalent, requiring extra synthetic steps to make the coupling partners. Some of these methods are shown in Figure 8.38.

In contrast, methylations catalysed by methyltransferases take place under mild, aqueous conditions without the need for harsh chemical reagents. They are typically more compatible with functional groups and do not produce any metal waste. Currently, only a small number of enzymes have been reported to catalyse this transformation, although researchers are actively looking to increase this number. Although the substrate range is very narrow, a few reactions of interest are detailed in Figure 8.39 to give you an idea of what is currently possible. In each case, the methyl group added by the enzyme is coloured in red.

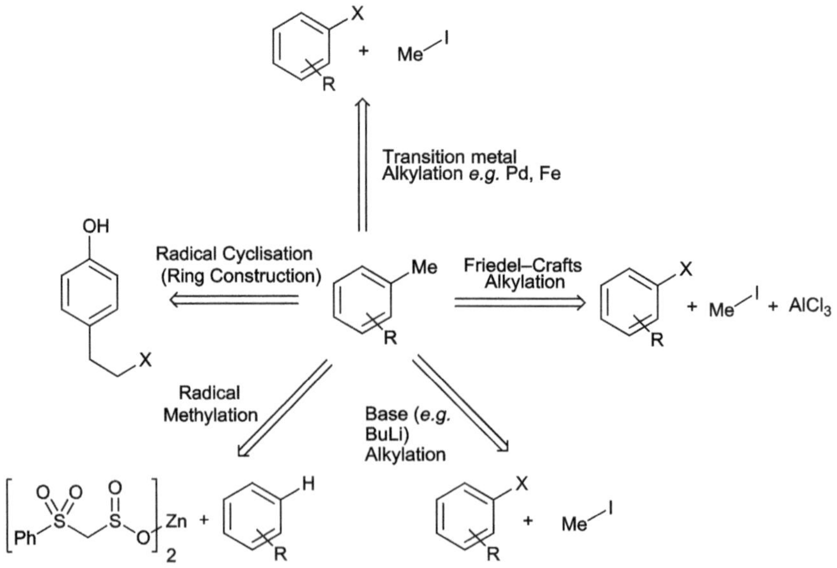

Figure 8.38 Selected examples of chemical methods for the alkylation of aromatic rings.

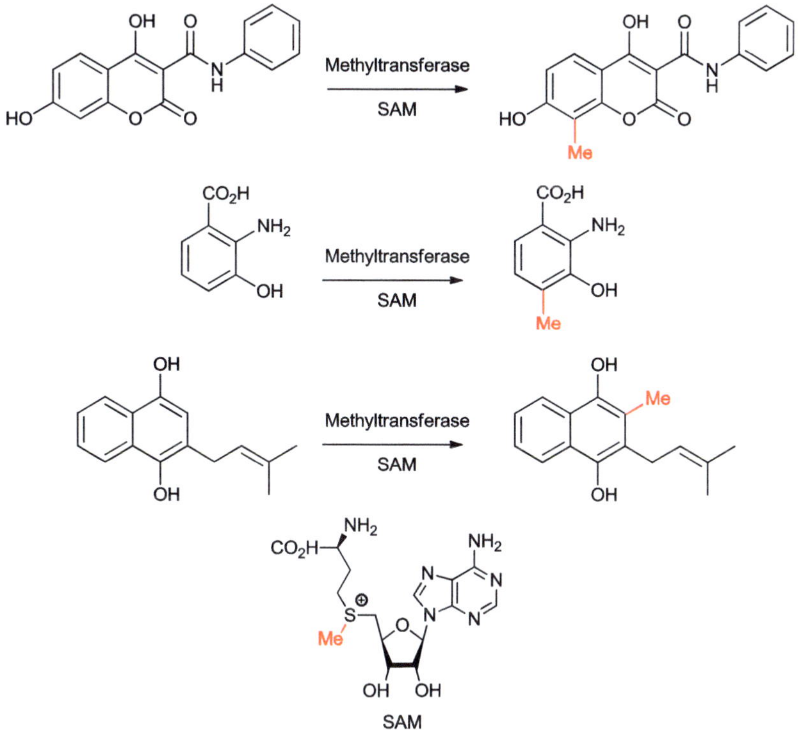

Figure 8.39 Selected examples of substrates that undergo aromatic methylation catalysed by methyltransferases.

8.7.2 Carbon–Carbon Bond Formation Using Prenyltransferases

A second family of transferase enzymes catalyse the addition of a dimethylallyl (or prenyl) carbon electrophile to aromatic substrates. As with methylation, prenylation is a common reaction in metabolic processes and the synthesis of natural products. Prenylated aromatic compounds have diverse activities, including antioxidant, anti-inflammatory, antiviral and anticancer properties. Aromatic prenyltransferases utilise the co-factors geranyl diphosphate or dimethylallyl diphosphate to catalyse the addition of a prenyl group to an aromatic ring, as shown in Figure 8.40.

Chemical methods for adding a prenyl group to an aromatic substrate are shown in Figure 8.41. Chemists traditionally use the Claisen rearrangement if an ortho alcohol substituent is present in

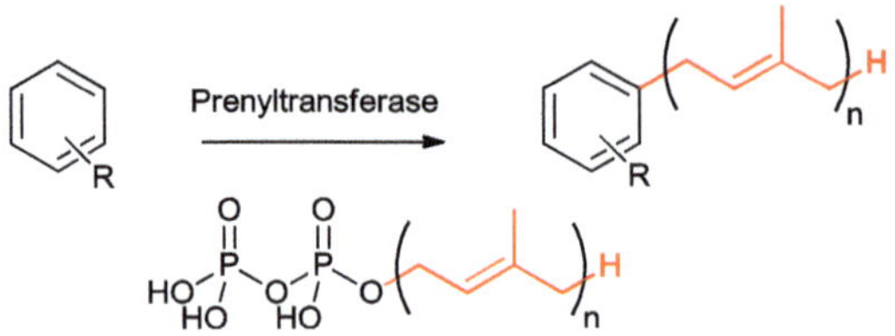

Figure 8.40 General transformation catalysed by prenyltransferases.

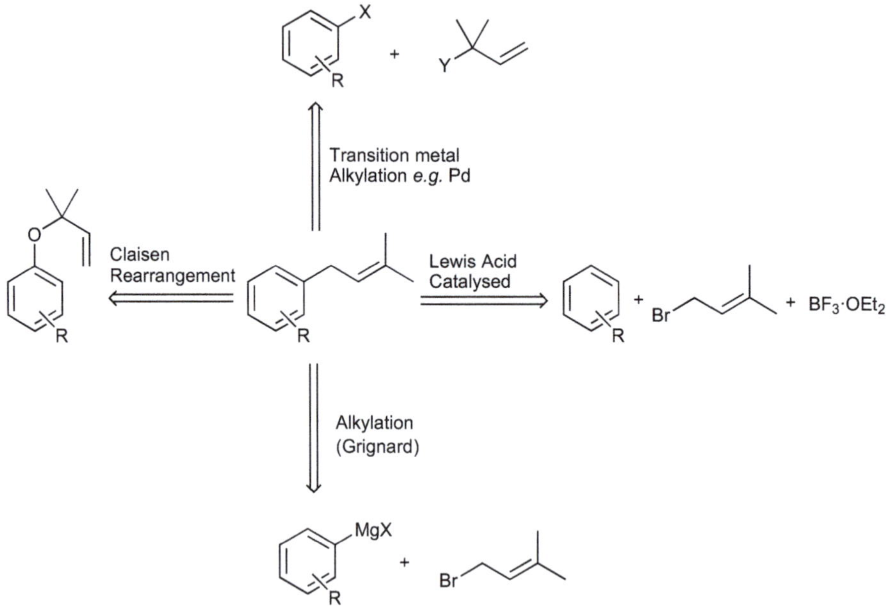

Figure 8.41 Selected examples of chemical methods for prenylation.

the substrate, although other approaches using strong bases or transition metal catalysts have been reported.

In comparison, prenyltransferase enzymes catalyse the reaction under mild conditions without the need for bases or metal catalysts. As with methyltransferases, these enzymes are highly regioselective, but individually they do not have a wide substrate scope (hence why they are highly selective). However, a range of enzymes are known leading to a wider range of possible substrates overall. Examples of products formed using prenyltransferases are shown in Figure 8.42. In each case, the bond formed in the reaction is coloured in red.

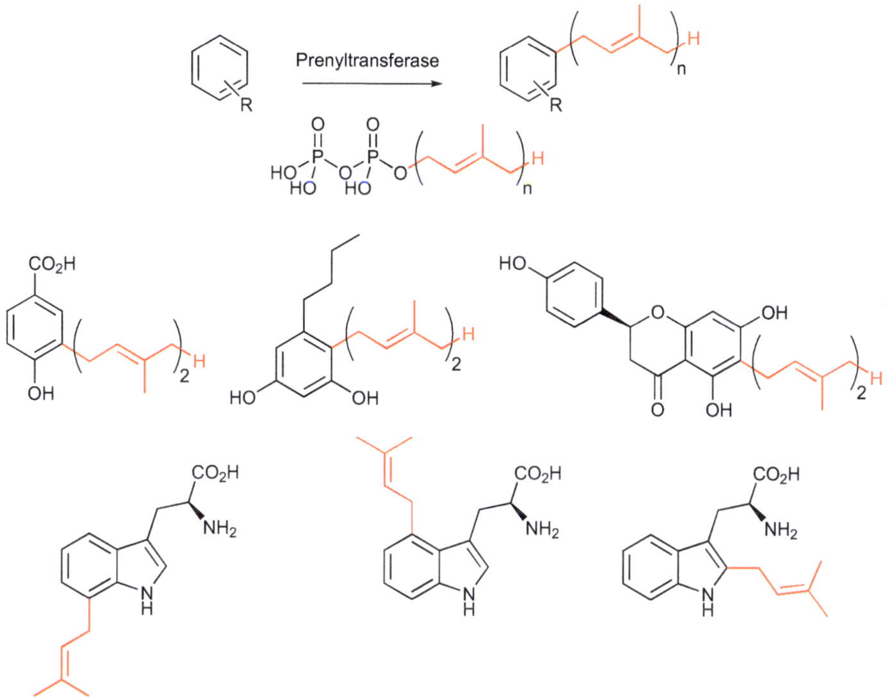

Figure 8.42 Selected examples of substrates that undergo prenylation catalysed by prenyltransferases.

8.8 Carbon–Carbon Bond Formation Using Carboxylases

Carboxylation is an important carbon–carbon bond forming re-action for both biosynthetic processes and organic chemistry. The addition of a carboxylate group to a molecule provides a group capable of hydrogen bonding with a target molecule or a functional handle to carry out further transformations. Carboxylations are found in a number of important biological pathways, such as purine biosynthesis and carbon fixation. A number of the enzymes used in these pathways have potential use as biocatalysts for or-ganic synthesis, especially if they act on a variety of substrates. The general reaction catalysed by carboxylase enzymes is shown in Figure 8.43.

A number of chemical methods exist for the carboxylation of organic compounds. These methods include lithium halogen exchange and then quenching with carbon dioxide, metal-catalysed

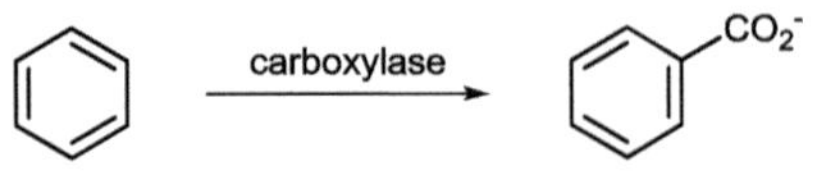

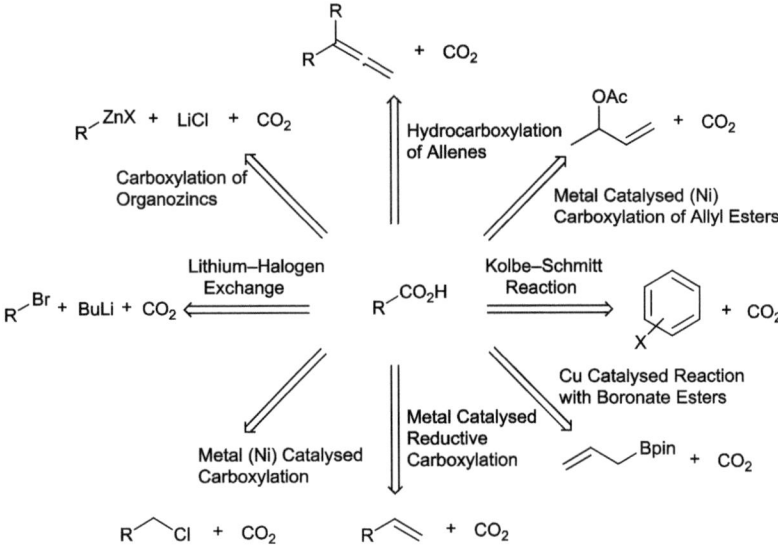

Figure 8.43 General transformation catalysed by carboxylases.

Figure 8.44 Selected examples of chemical methods for carboxylation.

reductive carboxylation or metal-catalysed cross-coupling of pseudohalides with carbon dioxide. These processes can require high temperatures or pressures and need metal catalysts and ligands to work. Carboxylation is usually carried out on an industrial scale using the Kolbe–Schmitt reaction, which requires high temperatures and pressures, and can also form a range of by-products that need to be separated from the desired compound. A range of chemical methods for carboxylation are shown in Figure 8.44.

Carboxylase enzyme-catalysed processes offer an alternative to the chemical methods above and don't require high temperatures or pressures and do not need metal catalysts or ligands. The reaction conditions are compatible with a range of functional groups and few, if any, by-products are formed. Carboxylase enzymes have only been studied recently for synthetic purposes and

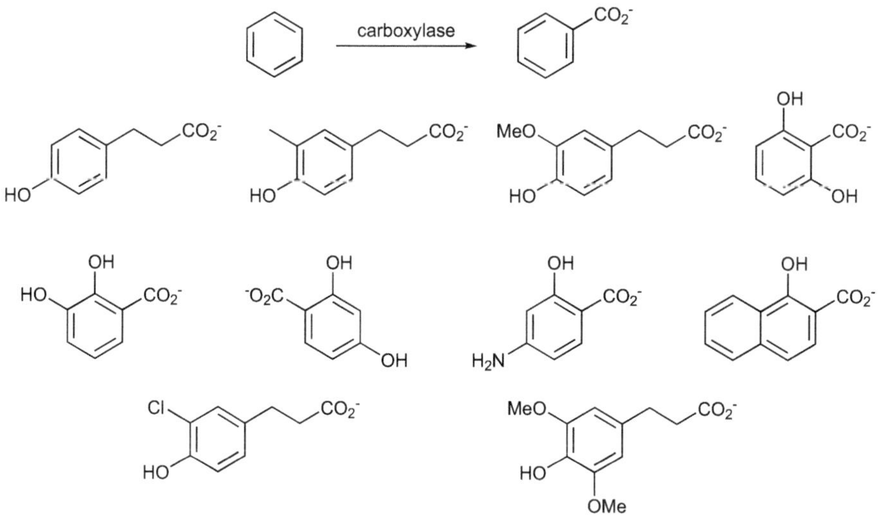

Figure 8.45 The scope of carboxylate products produced by carboxylase-catalysed reactions.

therefore the substrate scope of these biocatalysts is not as wide as some other enzymes. However, a range of phenol derivatives and other compounds can be carboxylated as shown in Figure 8.45. In each case the carboxylic acid product is shown. Access to one regioisomer or another is typically possible through the choice of an appropriate enzyme. It may also be possible to further increase the substrate scope of these enzymes through protein engineering.

8.9 Carbon–Carbon Bond Formation Using Terpene Cyclases

Terpenes and related compounds make up a very large family of natural products, which are found in both plants and animals. They can be further classed into subgroups depending upon how many molecules of isoprene, the monomer from which they are made, are present. For example, triterpenoids contain three isoprene units and are the precursors to steroids and sterols, which have an important role in biological function. In animals, steroids are important in cell membrane function and as hormones, whilst in plants they may also act as signalling molecules. Triterpene derivatives also have a wide range of uses in the food, health and biotechnology industries. The most studied transformation is the conversion of squalene into hopene, catalysed by the enzyme squalene hopene cyclase, as shown in Figure 8.46.

Once the mechanism of this reaction was discovered, chemists used this strategy to make triterpenes and steroids using strong Lewis acid catalysts. This methodology has been used in an elegant synthesis of a complex natural product, as shown in Figure 8.47. There are some drawbacks to this approach, however, as mixtures of products may be formed and the strong Lewis acids used may be incompatible with sensitive functional groups. In comparison, terpene cyclase enzymes are able to catalyse this transformation under mild reaction conditions with high selectivity.

A range of cyclisation reactions are catalysed by terpene cyclase enzymes, as shown in Figure 8.48. In addition to the cyclisation of triterpenoid precursors, simpler substrates with a range of functional groups also undergo reaction. The yields are not always high, but typically the reactions are very selective with a single product generated. In each case below, the product of the reaction is shown with the bonds made in the cyclisation coloured in red.

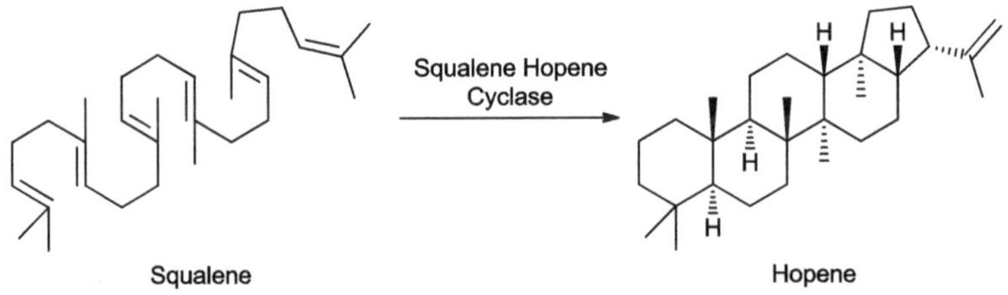

Figure 8.46 General transformation catalysed by squalene hopene cyclase.

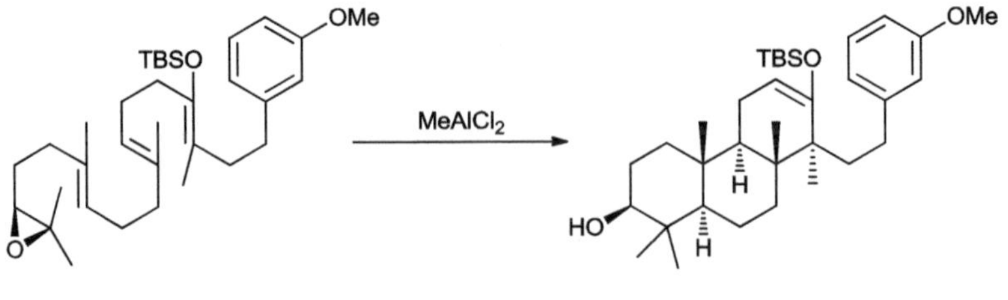

Figure 8.47 An example of chemical methods used to make squalene derivatives.

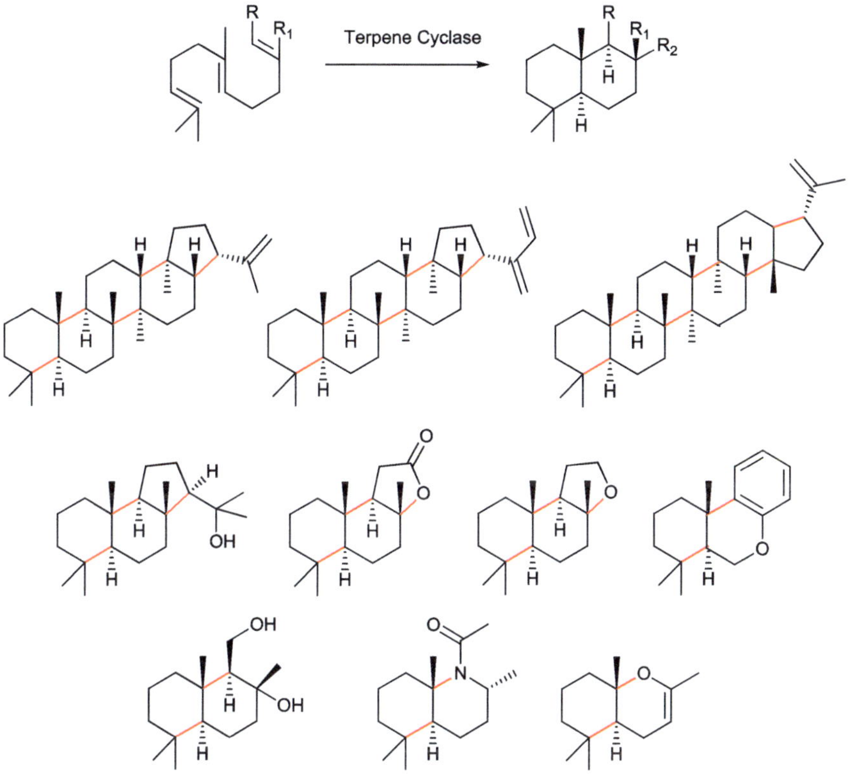

Figure 8.48 The scope of products made by squalene hopene cyclase-catalysed reactions.

The mechanism of squalene hopene cyclase with its natural substrate squalene is shown below. You do not have to know this mechanism, but it might be useful to help understand how these enzymes work or if you would like a more detailed understanding. The cyclisation reaction can be thought of as a series of reactions, each going *via* a carbocation intermediate. In order for the cyclisation reaction to begin, the substrate must adopt the all-chair conformation shown in Figure 8.49. The active site of the enzyme acts almost like a template, forcing the squalene to adopt this conformation, resulting in high stereochemical selectivity. The initial protonation of the first carbon–carbon double bond to start the reaction requires a relatively strong Brønsted acid. The enzyme achieves this through the use of an aspartic acid residue, which is further activated through a hydrogen bonding network involving histidine and tyrosine residues. Having carried out the initial

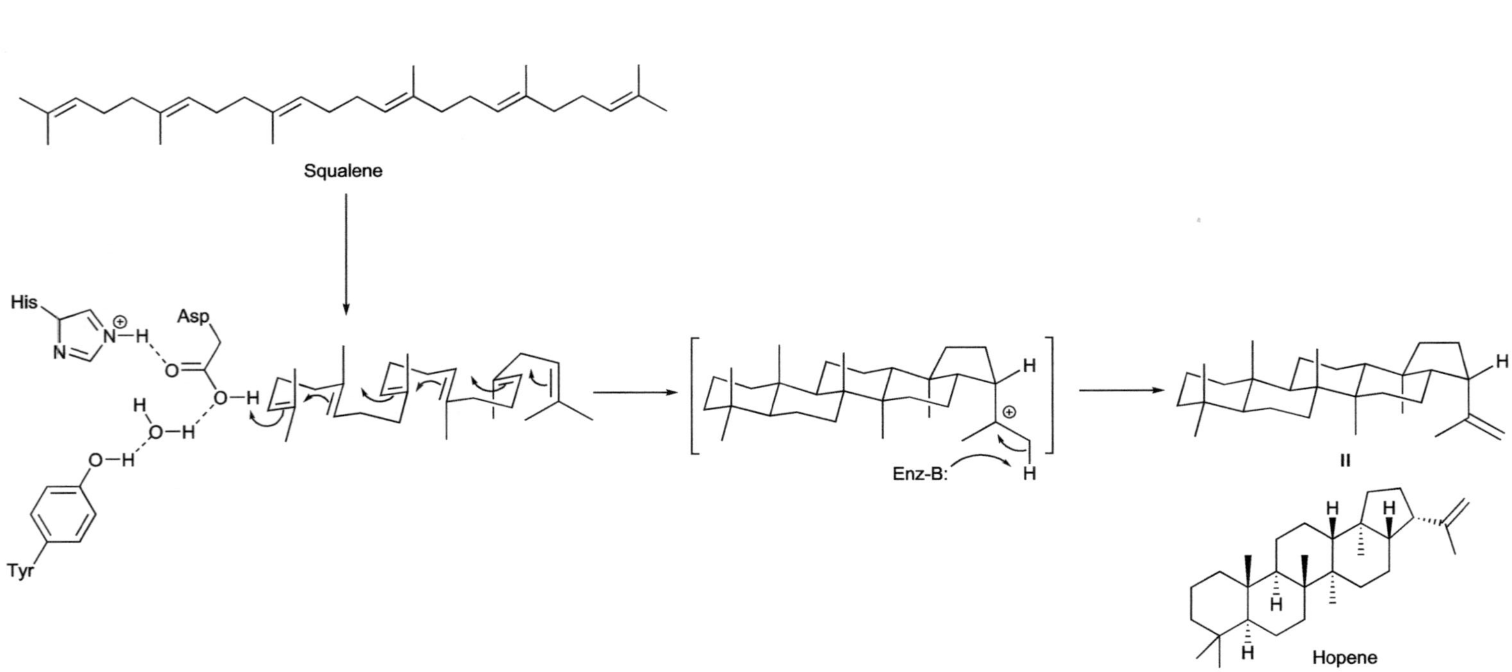

Figure 8.49 General mechanism of cyclisations catalysed by squalene hopene cyclase.

protonation to produce a carbocation, a series of cyclisation reactions then occurs. The enzyme active site shields each of the intermediate carbocations from attack from nucleophiles such as water. The final carbocation is then either deprotonated (as shown) to give hopene or reacts with a water molecule to produce hopanol.

9 Miscellaneous Biocatalysts

9.1 Introduction

This chapter will cover enzymes that do not readily fit into any of the previous eight chapters. Most of the enzymes covered here are known in the enzyme classification system as isomerases. Members of this group of enzymes are given the enzyme classification number 5, followed by a sub-group number depending on which type of bond they isomerise. This class of enzymes are called isomerases because, by definition, they catalyse geometric or structural changes within a molecule. Depending upon the type of isomerism, they are termed racemases, epimerases, *cis–trans* isomerases, mutases or cycloisomerases. We have already seen an example of a mutase enzyme, phenylammonia mutase (PAM), in Chapter 7. The transformations covered in more detail within this chapter are shown in Figure 9.1. Each of the reactions is discussed below in more detail, with a separate section for each one.

9.2 Racemisation and Epimerisation Processes

The origins of life on Earth continue to challenge researchers. One of the unanswered questions is why life evolved to use L-amino acids rather than the D-enantiomer. Unsurprisingly, as a result of this choice, the vast majority of enzymes have evolved to react with one enantiomer of a substrate over the other. This is seen in biological pathways and is the reason for the high stereoselectivities observed throughout the previous chapters. However, D-amino acids have been found in Nature and are of importance to several industries, such as

Biocatalysis in Organic Synthesis: The Retrosynthesis Approach
By Nicholas J. Turner and Luke Humphreys
© Nicholas J. Turner and Luke Humphreys 2018
Published by the Royal Society of Chemistry, www.rsc.org

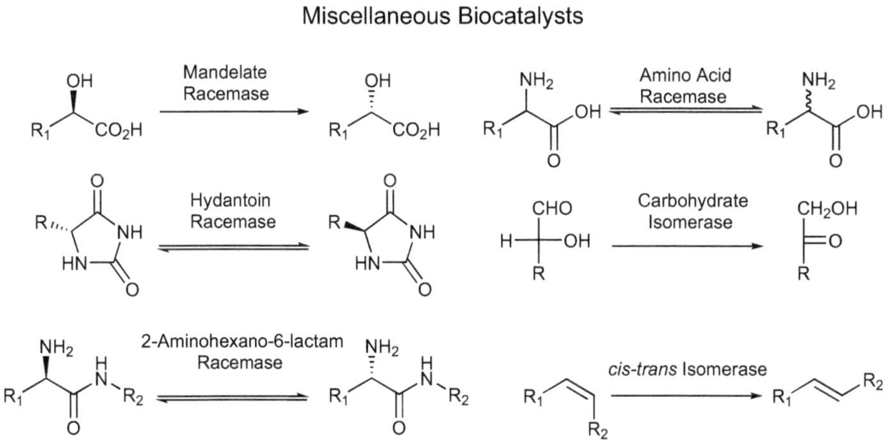

Miscellaneous Biocatalysts

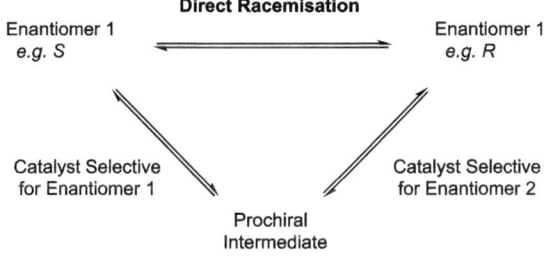

Figure 9.1 Overview of the transformations presented in this chapter.

Direct Racemisation

Enantiomer 1   Enantiomer 1
e.g. S   e.g. R

Catalyst Selective   Catalyst Selective
for Enantiomer 1   for Enantiomer 2

Prochiral
Intermediate

Two Step Process *via* Intermediate

Figure 9.2 The two possible methods of racemisation.

pharmaceuticals and the food industry, precisely because they may not be recognised in the manner that L-amino acids are. One way to produce the opposite stereoisomer of a particular compound is to produce a racemic mixture through a racemisation process. The racemate can then be separated or further reacted to isolate the desired compound. Most racemisation and epimerisation processes employ one of two strategies outlined in Figure 9.2. These are: a direct single-step racemisation or a two-step sequence in which one enantiomer is converted into the other *via* a prochiral intermediate. This second strategy typically involves a change in oxidation state, such as oxidation to a prochiral intermediate followed by selective reduction to produce the opposite stereoisomer of the starting material.

There are several benefits to carrying out a racemisation reaction. By accessing both stereoisomers of a compound we gain access to new chemical space, which we may not have been able to exploit previously. However, this is only useful if we can separate the stereoisomers. The most benefit is observed if we can link the racemisation to a second reaction that is selective for one enantiomer only. Without a racemisation, this would be a kinetic resolution reaction for which the maximum yield is 50%. However, with the unreacted enantiomer now being racemised, we can convert all of the starting material into the desired enantiomer so that the yield can now in theory be 100%. This is now a dynamic kinetic resolution reaction instead. Chemical methods of epimerisation and racemisation are carried out under acidic or basic conditions. This may require strong bases and harsh reaction conditions depending on the acidity of the proton that must be removed for racemisation to occur. As a result, other functional groups or chiral centres in the molecule maybe be adversely affected. Enzymatic methods of racemisation and epimerisation tend to occur under mild conditions, although they work most often on activated substrates containing acidic protons, such as protons adjacent to a carbonyl. The sections below detail a number of enzymes that are capable of racemising a range of substrates either directly or by a two-step sequence.

9.2.1 Racemisation of α-Hydroxy Carboxylic Acids

α-Hydroxy carboxylic acids are important chiral compounds as they are useful building blocks for synthesis. They can be prepared in a number of different ways and it is possible to chemically racemise the α-hydroxy substituent due to the acidity of the proton at that position. It is also possible to directly racemise α-hydroxy carboxylic acids using an enzyme from bacteria called mandelate racemase, as shown in Figure 9.3.

Mandelate racemase accepts a range of substrates, although the substituent attached to the carbon bearing the hydroxyl group must be able to stabilise a carbanion at this carbon. This is required to increase the acidity of the α-proton and help facilitate the racemisation. The range of substrates that undergo racemisation is shown in Figure 9.4. In this case, it is the *R*-enantiomer that is racemised.

Racemases have also been used in the synthesis of pharmaceutical compounds, such as angiotensin converting enzyme (ACE) inhibitors. These compounds are used to treat high blood pressure (hypertension)

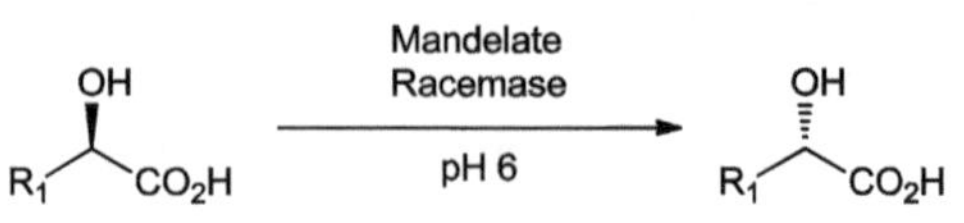

Figure 9.3 General transformation carried out by mandelate racemase.

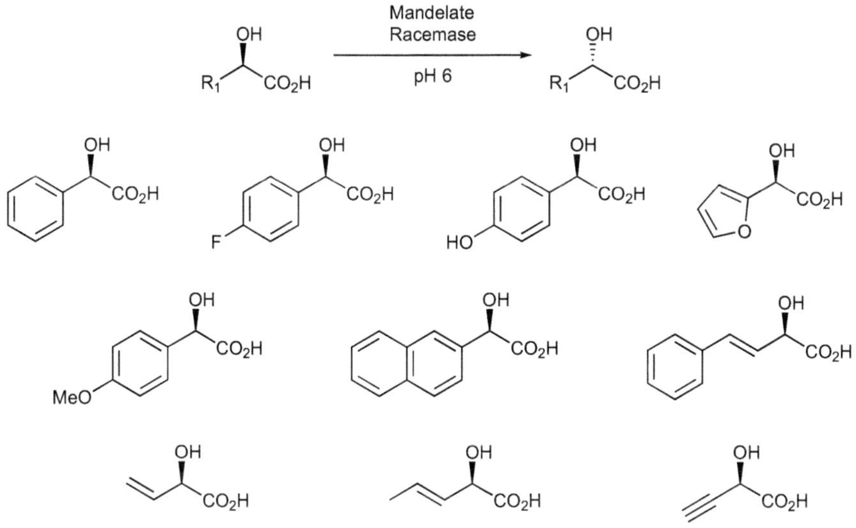

Figure 9.4 The substrate scope of racemisations carried out by mandelate racemase.

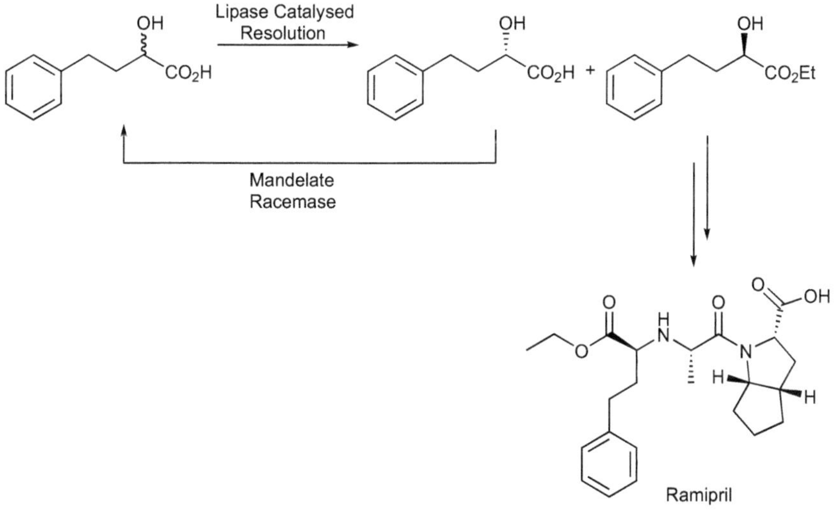

Figure 9.5 The synthesis of ramipril *via* a resolution reaction utilising mandelate racemase to recycle the unreacted substrate.

and heart failure. One class of ACE inhibitors, such as ramipril below, all incorporate the same α-hydroxy carboxylic acid building block. Mandelate racemase has been used as part of a deracemisation process to produce either enantiomer of the desired hydroxy carboxylic acid. As shown in Figure 9.5, a lipase-catalysed resolution is carried out and the unwanted enantiomer is then converted into a racemic mixture using mandelate racemase. After two or three cycles of this process, either enantiomer can be obtained in high yield and enantiomeric excess. This material can then be used to make compounds such as ramipril, depending on which enantiomer is required.

9.2.2 Racemisation of Hydantoins

Hydantoins are a useful motif in organic chemistry. They are found in a number of pharmaceutical drugs, particularly anticonvulsant medicines and also in agrochemicals. They can be formed by a number of different chemical methods, as shown in Figure 9.6. Several of these methods use toxic reagents or harsh reaction conditions, which may not be compatible with other functional groups.

Hydantoins that have a substituent at the carbon atom can be racemised by enzymes known as hydantoin racemases. These enzymes are found in microbes and are able to racemise the carbon atom alpha to the carbonyl under mild conditions, as shown in Figure 9.7.

A small number of substrates can be racemised using these enzymes and this reaction is used on a large scale to produce a number of different L- and D-amino acids. As shown in Figure 9.8, a

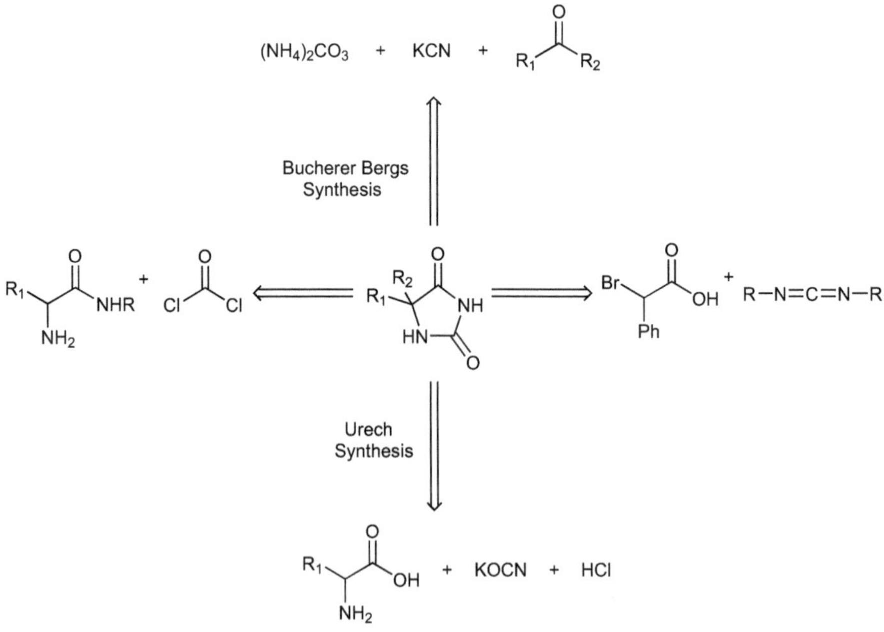

Figure 9.6 Selected chemical methods of hydantoin synthesis.

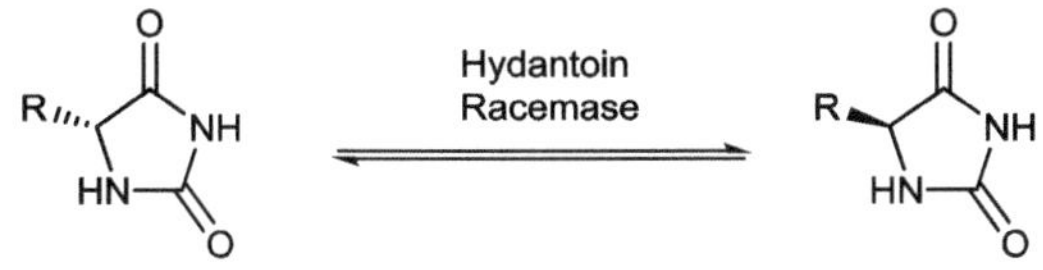

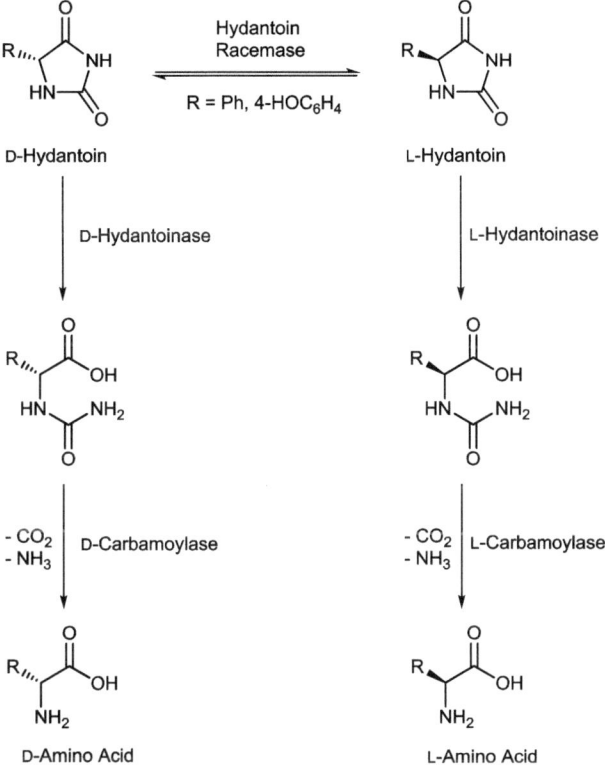

Figure 9.7 General transformation catalysed by hydantoin racemase.

Figure 9.8 The selective synthesis of either D- or L-amino acids utilising hydantoin racemase.

dynamic kinetic resolution reaction can be set up in which either hydantoin enantiomer can be reacted further by adding the appropriate hydantoinase and carbamoylase to the reaction mixture.

9.2.3 Racemisation of Amides

We have already mentioned in previous chapters how important amides are as a functional group in both naturally occurring

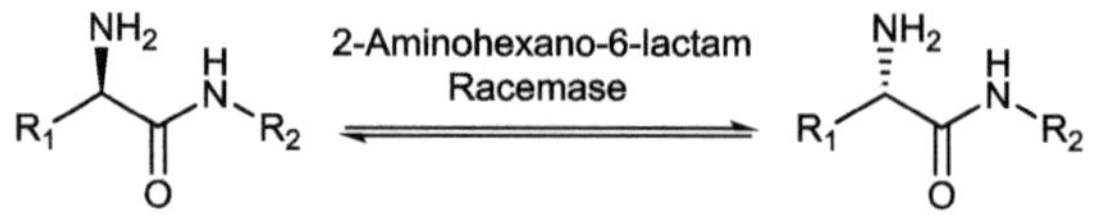

Figure 9.9 General transformation catalysed by 2-aminohexano-6-lactam racemase.

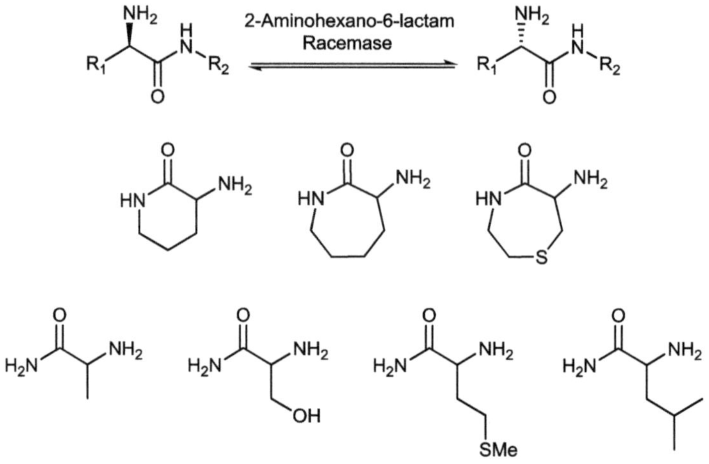

Figure 9.10 The substrate scope of 2-aminohexano-6-lactam racemase.

molecules and those which are man-made. They have been reported to be racemised by the enzyme 2-aminohexano-6-lactam racemase, which was isolated from a species of bacteria. As shown in Figure 9.9, this enzyme is capable of racemising the centre that is alpha to the carbonyl of an amide group.

The enzyme is pyridoxal phosphate-dependent and works on a small number of lactam and primary amide substrates, as shown in Figure 9.10.

Chemists have combined this racemase enzyme with amidase enzymes to synthesise a range of L- and D-amino acids, as shown in Figure 9.11. This dynamic kinetic resolution gives either enantiomer in very high yield and almost optical purity depending on the combination of enzymes used.

9.2.4   Racemisation of α-Amino Acids

We have already seen that amino acids are very important biological molecules and they have featured in a number of the previous chapters.

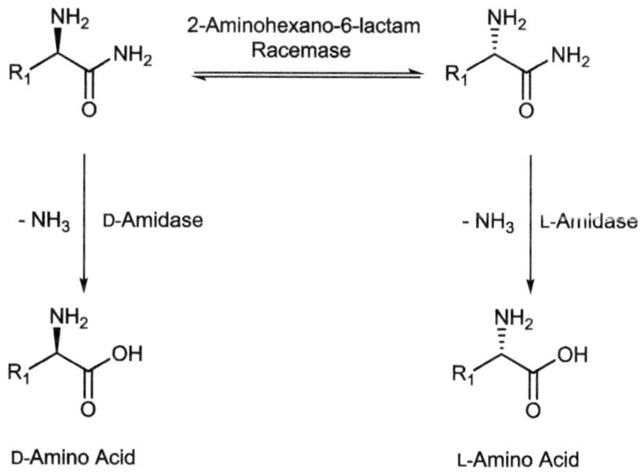

Figure 9.11 The selective synthesis of either D- or L-amino acids utilising 2-aminohexano-6-lactam racemase.

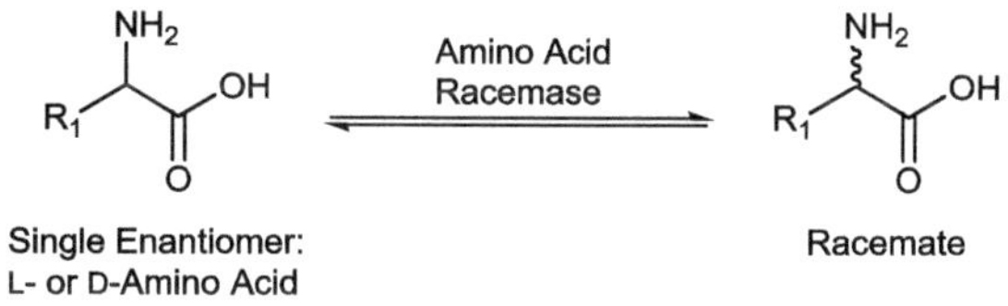

Figure 9.12 General transformation catalysed by amino acid racemases.

L-Amino acids play a vital role as the building blocks for peptides and proteins, whilst D-amino acids also play crucial roles. For example, D-glutamate is used in the biosynthesis of bacterial cell walls. The racemisation of amino acids is carried out by a family of enzymes known as amino acid racemases, as shown in Figure 9.12. These enzymes, which come from bacteria and plants, can be split into those which are pyridoxal phosphate (PLP)-dependent and those which are not. We will look at the PLP-dependent enzymes in more detail below.

Many amino acid racemases such as methionine, lysine and proline racemase accept a very limited substrate range, whilst others such as serine, alanine and ornithine racemase accept a few non-natural substrates. However, several racemases, such as isoleucine and arginine racemase along with two other enzymes isolated from bacteria, have a very broad substrate scope, as shown in Figure 9.13.

Using these enzymes, a wide range of amino acids and amino acid derivatives can be made. We saw in Chapter 7 how an alanine racemase was used to prepare D-alanine, which was then used as

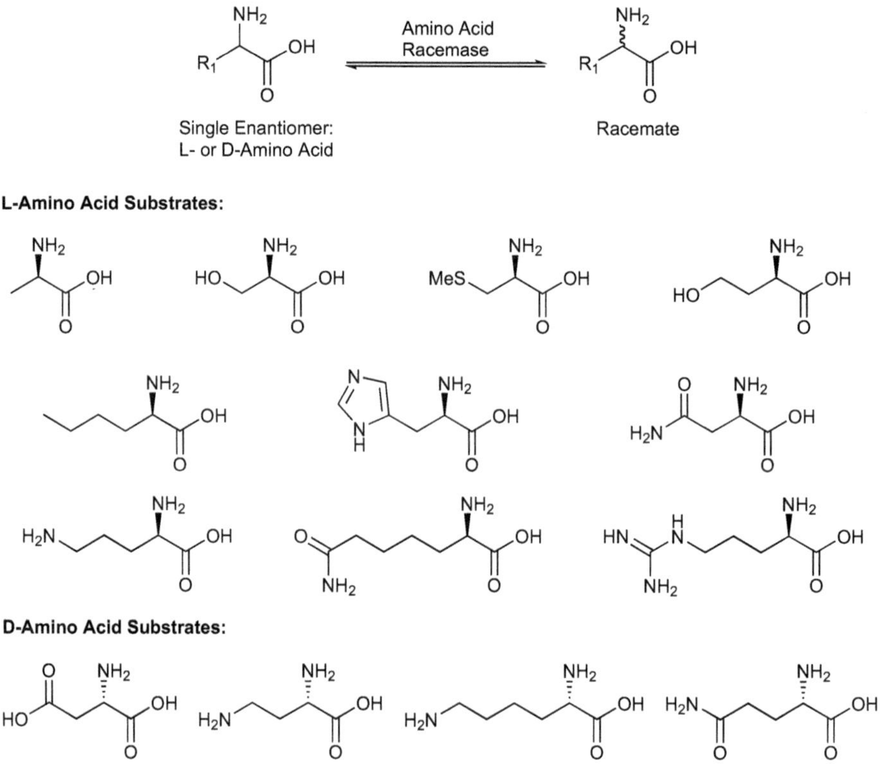

L-Amino Acid Substrates:

D-Amino Acid Substrates:

Figure 9.13   The substrate scope of amino acid racemases.

the amine donor in a transaminase-catalysed amination reaction to prepare *R*-amines. This process is shown again in Figure 9.14.

9.2.5   Epimerisation with Stereocomplementary Ketoreductases and Transaminases

As we mentioned earlier, one of the two possible strategies to epimerise stereocentres is to use more than one enzyme in a sequence involving oxidation and subsequent reduction. As shown in Figure 9.15, we start from a single enantiomer of the substrate and use an enzyme that selectively reacts with this enantiomer to produce a prochiral intermediate. The intermediate is then converted by a second enzyme back to the starting material. However, because the second enzyme has the opposite stereoselectivity to the first enzyme, it produces the enantiomer of the starting material. The whole process is equilibrium-controlled, which means that we produce a racemic mixture once equilibrium is reached.

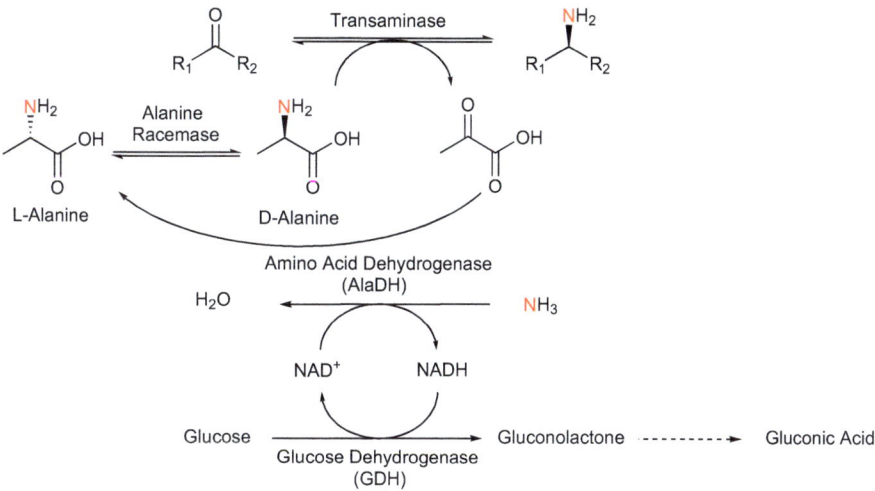

Figure 9.14 The use of alanine racemase to generate D-alanine as an amine donor for transaminase reactions.

Figure 9.15 General scheme of racemisation *via* a prochiral intermediate.

For these processes to work well, the reactivity of the two enzymes should be similar in order to minimise the reaction time. In addition, a minimal amount of co-factor should also be used to prevent a large amount of the prochiral intermediate being formed.

This process has been reported for the racemisation of secondary alcohols by a pair of complementary ketoreductases. As shown in Figure 9.16, the substrate alcohol enantiomer is oxidised by one ketoreductase to give a prochiral ketone intermediate. The ketone is reduced by the second ketoreductase to give the opposite enantiomer to that of the substrate. The amount of ketone is kept to a minimum by using a minimal amount of the nicotinamide co-factor. A range of secondary alcohols can be racemised as shown in Figure 9.16, in which the enantiomeric excess of the product is normally modest to good. In each case, the product enantiomer is shown, although access to either enantiomer is possible through the appropriate pair of enzymes.

A similar approach has been used with transaminase enzymes for the racemisation of amines. This time, stereocomplementary transaminases carry out a reversible amination of the prochiral ketone

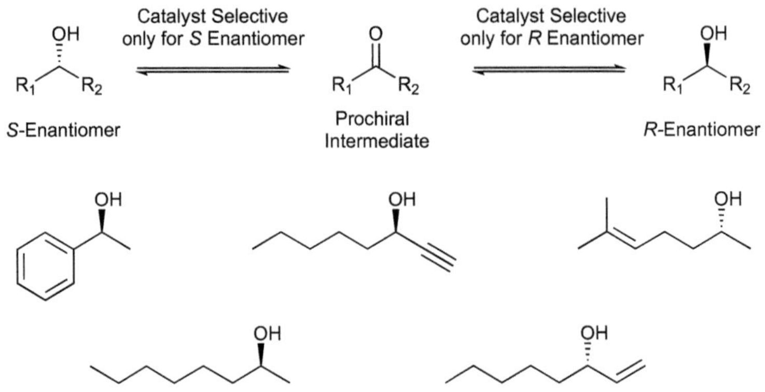

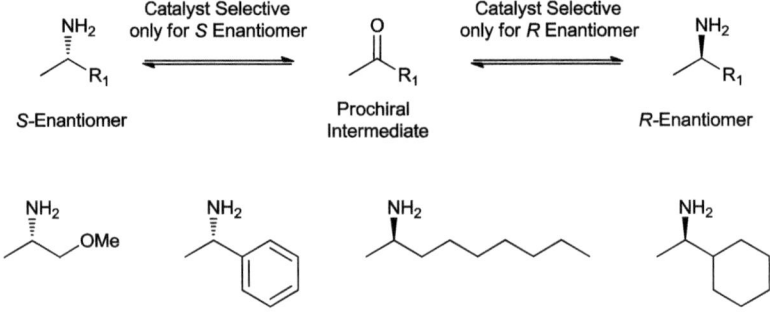

Figure 9.16 The substrate scope of alcohols that may be racemised using ketoreductases.

Figure 9.17 The substrate scope of amines that may be racemised using transaminases.

intermediate with pyridoxal phosphate as the co-factor. The rate of racemisation can be sped up by the addition of an achiral amine donor or acceptor to help shift the equilibrium position of the reaction. This process works for a small number of secondary amines, as shown in Figure 9.17. In each case, the product amine shown is formed in moderate to very good enantiomeric excess. It is also possible to obtain the other enantiomer of the product shown in Figure 9.17 by using a different pair of transaminase enzymes.

9.3 Isomerisation Processes

Isomerisation processes are similar to those we have seen already in this chapter, typically involving a change in the structure of the substrate without any change in the molecular formula. By making a change to the molecule's structure, we can change the properties of the molecule, including its reactivity, sterics or physical properties such as boiling point. The advantages of this type of process are that we may be able to access structures that are difficult to make *via* other means. Isomerisation also allows several molecules to be made by the same biological pathway rather than having multiple pathways each making an individual molecule. In practice this means that from a common precursor, we could use an isomerase to make two different molecules. In the sections below, we will consider several classes of enzymes that are capable of carrying out isomerisations.

9.3.1 Isomerisation of Carbohydrates

Sugars and sugar derivatives are an important family of biological molecules with a range of biological activities. They are found in plants and simple sugars, such as glucose, and are extracted on a million tonne scale for the food industry every year. Simple and more complex sugars are found in most foodstuffs and are a source of energy for animals and humans alike. Sugars are constituent parts of vital biological compounds, such as RNA and DNA, and are also used as the monomer unit for oligosaccharides, which are used in cell structure or for the storage of monomers. Carbohydrate chemistry is a huge area of research and is beyond the scope of this book; however, one enzyme type involved in sugar chemistry is the carbohydrate isomerases. In general, these enzymes catalyse the isomerisation of the carbonyl present in sugars from the aldose form (C-1) to the ketose form (C-2), as shown in Figure 9.18 for the conversion of D-glucose to D-fructose.

Chemical methods also exist for the isomerisation of sugars, but they typically require the use of bases which may not be compatible

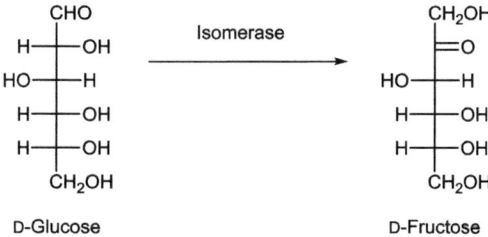

Figure 9.18 The conversion of D-glucose to D-fructose using an isomerase.

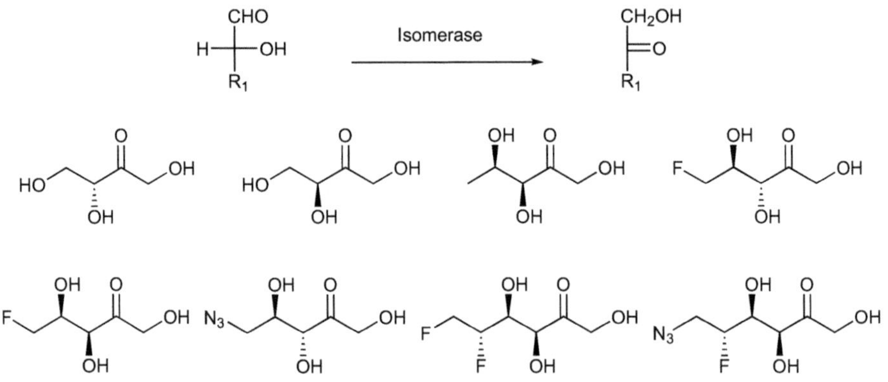

Figure 9.19 Selected examples of substrates that undergo isomerisation catalysed by carbohydrate isomerases.

with other functional groups present in the molecule. A number of different carbohydrate isomerases exist with most acting upon a limited substrate scope. However, some carbohydrate isomerases, such as D-xylose isomerase and L-rhamnose isomerase, display a remarkably broad substrate scope. As shown in Figure 9.19, these enzymes work on a range of natural and non-natural substrates. The yields and selectivities range from good to very good and in each case the product is displayed in the ketose form.

9.3.2 Isomerisation of Alkenes

We have seen in previous chapters that alkenes are important groups as they can be transformed into a range of other functional groups. Alkenes exist in two different isomeric forms, *cis* and *trans*. The alkene geometry can affect the physical properties and sterics of a molecule. This can result in differences, such as a change in the boiling point of a molecule or how it interacts with a biological target. Although many selective methods of alkene synthesis exist, one way of making alkenes with a defined geometry is to start from an alkene of the opposite geometry and isomerise it to the desired one. The isomerisation of alkenes is catalysed by the enzyme linoleate *cis–trans* isomerase. As shown in Figure 9.20, *cis*-alkenes are isomerised to the more stable *trans*-isomers.

The isomerisation of alkenes can be carried out using a number of chemical methods. Traditionally, alkenes have been isomerised using heat or light, which may not be compatible with other sensitive functional groups in the molecule. More modern methods use

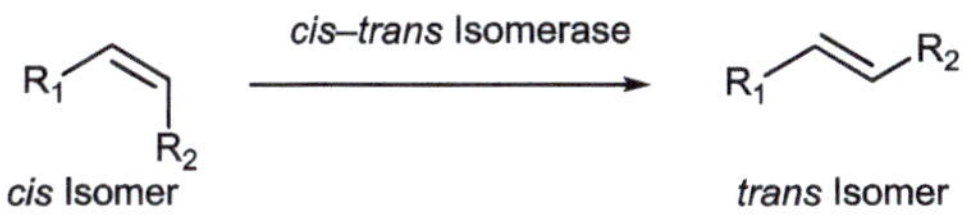

Figure 9.20 General transformation catalysed by *cis–trans* isomerases.

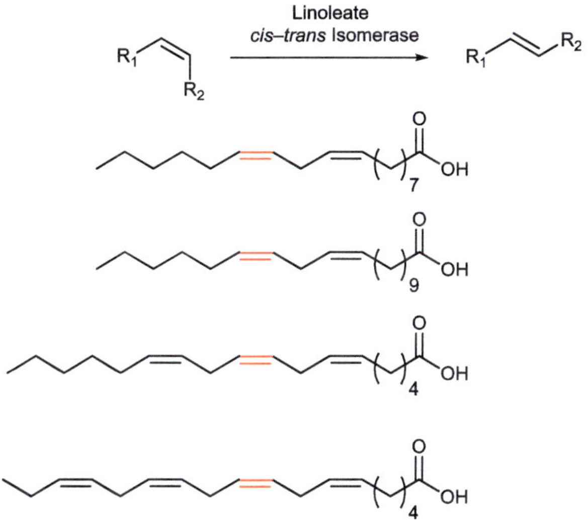

Figure 9.21 The substrate scope of linoleate *cis–trans* isomerase.

transition metal catalysts and ligands that must be disposed of at the end of the reaction, which can be problematic on a large scale. In comparison, enzymatic methods are very regioselective and occur under mild conditions. A small range of polyunsaturated acids are substrates for linoleate *cis–trans* isomerase, as shown in Figure 9.21. Despite the limited substrate scope, these enzymes are very regioselective as only the alkene in red undergoes isomerisation.

10 Biocatalytic Disconnections and Functional Group Interconversions

Introduction

Biocatalytic retrosynthesis is based on identifying potential discon-nections with the knowledge that the forward reaction is synthetically possible. In Chapters 3–9 we have provided a comprehensive de-scription of both the major enzyme classes, and importantly the in-dividual biocatalysts within these classes that are finding applications in organic synthesis. In this chapter, we now present a more formal and structured approach to 'biocatalytic retrosynthesis' in which the approach taken is 'how can I select a biocatalyst to prepare a par-ticular type of functional group or combination of functional groups that occurs in my target molecule?' This chapter presents all of the most important disconnections that are possible for acyclic, cyclic and aromatic systems, in order to gain an understanding of where biocatalysts can be applied in organic synthesis. The information has been organised into 6 basic sections. Section 10.1 covers basic functional groups, *e.g.* if you want to synthesise a secondary alcohol (Section 10.1.1) or tertiary amine (Section 10.1.2) these sections present all of the possible methods to achieve the conversion, ir-respective of which particular enzyme catalyses the transformation. Section 10.2 deals with target molecules containing two or more functional groups (*e.g.* 1,3-hydroxy ketone or 1,2-amino alcohol) and is organised according to the type of synthetic transformation

Biocatalysis in Organic Synthesis: The Retrosynthesis Approach
By Nicholas J. Turner and Luke Humphreys
© Nicholas J. Turner and Luke Humphreys 2018
Published by the Royal Society of Chemistry, www.rsc.org

employed, *i.e.* C–C or C–N bond formation, redox methods or those based on hydrolysis. Section 10.3 describes biocatalytic methods for preparing carbocyclic and heterocyclic ring systems and Section 10.4 provides an overview of how to prepare substituted aromatic and heteroaromatic target molecules, including enzymes that modify both the aromatic ring directly and also substituents on the ring. Finally, Sections 10.5 and 10.6 deal with specific methods for carbohydrates and nucleosides, respectively. For all of the sections, a scheme giving an overview of the various disconnections is provided together with selected examples of where those particular transformations have been applied in target molecule synthesis. To enable the reader to quickly find further examples of a desired transformation, there is a specific reference that gives a link to the relevant chapter in *Science of Synthesis: Biocatalysis in Organic Synthesis*.[115–117]

Chapter 10 should be read in conjunction with the preceding Chapters 3–9 since to some extent it presents very similar material but from a different perspective. Indeed, we have deliberately arranged for some of the examples to occur in both parts of the book, in order to emphasise the application of a particular biocatalyst in both a 'synthetic' and 'retrosynthetic' sense. In order to gain familiarity with the application of biocatalysts it is important to know exactly what types of reactions they can catalyse together with an appreciation of where they might be used. The aim of Chapter 10 therefore is to introduce a greater level of planning into using enzymes by application of retrosynthetic disconnections.

Inevitably during the preparation of this chapter, some decisions had to made regarding either inclusion or exclusion of specific disconnections. In some cases, this decision was made on the basis that although the specific enzyme-catalysed reaction has been reported, it has not yet been applied in a synthetic context. Also excluded from this chapter are reactions catalysed by 'synthetic' or 'artificial' enzymes. This field is rapidly advancing and undoubtedly expanding the tool-box of available new biocatalysts, and would represent an excellent addition to a future update of guidelines for 'biocatalytic retrosynthesis'.

10.1 Functional Group Interconversions

10.1.1 1°, 2° and 3° Alcohols

Primary alcohols can be prepared *via* a number of different approaches, the most important of which involve the reduction of an aldehyde (1.1.1) using a ketoreductase[116n] or the hydrolysis of an ester (1.1.2) with a lipase or esterase[115b] (Scheme 10.1). Other approaches involve hydrolysis of a primary alkyl halide (1.1.3) using a halohydrin dehalogenase (HHD),[116q] oxidation of a terminal methyl group with a P450 monooxygenase in the presence of molecular oxygen[117b] (1.1.4) or the equivalent transformation using an unspecific peroxygenase (UPO) (1.1.5) in the presence of hydrogen peroxide.[117c]

As an example of the use of an alcohol dehydrogenase, the synthesis of (S)-ketoprofenol was achieved *via* reduction of the corresponding racemic aldehyde, with dynamic kinetic resolution, using horse liver alcohol dehydrogenase (HLADH) with co-factor recycling (Figure 10.1).[1]

Enantiomerically pure secondary alcohols are frequently used building blocks in organic synthesis and arguably the application of enzyme-mediated transformations for their preparation represents one of the major achievements of biocatalysis in the past ten years. A number of biocatalytic methods are now available for their synthesis and indeed some of these are used on a large scale in industrial processes (Scheme 10.2). The two most popular approaches are based on ketone reduction (1.1.6), using either a ketoreductase (KRED) or

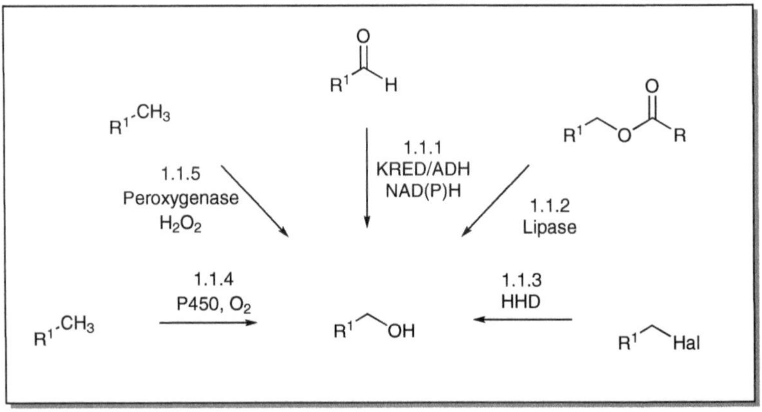

Scheme 10.1   1° alcohols.

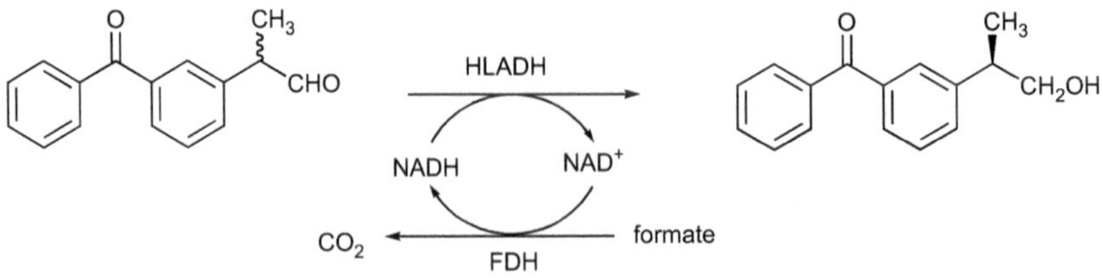

Figure 10.1   Enantioselective reduction of an aldehyde with DKR.

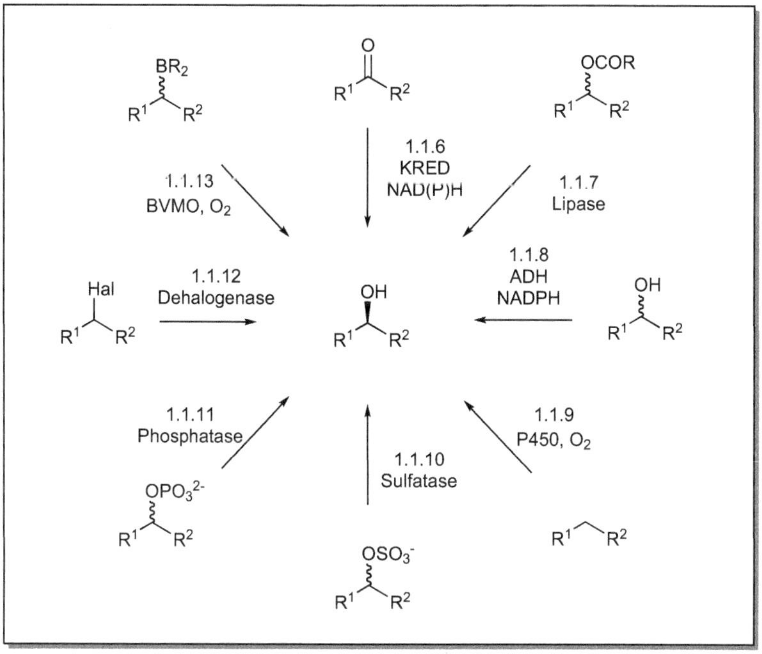

Scheme 10.2 2° alcohols.

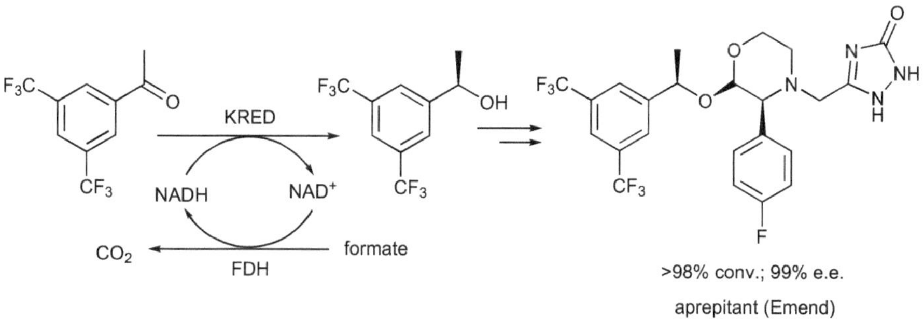

>98% conv.; 99% e.e.

aprepitant (Emend)

Figure 10.2 Asymmetric reduction of a ketone using a KRED.

alcohol dehydrogenase,[116n] or the kinetic resolution of a racemic ester (1.1.7) using a lipase or esterase.[115b]

As an example of the former, scientists at Merck have developed an (immobilised) ketoreductase (KRED) for the asymmetric reduction of a ketone to provide the key chiral alcohol building block for the synthesis of the active pharmaceutical ingredient (API) aprepitant (Emend) in high conversion and high enantiomeric excess (e.e.) (Figure 10.2).[2]

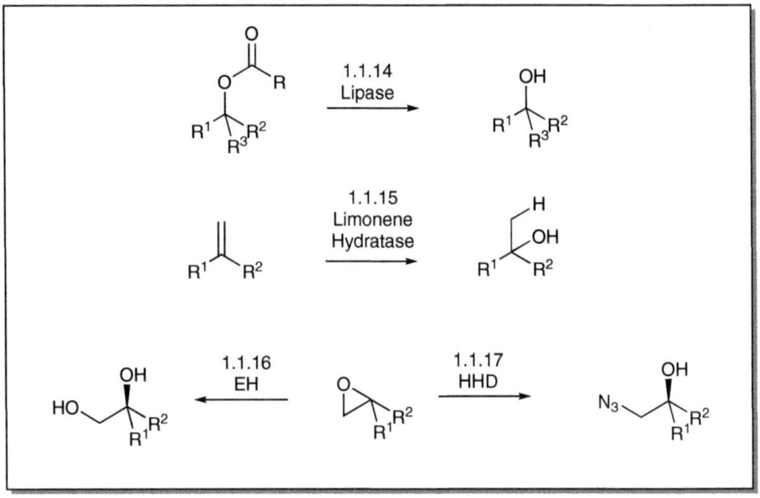

Scheme 10.3 3° alcohols.

It is also possible to deracemise a racemic secondary alcohol (1.1.8) using a whole cell system containing enantiocomplementary ADHs.[3] Other approaches are based on oxidation (1.1.9) using a P450 monooxygenase or unspecific peroxygenase,[117b,117c] hydrolysis of a sulfate (1.1.10) with a sulfatase,[4] phosphatase-mediated hydrolysis of a phosphate (1.1.11)[115d] and hydrolysis of an alkyl halide (1.1.12) with a HHD.[116q] An unusual option is oxidation of a carbon–boron bond using BVMO (1.1.13).[5]

Tertiary alcohols present specific challenges in view of the sterically hindered nature of these compounds (Scheme 10.3). The most general method involves the resolution of the corresponding ester (1.1.14) using a specific lipase or esterase containing a GGGX motif at the active site of the enzyme.[6] Limonene hydratase catalyses regiospecific hydration of a terminal alkene (1.1.15) to generate the corresponding tertiary alcohol.[117b] Alternative approaches include ring opening of an epoxide using an epoxide hydrolase (1.1.16) to give the diol[116r] or HHD in the presence of azide (1.1.17) to produce the azido alcohol.[116q]

10.1.2 1°, 2° and 3° Amines

Chiral amines, including amino acids, are estimated to comprise *ca.* 40% of all chiral intermediates involved in the production of active pharmaceutical ingredients (APIs). Although transitional metal-based methods have traditionally been used for the synthesis of chiral amines, there are often challenges to be overcome in terms of contamination of the amine product with trace amounts of metal. In addition, the use of a costly metal on a large scale may militate against the sustainability of the process in the longer term.

For these reasons, there has been an intense effort during the past 10 years to develop a range of biocatalytic methods for the preparation of chiral amines, particularly on a large scale. For 1° amines, the various options available are shown in Scheme 10.4. Traditionally, enantiomerically pure chiral primary amines have been produced by hydrolysis of a racemic amide using an acylase[115g] (1.2.1) or the reverse process in which a lipase is used to catalyse acylation in the presence of an acyl donor such as methoxyethyl acetate.[115c] This latter approach was ultimately commercialised on a large scale by BASF for the manufacture of specific chiral amines.[7] More recently, the use of ω-transaminases (ω-TAs) for the asymmetric amination of prochiral ketones (1.2.2) using a suitable amine donor (*e.g.* isopropyl amine, alanine) has been developed as a versatile approach[116m] (Figure 10.3). Transaminases have been engineered to be stable and are now widely available for the synthesis of both (*S*)- and (*R*)-amines in high optical purity. For example, the ω-TA from *Paracoccus denitrificans* was

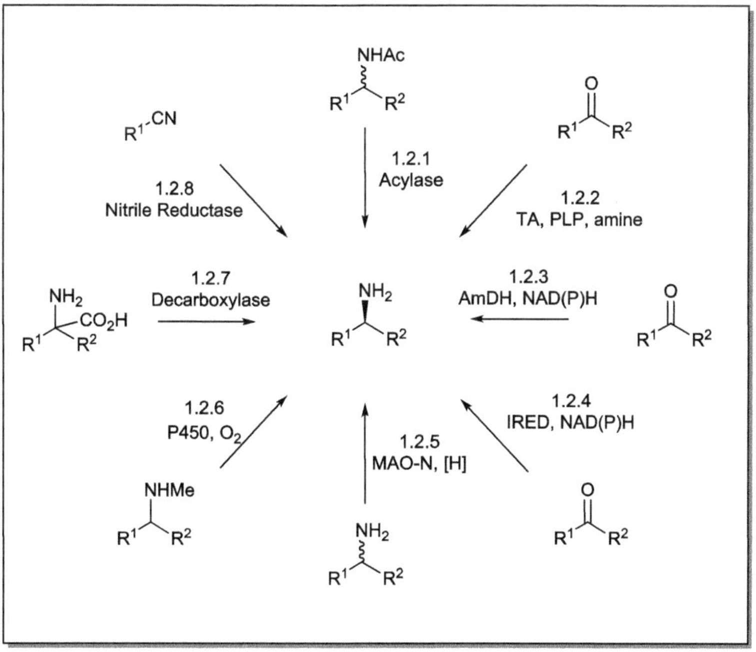

Scheme 10.4 1° amines.

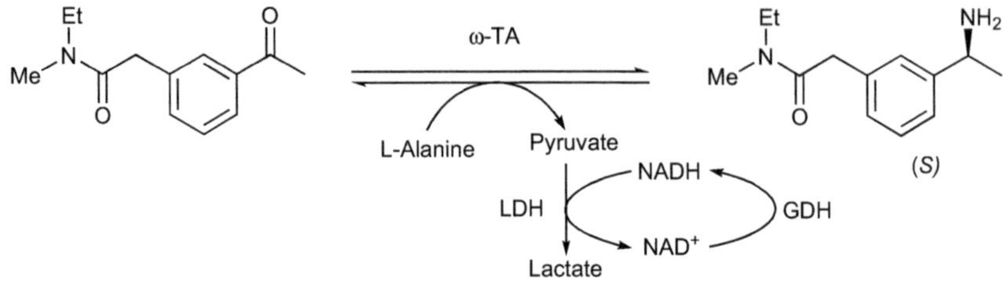

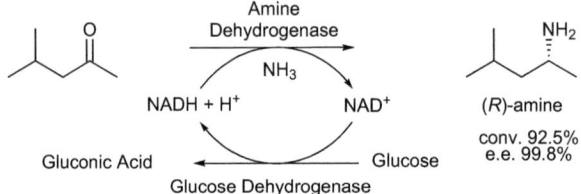

Figure 10.3   ω-TA-mediated amination of a ketone using L-alanine as the amine donor.

Figure 10.4   Reductive amination of 4-methyl-butano-2-one using an amine DH.

used for the asymmetric enzymatic transamination of a precursor ketone for the synthesis of (S)-rivastigmine in four steps (66% overall yield).[8]

Another recently developed method is the reductive amination of a ketone using either an engineered amine dehydrogenase[116k] (1.2.3) or an imine reductase[116l] (1.2.4). Both systems require ammonia as the amine source and also NAD(P)H as a co-factor (Figure 10.4). L-Leucine dehydrogenase from *Bacillus stearothermophilus* has been engineered for activity towards ketones not possessing a carboxylic acid. After eleven rounds of mutagenesis, an amine dehydrogenase with activity towards 4-methylpentanone was identified.[9]

These AmDHs have also been used as part of a redox neutral cascade to convert alcohols to amines in the presence of ammonia and catalytic NADH *via* 'hydrogen borrowing' (Figure 10.5).[10]

A conceptually different approach is based on deracemisation of a racemic amine (1.2.5) using an enantioselective monoamine oxidase in the presence of a non-selective chemical reducing agent (*e.g.* ammonia borane).[117h] Other less frequently used methods include dealkylation of a secondary amine (1.2.6) with a P450 monooxygenase,[11] decarboxylation of an α-amino acid (1.2.7) with a decarboxylase[116d] and reduction of a nitrile (1.2.8) with a nitrile reductase.[12]

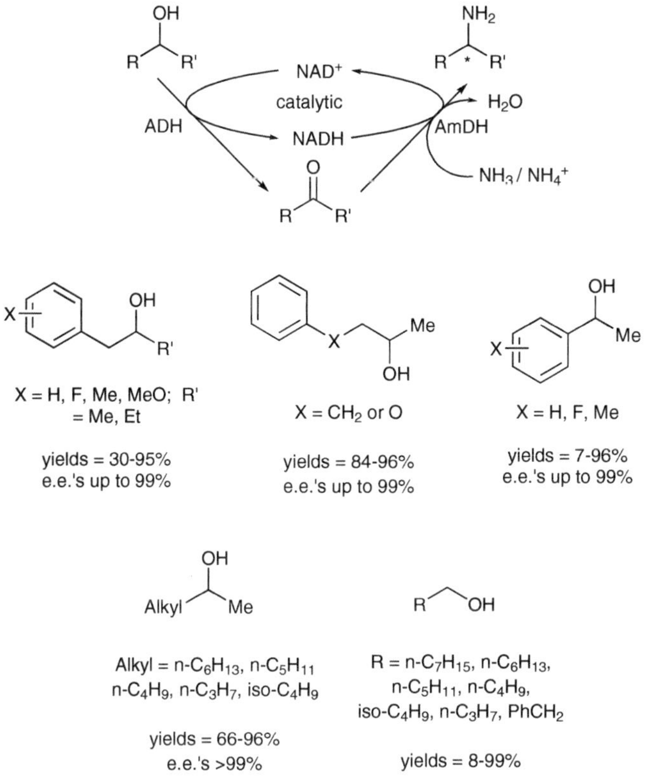

Figure 10.5 Conversion of alcohols to amines *via* biocatalytic 'hydrogen borrowing'.

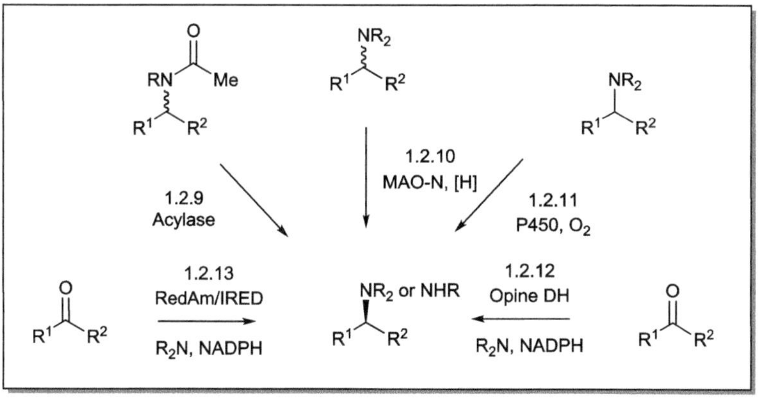

Scheme 10.5 2° and 3° amines.

Comparatively fewer methods are available for the synthesis of 2° and 3° amines (Scheme 10.5). By analogy with the synthesis of a 1° amine, acylase-catalysed hydrolysis of an amide[115g] (1.2.9) yields a secondary amine. Monoamine oxidase (MAO-N) can be used for deracemisation of (cyclic) 2° and 3° amines[117h] (1.2.10). This process has broad substrate scope as a result of extensive engineering of the monoamine oxidase from *Aspergillus niger*.

As an example, MAO-N was combined with nanoscale bioreduced Pd(0) particles (Pd(0)MAO-N) to catalyse the deracemisation of 1-methyltetrahydroisoquinoline leading to the accumulation of the (R)-enantiomer with high enantiomeric excess (96%) after 5 cycles (Figure 10.6).[13]

P450-mediated dealkylation of a 3° amine (1.2.11) has been reported for 2° amine synthesis.[14] Emerging methods that hold real promise are those based on reductive amination using either opine dehydrogenase (1.2.12) or a reductive aminase (RedAm)/imine reductase (IRED)[15] (1.2.13). For example, a range of different secondary amines can be synthesised with high efficiency using the reductive aminase from *Aspergillus oryzae* (AspRedAm).[16] Tertiary amines can be prepared *via* reductive amination using opine dehydrogenases (Figure 10.7).[17]

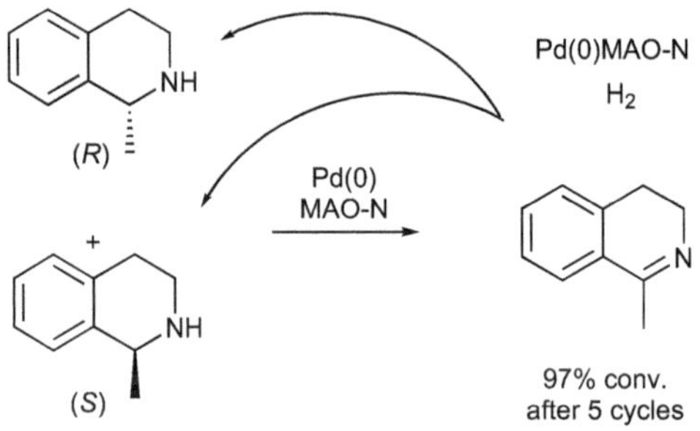

Figure 10.6   Deracemisation of 1-methyltetrahydroisoquinoline using MAO-N with Pd/formate.

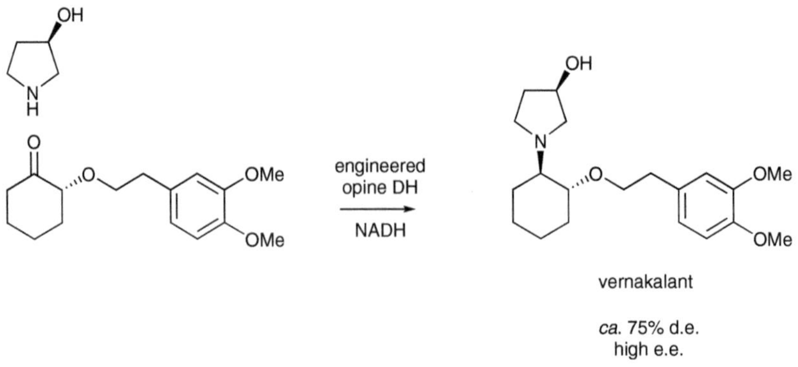

Figure 10.7   Reductive amination using an engineered opine dehydrogenase.

10.1.3 Carboxylic Acids/Esters/Amides

Carboxylic acids, like alcohols and amines, are widely used as synthetic intermediates and are useful precursors of amides, esters and aldehydes. They can be prepared under hydrolytic conditions (Scheme 10.6) from the corresponding ester (1.3.1) using a lipase/esterase,[115b] from the carboxamide (1.3.2) using an amidase,[115g] from a general amide (1.3.3) using a protease,[115g] and from the precursor nitrile (1.3.4) using a nitrilase.[115f]

An example of the lipase-mediated approach is the synthesis of a series of 3-aryl alkanoic acids *via* kinetic resolution of the corresponding ethyl esters. (±)-Ethyl 3-phenylpentanoate was hydrolysed to its corresponding acid with an e.e. of 94% and its ester with lipase B from *Candida antarctica* (Figure 10.8).[18]

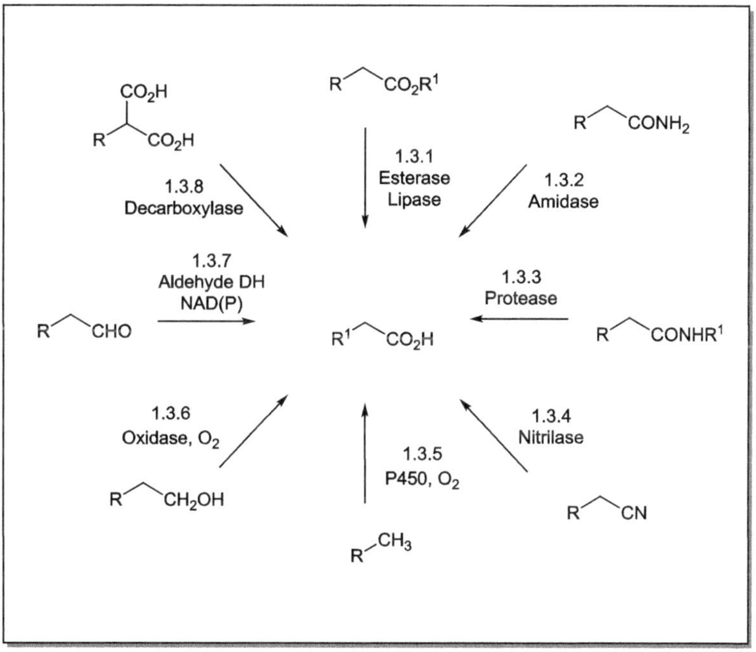

Scheme 10.6 Carboxylic acids.

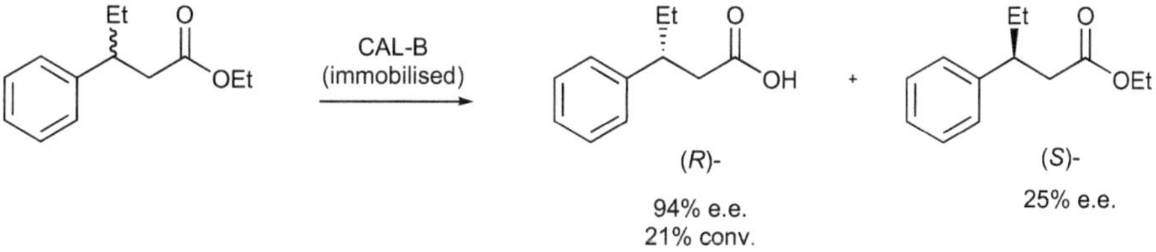

Figure 10.8 Kinetic resolution of an ester using immobilised CAL-B.

Carboxylic acids can also be synthesised by a range of oxidation-mediated processes, *i.e.* by oxidation of an alkane (1.3.5) using a P450 monooxygenase,[117b] from a primary alcohol or aldehyde (1.3.6) using an oxidase[17f] or laccase,[117e] and from an aldehyde (1.3.7) *via* an aldehyde dehydrogenase.[117d] For example, a range of carboxylic acids were produced by aldehyde oxidase (PaoABC)-catalysed oxidation of the corresponding aldehyde. The aldehydes were generated *in situ* by galactose oxidase-catalysed oxidation of the alcohol (Figure 10.9).[19]

A range of aldehyde dehydrogenases (AldDHs) have been reported for aldehyde oxidation. Regeneration of NAD$^+$ was achieved using an H$_2$O-forming NAD(P)H-oxidase (NOX) from *Lactobacillus sanfranciscensis*. Screening of a broad range of substrates demonstrated that both the oxidative dynamic resolution and 'through oxidation' from the alcohol was feasible, with good e.e.'s and yields of the aldehyde (Figure 10.10).[20]

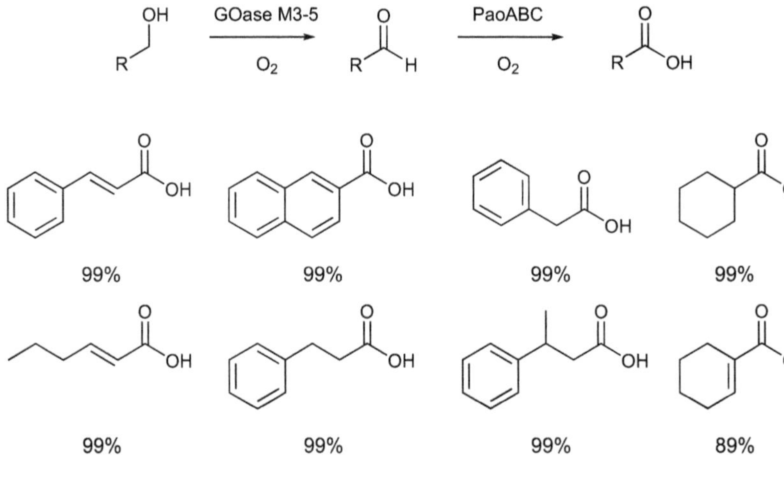

Figure 10.9 Conversion of primary alcohols to carboxylic acids *via* the aldehyde using GOase with PaoABC in the presence of oxygen.

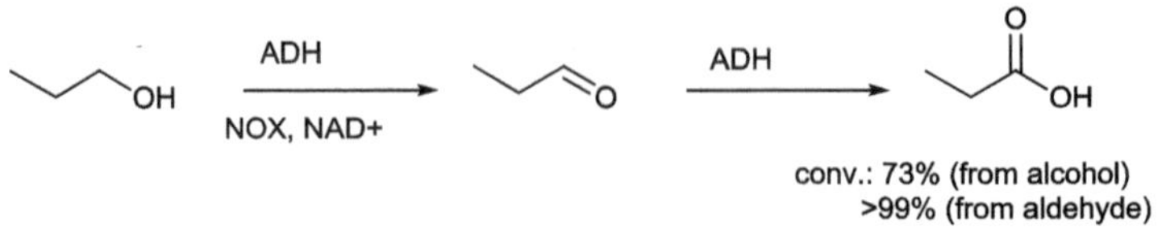

conv.: 73% (from alcohol)
>99% (from aldehyde)

Figure 10.10 Sequential oxidation of a primary alcohol to the aldehyde and acid using an aldehyde dehydrogenase with co-factor recycling.

Finally, for specific carboxylic acids, namely arylmalonic acid derivatives, it is also possible to decarboxylate a malonic acid derivative (1.3.8) using a decarboxylase.[116d]

Carboxylic esters can be derived simply by esterification (or transesterification) of the corresponding acid/ester (Scheme 10.7, 1.3.9) in the presence of an alcohol using a lipase/esterase/protease,[115c] from the carboxamide (1.3.10) using an amidase and from a general amide (1.3.11) using a protease.[115g]

The transesterification approach using the protease subtilisin A from *Bacillus licheniformis* was exploited in the synthesis of valaciclovir from aciclovir and L-valine methyl ester (Figure 10.11).[21]

Alcalase catalyses the conversion of *C*-terminal α-carboxyamides into the corresponding primary alkyl ester, the highest conversion being observed with methanol (98%) (Figure 10.12).[22]

Two other approaches to carboxylic esters are *via* BVMO-mediated oxidation of an acyclic ketone[117g] (1.3.12) and reduction of an unsaturated ester (1.3.13) using an ene reductase[116] (ERED). For

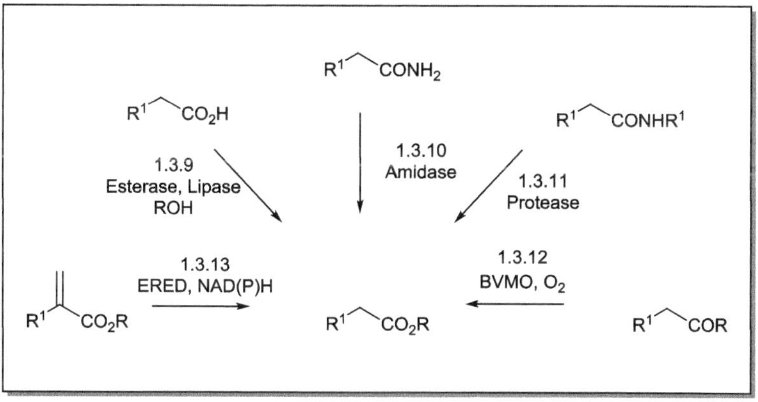

Scheme 10.7 Carboxylic esters.

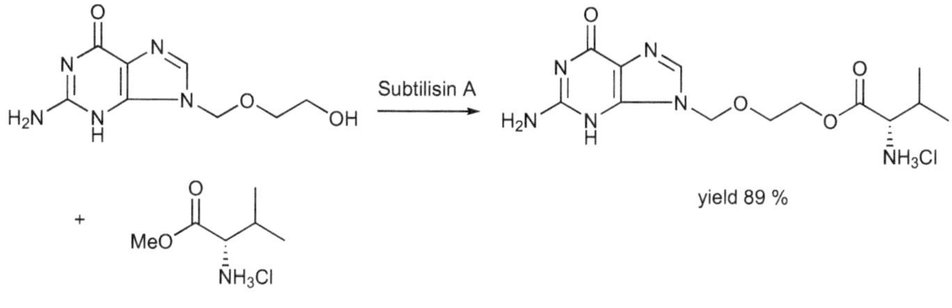

Figure 10.11 Synthesis of valaciclovir *via transesterification* using subtilisin.

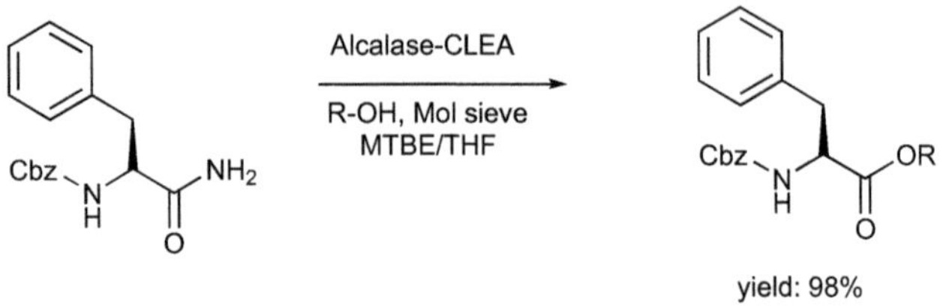

Figure 10.12   Alcalase-catalysed conversion of a carboxamide to an ester.

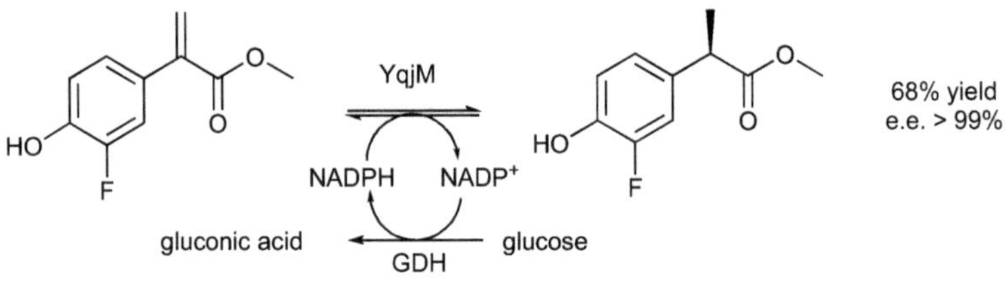

Figure 10.13   Asymmetric reduction of an α,β-unsaturated ester using an ERED.

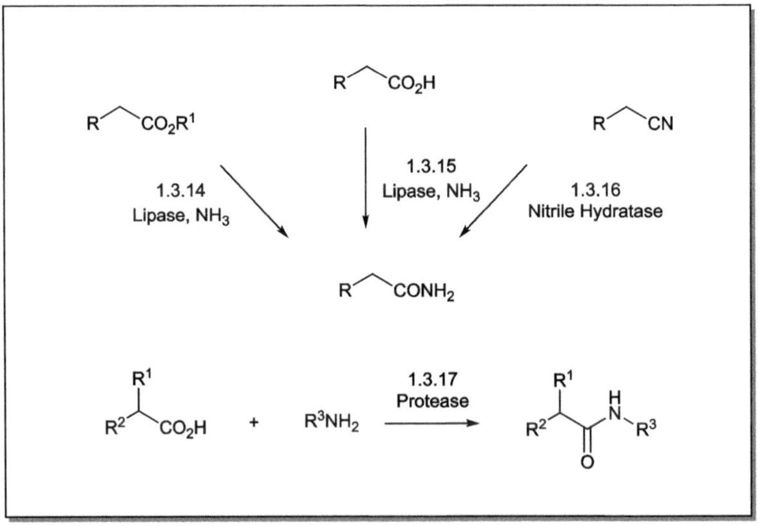

Scheme 10.8   Carboxylic acid amides.

example, an ERED from *Bacillus subtilis* (YqjM) catalyses the formation of (*R*)-flurbiprofen methyl ester, with a glucose dehydrogenase (GDH) co-factor recycling system, in high enantiomeric purity (Figure 10.13).[23]

Carboxylic acid amides are typically obtained from the corresponding ester (Scheme 10.8, 1.3.14) using a lipase,[115c] from the acid

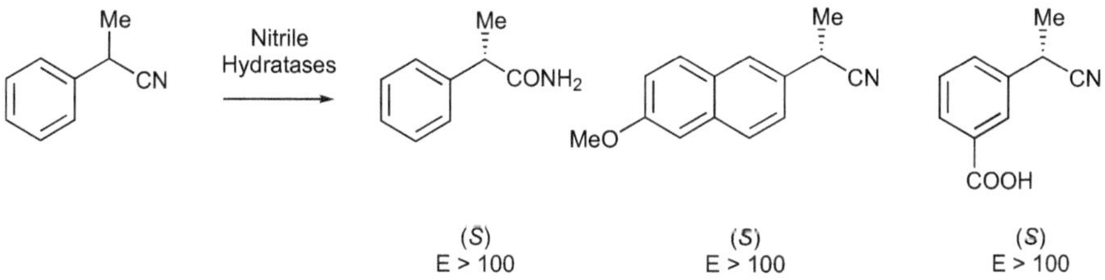

Figure 10.14 Nitrile hydratase-catalysed resolution of a racemic nitrile.

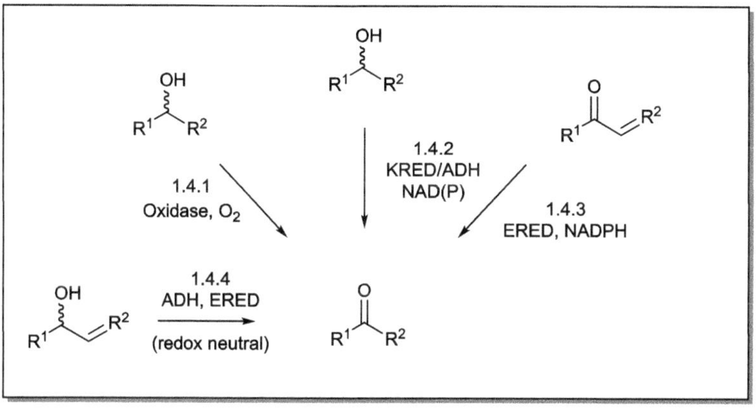

Scheme 10.9 Ketones.

(1.3.15) using either a lipase or protease[115h] and from a nitrile (1.3.16) using a nitrile hydratase[115e] Four cobalt-containing nitrile hydratases were screened for enantioselectivity against a panel of chiral nitriles, resulting in the identification of several enzymes that were (S)-selective and with E values >100 (Figure 10.14).[24]

10.1.4 Ketones/Aldehydes/Imines/Iminiums

Ketones can be prepared *via* direct oxidation of an alcohol using either an oxidase (Scheme 10.9, 1.4.1) in the presence of molecular oxygen[117f] or a KRED/ADH (1.4.2) in the presence of NAD(P) with an appropriate co-factor recycling system.[117d]

A variant of galactose oxidase (M3-5) was shown to have broad specificity and high enantioselectivity towards a range of benzylic secondary alcohols that underwent oxidation to the corresponding ketone with concomitant kinetic resolution (Figure 10.15).[25]

Ketones can also be accessed *via* reduction of an enone using an ERED[116g] (1.4.3) or *via* an interesting redox neutral isomerisation of an allylic alcohol (1.4.4). Thus, the NAD$^+$-dependent alcohol

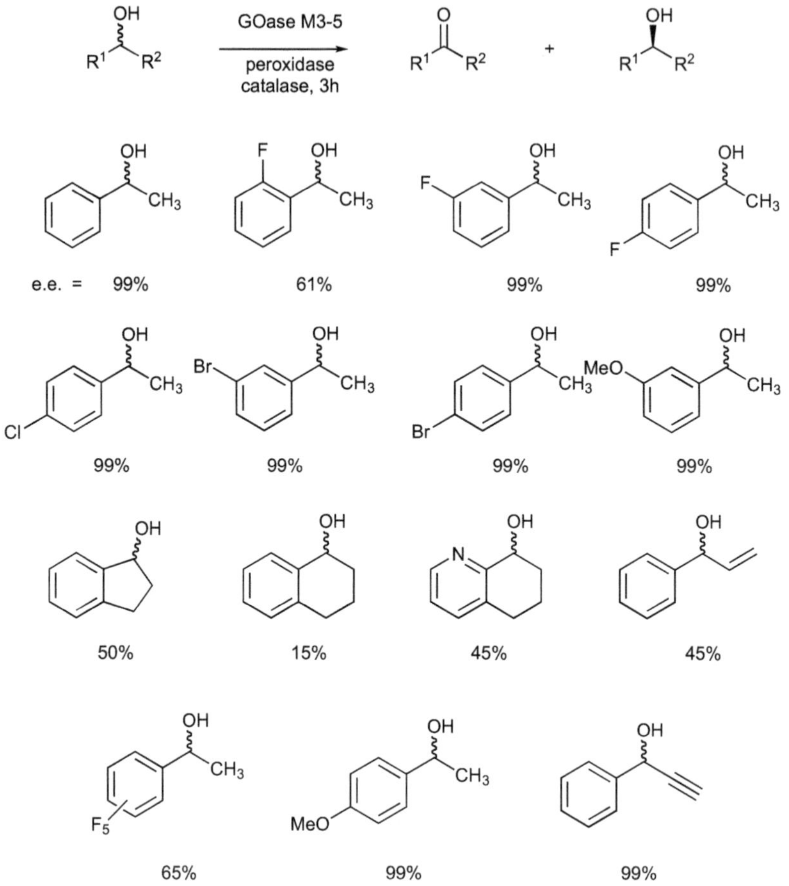

Figure 10.15   Galactose oxidase-catalysed oxidation of secondary alcohols.

dehydrogenase from *Thermus* sp. ATN1 (TADH) catalyses the oxidation of cyclohexenol to cyclohexanone, followed by a reduction to cyclohexanone catalysed by the NADH-dependent enoate reductase from *Thermus scotoductus* SA-01 (TsER) (Figure 10.16).[26]

Aldehydes can be similarly accessed *via* oxidation of a primary alcohol (Scheme 10.10, 1.4.5) using an oxidase[117f] or using a KRED/ADH[117d] (1.4.6). A general approach is reduction of the corresponding carboxylic acid using CAR in the presence of NADPH and ATP[116o] (1.4.7). Pyruvate decarboxylase is a TDP-dependent enzyme that catalyses decarboxylation of an α-keto acid (1.4.8) to the corresponding aldehyde.[116d]

Finally, imines or iminiums can be prepared using two complementary approaches, namely oxidation of an amine (Scheme 10.11,

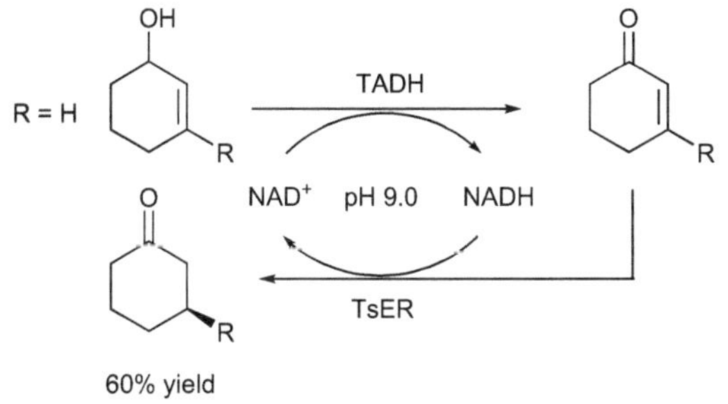

Figure 10.16 Redox neutral conversion of an allylic alcohol to a β-substituted ketone.

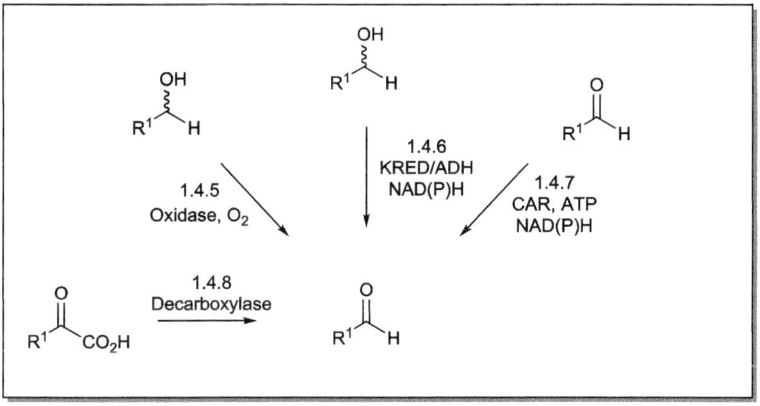

Scheme 10.10 Aldehydes.

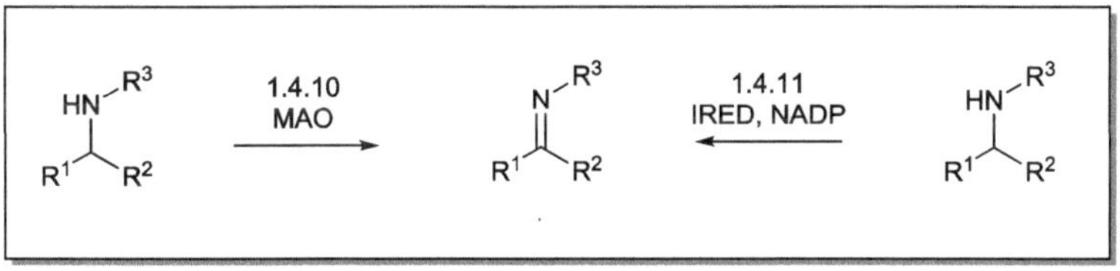

Scheme 10.11 Imines/iminiums.

1.4.10) using an amine oxidase[117h] or the equivalent transformation (1.4.11) using the reverse activity of an imine reductase (IRED).[116l]

Monoamine oxidase was used for the oxidative desymmetrisation of 6,6-dimethyl-3-azabicyclo[3.1.0]-hexane to yield the corresponding amino sulfonate bicyclic [3.1.0] proline moiety in high yield and e.e. (Figure 10.17). This compound is a key intermediate in the synthesis of boceprevir, a drug treatment for hepatitis C.[27]

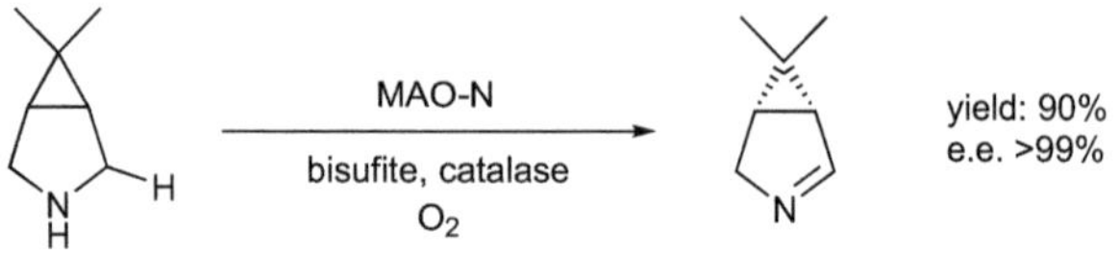

Figure 10.17 Oxidative desymmetrisation of a prochiral amine using MAO-N to give a chiral imine.

10.1.5 Other Functional Groups

Section 10.1.5 deals with the synthesis of all other functional groups as shown in Scheme 10.12. Epoxides can be prepared starting from either an alkene (1.5.1) with a monooxygenase[116p] or from the corresponding halohydrin (1.5.2) using an HHD.[116q]

Oxidation of a sulfide to a sulfoxide (1.5.3) can be carried out using a P450 monooxygenase or BVMO.[117] The flavin-containing monooxygenase (FMO) from *Methylophaga* sp. was used for enantioselective sulfoxidation of aryl sulfides. Reaction of the *para*-Cl sulfide proceeded in high conversion and e.e., whereas *meta*-substitution on the aromatic ring together with electron-withdrawing groups had a negative effect on the selectivity (Figure 10.18).[28]

Haloalkanes can be prepared directly from alcohols (1.5.4) using HHD.[116q] Fluorinase is a somewhat specific enzyme for preparing a small number of fluorine-containing compounds using F^- (1.5.5).[117j]

Oxidation of an *O*-alkylhydroxylamine with MAO-N yields an oxime in a reaction that is both enantioselective with respect to the substrate and specific for one geometric isomer of the product[117h] (1.5.6). Oxidation of an amine with a flavin monooxygenase (FMO) yields an *N*-oxide (1.5.7). For example, human flavin monooxygenases (hFMOs) catalyse the oxidation of moclobemide to moclobemide-*N*-oxide, which can then be isolated in 55% yield from a 1 L scale reaction (Figure 10.19).[29]

Hydroxamic acids can be obtained from an aldehyde and nitroso compound using a thiamine diphosphate (TDP)-dependent lyase (1.5.8). Benzaldehyde lyase (BAL) from *Pseudomonas fluorescens* catalyses the formation of *N*-arylhydroxamic acids from aldehydes and nitrosobenzene in yields ranging from 65% to 94% (Figure 10.20). Kinetic resolution of benzoin derivatives produces (*S*)-benzoins (85–95% e.e.) and arylhydroxamic acids (90–97% conv.) under the same conditions.[30]

Oxidation of an amine with a P450 monooxygenase gives rise to a nitro-containing compound[117b] (1.5.9).

An alcohol can be phosphorylated (1.5.10) using a kinase/ATP or engineered phosphatase[115d] and sulfated (1.5.11), particularly in the context of carbohydrate chemistry, using a sulfotransferase and the co-factor 3-phosphoadenosine 5-phosphosulfate (PAPS).[31] The acid

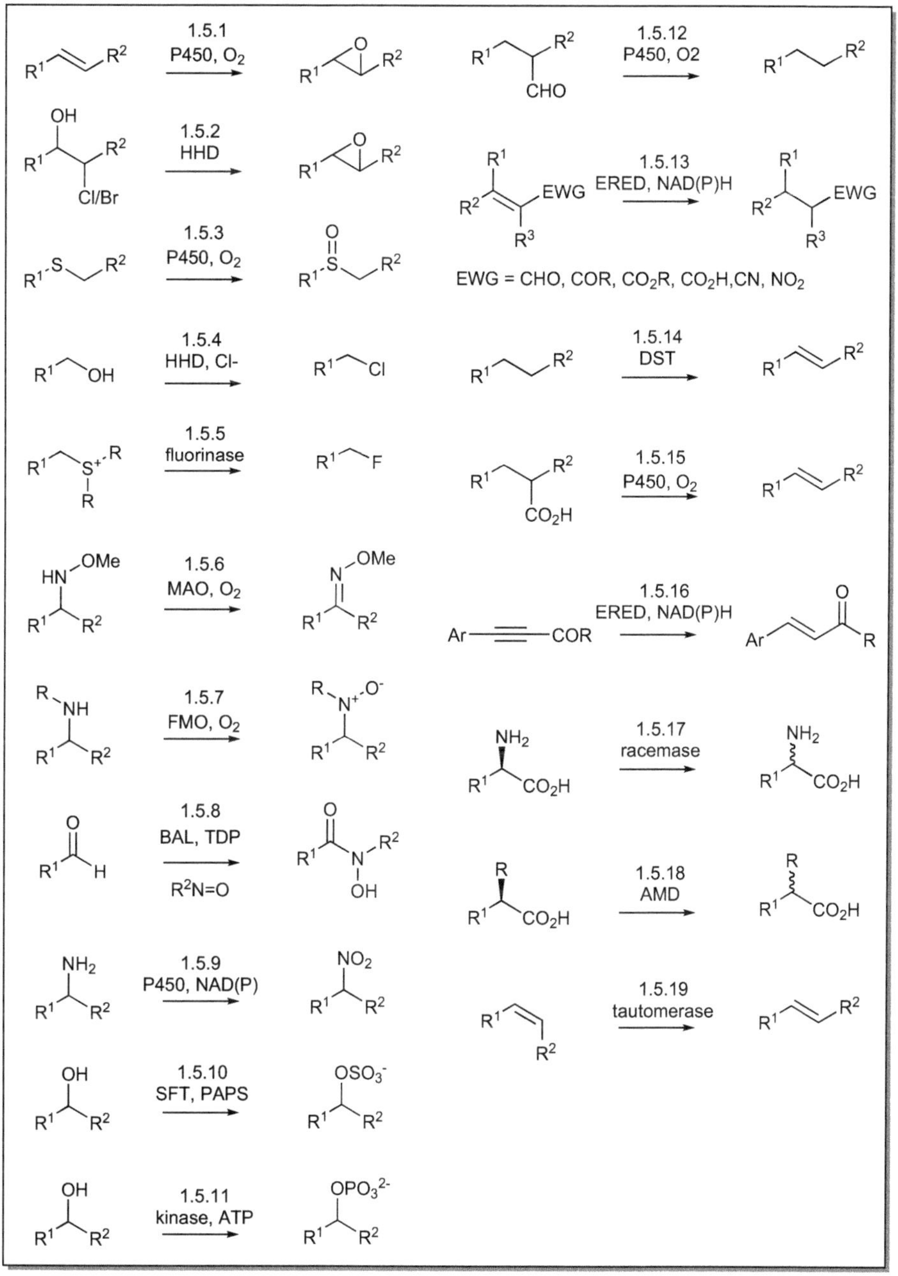

Scheme 10.12 Other functional groups.

phosphatase from *Shigella flexneri* (PhoN-Sf) catalyses the transfer of a phosphate group from pyrophosphate to alcohols. This reaction has been used to produce D-glucose-6-phosphate as well as N-acetyl-D-glucosamine-6-phosphate, allyl phosphate, glycerol-L-phosphate and inosine-5-monophosphate in high yields (Figure 10.21).[32]

Figure 10.18 Asymmetric oxidation of an arylsulfide to give a chiral sulfoxide.

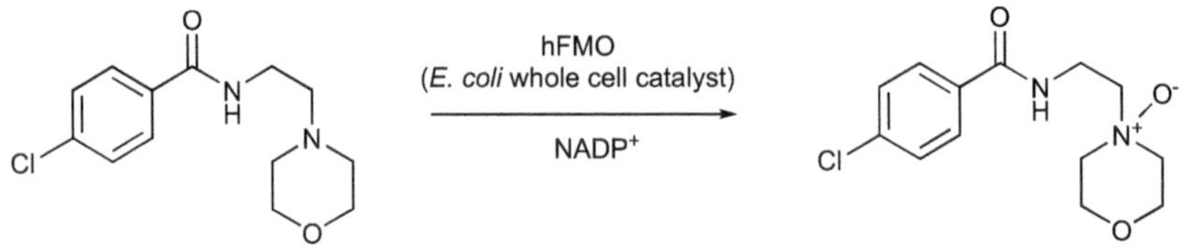

Figure 10.19 Oxidation of an amine to an amine oxide using a flavin monooxygenase (FMO).

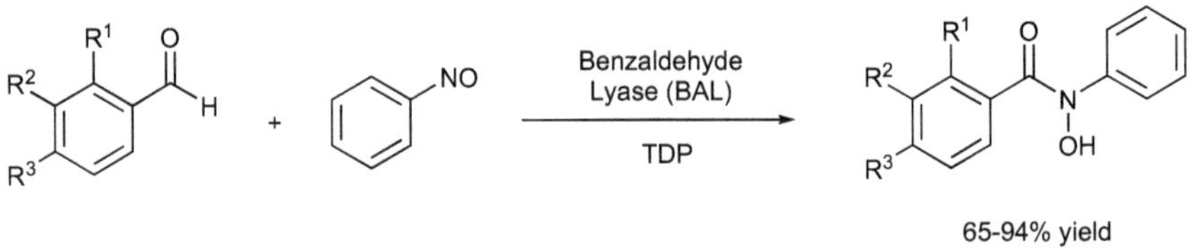

65-94% yield

Figure 10.20 Benzaldehyde lyase (BAL)-mediated condensation of an aldehyde and nitro-compound.

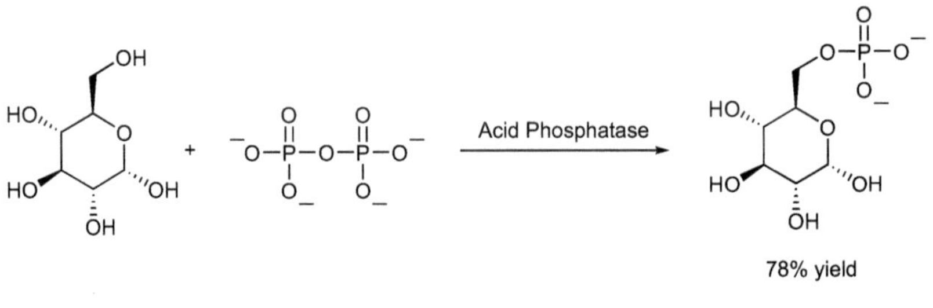

78% yield

Figure 10.21 Phosphorylation of a sugar using an acid phosphatase and pyrophosphate as the phosphate donor.

Decarboxylation of an aldehyde gives rise to an alkane (1.5.12). For example, cyanobacterial aldehyde decarbonylase catalyses the conversion of fatty aldehydes (RCHO) to alka(e)nes (R–H) and formate as shown in the example in Figure 10.22 with 1-octadecanal.[33]

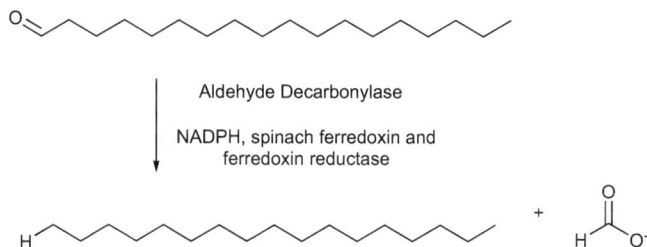

Figure 10.22 Oxidative decarbonylation of an aldehyde to an alkane using an aldehyde decarbonylase.

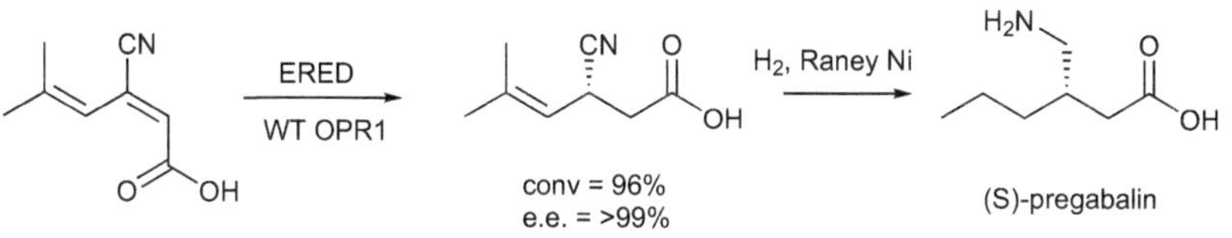

Figure 10.23 ERED-mediated reduction of an alkene as a route to (*S*)-pregabalin.

Ene reductases (EREDs) carry out the reduction of a wide range of activated alkenes to yield substituted alkanes[116g] (1.5.13) as shown for the synthesis of pregabalin *via* reduction of the precursor cyano alkene (Figure 10.23).[34]

Alkenes can be synthesised by either desaturation of an alkane (1.5.14) using a desaturase[35] or oxidative decarboxylation (1.5.15) using a decarboxylase.[116d] Reduction of an activated alkyne with an ERED yields an *E*-alkene[116g] (1.5.16).

Racemase enzymes are able to convert enantiomers of amino acids[115j] (1.5.17). In addition, engineered AMD enzymes have been shown to racemise α-substituted carboxylic acids (1.5.18). Several active site variants of the arylmalonate decarboxylase (AMDase) from *Bordetella bronchiseptica* have been shown to possess racemisation activity. The G74C/V43A variant mediated the deuteration of a number of aryl, pyridyl and vinylcarboxylic acids (Figure 10.24).[36]

Tautomerases catalyse the interconversion of *E*- and *Z*-alkenes[115j] (1.5.19), *e.g.* the conversion of *Z*-nitro-styrene to the corresponding *E*-isomer (Figure 10.25).[37]

10.2 Synthesis of Target Molecules Containing Two or More Functional Groups

In this section, we examine the biocatalytic disconnections that lead to the formation of target molecules that contain two or more functional

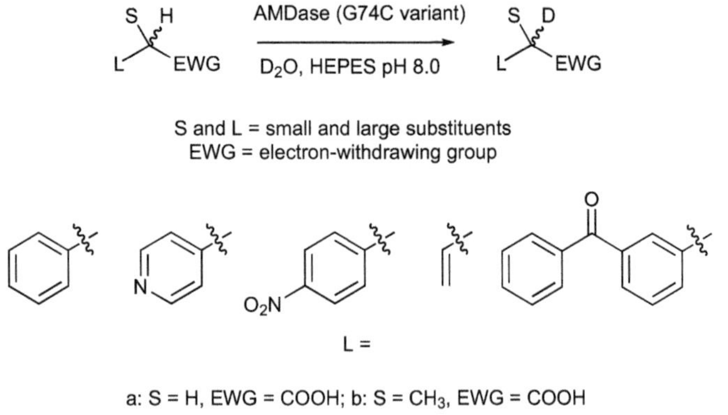

S H
L EWG
AMDase (G74C variant)
D_2O, HEPES pH 8.0
S D
L EWG
S and L = small and large substituents
EWG = electron-withdrawing group
N
O_2N
O
L =
a: S = H, EWG = COOH; b: S = CH_3, EWG = COOH

Figure 10.24 Racemisation of substituted carboxylic acids using an engineered arylmalonate decarboxylase (AMD).

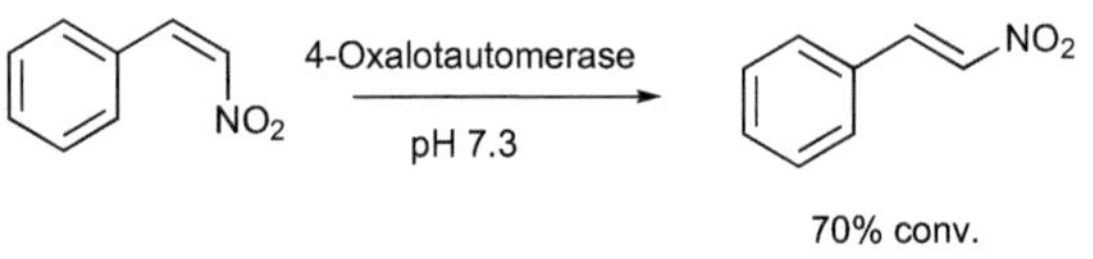

NO_2
4-Oxalotautomerase
pH 7.3
NO_2
70% conv.

Figure 10.25 Interconversion of (*Z*)- and (*E*)-alkenes using 4-oxalotautomerase.

groups. In some cases, this analysis leads to the identification of enzymes that have already been discussed in Section 10.1 and which display broad substrate tolerance, which allows for the introduction of other functional groups into the substrates. However, in other cases, it will be seen that there are specific classes of enzymes that catalyse C–C, C–O and C–N bond forming reactions, which result in the formation of products with multiple functional groups, and it is these enzymes which offer increasing opportunities for disconnecting target molecules in ways which are either unique to biocatalysis or for which the application of a biocatalyst is the optimal way to synthesise the target molecule.

10.2.1 Disconnections Involving C–C Bond Formation (Aldolases & Related Enzymes)

Aldolases are a broad class of enzymes able to catalyse asymmetric C–C bond forming reactions between aldehydes and a wide range of different nucleophilic donors.[116b] A number of members of this family generate 1,2-diols and they can be divided into those that use dihydroxyacetone phosphate (DHAP) as the nucleophile (Scheme 10.13, 2.1.1) and those that use non-phosphorylated ketones such as

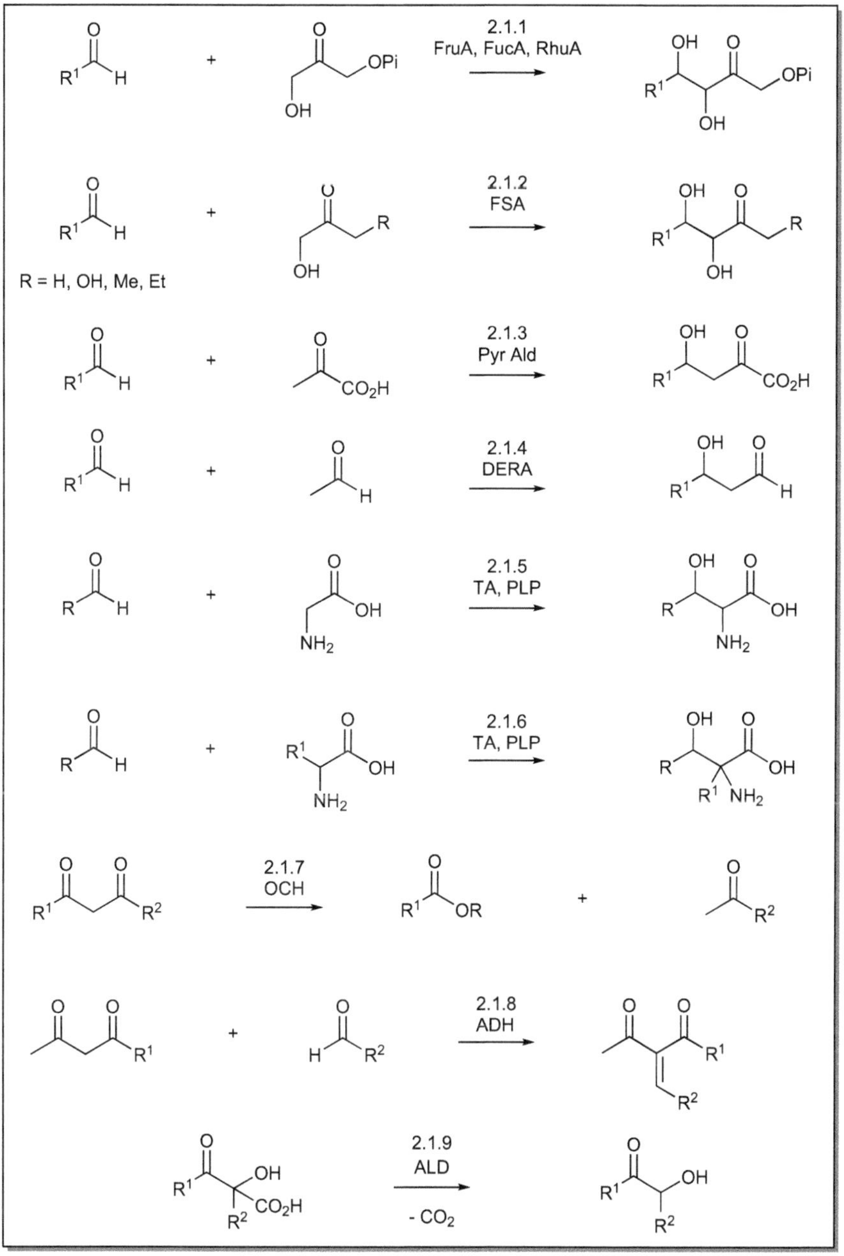

Scheme 10.13   C–C bond formation using aldolases and related enzymes.

dihydroxyacetone (2.1.2). The DHAP-dependent aldolases generate complementary stereochemistry of the 1,2-diol depending upon their origin. Amongst the latter class, the enzyme fructose-6-phosphate aldolase (FSA) displays broad donor-substrate specificity and is able to utilise hydroxyacetone, hydroxybutanone and glycolaldehyde. For example, FSA mutant A129S/A165G catalyses the aldol reaction between N-Cbz-glycinal and dihydroxyacetone (Figure 10.26). The aldol product can be subsequently converted to a valuable intermediate for the synthesis of a wide range of 2-aminomethyl DAB derivatives.[38]

Pyruvate-dependent aldolases (2.1.3) use pyruvic acid as the nucleophile and accept a wide range of aldehydes. These aldolases are involved in the biosynthesis of neuraminic acid[39] and have been used in the synthesis of the neuraminidase inhibitor zanamivir (Figure 10.27).

2-Keto-3-deoxy-6-phosphogluconate (KDPG) aldolases from *Thermotoga maritima* and *Escherichia coli* were evolved for retro-aldol cleavage of 2-keto-4-hydroxyoctanoate (Figure 10.28). Mutations were

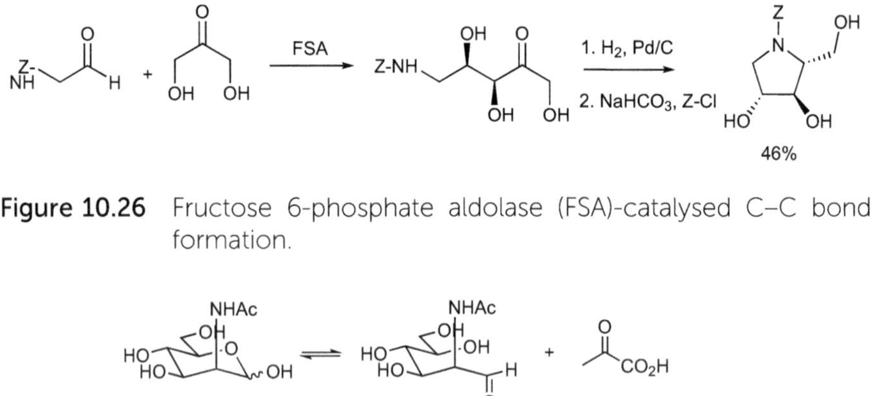

Figure 10.26 Fructose 6-phosphate aldolase (FSA)-catalysed C–C bond formation.

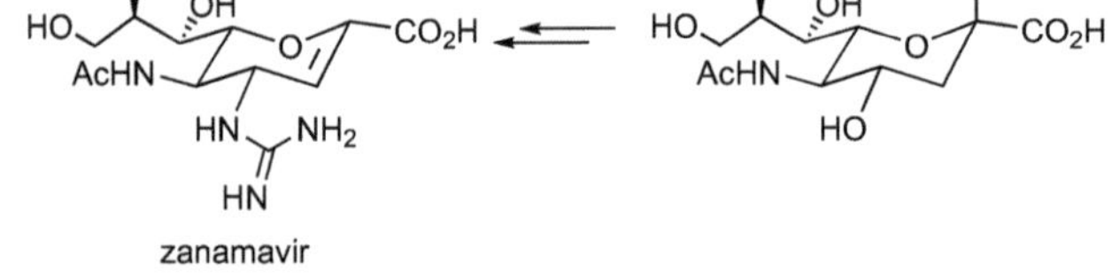

Figure 10.27 Synthesis of an intermediate for zanamivir using neuraminic acid aldolase (NANA).

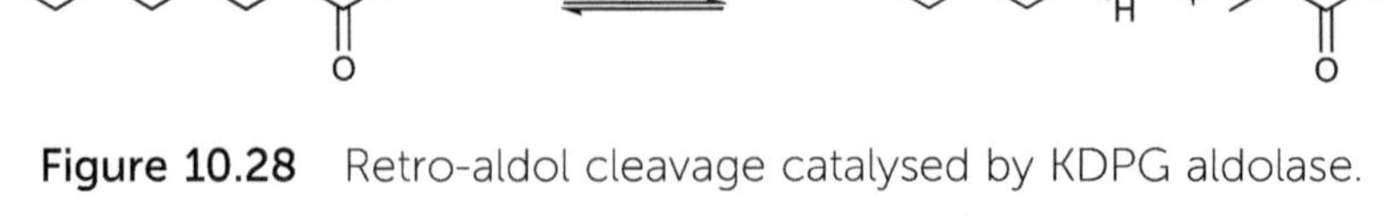

Figure 10.28 Retro-aldol cleavage catalysed by KDPG aldolase.

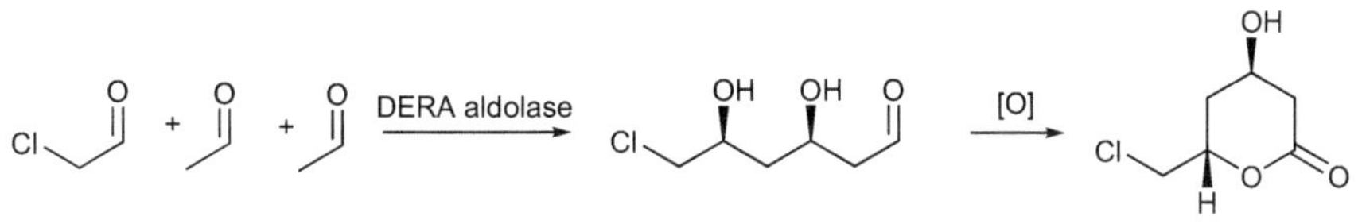

Figure 10.29 DERA-mediated asymmetric C–C bond formation to provide an intermediate for Lipitor synthesis.

identified that lower the K_M value up to 100-fold (*E. coli* KDPG aldolase), and that enhance the efficiency of retro-aldol cleavage by up to 25-fold (*T. maritima* KDPG aldolase).[40]

Deoxyribose phosphate aldolase (DERA) catalyses the condensation of acetaldehyde and glycerol-3-phosphate, but is able to utilise other aldehyde acceptors to generate various β-hydroxyaldehydes (2.1.4). DERA can also catalyse a double aldol condensation between chloro-acetaldehyde and two molecules of acetaldehyde, a process that has been used to generate a key building block for the manufacture of Lipitor (Figure 10.29).[41]

D- and L-threonine aldolases catalyse C–C bond formation between glycine and acceptor aldehydes (2.1.5). They can also use alanine as the donor leading to quaternary α-amino acids (2.1.6). A related PLP-dependent enzyme with broad application is serine hydroxymethyl-transferase (SHMT), which is able to utilise glycine, alanine and serine. Finally, 6-oxo-camphor hydrolase (OCH) catalyses an asymmetric Claisen ester-type cleavage of a β-diketone (2.1.7),[42] whereas the enzyme aceto-acetyl dehydratase (AcetDH) can catalyse aldol condensation followed by dehydration[116b] (2.1.8), and the enzyme acetolactate decarboxylase (ALD) catalyses decarboxylation of an activated β-keto acid[116d] (2.1.9).

10.2.2 Disconnections Involving C–C Bond Formation (TDP-dependent Lyases)

A second large family of enzymes that mediate C–C bond formation are the thiamine diphosphate lyases (TDP-lyases).[116c] These enzymes utilise TDP to generate enzyme-bound species that are equivalent to acyl anions and then condense these species with a wide range of different electrophiles including aldehydes, ketones, keto acids and enones to give a very broad range of substitution patterns. Use of an aldehyde acceptor can lead to either a phenylacetylcarbinol (PAC) product (Scheme 10.14, 2.2.1) or hydroxypropanone (HPP) (2.2.2) or even a benzoin (2.2.3) if both aldehyde substrates are aromatic. Benzaldehyde lyase (BAL) from *Pseudomonas fluorescens* catalyses the

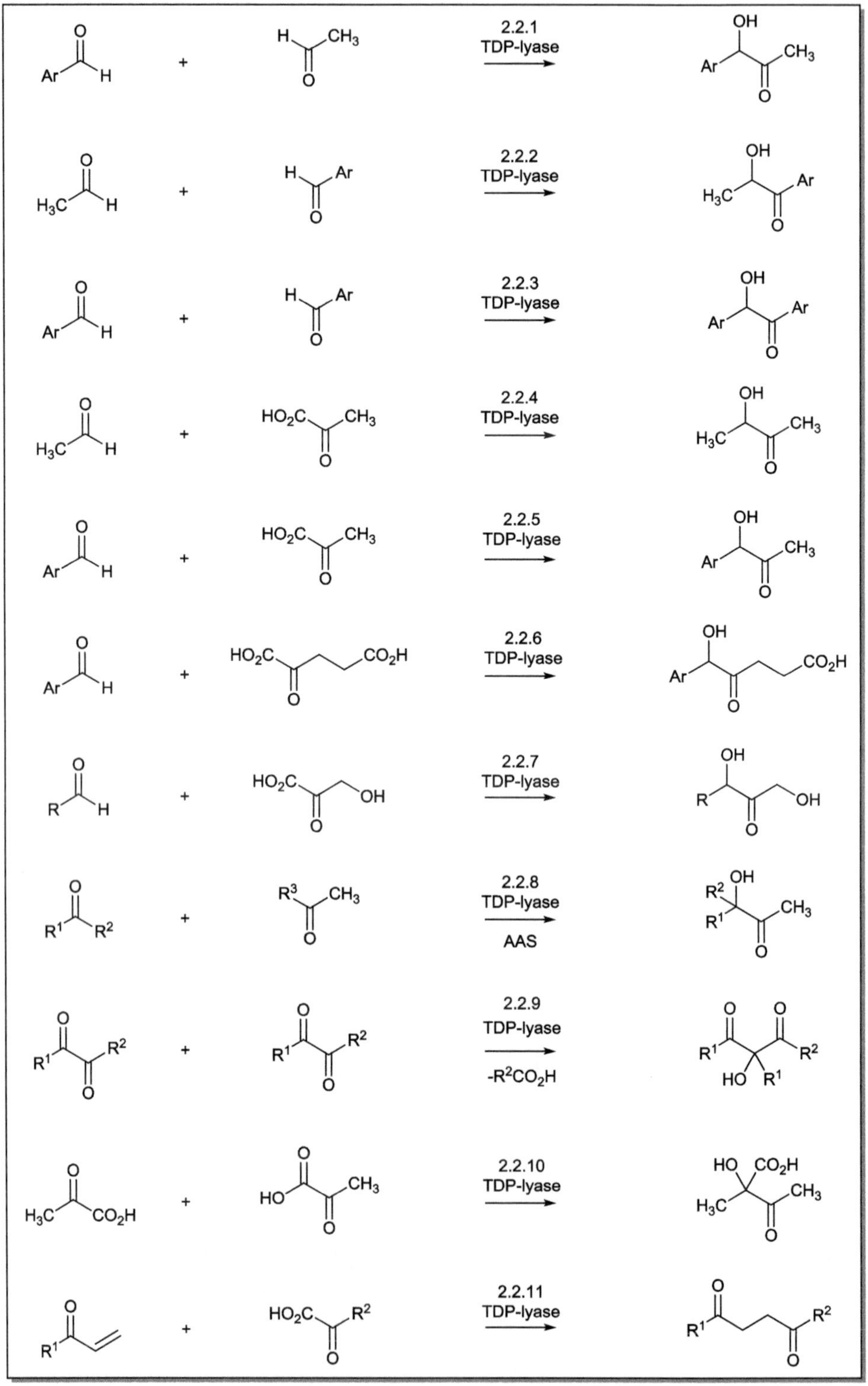

Scheme 10.14 C–C bond formation catalysed by TDP-dependent lyases.

carboligation of benzaldehyde 2-methyl aldehyde as an acceptor aldehyde with simultaneous kinetic resolution to afford a highly diastereoselective α-hydroxy ketone (both diastereoisomers >99% e.e.) (Figure 10.30).[43]

Pyruvate can also function as a donor in a reaction with acetalde-hyde (2.2.4) leading to acetoin or with an aromatic aldehyde (2.2.5) providing another route to PAC derivatives. Longer chain α-keto acids can also be tolerated (2.2.6). Transketolase utilises hydroxypyruvate as a donor (2.2.7) leading to a range of keto diol products. A particularly powerful application of TDP-dependent lyases is in the generation of tertiary alcohols *via* C–C bond formation. Various types of ketones can be used as acceptors including simple ketones (2.2.8), diketones (2.2.9) and α-keto acids (2.2.10).

Finally, the use of an enone as an acceptor leads to the synthesis of a 1,4-diketone *via* a 'Stetter'-type reaction (2.2.11) (Figure 10.31).[44]

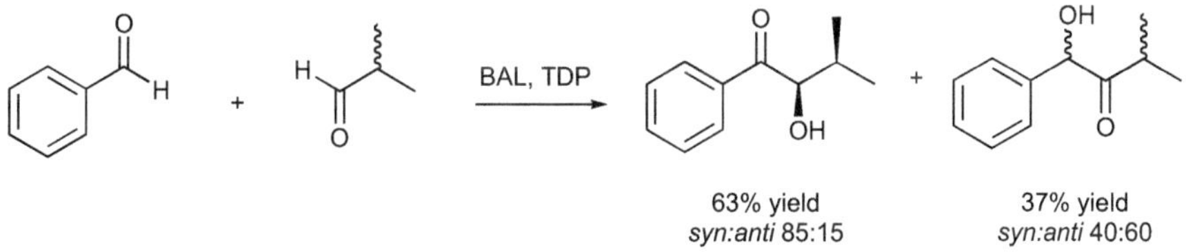

63% yield
syn:anti 85:15

37% yield
syn:anti 40:60

Figure 10.30   C–C bond formation using benzaldehyde lyase (BAL).

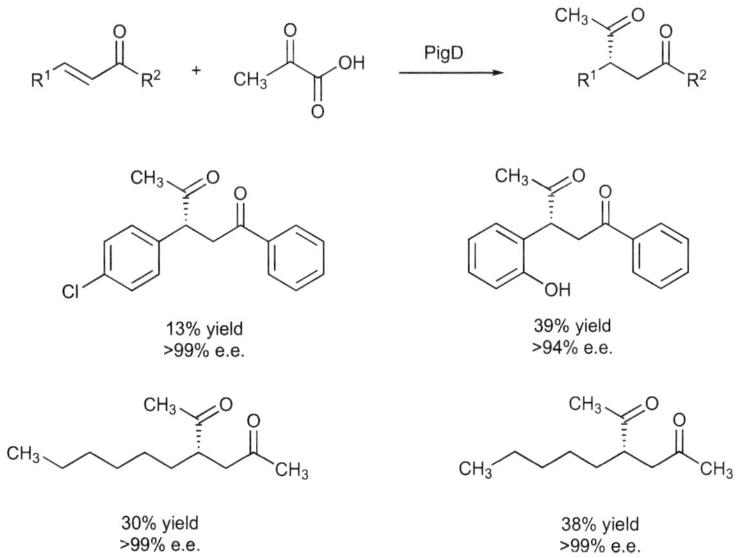

13% yield
>99% e.e.

39% yield
>94% e.e.

30% yield
>99% e.e.

38% yield
>99% e.e.

Figure 10.31   'Stetter' reaction catalysed by the TDP-dependent lyase PigD.

10.2.3 Disconnections Involving C–C Bond Formation (Other Enzymes)

Three other types of biocatalyst can be used for C–C bond formation, namely hydroxynitrile lyase (HNL), halohydrin dehalogenase[116q] (HHD) and the promiscuous activity of a lipase.[116j] HNL enzymes catalyse the addition of a cyanide ion to aromatic and aliphatic aldehydes to give both (R)- and (S)-cyanohydrins (Scheme 10.15, 2.3.1). An engineered HNL was used to convert *ortho*-Cl benzaldehyde to the corresponding cyanohydrin, which was then hydrolysed to give *ortho*-Cl mandelic acid (Figure 10.32).[45]

HNLs are also able to accept ketone substrates (2.3.2), although with lower activities, as well as use nitroalkanes as donors (2.3.3) in a Henry-type reaction. The HNL from *Hevea brasiliensis* catalyses the addition of nitroethane to benzaldehyde yielding 2-nitro-1-phenyl-propanol in good yield and high e.e. (Figure 10.33).[46]

HHD biocatalysts can also catalyse cyanide addition to either an epoxide (2.3.4) or a chlorohydrin (2.3.5) to generate a β-hydroxynitrile. Lipase enzymes, which principally catalyse hydrolytic reactions, can mediate a wide range of C–C bond forming processes on account of their ability to stabilise negatively charged products in the 'oxyanion hole'. Thus, they can catalyse aldol-type processes (2.3.6), Michael reactions with nitroalkenes (2.3.7 & 2.3.8 & 2.3.9), Mannich reactions (2.3.10) and conjugate addition to enones (2.3.11 & 2.3.12). Porcine pancreas lipase, type II (PPL II) catalyses the direct asymmetric aldol reaction of heterocyclic ketones with aromatic aldehydes (Figure 10.34). Enantioselectivities of up to 87% and diastereoselectivities of up to 83 : 17 (*anti* : *syn*) can be achieved.[47]

A co-catalyst system was developed using lipase B from *Candida antarctica* (CAL-B) and acetamide for the Michael addition of less activated ketones and nitroolefins. A range of structurally diverse nitroolefins and less activated cyclic and acyclic monoketones were tested with this co-catalyst system to produce product yields between 25–68%. The highest yield of 72% was for the product from *trans*-β-nitrostyrene and cyclohexanone (Figure 10.35).[48]

10.2.4 Disconnections Involving C–N Bond Formation or Cleavage

Amino acid oxidases catalyse the oxidation of both L- and D-amino acids to the corresponding α-keto acid *via* the intermediate imine, which

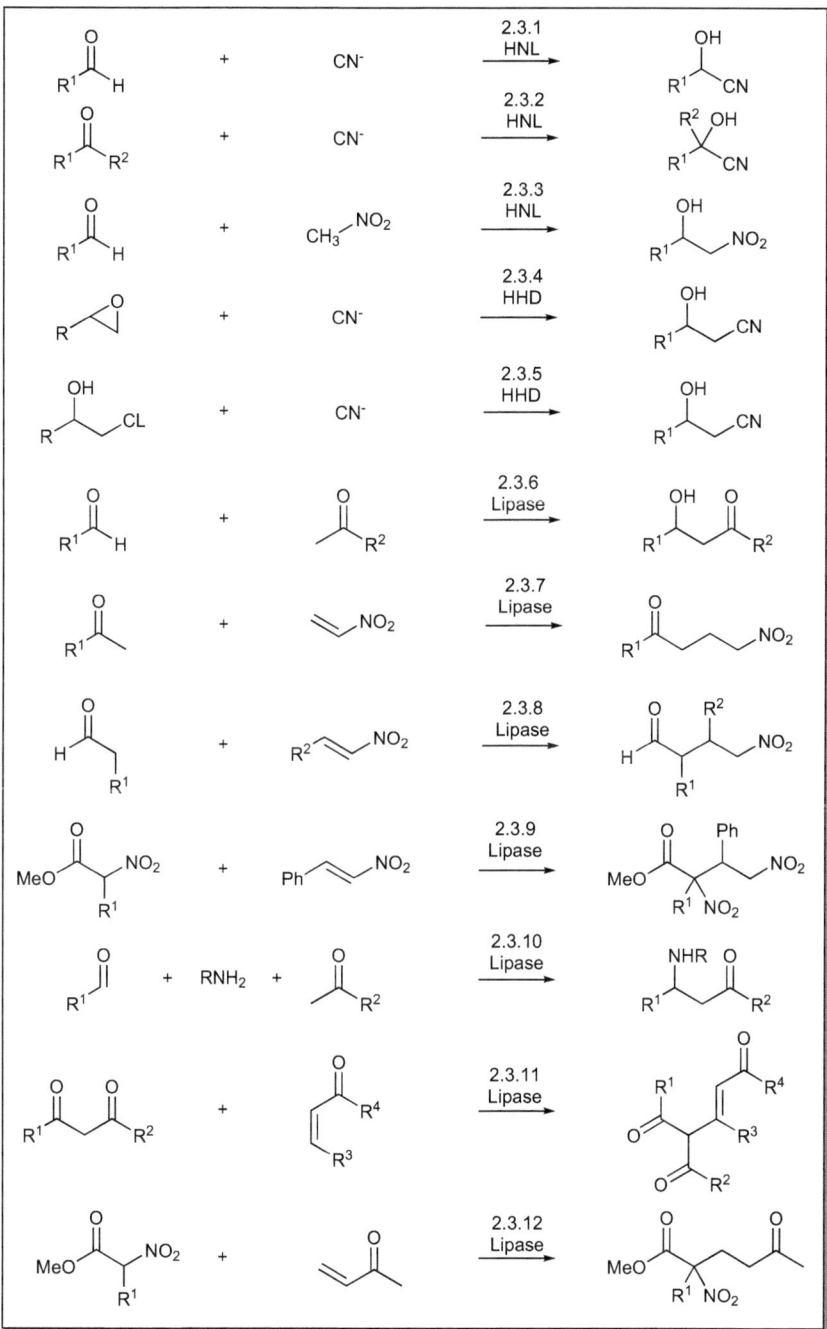

Scheme 10.15   C–C bond formation catalysed by other enzymes.

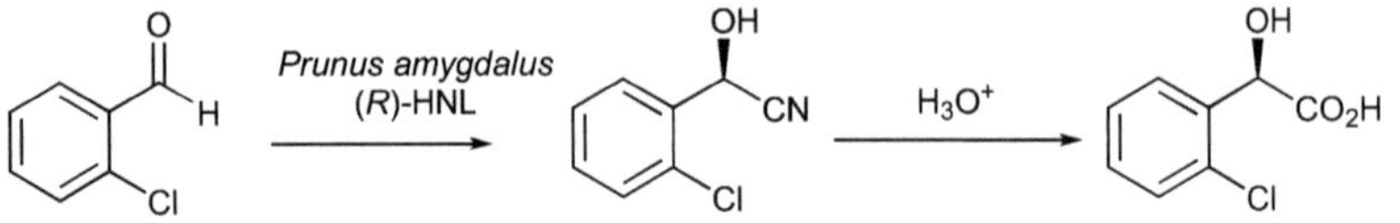

Figure 10.32 Enantioselective cyanohydrin synthesis with a hydroxynitrile lyase (HNL).

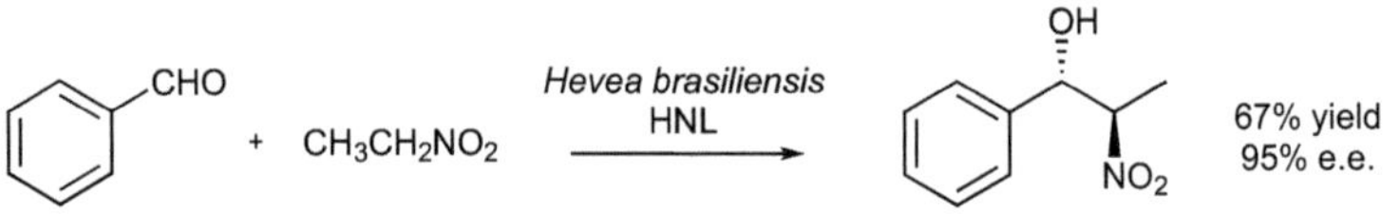

Figure 10.33 Asymmetric Henry reaction using an HNL.

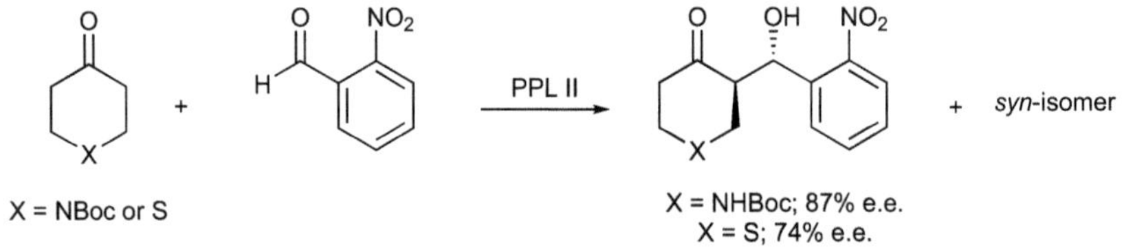

Figure 10.34 Lipase-catalysed C–C bond formation.

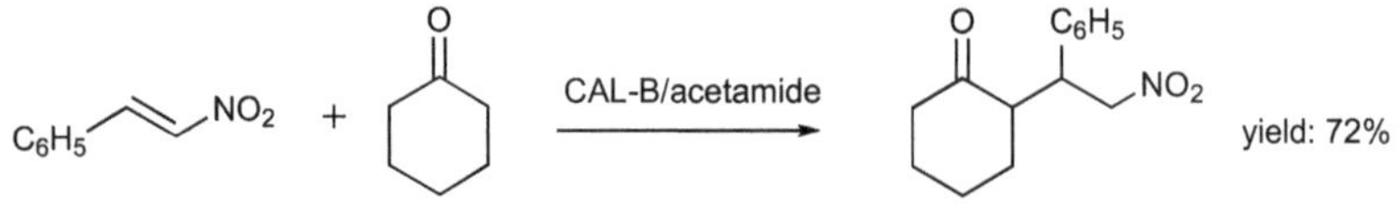

Figure 10.35 CAL-B-catalysed Michael-type reaction.

undergoes spontaneous hydrolysis[117h] (Scheme 10.16, 2.4.1). The addition of an ω-transaminase (ω-TA) allows recycling of the α-keto acid back to the enantiomerically pure amino acid, which overall results in deracemisation (2.4.2). Using this approach, deracemisation of racemic homoalanine was achieved using the D-amino acid oxidase from *Rhodotorula gracilis* (fused to *Vitreoscilla* haemoglobin to give VHb-DAAO) and the ω-transaminase from *Vibrio fluvialis* JS17 (ω-TA) (Figure 10.36). Conversion of 500 mM substrate to 485 mM product (>99% e.e.) was achieved in a biphasic system with benzylamine.[49]

ω-TA enzymes in general have broad substrate acceptance and hence are able to generate a range of different amine products containing additional functionality.[116m] Thus, they can be used to generate a 1,2-amino alcohol (2.4.3), 1,3-amino alcohol (2.4.4), 1,2-amino ester (2.4.5), 1,3-amino ester (2.4.6), 1,2-amino ether (2.4.7) and

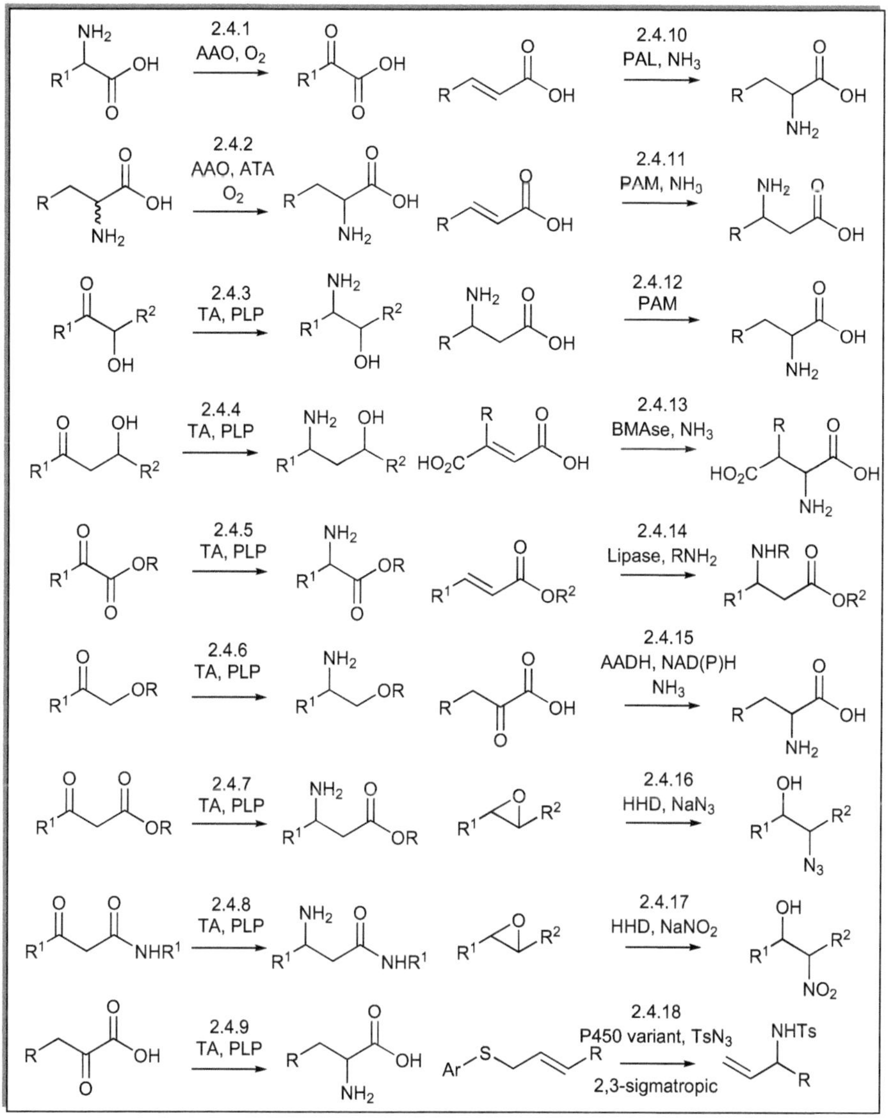

Scheme 10.16 Enzymes catalysing C–N bond formation and cleavage.

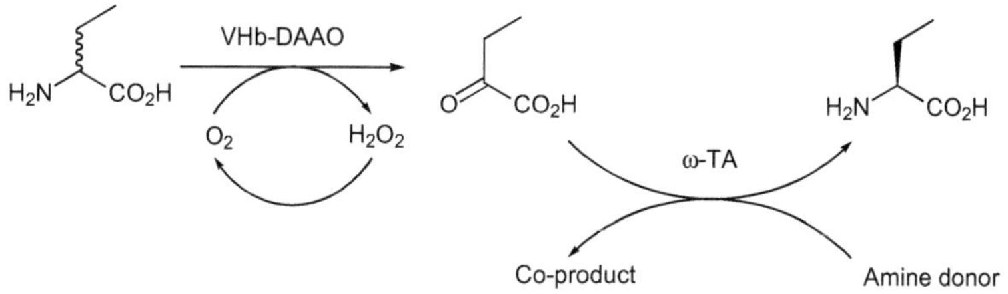

Figure 10.36 Deracemisation of 2-amino-butyrate using a combination of DAAO and ω-TA.

1,3-amino amide (2.4.8). In addition, α-TA enzymes catalyse amination of α-keto acids to yield α-amino acids (2.4.9). Three (S)-selective ω-TAs from *Paracoccus denitrificans* (Strep-PD ω-TA), *Pseudomonas fluorescens* (PF ω-TA) and *Vibrio fluvialis* (His-Vf ω-TA) were employed for the asymmetric reductive amination of selected prochiral ketones. His-tagged ω-TA from *Vibrio fluvialis* converted ethyl acetoacetate to the corresponding amine with high e.e. (>99%) (Figure 10.37).[50]

An example of amination of a β-keto amide is given by the use of an ω-TA, in the presence of isopropylamine as an amine donor, as the key step in the synthesis of the API for sitagliptin, an important drug used for the treatment of diabetes (Figure 10.38).[51]

The addition of ammonia to α,β-unsaturated carboxylic acids is catalysed by ammonia lyases and can lead to both α-(2.4.10) and β-(2.4.11) amino acids.[116i] The related aminomutase enzymes catalyse interconversion of α- and β-amino acids (2.4.12). Phenylalanine ammonia lyase was used to catalyse amination of *ortho*-bromo cinnamic acid as the key step in the synthesis of the anti-hypertensive drug perindopril (Figure 10.39).[52]

β-Methylaspartase catalyses the addition of ammonia to β-methylaspartate and some analogues (2.4.13). Engineered active site mutants of β-methylaspartase show broadened substrate specificity. For example, L384A exhibits a broad electrophile scope including

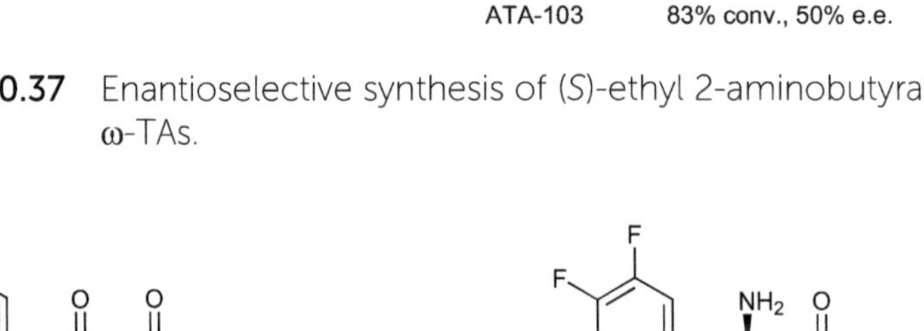

Strep-PD ω-TA	99% conv., >99% e.e.
PF ω-TA	90% conv., 90% e.e.
His-Vf ω-TA	99% conv., >99% e.e.
ATA-103	83% conv., 50% e.e.

Figure 10.37 Enantioselective synthesis of (S)-ethyl 2-aminobutyrate using ω-TAs.

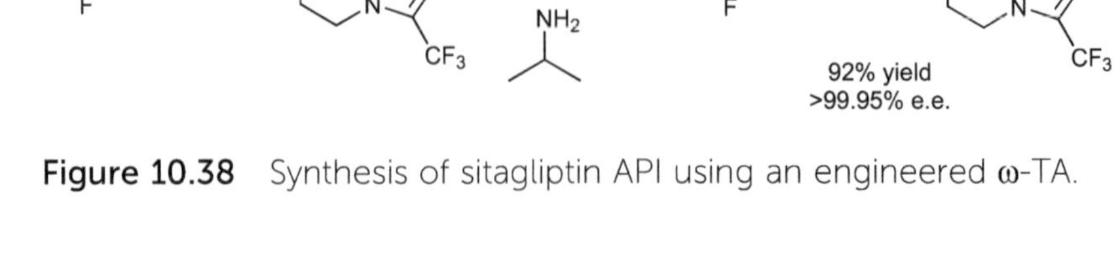

Figure 10.38 Synthesis of sitagliptin API using an engineered ω-TA.

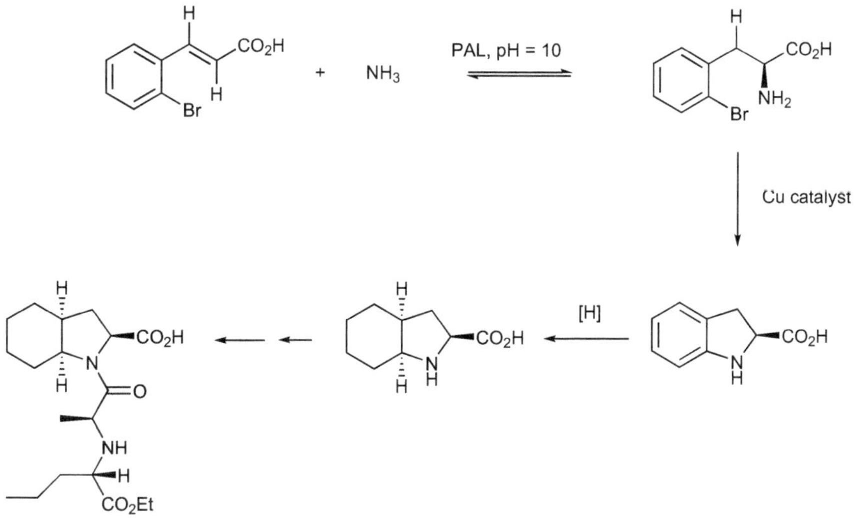
PAL, pH = 10
NH3
Cu catalyst
[H]
CO2H
NH2
Br
CO2Et

perindopril

Figure 10.39 Phenylalanine ammonia lyase (PAL)-catalysed α-amination of *o*-bromo-cinnamic acid.

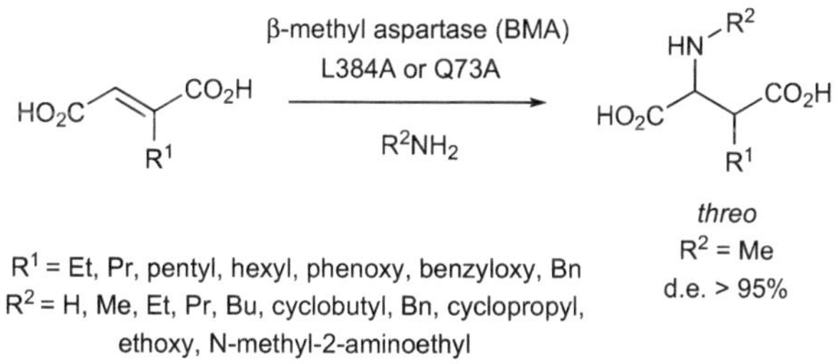
β-methyl aspartase (BMA)
L384A or Q73A
R2NH2
threo
R2 = Me
d.e. > 95%
R1 = Et, Pr, pentyl, hexyl, phenoxy, benzyloxy, Bn
R2 = H, Me, Et, Pr, Bu, cyclobutyl, Bn, cyclopropyl,
ethoxy, N-methyl-2-aminoethyl

Figure 10.40 Synthesis of L-aspartate derivatives using β-methylaspartase.

fumarate derivatives with alkyl, aryl, alkoxy, aryloxy, alkylthio and arylthio substituents at the C-2 position. Q73A exhibits a wide nucleophile scope including structurally diverse linear and cyclic alkylamines (Figure 10.40).[53]

Engineered variants of the phenylalanine amino mutase EncP were used to regioselectively aminate substituted cinnamic acids to yield β-amino acids in high enantiomeric excess (Figure 10.41).[54]

Conjugate addition of ammonia to acrylate derivatives can also be catalysed using a lipase[116j] (2.4.14). The synthesis of *N*-substituted β-amino esters has been achieved using the aza-Michael addition of mono- and bifunctional amines to acrylates using the lipase from

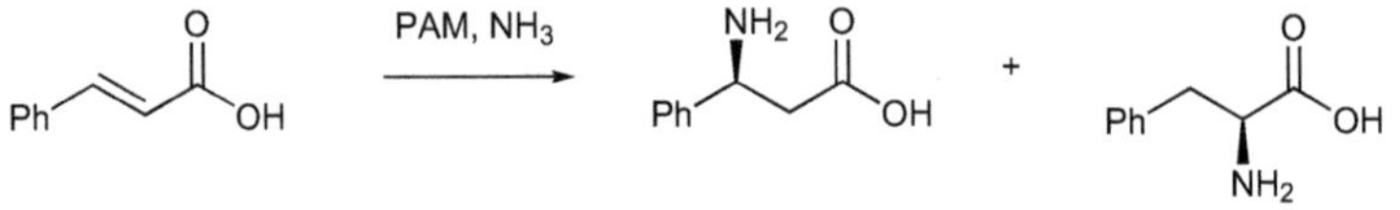

EncPwt	56	:	44
EncPR299K	88	:	12
EncPE293Q	43	:	57
EncPE293M	18	:	82

Figure 10.41 Amination of cinnamic acid using engineered PAM variants.

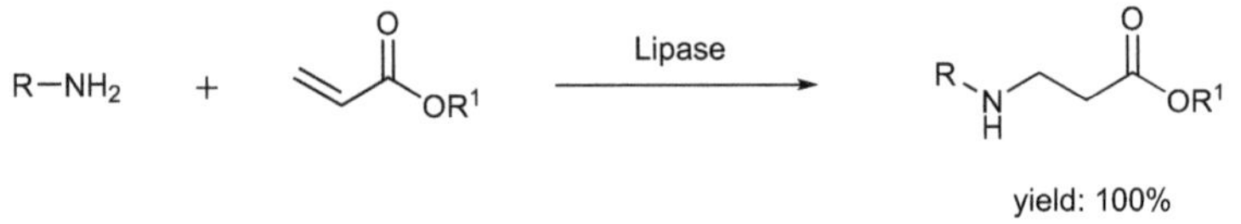

yield: 100%

Figure 10.42 Lipase-catalysed conjugate addition of amines to α,β-unsaturated esters.

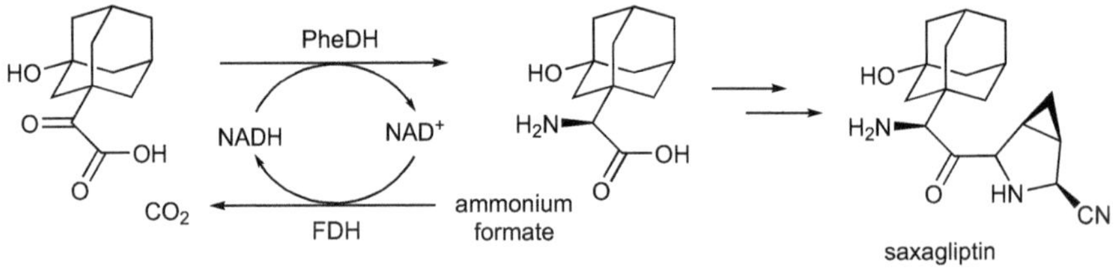

saxagliptin

Figure 10.43 Reductive amination of a keto acid using PheDH to provide a key building block for the synthesis of saxagliptin.

Rhizomucor miehei. A total of 22 *N*-substituted β-amino esters and 2 *N,N*-disubstituted double Michael adducts were obtained. The product yields ranged from 45–100%, with the best yields obtained when alkanolamines and diamines were used as the donors (Figure 10.42).[55]

Amino acid dehydrogenases (AADHs) are NAD(P)H-dependent enzymes that convert α-keto acids to predominantly L-α-amino acids although there are also examples of production of D-amino acids[116k] (2.4.15). An engineered phenylalanine dehydrogenase (PheDH) from *Thermoactinomyces* sp. was used for the enantioselective reductive amination of an adamantly-containing α-keto acid to yield the corresponding α-amino acid in high e.e. This amino acid was used as a building block for the anti-diabetic drug saxagliptin (Figure 10.43).[56]

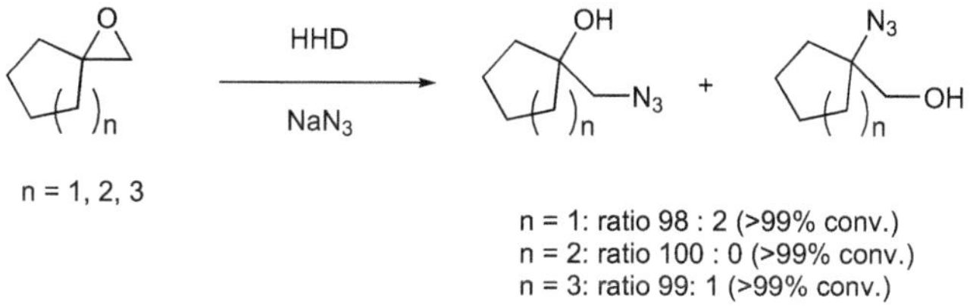

n = 1, 2, 3

n = 1: ratio 98 : 2 (>99% conv.)
n = 2: ratio 100 : 0 (>99% conv.)
n = 3: ratio 99: 1 (>99% conv.)

Figure 10.44 Epoxide ring-opening with an azide catalysed by a HHD.

HHD enzymes are able to catalyse the ring opening of an epoxide with either an azide (2.4.16) or a nitrate ion[116q] (2.4.17). Halohydrin dehalogenase from *Arthrobacter* sp. (HheA) catalyses the highly regioselective azidolysis of spiroepoxides with 5-, 6- and 7-membered alkane rings. The enzyme from *Agrobacterium radiobacter* (HheC) displayed moderate to high enantioselectivity with racemic 1-oxaspiro[2.5]octanes (Figure 10.44).[57]

Finally, a P450 variant has recently been used for the synthesis of enantiomerically enriched allylic amines *via* conversion of an allylic sulfide to a sulfinimine followed by [2,3]-sigmatropic rearrangement (2.4.18).[58]

10.2.5 Disconnections Involving Hydrolysis Reactions (Non-amino Acid)

HHD biocatalysts can catalyse hydrolysis of α-halo acids (Scheme 10.17, 2.5.1) and α-halo amides (2.5.2) to the corresponding hydroxyl derivatives.[116q] A *Pseudomonas* sp. containing a HHD has been used to resolve racemic chloropropionic acid to give a building block for the synthesis of the herbicide (R)-fluazifop (Figure 10.45).[59]

Six haloalkane dehalogenases were expressed and tested for their ability to perform kinetic resolution of racemic α-bromoamides. Activity towards eight substrates was found, with a preference for the (R)-α-bromoamide in all cases. For preparative scale conversions, DbjA (*B. japonicum*) and LinB (*S. paucimobilis* UT26) were selected, giving products of high optical purity (87–99% e.e. or d.e.) in yields from 31% to 50% (Figure 10.46).[60]

The HHDs LinB (*Sphingomonas paucimobilis* UT26) and DhaA31 (variant of DhaA from *Rhodococcus rhodochrous* NCIMB 13064) were shown to catalyse the hydrolysis of dibromide substrates (Figure 10.47).[61]

The same enzymes catalyse the ring opening of epoxides with a halide ion to yield a chlorohydrin (2.5.3).

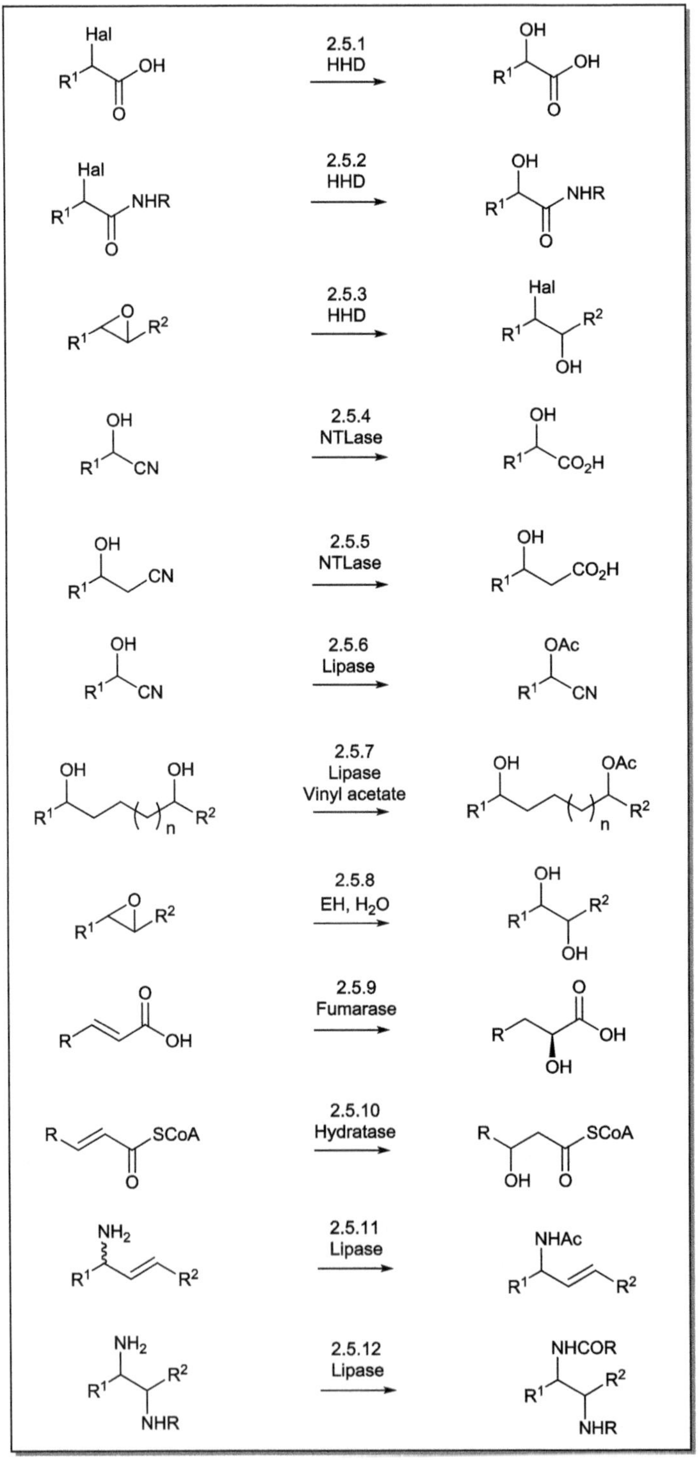

Scheme 10.17   Disconnections involving hydrolysis reactions (non-amino acid).

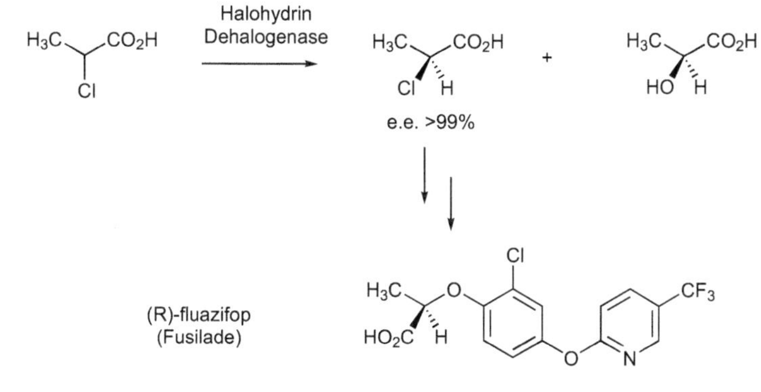

Figure 10.45 HHD-catalysed resolution of racemic 2-chloropropionic acid.

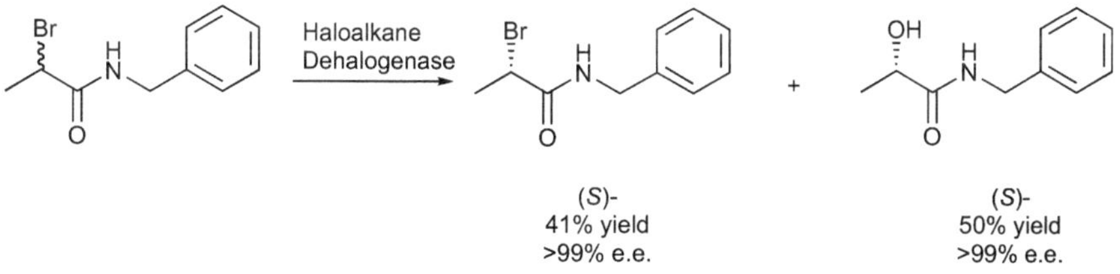

Figure 10.46 Kinetic resolution of an α-bromo-amide using HHD.

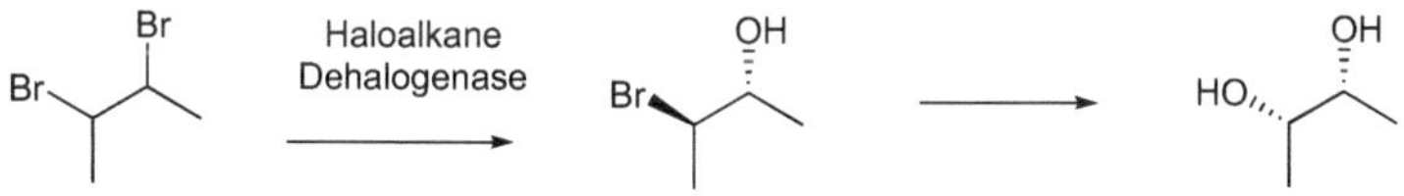

Figure 10.47 Conversion of a 1,2-bromoalkane to a 1,2-diol.

Nitrilase enzymes are able to hydrolyse a broad range of different nitrile containing substrates.[115f] A nitrilase (nitrilase III) was identified from a metagenomic library and applied to the asymmetric hydrolysis of 3-hydroxyglutaronitrile to yield the β-hydroxy acid product in 95% yield and >90% e.e. Esterification of the acid provided an important building block for the cholesterol lowering agent Lipitor (Figure 10.48).[62]

Nitrilases are also able to hydrolyse both cyanohydrins (2.5.4) and β-hydroxynitriles (2.5.5) to the corresponding hydroxy carboxylic acids. (S)-Hydroxynitrile lyase (MeHNL from *Manihot esculenta*) and aryl-acetonitrilase (NLase from *Pseudomonas fluorescens*) were co-expressed in *E. coli* and suspensions of the resting cells in buffer were then used to convert benzaldehyde and cyanide into mandelic acid and mandeloamide. The two-phase system produced the (S)-acid and

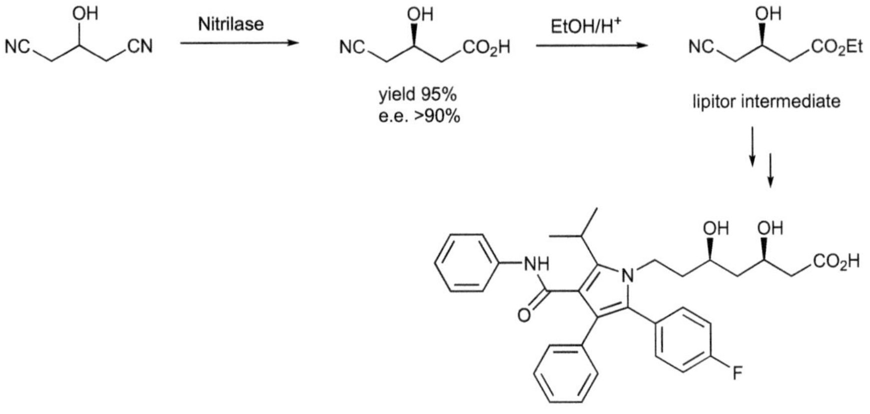

Figure 10.48 Hydrolysis of 3-hydroxyglutaronitrile using a nitrilase to provide an intermediate for the synthesis of Lipitor.

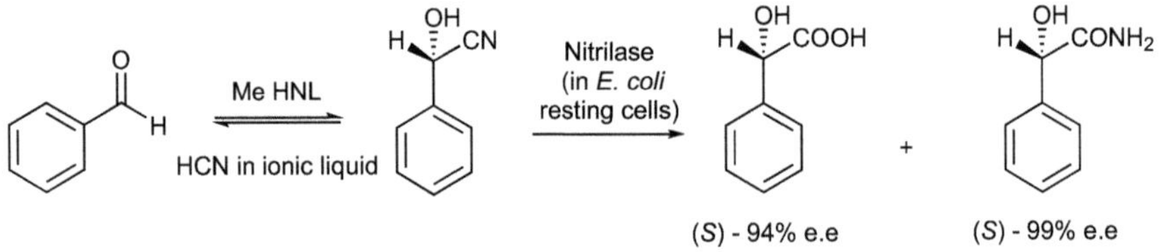

(S) - 94% e.e   (S) - 99% e.e

Figure 10.49 Combined HNL and nitrilase-catalysed synthesis of enantiomerically enriched α-hydroxy acids.

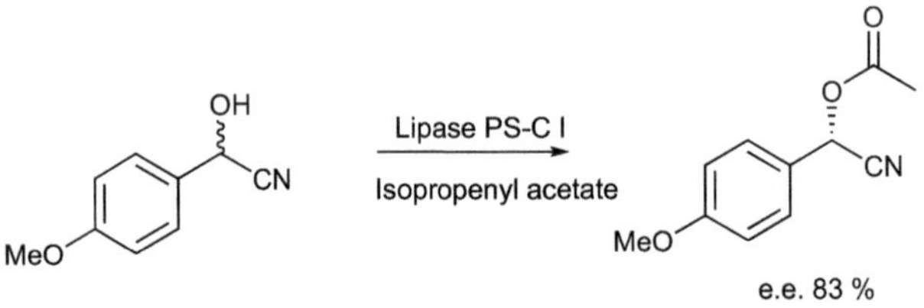

e.e. 83 %

Figure 10.50 Lipase-catalysed acylation of a cyanohydrin.

(S)-amide in 94% e.e. with combined yields of 87–100% (Figure 10.49).[63]

Lipases can both acetylate[115c] and deacetylate[115b] both cyanohydrins (2.5.6) and also a wide range of 1,n-diols/hydroxyl esters, and diesters (2.5.7). An immobilised lipase on a ceramic support (PS-C I) was used as the biocatalyst for the acetylation of a racemic cyanohydrin to the acetylated (R)-ester, which was produced in moderate e.e. alongside other substituted cyanohydrins (Figure 10.50).

Porcine pancreas lipase (PPL) was used for the desymmetrisation of a *meso*-diol to give the monoacetate and the corresponding *meso*-diacetate, which was converted back to the *meso*-diol with sodium methoxide in methanol. The monoacetate then served as an intermediate in the synthesis of (+)-bourgeanic acid (Figure 10.51).[64]

Epoxide hydrolases[116r] mediate ring opening of epoxides to yield diols (2.5.8). For example, the epoxide hydrolase HXN-200 catalyses ring opening and desymmetrisation of 5- and 6-membered ring epoxides (Figure 10.52).[65]

Finally, hydration of C=C can be catalysed by either fumarase (2.5.9) or enoyl CoA hydratase (2.5.10), amongst other enzymes.[116h]

Lipases can catalyse acylation of allylic amines (2.5.11) and di-amines[115c] (2.5.12). 1-(4-Methylnaphthalen-1-yl)-prop-2-en-1-amine was prepared *via* the *Candida antarctica* lipase B (CaLB)-catalysed resolution of racemic 4-methylnaphthalen-1-yl in excellent enantio-meric excess (>99.9%) (Figure 10.53).[66]

CALB was also used for the kinetic resolution of enantiopure 2-(1*H*-imidazol-1-yl)cycloalkanamines to produce optically pure imi-dazole derivatives. Racemic (+)-*trans* cycloalkanamine was reacted with ethyl acetate with CALB to yield enantiopure (1*R*,2*R*) *trans*-acetamide and (1*S*, 2*S*) *trans*-amine (Figure 10.54).[67]

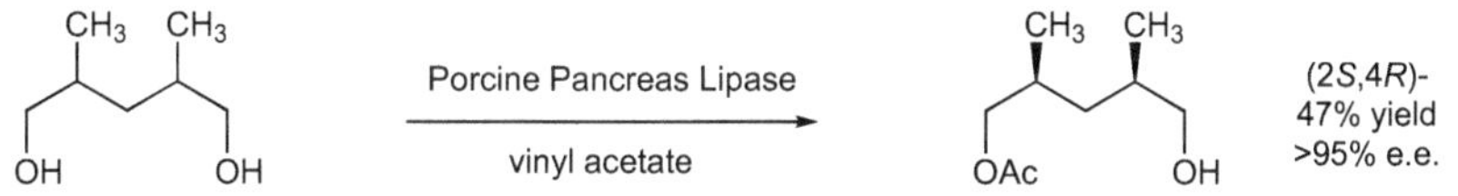

Figure 10.51   PPL-mediated desymmetrisation of a *meso*-1,5-diol.

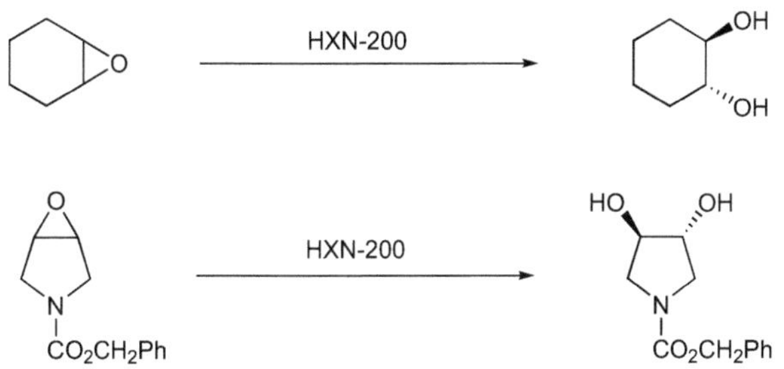

Figure 10.52   Desymmetrisation of prochiral epoxides using the epoxide hydrolase HXN-200.

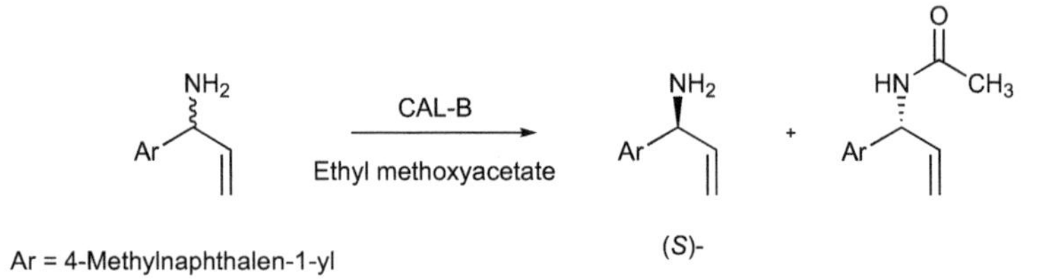

Ar = 4-Methylnaphthalen-1-yl

Figure 10.53   CAL-B-mediated resolution of an allylic amine.

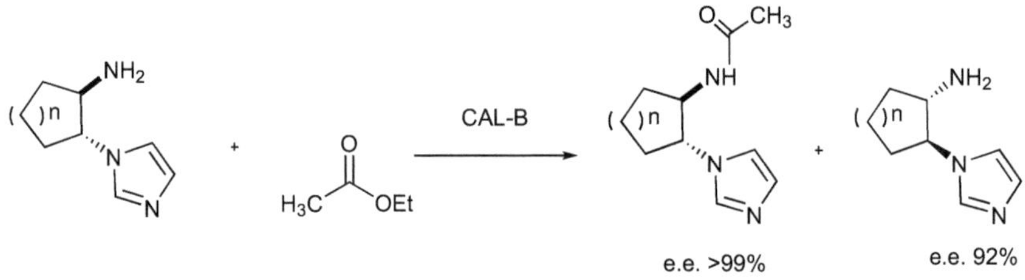

e.e. >99%   e.e. 92%

Figure 10.54   CALB acylation of an imidazole derivative.

10.2.6   Disconnections Involving Hydrolysis Reactions (Amino Acid)

In this section, we look at enzymes that catalyse the hydrolysis of substrates leading to amino acids or their derivatives. Many of these biocatalysts are involved in commercially important processes for the production of amino acids on a manufacturing scale. α-Amino acids can be prepared by hydrolysis of an α-amino nitrile using a nitrilase[115f] (Scheme 10.18, 2.6.1) or an N-acyl amino acid using an acylase (2.6.2), or from an N-carbamoyl derivative using a carbamoylase[115i] (2.6.3).

Proteases are able to catalyse the hydrolysis of N-substituted amino acid thioesters under DKR conditions (2.6.4) and amidases can catalyse the hydrolysis of amino acid amide derivatives[115g] (2.6.5). The protease subtilisin was used for the dynamic kinetic resolution of racemic N-protected-naphthylaminoacid thioesters. Of the four compounds tested, all produced their L-enantiomers in excellent yield (90–93%) and almost complete stereoselectivity (Figure 10.55).[68]

Resting whole cells of *Rhodococcus erythropolis* AJ270 can catalyse the enantioselective hydrolytic desymmetrisation of a number of prochiral α-substituted α-aminomalonamides to functionalised α-tetrasubstituted α-amino acids (Figure 10.56).[69]

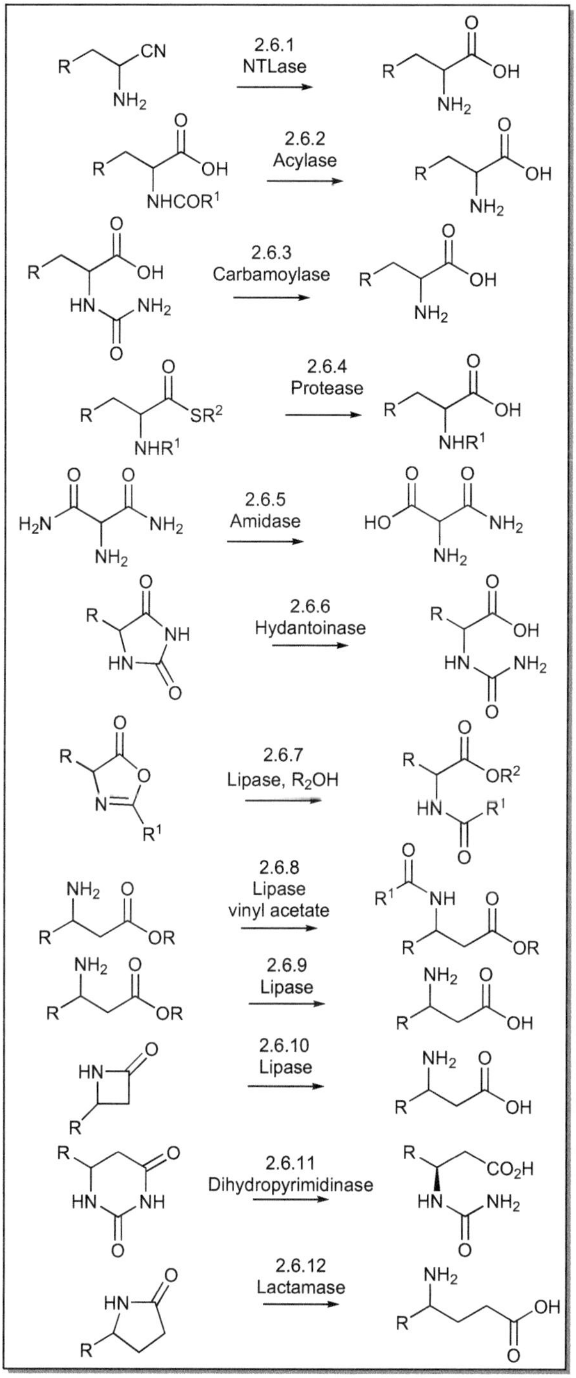

Scheme 10.18 Disconnections involving hydrolysis reactions (amino acid).

Figure 10.55   Protease-catalysed DKR of an amino acid thioester.

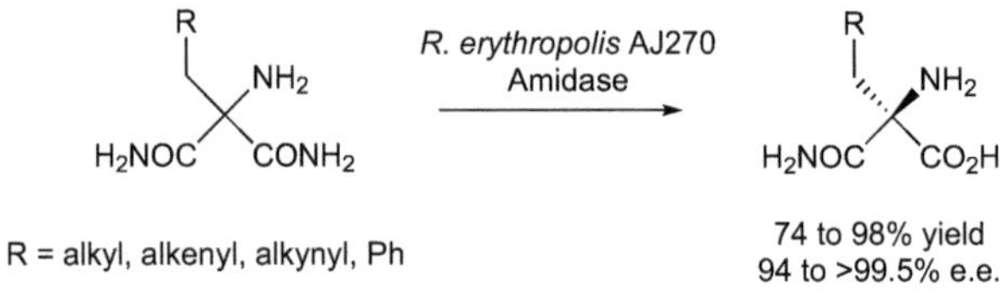

Figure 10.56   Amidase-catalysed hydrolysis of a 1,3-diamide.

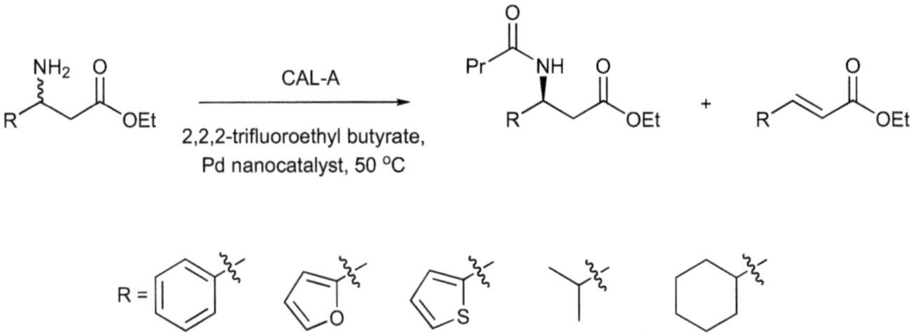

Figure 10.57   Lipase-catalysed kinetic resolution of a β-amino ester.

Other approaches to α-amino acid derivatives are based on hydrolysis of a hydantoin using a hydantoinase (2.6.6) and ring-opening of an azlactone using a lipase in the presence of an alcohol (2.6.7); both of these reactions also operate under DKR conditions.[115i] For the corresponding β-amino acids, approaches are based on lipase-catalysed acylation of amino esters[115c] (2.6.8), hydrolysis of N-acyl amino acids[115g] (2.6.9), lipase-catalysed ring opening of β-lactams[115c] (2.6.10) or hydrolysis of a dihydropyrimidine using a dihydropyrimidinase[115i] (2.6.11).

Candida antarctica lipase A (CALA) was used for the acylation of an amino ester in the presence of a palladium nanocatalyst to promote racemisation of the substrate. A temperature reduction of 50 °C proved most effective, reducing the eliminated by-product, and yielding the product in 99% yield and 96% e.e. with CALA/GAmP-MCF (Figure 10.57). Various β-amino esters were similarly resolved with >97% yield and 96–99% e.e.[70]

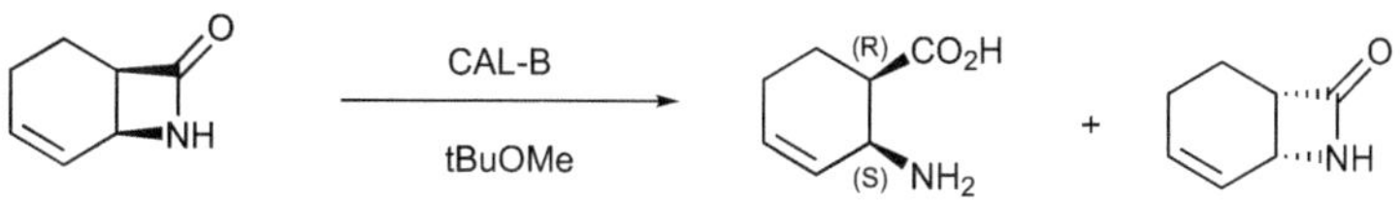

Figure 10.58 CAL-B-mediated β-amino acid synthesis.

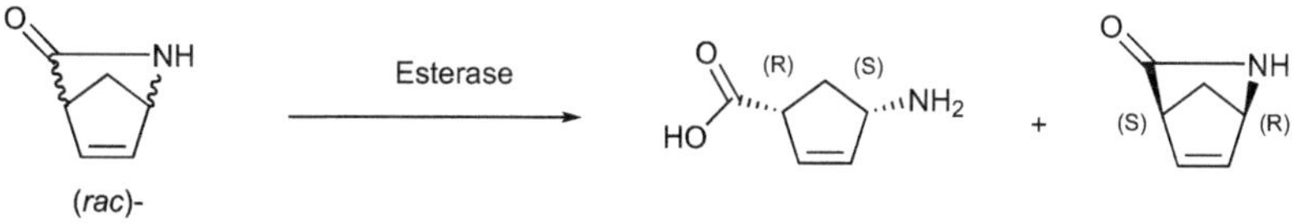

Figure 10.59 γ-Lactam synthesis using an esterase.

As the first step in the enantioselective synthesis of 2-aminocyclo-hexanecarboxylates, an enzymatic resolution of a racemic bicyclic β-lactam with CAL-B was employed (Figure 10.58).[71]

Finally, γ-amino acids can be accessed from the corresponding γ-lactam using a lactamase (2.6.12). *Pseudomonas fluorescens* esterase I (PFEI) was found to have promiscuous (−)-γ-lactamase activity (Figure 10.59). In the kinetic resolution of the Vince lactam, the (−)-lactam was preferentially hydrolysed ($E > 100$).[72]

10.2.7 Disconnections Involving Redox Reactions (KREDs & Other DHs)

KREDs have very broad substrate tolerance and hence are very useful biocatalysts for the generation of fragments containing two or more functional groups.[116n] For example, they can be used to generate a 1,2-diol from either a hydroxyketone (Scheme 10.19, 2.7.1) or a diketone (2.7.2), or a hydroxyketone (2.7.3). Baker's yeast has been used to reduce the following diketone to the corresponding *cis-* and *trans-*diols with high e.e. (Figure 10.60).[73]

CgKR2 from *Candida glabrata* was cloned and expressed in *E. coli* and used for the reduction of ethyl 2-oxo-4-phenylbutyrate (OPBE) to ethyl (R)-2-hydroxy-4-phenyl-butyrate ((R)-HPBE) in high e.e., which is an important chiral precursor for ACE inhibitors (Figure 10.61).[74]

KREDs can also be employed as biocatalysts to prepare a 1,3-diol (2.7.4), 1,3-hydroxyester (2.7.5), halohydrin (2.7.6) and β-hydro-xynitrile (2.7.7). Scientists at Bristol-Meyers Squibb screened a panel of 100 micro-organisms and identified *Streptomyces nodosus* as the ideal catalyst for reduction of a protected amino acid derived

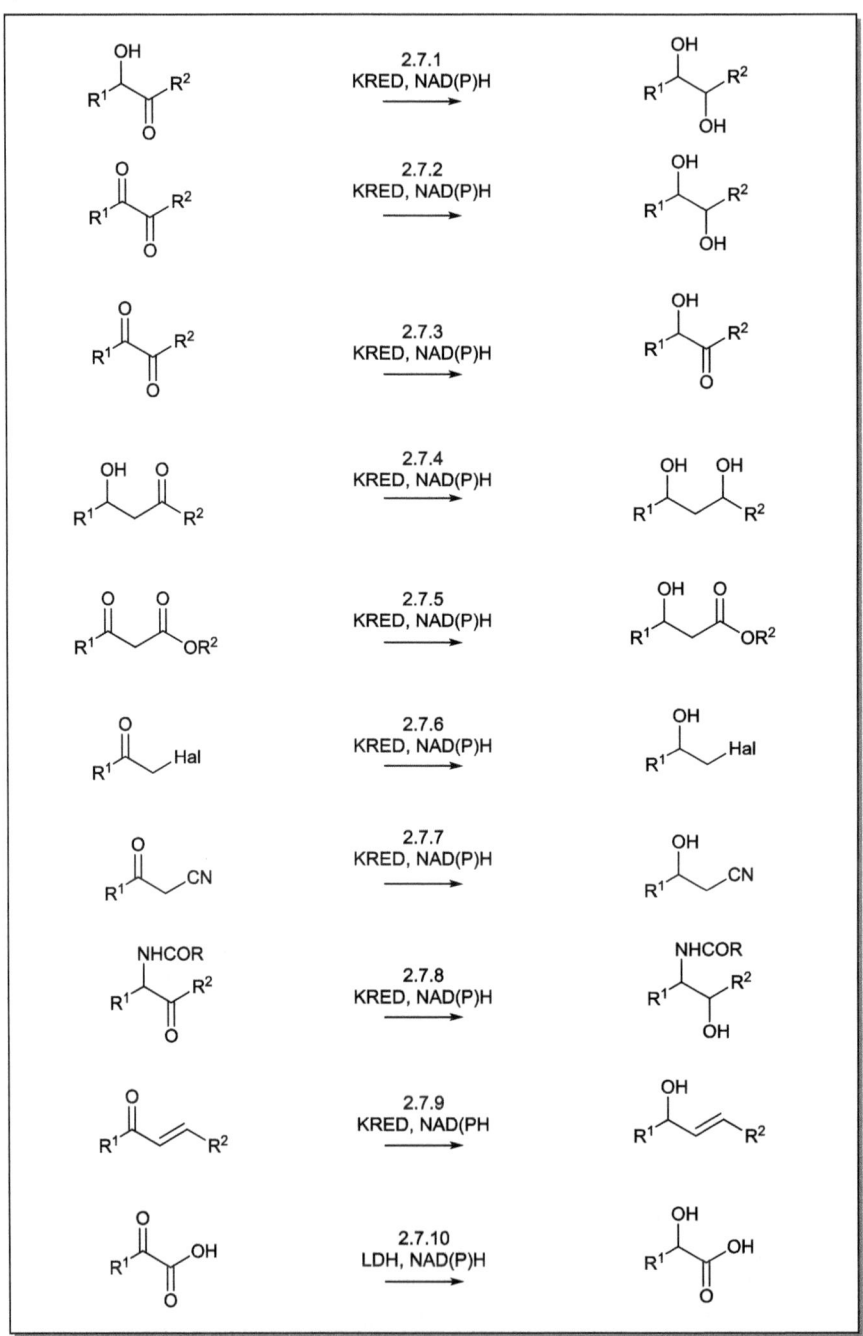

Scheme 10.19 Disconnections involving redox reactions (KREDs and other ADHs).

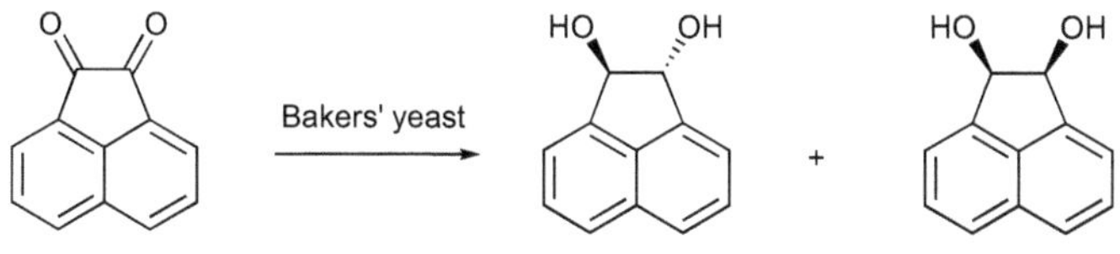

Figure 10.60   Baker's yeast reduction of a 1,2-diketone.

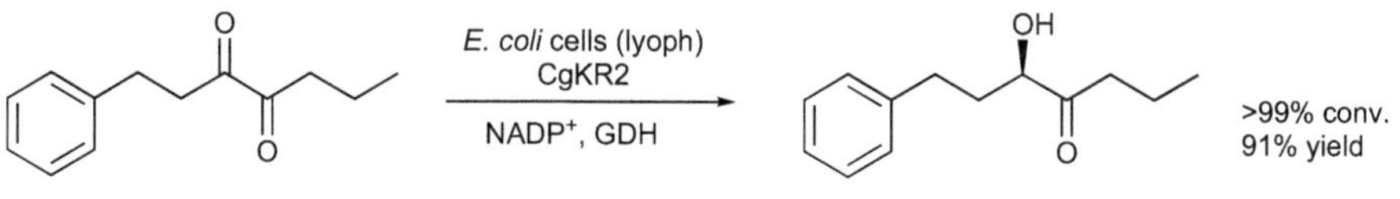

Figure 10.61   Reduction of a 1,2-diketone to an α-hydroxy ketone.

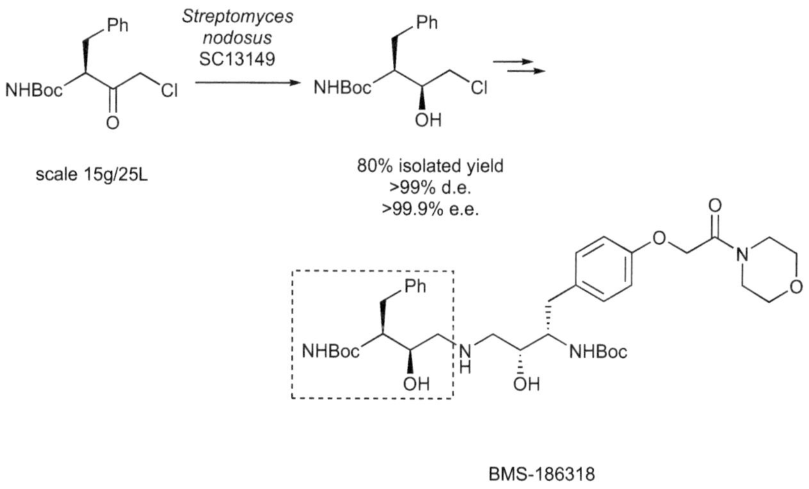

scale 15g/25L

80% isolated yield
>99% d.e.
>99.9% e.e.

BMS-186318

Figure 10.62   Whole cell-mediated reduction of an α-chloroketone.

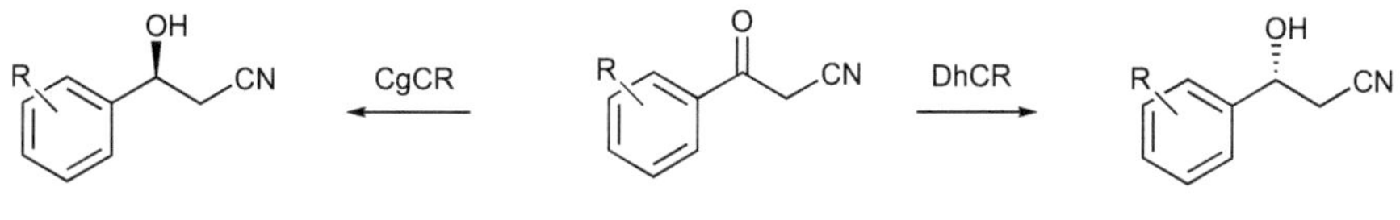

Figure 10.63   Synthesis of β-hydroxyketones *via* ketone reduction.

α-haloketone (Figure 10.62). The chlorohydrin product was used as an intermediate for an HIV protease inhibitor.[75]

KREDs were identified through genome mining that were able to catalyse the reduction of substituted β-ketonitriles to either enantiomer of the corresponding β-hydroxynitriles (Figure 10.63).[76]

A recombinant *E. coli* whole cell biocatalyst was developed in which diketoreductase from *Acinetobacter baylyi* was co-expressed with glucose dehydrogenase (GDH) from *Bacillus megaterium* to enable efficient co-factor recycling (Figure 10.64).[77]

Finally, KREDs and ADHs can reduce amino ketones (2.7.8) and enones to allylic alcohols (2.7.9). A prochiral enone was reduced using the ADH from *Thermoanaerobacter* sp. (ADH-T) to yield the (*R*)-enantiomer of ethyl 5-hydroxyhept-6-enoate, whereas use of the ADH from *Lactobacillus* sp. (ADH-LB) produced the (*S*)-enantiomer (Figure 10.65).[78]

For the reduction of α-keto acids, lactate dehydrogenase (LDH) is a highly versatile and useful biocatalyst (2.7.10). The enzymatic reduction of phenyl pyruvate to the corresponding substituted phenyl lactic acid was achieved in 95% yield (Figure 10.66).[79]

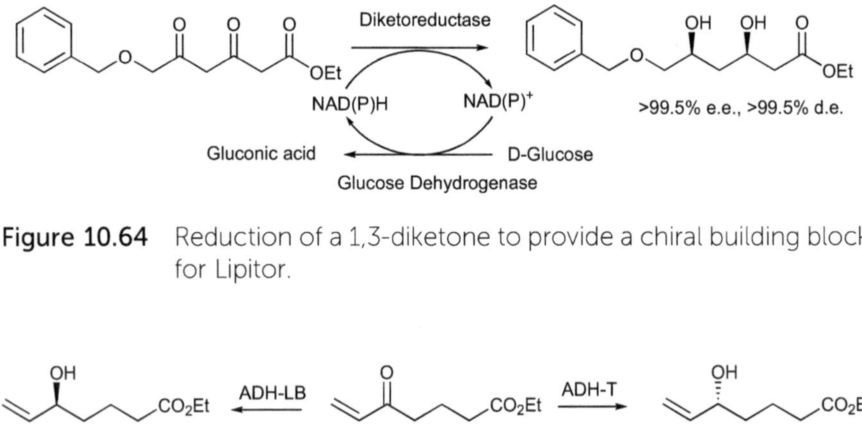

Figure 10.64   Reduction of a 1,3-diketone to provide a chiral building block for Lipitor.

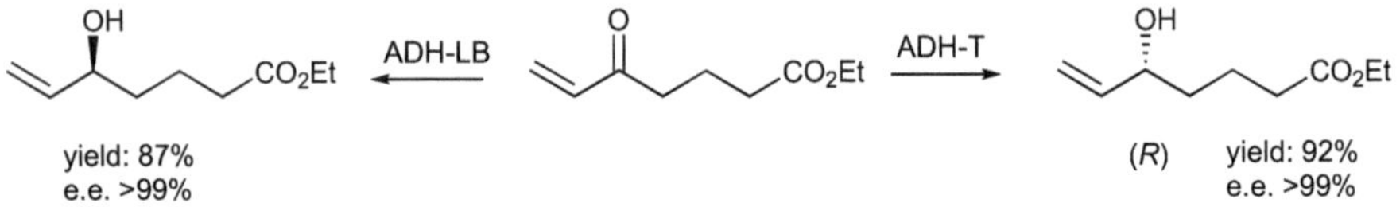

Figure 10.65   Reduction of an enone to either enantiomer of an allylic alcohol.

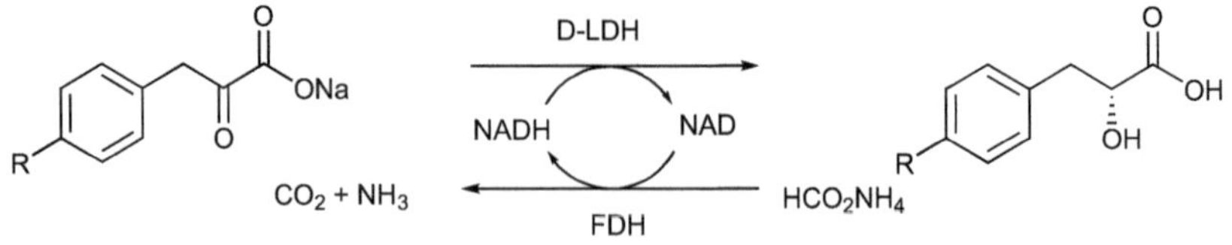

Figure 10.66   Lactate dehydrogenase (LDH)-mediated reduction of an α-keto acid.

10.2.8 Disconnections Involving Redox Reactions (EREDs)

Ene reductases (EREDs) are flavin mononucleotide (FMN)- and NAD(P)H-dependent enzymes that reduce substrates containing a C=C bond activated by a suitable substituent.[116g] The electron-withdrawing group provides not only polarization of the C=C bond but also the necessary binding of the substrate at the active-site of the enzyme by formation of hydrogen bonds with a His/Asp or His/His pair. EREDs are often highly chemo-, regio- and stereoselective and hence offer a versatile method for creating molecules with multiple functionalities and predetermined relative and absolute stereochemistries. In common with KREDs, EREDs also have broad substrate tolerance and are able to catalyse C=C reduction of all of the substrate classes shown in Scheme 10.20, 2.8.1–2.8.14. These reactions proceed *via trans*-addition of hydrogen and mechanistically bear a strong resemblance to Michael-type reactions, hence the requirement for activation of an alkene, often with two or more electron-withdrawing substituents. A wide range of different substrate types can be accommodated as shown by the representative set of C=C bond-containing motifs in Figure 10.67.[80]

10.2.9 Disconnections Involving Oxidation Reactions

Finally, we look at biocatalysts that generate target molecules containing two or more functional groups *via* oxidation processes. Many of these transformations involve the introduction of one or more oxygen atoms into the product *via* a mono- or dioxygenase. Naphthalene dioxygenase (NDO) catalyses the dihydroxylation of an alkene to give a 1,2-diol[117a] (Scheme 10.21, 2.9.1). Carboxylic acids can be converted to the corresponding α-hydroxy acid with a P450 monooxygenase[117b] (2.9.2) or to the α-peroxy acid with a peroxygenase[117c] (2.9.3). P450 monooxygenases also oxidise alkenes to allylic alcohols (2.9.4), whereas lipoxygenases catalyse oxidation of a skipped diene to a conjugated diene hydroperoxide[117c] (2.9.5). P450-BM3 mutant F87L/A328V catalyses the (*R*)-selective hydroxylation of cyclohexene-1-carboxylic acid methyl ester (Figure 10.68), whereas the A328S variant catalyses the (*S*)-selective hydroxylation. Both mutants were developed using rational structure-guided directed evolution and were highly regio- and enantioselective.[81]

Styrene monooxygenase catalyses oxidation of an allylic alcohol (2.9.6) and allylic ether[117a] (2.9.7). BVMOs can catalyse oxidation of a β-amino ketone to the corresponding β-amino acid ester (2.9.8) and

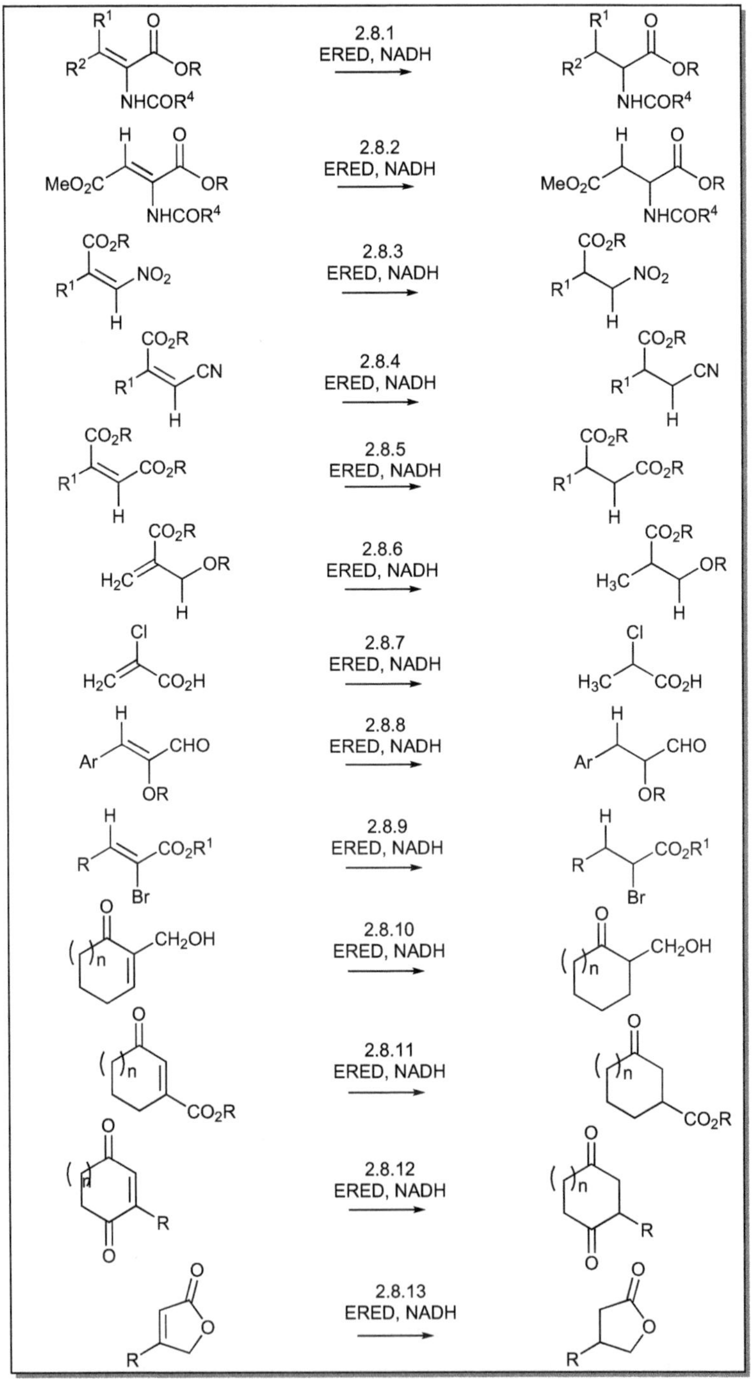

Scheme 10.20   Disconnections involving redox reactions (EREDs).

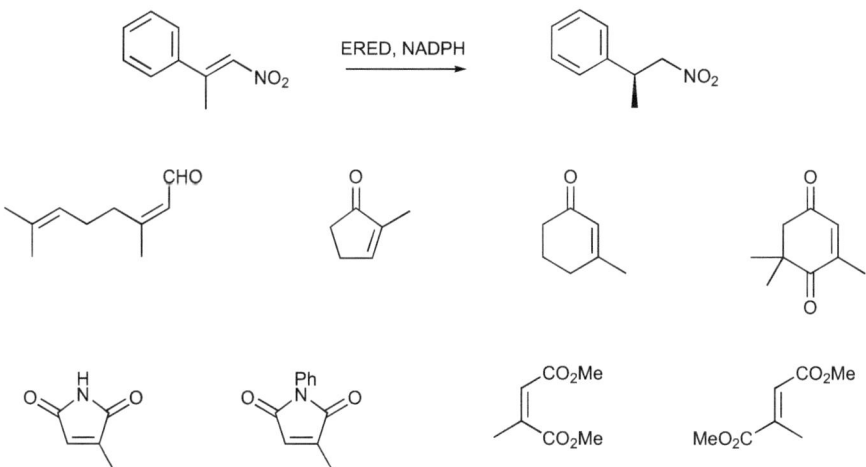

Figure 10.67 Examples of different C=C bond-containing substrates that undergo reduction by an ERED.

also a 1,3-ketone ester to an α-acyl ester[117g] (2.9.9). Styrene monooxygenase (SMO) has been used for the epoxidation of an allylic alcohol to yield the corresponding epoxide in high yield and selectivity (Figure 10.69). The SMO enzyme showed a preference for *para-* and *meta*-substituted substrates, with no activity with *ortho*-substituents. The reactivity generally followed the trend of H > F > Cl > Br.[82]

Finally, a haloperoxidase converts an alkene to a halohydrin[117c] (2.9.10) and asparagine hydroxylase catalyses hydroxylation of asparagine at the β-position (2.9.11). L-Asparagine hydroxylase is a member of the 2-oxoglutarate-dependent hydroxylase family, which also includes L-proline hydroxylase, and was shown to catalyse the regio- and stereoselective hydroxylation of L-asparagine (Figure 10.70).[83]

Vanadium bromoperoxidase (V-BrPO) from *Delisea pulchra* catalyses the bromination of 3-oxo-hexanoylhomoserine lactone to form dibromo-3-oxo-hexanoylhomserine lactone (Figure 10.71).[117c] V-BrPO also catalyses the bromolactonisation of 4-pentynoic acid to give bromofuranone.[84]

10.3 Cyclic Systems

10.3.1 Cyclic Systems: 3-, 4- and 5-membered Rings

Ring-forming reactions are relatively under-represented in biocatalysis, certainly compared to traditional organic synthesis where there

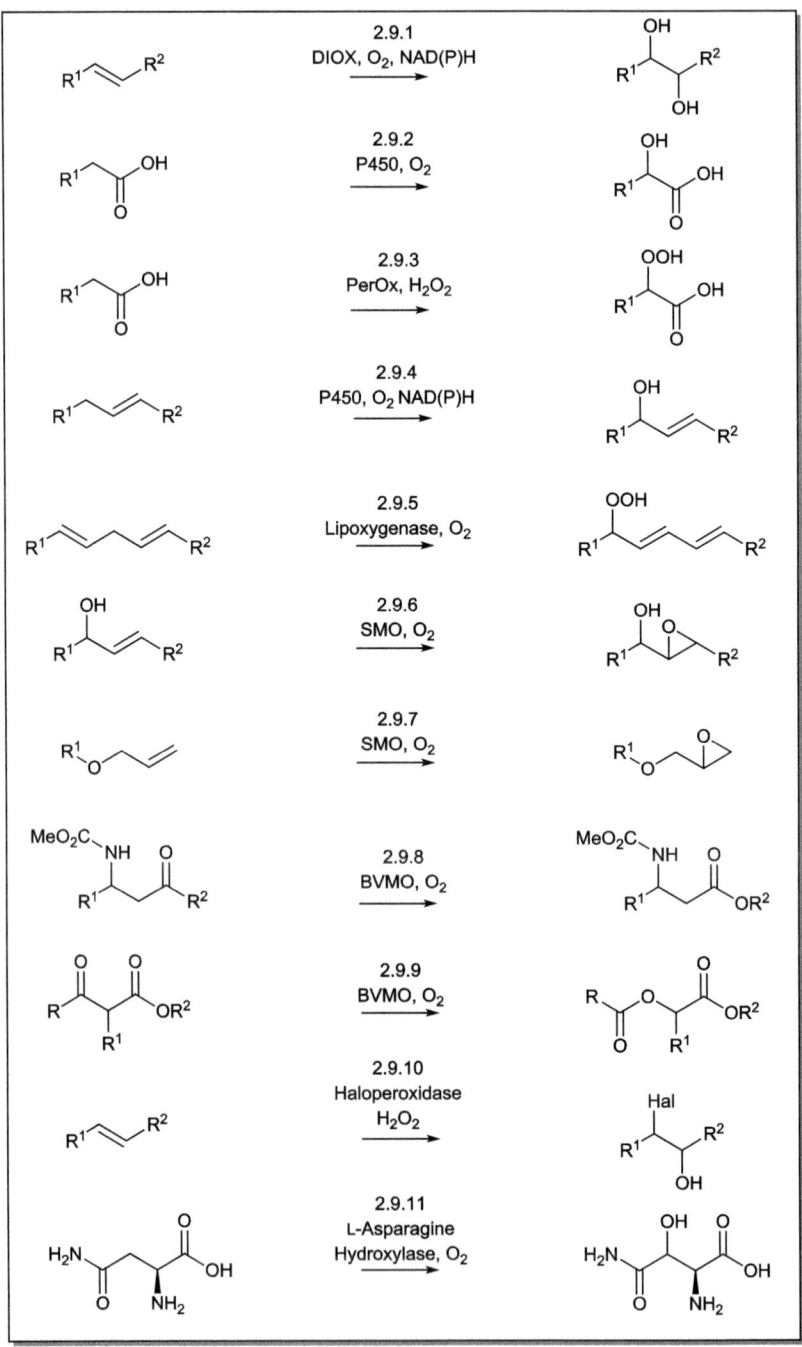

Scheme 10.21 Disconnections involving oxidation reactions.

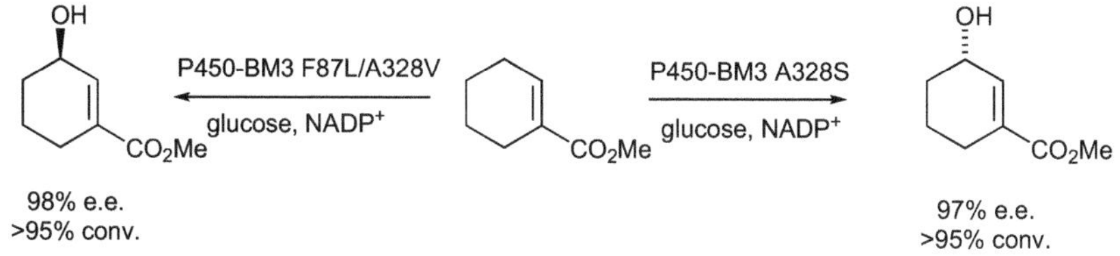

Figure 10.68 Allylic oxidation of an alkene using P450-BM3.

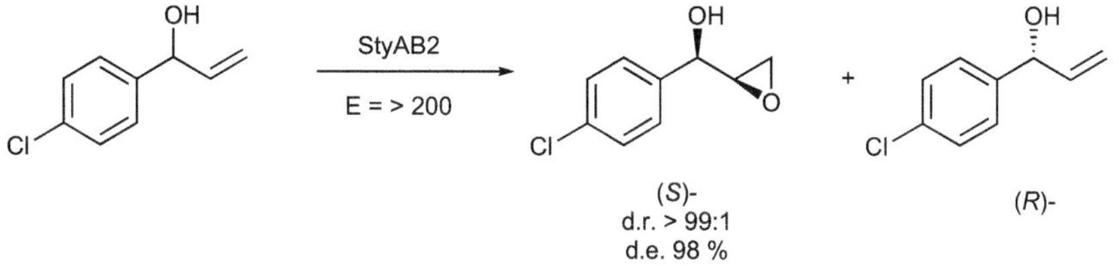

Figure 10.69 Epoxidation of an allylic alcohol using styrene monooxygenase (SMO).

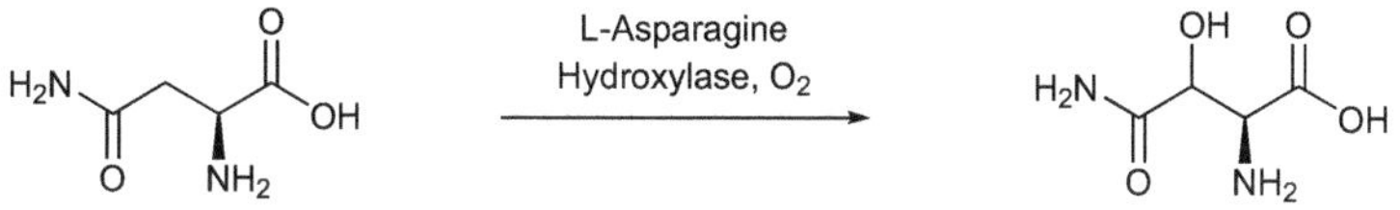

Figure 10.70 Hydroxylation of L-asparagine using a hydroxylase.

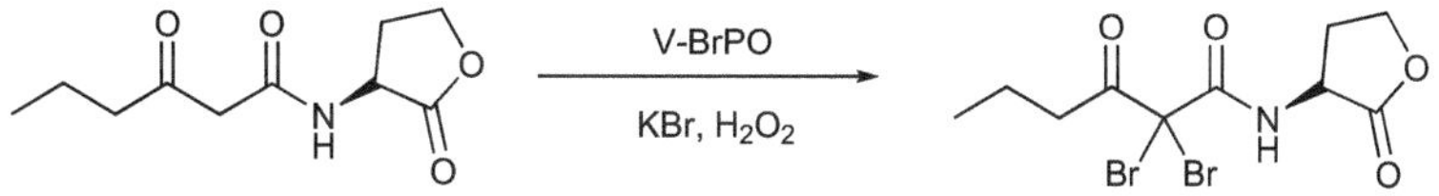

Figure 10.71 Dibromination of a β-keto amide with a vanadium-dependent peroxidase.

are multiple ways of creating cyclic target molecules from acyclic precursors. A recent review provides an excellent overview on the use of biocatalysts for the formation of 3- to 6-membered carbocyclic and heterocyclic rings.[85] A recent and important development is the discovery that certain P450 variants are able to catalyse the enantioselective formation of substituted cyclopropanes using diazo esters and styrene derivatives as precursors (Scheme 10.22, 3.1.1)[86] as well as the formation of aziridines from styrenes using tosyl azide (3.1.2) (Figure 10.72).[87]

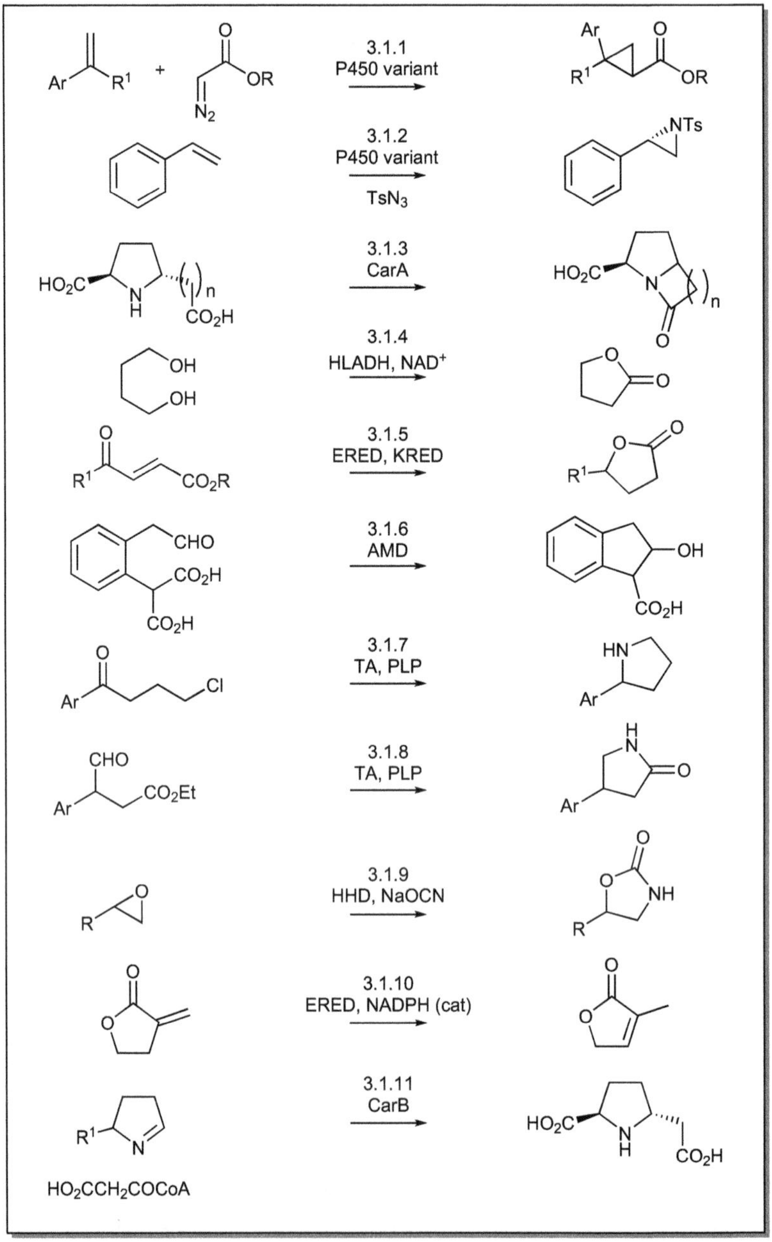

Scheme 10.22   Synthesis of 3-, 4- and 5-membered rings.

CarA, an enzyme involved in the biosynthesis of clavulanic acid, is able to catalyse the formation of a β-lactam from a β-amino acid (3.1.3). 1,4-Diols can be converted to 5-membered ring lactones *via* sequential oxidation *via* a lactol using HLADH[117d] (3.1.4).

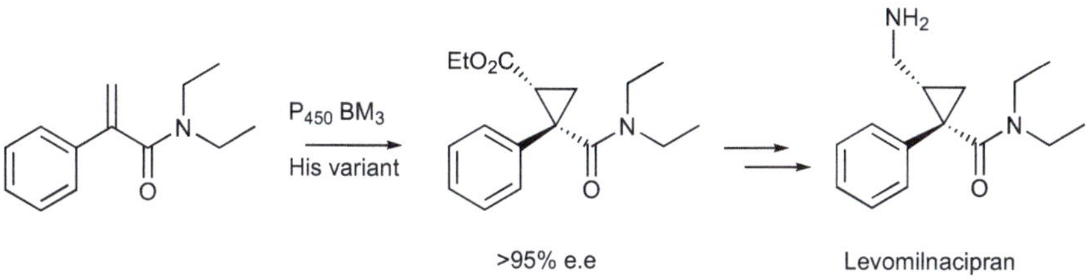

Figure 10.72 Asymmetric cyclopropanation catalysed by a P450 variant.

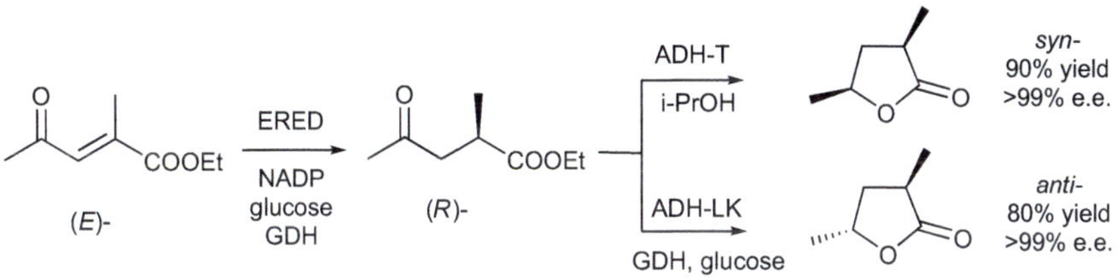

Figure 10.73 Tandem ERED and ADH transformation as a route to chiral δ-lactones.

The combination of an **ERED** and **ADH** converts an enone ester to the butyrolactone (3.1.5). A one-pot method for the preparation of lactones was developed involving initial reduction by OYE1 (Old Yellow Enzyme 1, *Saccharomyces carlsbergensis*) before adding an alcohol dehydrogenase to give the hydroxyester, which underwent spontaneous lactonisation. The use of ADH-T (*Thermoanaerobacter* sp.) or ADH-LK (*Lactobacillus kefir*) gave the *syn-* and *anti-*products, respectively, in high yield and >99% e.e. (Figure 10.73).[88]

Arylmalonate decarboxylase (AMD) catalyses C–C decarboxylation followed by C–C bond formation to yield substituted indanes[116d] (3.1.6). ω-TA enzymes can be used to convert chloroketones to substituted pyrrolidines (3.1.7) and also ester aldehydes to 5-membered ring lactams[116m] (3.1.8). Treatment of an epoxide with HHD and NaOCN yields an oxazolidinone[116q] (3.1.9). Exomethylene 5-membered ring lactones can be isomerised to the endocyclic isomers using an ERED and catalytic NADPH[116g] (3.1.10). Finally, CarB catalyses the addition of a malonyl-derived C2 unit to a pyrroline (3.1.11).

10.3.2 Cyclic Systems: 5- or 6-membered Carbocyclic Rings

In this short section, we look at transformations that operate equally well on 5- or 6-membered rings. IRED biocatalysts catalyse the reduction of a wide range of cyclic imines to the corresponding cyclic amines using NADPH[116l] (Scheme 10.23, 3.2.1). For example, a range of 5-, 6- and 7-membered cyclic imines were reduced to the corresponding cyclic amines using an NADPH-dependent imine reductase from *Streptomyces* sp. The reactions could be carried out either using whole cells or alternatively the purified enzyme with a co-factor recycling system (Figure 10.74).[89]

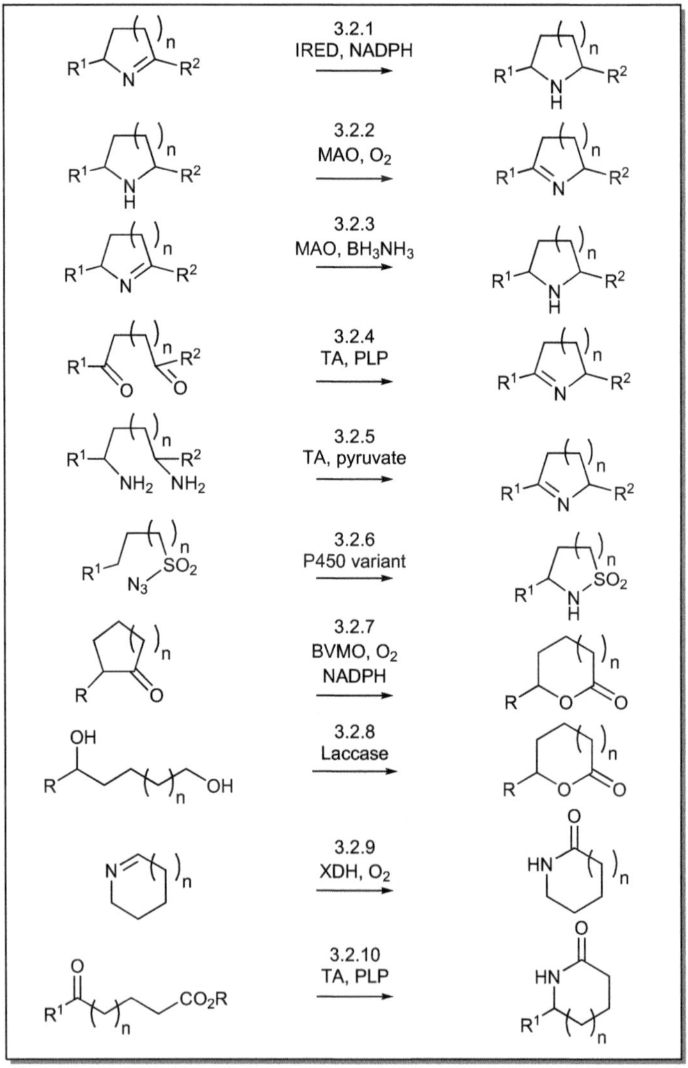

Scheme 10.23 Synthesis of either 5- or 6-membered rings.

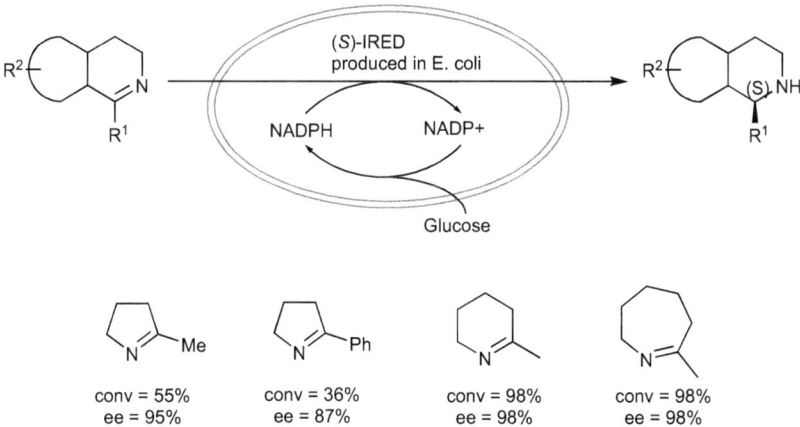

Figure 10.74 Asymmetric reduction of imines using NADPH-dependent IREDs.

In contrast, monoamine oxidase (MAO-N) catalyses the oxidation of cyclic amines to cyclic imines[117h] (3.2.2). In combination with ammonia borane, this system can be used to convert imines to cyclic amines (3.2.3). 1,*n*-diketones can be converted to cyclic imines *via* regioselective amination followed by spontaneous cyclisation (3.2.4). 2,5-Disubstituted pyrrolidines were prepared *via* oxidation/reduction of preformed imines, which themselves were generated from *trans*aminase-mediated asymmetric amination of the corresponding 1,4-diketones (Figure 10.75).[90]

Alternatively, 1,*n*-diamines can be converted to the same target molecule *via* regioselective deamination followed by cyclisation using putrescine transaminase in the presence of pyruvate as the amine acceptor[116m] (3.2.5).

A P450 variant has been used to generate a sulfonyl azide-derived nitrene, which then undergoes C–H bond insertion to yield a γ-sultam (3.2.6) (Figure 10.76).[91]

BVMOs catalyse the conversion of a wide range of cyclic ketones to the corresponding lactone[117g] (3.2.7). An FAD-containing mono-oxygenase from *Stenotrophomonas maltophilia* (SMFMO) catalyses the regioselective Baeyer–Villiger oxidation of bicyclo[3.2.0]hept-2-en-6-one (Figure 10.77). Quantitative conversion to the 2-oxa and 3-oxa lactone products occurred in a 5 : 1 ratio, although with low enantioselectivity. The enzyme was able to employ either NADH or NADPH as a co-factor.[92]

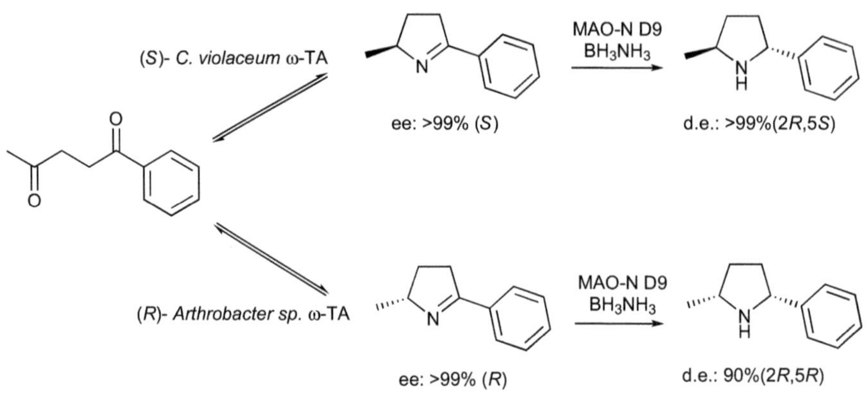

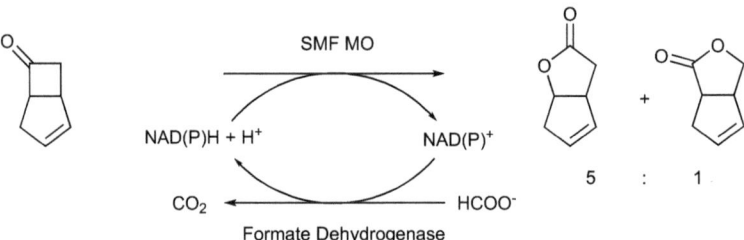

Figure 10.75 Combination of ω-TA with MAO-N/BH₃ for conversion of 1,5-diketones to 2,5-disubstituted pyrrolidines.

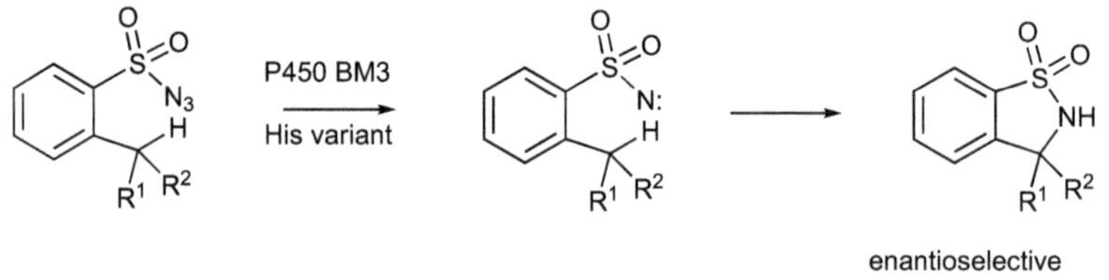

Figure 10.76 P450 variant-mediated nitrene formation and insertion into a C–H bond provides a route to sultams.

Figure 10.77 BVMO-catalysed formation of bicyclic lactones.

5- and 6-Membered ring lactones can also be prepared from diols using laccase[117c] (3.2.8). Oxidation of cyclic imines using xanthine dehydrogenase (XDH) gives rise to lactams (3.2.9) (Figure 10.78).[93]

Finally, amination of a keto ester using a TA followed by spontaneous cyclisation also provides a route to 5- and 6-membered ring lactams (3.2.10).

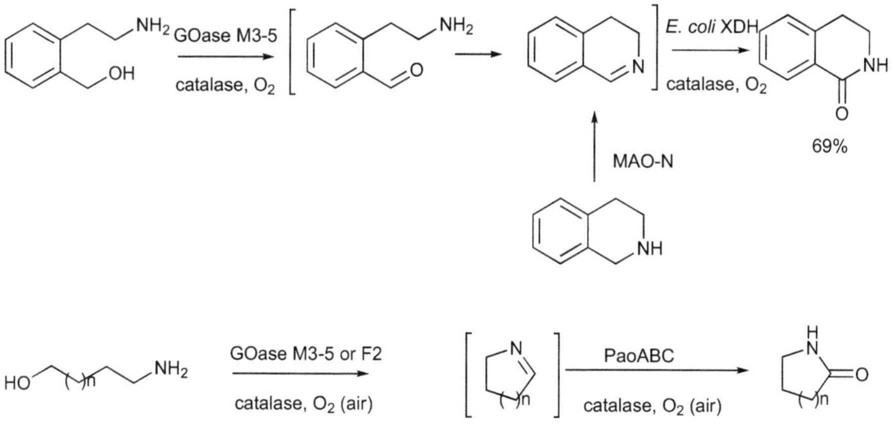

Figure 10.78 Synthesis of 5- and 6-membered ring lactams by tandem oxidation of amino alcohols using GOase together with either XDH or PaoABC.

10.3.3   Cyclic Systems: 6- and 7-membered Carbocyclic Rings

A wide variety of substituted carbocyclic 6-membered rings can be prepared *via* squalene hopane cyclase (SHC)-catalysed cyclisation of 1,5-dienes in the presence of different internal nucleophiles X (Scheme 10.24, 3.3.1). By the judicious choice and placement of the pendant nucleophile X, it is possible to trap the initially generated cyclic carbonium ion to access a wide range of functional groups including lactones, enol ethers, cyclic ethers and nitrogen-containing rings (Figure 10.79).[94]

In a rather specialised transformation, laccase catalyses the oxidative rearrangement of substituted furans to 6-membered ketolactols[117e] (3.3.2). Cyclic enamines equilibrate with enamines in the presence of IREDs to generate stereochemically defined 2,3-disubstituted piperidines (3.3.3). 2,3-Disubstituted piperidines were prepared *via* imine reductase-mediated reduction of cyclic imines, which are in equilibrium with the enamine form allowing two adjacent stereogenic centres to be established in one step (Figure 10.80).[95]

Norcoclaurine synthase (NCS) catalyses a Pictet–Spengler cyclisation on an *in situ* formed imine or iminium ion to yield a tetrahydroisoquinoline[116e] (3.3.4). The analogous transformation on tryptamine derivatives catalysed by strictosidine synthase (STS) yields β-carbolines (3.3.5) and isomers (3.3.6). The enzyme norcoclaurine synthase (NCS) from *Thalictrum flavum* catalysed C–C bond formation

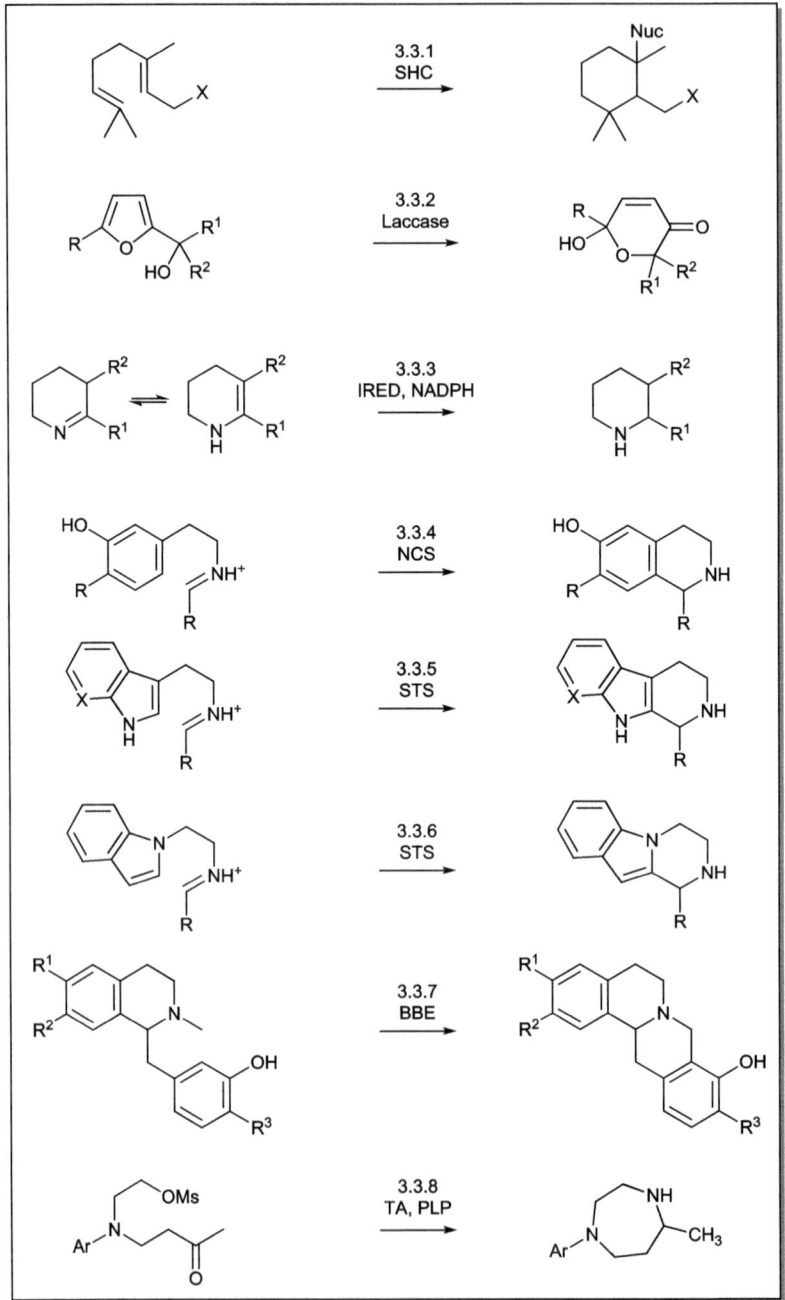

Scheme 10.24 Synthesis of cyclic 6- or 7-membered rings.

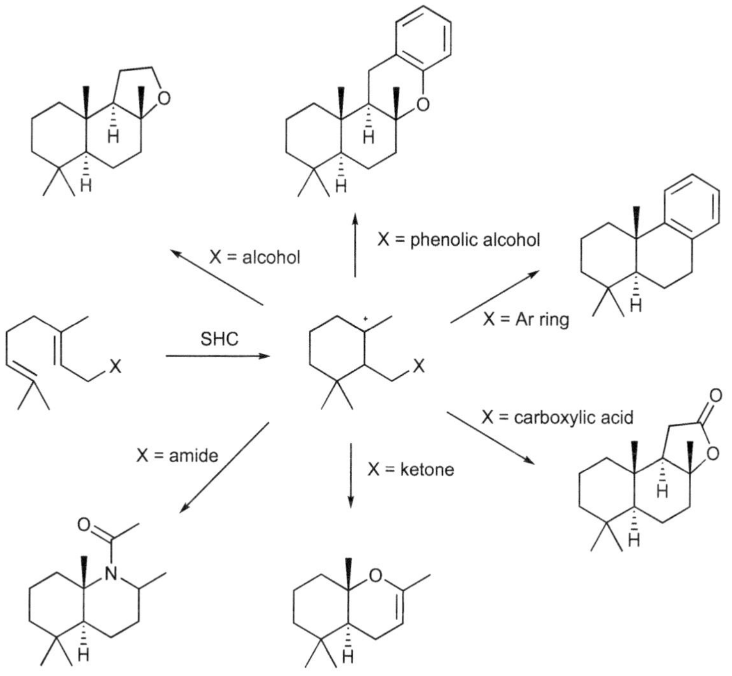

Figure 10.79 Range of different products available from SHC-catalysed cyclisation of a carbocation derived from a 1,5-diene.

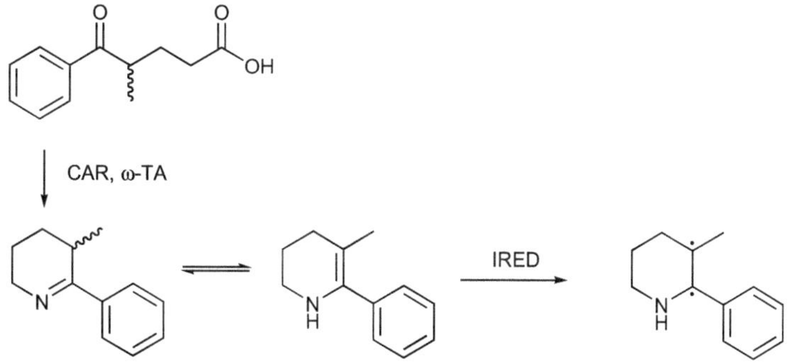

Figure 10.80 Synthesis of a 2,3-disubstituted piperidine using a cascade process involving CAR/ω-TA/IRED.

between 4-hydroxyphenylacetaldehyde (4-HPAA) and dopamine to afford (*S*)-norcoclaurine (Figure 10.81). A variety of additional aldehydes were also shown to be substrates for this enzyme, producing an array of substituted tetrahydroisoquinolines, with conversions

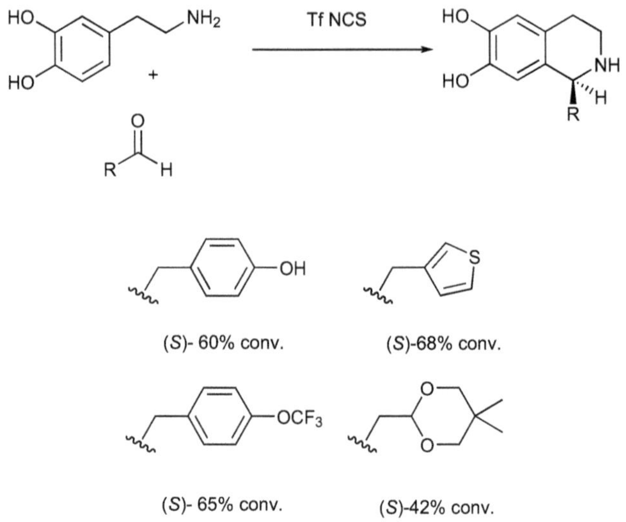

Figure 10.81 Synthesis of 1-substituted tetrahydroisoquinolines using norcoclaurine synthase (NCS).

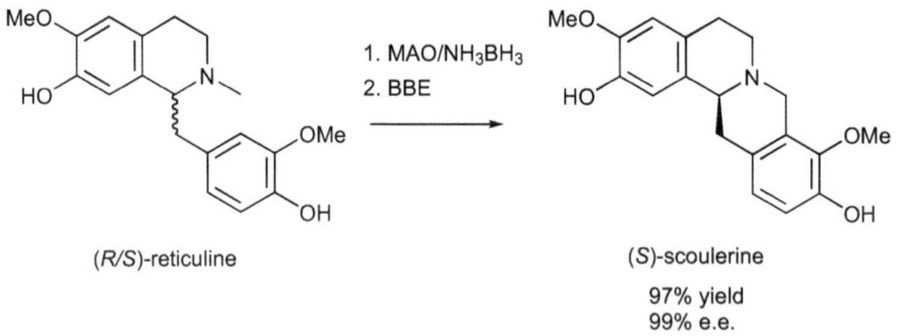

Figure 10.82 Conversion of racemic reticuline to (S)-scoulerine using MAO-N/ammonia-borane/BBE.

between 42% and 71%. In contrast to the relaxed aldehyde specificity, NCS appears to have a strict requirement for the amine substrate dopamine.[96]

Berberine bridge enzyme (BBE) catalyses the conversion of (S)-reticuline to (S)-scoulerine (3.3.7). BBE was combined with MAO-N and ammonia borane to carry out an *in situ* deracemization followed by C–C bond formation leading to the synthesis of scoulerine and its analogues (Figure 10.82).[97]

Finally, TA-mediated amination followed by base-mediated cyclisation generates 7-membered ring diazepanes[116m] (3.3.8).

10.4 Aromatic and Heteroaromatic Rings

10.4.1 Direct Modification of Aromatic and Heteroaromatic Rings

This section deals with transformations in which an aromatic or heteroaromatic ring undergoes direct modification by either substitution or oxidation. Hydroxylation of an aromatic ring catalysed by a P450 monooxygenase (Scheme 10.25, 4.1.1) leads to the phenol whereas a phenol can be oxidised to a catechol using either a P450 or

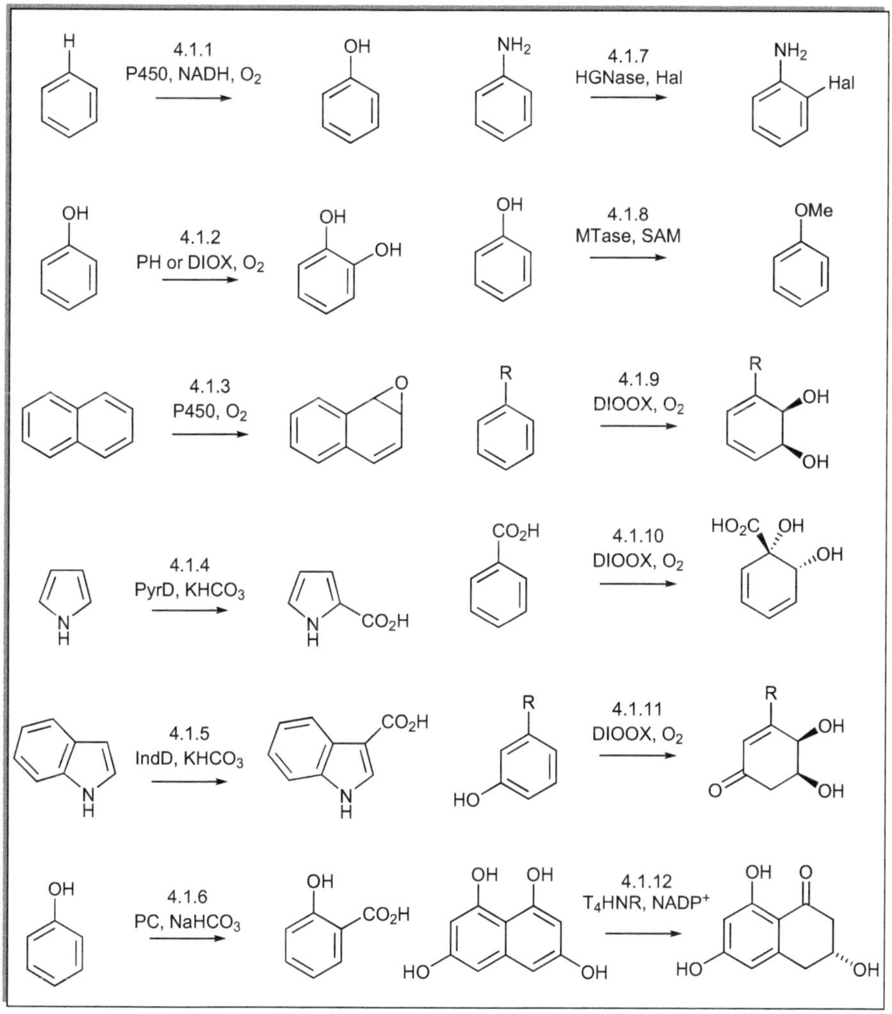

Scheme 10.25 Direct modification of aromatic and heteroaromatic rings.

peroxidase[117a,e] (4.2.2). For example, horse radish peroxidase (HRP), immobilised inside the pores of ordered mesoporous silica (OMS), was used to convert L-tyrosine to L-DOPA with *ca.* 45% conversion (Figure 10.83).[98]

Naphthalene can also be converted to the corresponding epoxide using a dioxygenase[117a] (4.1.3). Carboxylation of aromatic and heteroaromatic systems can be catalysed by using carboxylase enzymes in the presence of high concentrations of bicarbonate.[116d] Pyrrole carboxylase converts pyrrole to pyrrole-2-carboxylic acid (4.1.4) and indole carboxylase catalyses the analogous transformation of the indole (4.1.5). Benzoic acid decarboxylase converts phenol derivatives to *ortho*-carboxylic acid derivatives in the presence of potassium hydrogen carbonate (4.1.6) (Figure 10.84).[99]

Halogenation of aromatic compounds is a general and important transformation in Nature for the production of a wide range of chlorinated and brominated natural products.[117j] Two distinct classes of enzymes are responsible for this transformation, namely peroxidases and flavin-dependent halogenases (4.1.7). For example, the enzyme VBrPO(AnI), a vanadate(v)-dependent bromoperoxidase I from the brown alga *Ascophyllum nodosum*, catalyses mono-, di- and tri-bromination of substituted pyrroles (Figure 10.85). Combined yields of all products of up to 91% can be obtained by continuous addition of hydrogen peroxide and sodium bromide.[100]

Halogenation and bromination of tryptophan and its derivatives is mediated by the halogenases PrnA and RebH in the presence of either a chloride or bromide ion, respectively (Figure 10.86).[101]

Another broad class of enzymes are the SAM-dependent alkyl transferases,[116f] which catalyse methylation of a phenol to the

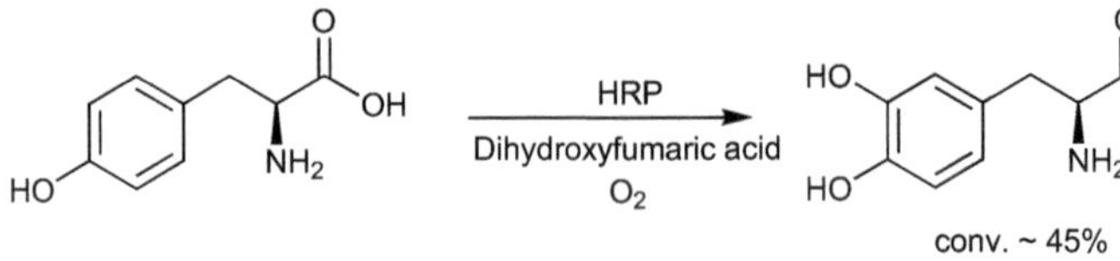

Figure 10.83 Conversion of L-tyrosine to L-DOPA using horse radish peroxidase (HRP).

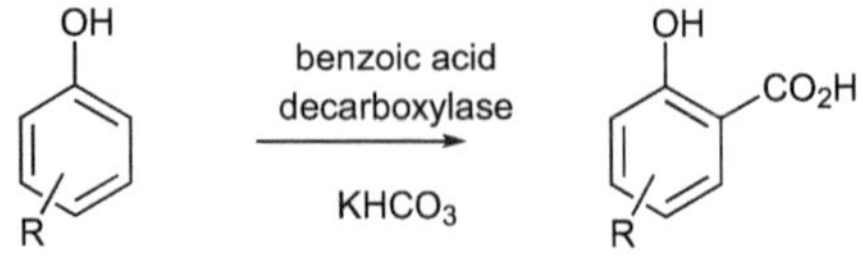

Figure 10.84 A biocatalytic 'Kolbe–Schmidt' reaction.

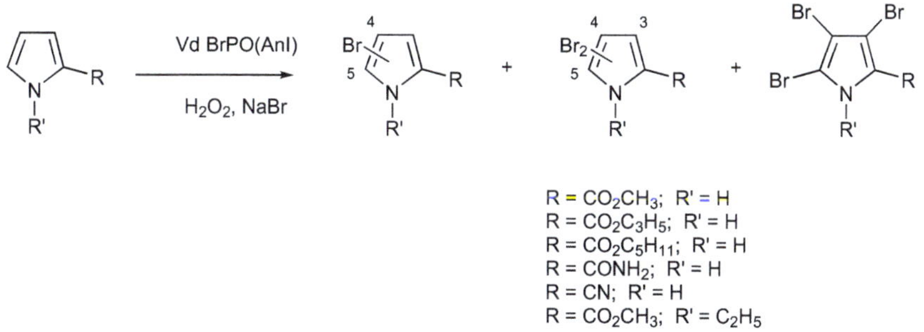
Vd BrPO(AnI)
H_2O_2, NaBr
R = CO_2CH_3; R' = H
R = $CO_2C_3H_5$; R' = H
R = $CO_2C_5H_{11}$; R' = H
R = $CONH_2$; R' = H
R = CN; R' = H
R = CO_2CH_3; R' = C_2H_5

Figure 10.85 Bromination of pyrrole derivatives using a vanadium-dependent bromoperoxidase.

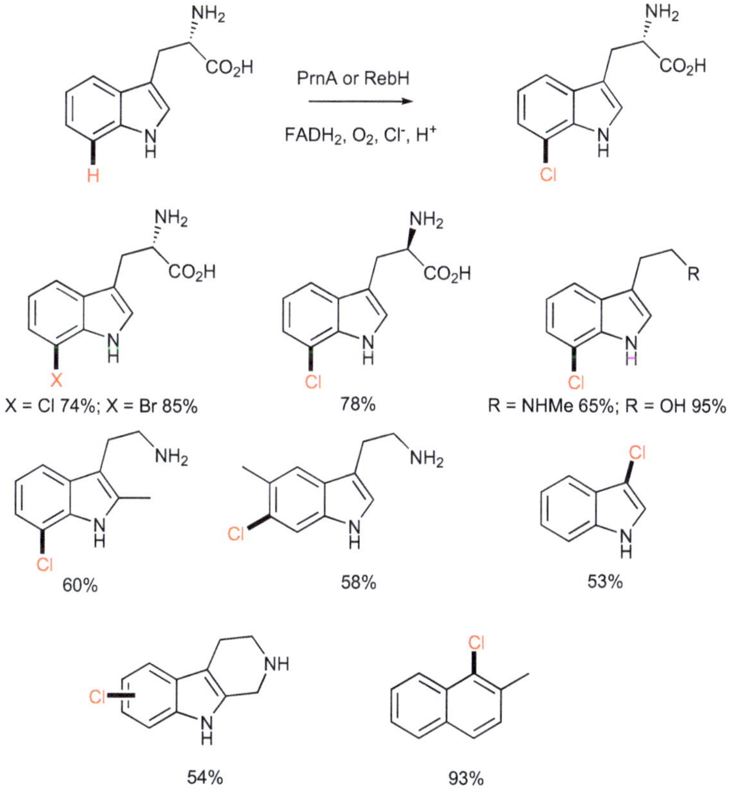
PrnA or RebH
$FADH_2$, O_2, Cl^-, H^+
X = Cl 74%; X = Br 85%
78%
R = NHMe 65%; R = OH 95%
60%
58%
53%
54%
93%

Figure 10.86 Halogenation tryptophan derivatives using the halogenases PrnA or RebH.

corresponding aryl methyl ether (4.1.8). These enzymes often exhibit regioselectivity, for example in the methylation of 3,4-dihydroxybenzoic acid to give the 3-methyl ether (Figure 10.87).[102]

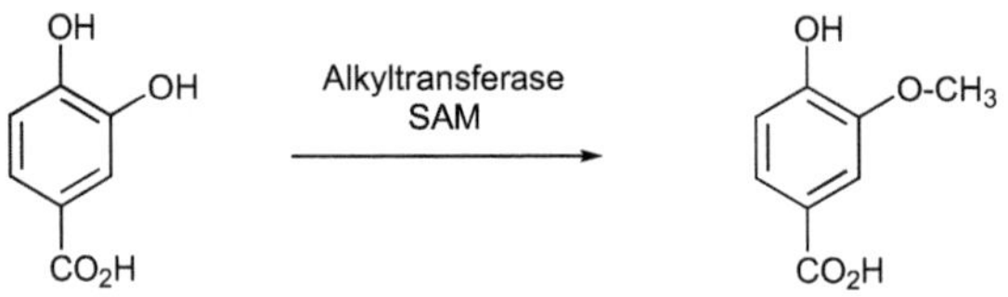

Figure 10.87 Regioselective methylation of a catechol using a SAM-dependent alkyltransferase.

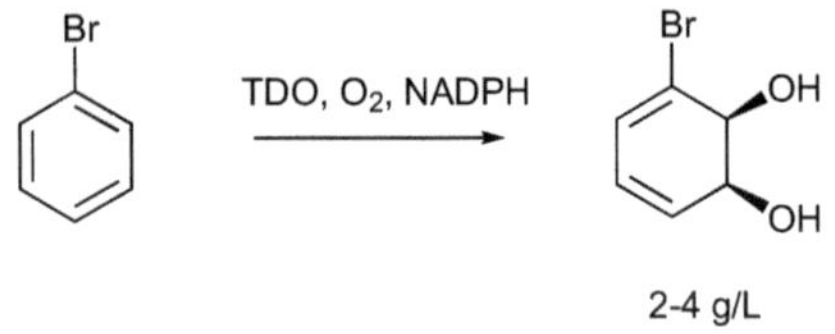

2-4 g/L

Figure 10.88 Oxidation of bromobenzene to a chiral cyclohexadiendiol using toluene dioxygenase (TDO).

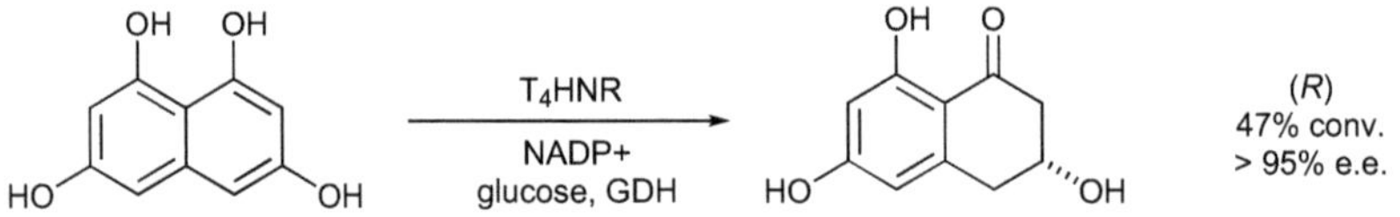

(R)
47% conv.
> 95% e.e.

Figure 10.89 Reduction of a 1,3-diphenol to the corresponding β-hydroxyketone.

Dioxygenases catalyse the conversion of a broad range of substituted arenes to the corresponding cyclohexadiene 2,3-diols[117a] (4.1.9). For example, toluene dioxygenase (TDO) from *Pseudomonas putida* was used as the biocatalyst for the conversion of bromobenzene to bromocyclohexadienediol. A 2 L fermentation gave access to 2–4 g L^{-1} of the product, which was then subjected to further chemistry to afford pancratistatin analogues in high yield (Figure 10.88).[103]

The use of a benzoic acid leads to a change in regioselectivity to give the 1,2-diol (4.1.10), where a phenol generates the corresponding enone (4.1.11). Finally, 1,3-dihydroxynaphthalenes can be reduced to β-hydroxyketones (4.1.12). Tetrahydroxynaphthalene reductase (T4HNR) selectively reduces 1,3,6,8-tetrahydroxynaphthalene to scytalone with high enantiomeric purity (Figure 10.89). T4HNR has a broad substrate range and can tolerate substrates with other hydroxy and methoxy substituents, as well as tetralone derivatives.[104]

Tryptophan synthase (Scheme 10.26, 4.1.14) is a PLP-dependent enzyme that catalyses C–C bond formation between indole and serine (Figure 10.90).[116j] Biofilms of *E. coli* PHL644 (harbouring an expression plasmid with the tryptophan synthase gene from *Salmonella enterica*)

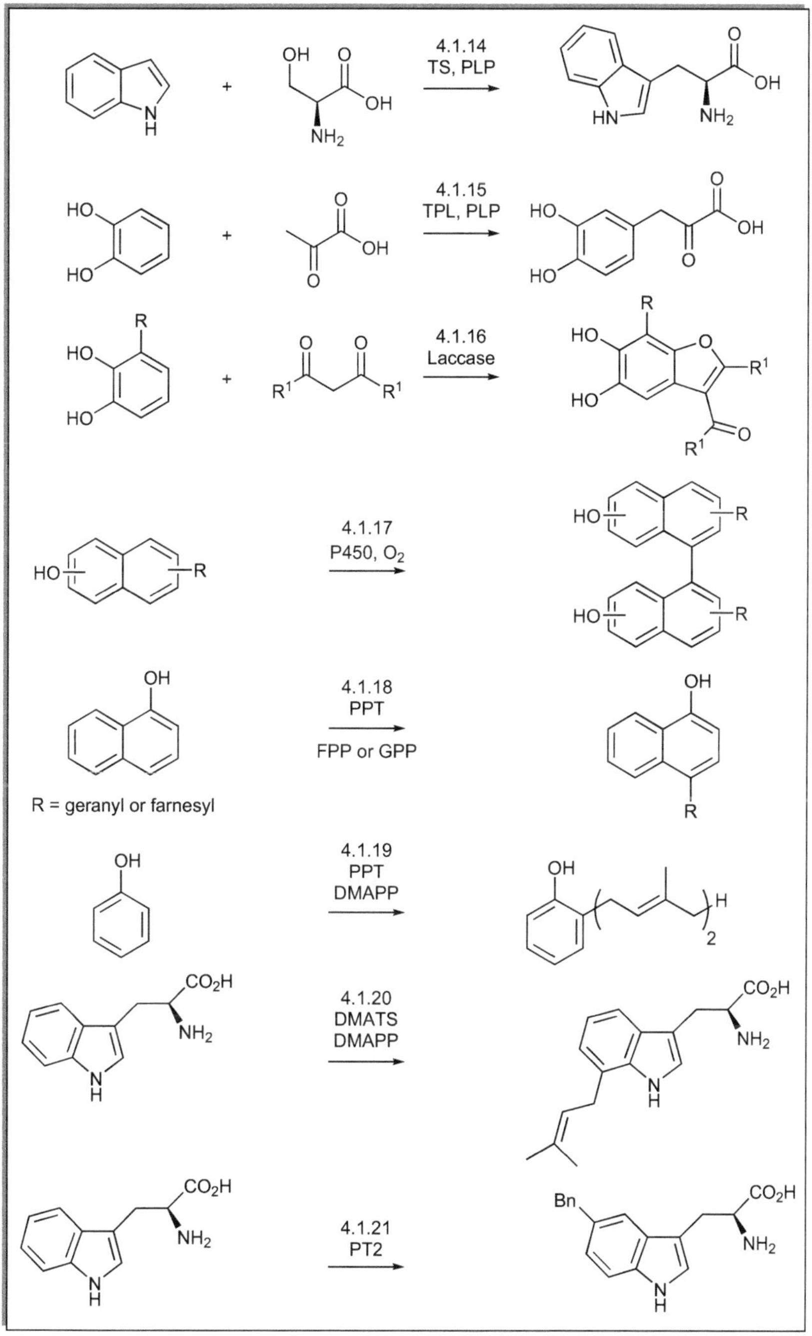

Scheme 10.26   Direct modification of aromatic and heteroaromatic rings.

were generated by spin-coating cells onto a poly-L-lysine-coated glass slide and maturing for up to 7 days. In biotransformations with indole analogues, the engineered biofilm was found to produce higher amounts of halotryptophans than the purified protein or a cell-free lysate.[105]

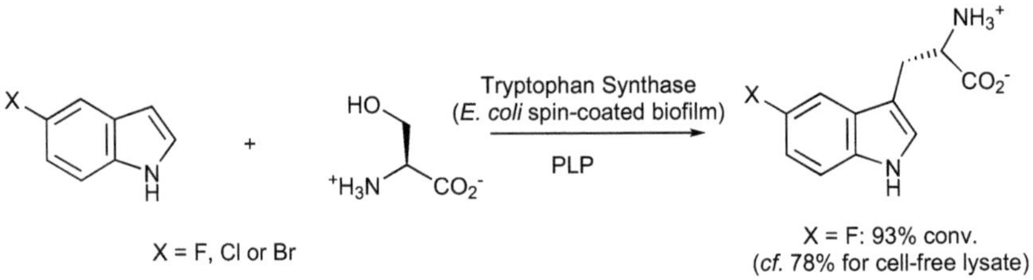

Figure 10.90 Tryptophan synthase-catalysed synthesis of tryptophan analogues.

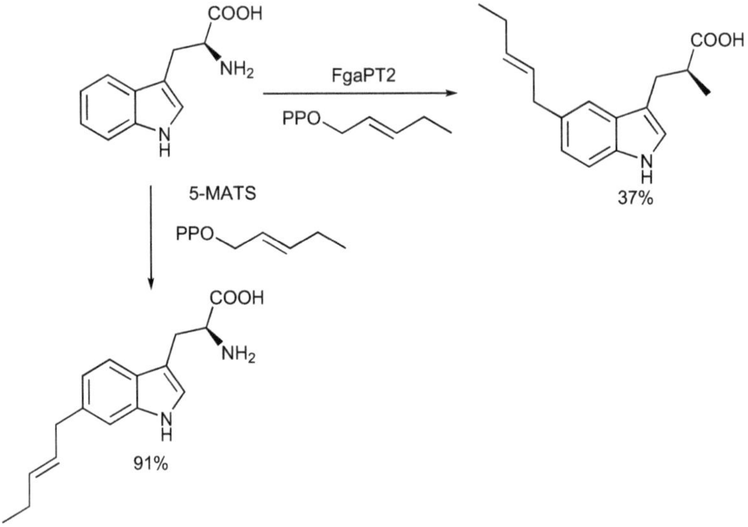

Figure 10.91 Alkylation of tryptophan at either C-4 or C-5.

A related PLP enzyme is TP lyase, which catalyses C–C bond formation between a catechol and pyruvic acid to give an aryl pyruvic acid (4.1.15). Laccase mediates C–C bond formation on furan derivatives[117e] (4.1.16). Two naphthol units can be ligated using a P450 monooxygenase to give chiral, non-racemic, binaphthol derivatives (4.1.17).[106]

Finally, we look at a series of enzymes that catalyse C–C bond formation of arenes using electrophiles derived from farnesyl pyrophosphate (FPP), geranyl pyrophosphate (GPP) or dimethylallyl pyrophosphate[116f] (DMAPP). The enzyme PPT can use either a naphthol (4.1.18) or a phenol (4.1.19) as the nucleophile in this reaction. Tryptophan is a substrate for both DMATS (4.1.20) and PT2 (4.1.21) enzymes.

FgaPT2 from *Aspergillus niger* and 5-DMATS from *Aspergillus clavatus* catalyse the prenylation of L-tryptophan using dimethylallyl phosphate (DMAPP) at C-4 and C-5, respectively (Figure 10.91).[107]

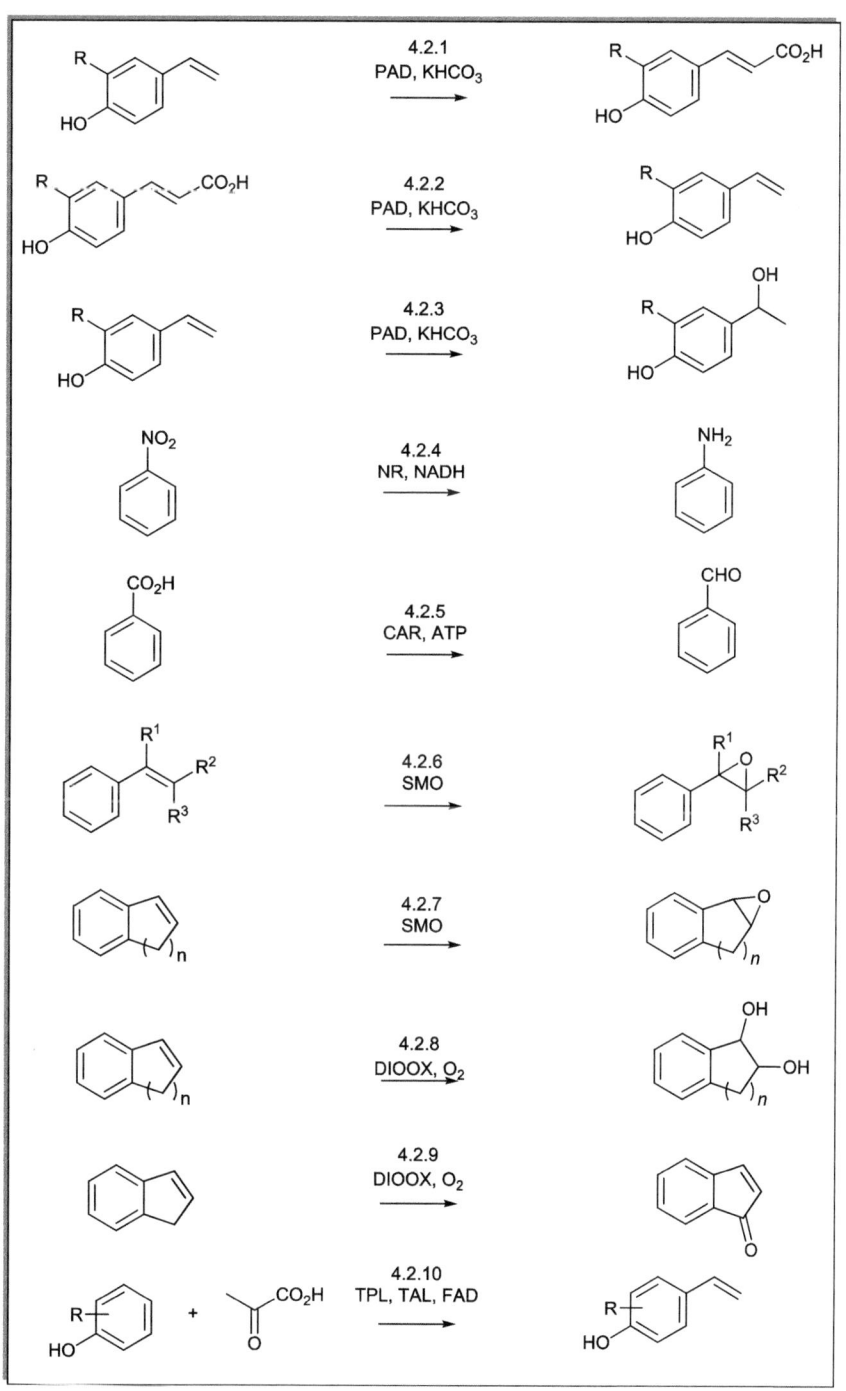

Scheme 10.27 Indirect modification of aromatic and heteroaromatic rings.

10.4.2 Indirect Modification of Aromatic and Heteroaromatic Rings

Members of the PAD enzyme family catalyse the carboxylation of *para*-hydroxy styrene derivatives (Scheme 10.27, 4.2.1) as well as the reverse reaction[116d] (4.2.2). The enzyme also exhibits a promiscuous activity, namely the enantioselective hydration of styrene derivatives to give phenethanols (4.2.3).[108] As an example of carboxylation activity, 2,3-dihydroxybenzoate decarboxylase (*Aspergillus oryzae*) (2,3-DHBD Ao) produces the corresponding *O*-hydroxybenzoic acid derivative from the starting styrene derivative in moderate conversion, whilst the phenolic acid decarboxylase PAD Lp (*Lactobacillus plantarum*) carboxylates selectively at the β-carbon atom to produce the corresponding *E*-cinnamic acid (Figure 10.92).[109]

Nitroreductases catalyse the reduction of substituted electron-deficient nitroarenes to give nitroso derivatives (Figure 10.93), which in some cases can be isolated, but often undergo further rapid reduction to the hydroxylamine and aniline (4.2.4).[110]

Carboxylic acid reductase (CAR) catalyses the reduction of a wide range of substituted benzoic acid derivatives to the corresponding benzaldehydes in high conversion with no over-reduction to the alcohol[116o] (4.2.5) (Figure 10.94).

There are a group of oxidation enzymes that can modify styrenes and indenes.[117a] Styrene monooxygenase (SMO) catalyses epoxidation of styrene derivatives (4.2.6). Indenes can undergo three types of

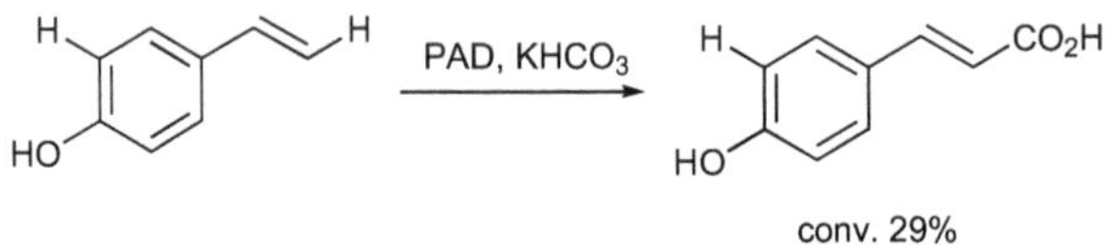

conv. 29%

Figure 10.92 Carboxylation of 4-hydroxy-styrene using phenolic acid decarboxylase in the presence of potassium hydrogen carbonate.

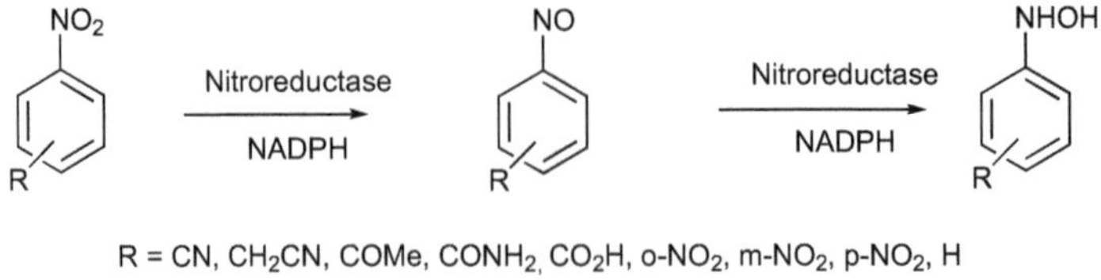

R = CN, CH₂CN, COMe, CONH₂, CO₂H, o-NO₂, m-NO₂, p-NO₂, H

Figure 10.93 Nitroreductase-mediated reduction of nitroarenes.

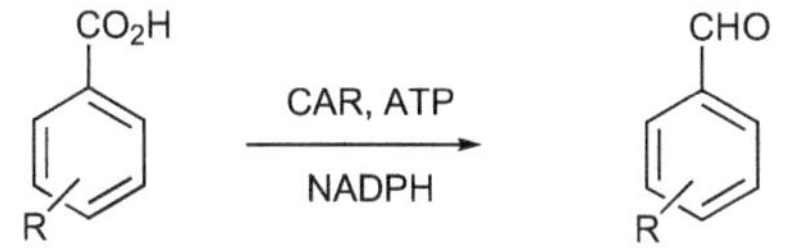

Figure 10.94   CAR-catalysed reduction of benzoic acids to benzaldehydes.

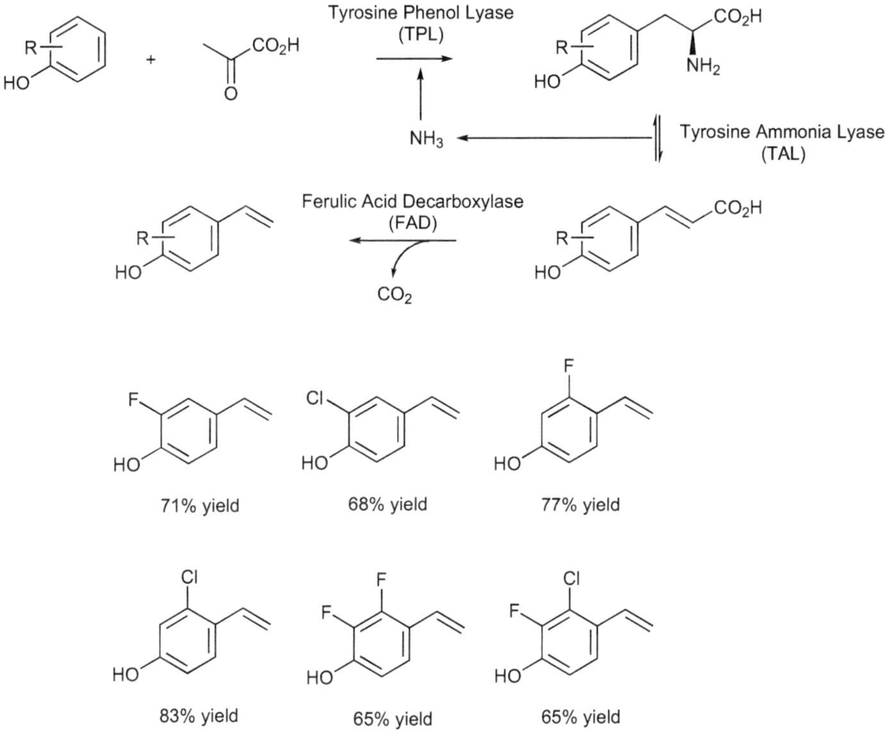

Figure 10.95   Three-enzyme system for the overall *para*-vinylation of substituted phenols.

oxidation process to the epoxide (4.2.7), the diol (4.2.8) and the corresponding enone (4.2.9).

Finally, an interesting cascade has been reported, which results in overall *para*-vinylation of substituted phenols (4.2.10). Initial C–C bond formation between the phenol and pyruvate in the presence of ammonia catalysed by tyrosine phenol lyase (TPL) leads to the substituted tyrosine, which then undergoes deamination catalysed by tyrosine ammonia lyase (TAL). Finally, decarboxylation to the styrene derivative is achieved by using ferulic acid decarboxylase (FAD) (Figure 10.95).[111]

10.5   Biocatalysts Acting on Carbohydrates

The use of enzymes for carbohydrate synthesis is a rapidly developing area, which offers great promise and application in synthesis, primarily in view of the control of stereochemistry and the lack of requirement for protecting groups. The principal type of biocatalyst acting on carbohydrates are glycosidases (Scheme 10.28, 5.1.1), which catalyse hydrolysis of glycosides to the parent carbohydrate and an

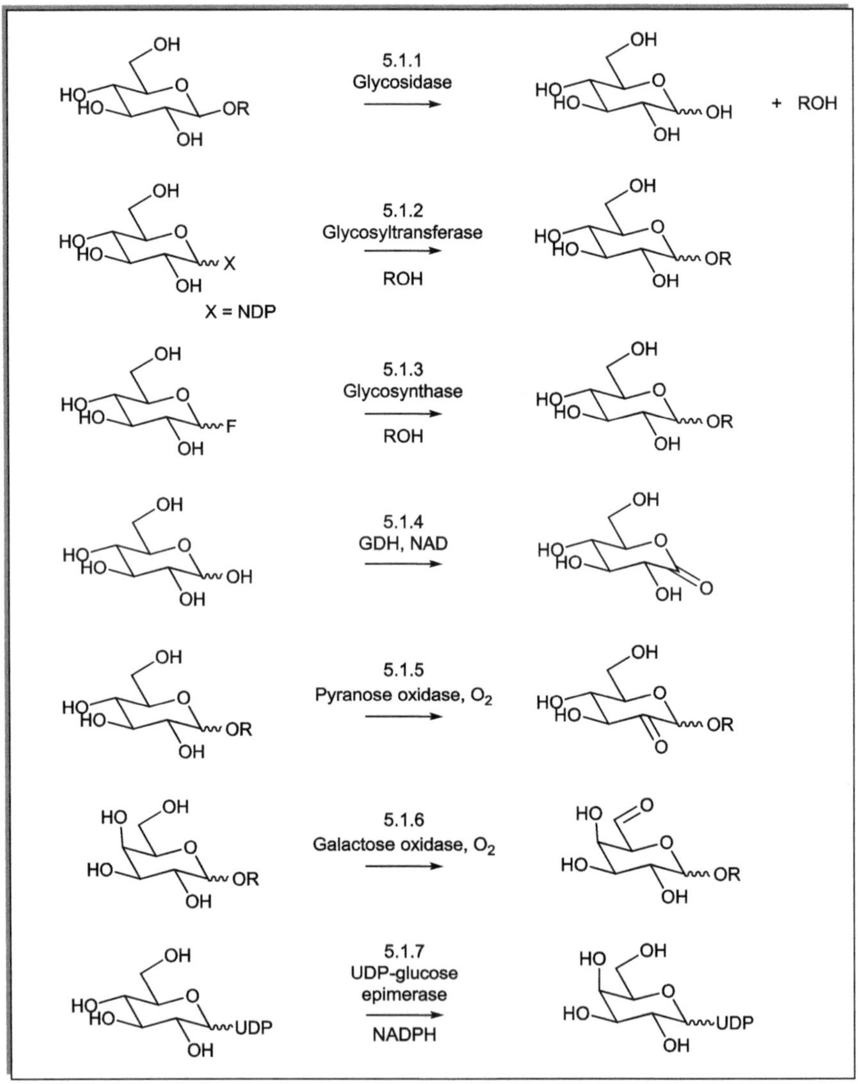

Scheme 10.28   Biocatalysts acting on carbohydrates.

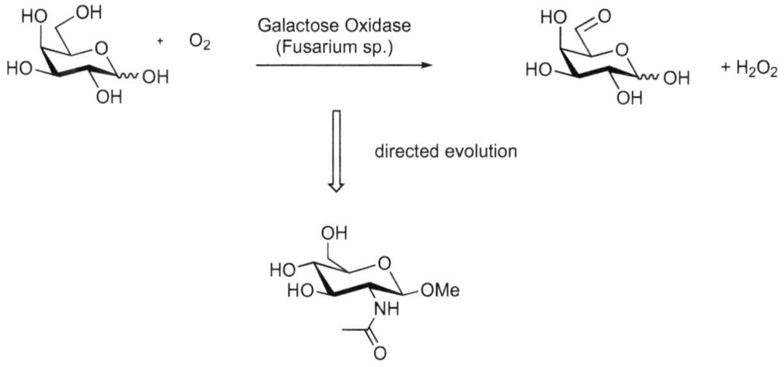

Figure 10.96 Directed evolution of GOase to enable oxidation of carbohydrates other than galactose at C-6.

alcohol,[115k] and secondly glycosyl transferases (5.1.2), which catalyse
the formation of a wide range of different carbohydrate linkages
using NDP-activated sugars as glycosyl donors.[115l] More recently, glycosynthases[115k] have been engineered that are able to use glycosyl
fluorides as glycosyl donors (5.1.3). Glucose dehydrogenase, which is
employed in the recycling of the NADH co-factor, catalyses the oxidation of glucose at C-1 to yield gluconolactone (5.1.4). Pyranose oxidase, by contrast, oxidises sugars at C-2 to the ketone (5.1.5). Galactose
oxidase is specific for galactose and related sugars and oxidises at C-6
to the aldehyde (5.1.6). However, this enzyme has also been engineered
to accept other sugars such as *N*-acetylglucosamine (Figure 10.96).[112]

  Finally, UDP-glucose epimerase interconverts UDP-glucose with
UDP-galactose (5.1.7).

10.6 Biocatalysts Acting on Nucleosides

Although lipases, esterases and proteases are widely used in the
selective protection, deprotection and modification of nucleosides,
especially the sugar component, this section focuses only on those
biocatalysts that have a particular application for this class of compound. For excellent overviews, see ref. 113 and 114. Adenosine
deaminase (ADA) catalyses the deamination of 2'-deoxynucleosides
containing 1,3-diamino purines (Scheme 10.29, 6.1.1). Adenylate
deaminase (ADADA) shows much wider substrate tolerance and
catalyses an analogous process on a broad range of unnatural nucleosides containing, for example, amine, methoxy and chloro

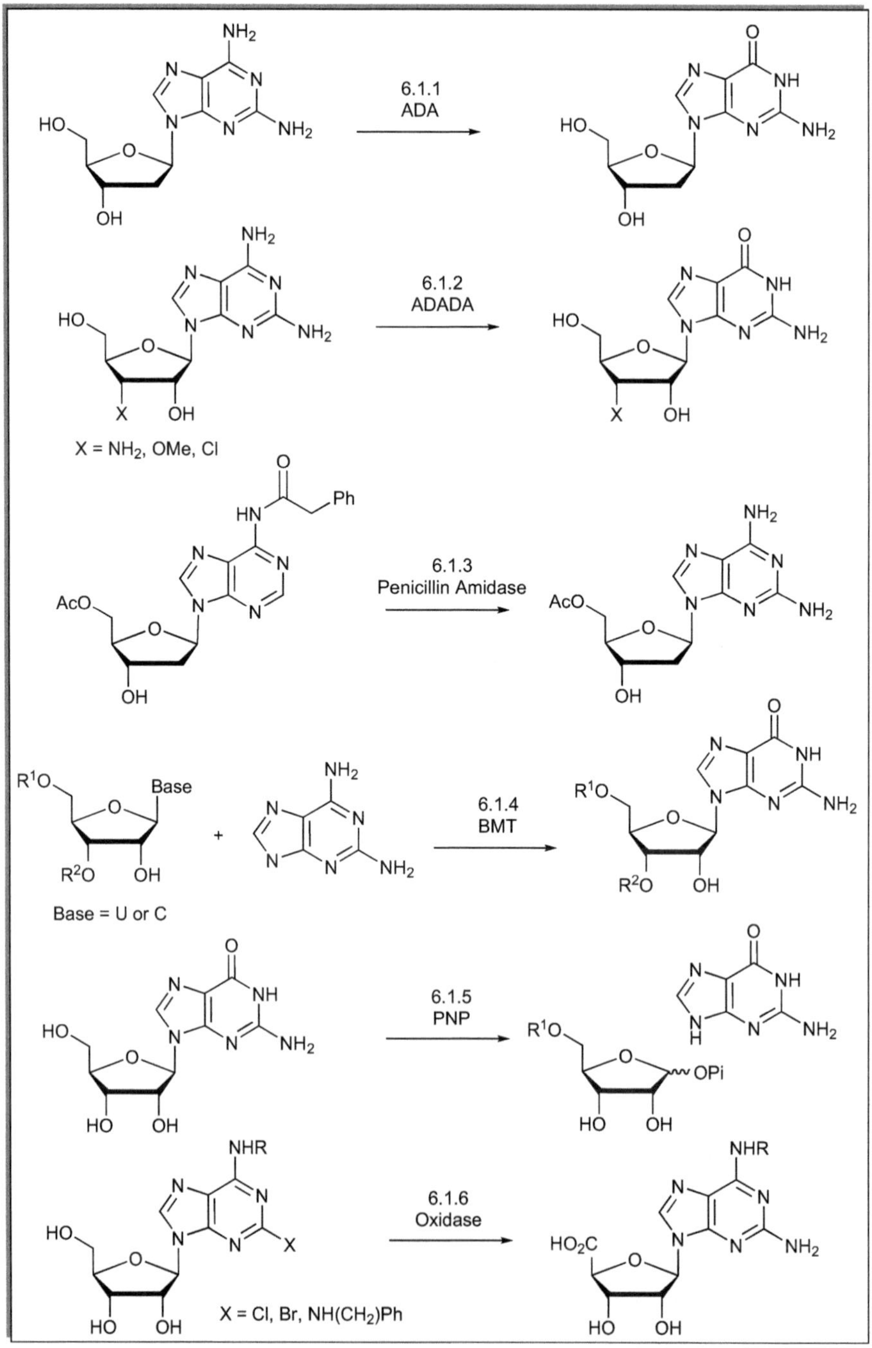

Scheme 10.29   Biocatalysts acting on nucleosides.

functional groups at C-3 (6.1.2). Penicillin amidase hydrolyses protected nucleosides containing phenylacetamide groups (6.1.3). Nucleoside phosphorylases (NPs) are a family of enzymes that catalyse the reversible phosphorolysis of nucleosides and the transferase reaction involving purine or pyrimidine bases. *E. coli* BMT is an enzyme with broad activity that uses uridine or cytosine nucleosides as substrates and introduces alternative purine bases (6.1.4). Purine nucleoside phosphorylase (PNP) is also able to catalyse exchange of the base from purine nucleosides (6.1.5). Finally, nucleoside oxidase catalyses the oxidation of the C-5–OH of some nucleosides to the corresponding carboxylic acid (6.1.6).

Abbreviations

AADH	amino acid dehydrogenase
ADH	alcohol dehydrogenase
AldDH	aldehyde dehydrogenase
AcetDH	acetolactate dehydratase
ALD	acetolactate decarboxylase
AmDH	amine dehydrogenase
AMDase	arylmalonate decarboxylase
AO	alcohol oxidase
ATP	adenosine triphosphate
BAL	benzaldehyde lyase
BMA	β-methylaspartase
BVMO	Baeyer–Villiger monooxygenase
CAL-B	*Candida antarctica* lipase B
CAR	carboxylic acid reductase
CLEA	cross-linked enzyme aggregate
DAAO	D-amino acid oxidase
DKR	dynamic kinetic resolution
DST	desaturase
DERA	deoxyribose aldolase
EH	epoxide hydrolase
ERED	ene reductase
FAD	ferulic acid decarboxylase
FDH	formate dehydrogenase
FMO	flavin monooxygenase
FSA	fructose-6-phosphate aldolase
GOase	galactose oxidase

GDH	glucose dehydrogenase
IRED	imine reductase
KRED	ketoreductase
HHD	halohydrin dehalogenase
HLADH	horse liver alcohol dehydrogenase
HNL	hydroxynitrile lyase
HPO	horse radish peroxidase
HYDase	hydantoinase
KIN	kinase
LDH	lactate dehydrogenase
LIP	lipase
NAD^+	nicotinamide adenine dinucleotide (oxidised form)
NADH	nicotinamide adenine dinucleotide (reduced form)
$NADP^+$	nicotinamide adenine dinucleotide phosphate (oxidised form)
NADPH	nicotinamide adenine dinucleotide phosphate (reduced form)
MAO-N	monoamine oxidase N
NHase	nitrile hydratase
NTLase	nitrilase
NITRase	nitrile reductase
NOX	NADH oxidase
NRase	nitroreductase
OCH	6-oxo-camphor hydrolase
OpDH	opine dehydrogenase
PAD	*para*-hydroxy cinnamic acid decarboxylase
PAL	phenylalanine ammonia lyase
PAM	phenylalanine aminomutase
PaoABC	periplasmic alcohol oxidase
PAPS	3-phosphoadenosine 5-phosphosulfate
PLP	pyridoxal phosphate
P450	P450 monooxygenase
PPase	phosphatase
PPL	porcine pancreatic lipase
SFT	sulfotransferase
TA	threonine aldolase
TAL	tyrosine ammonia lyase
TPL	tyrosine phenol lyase
ω-TA	ω-transaminase
TDP	thiamine diphosphate
UPO	unspecific peroxygenase

References

1. P. Galletti, E. Emer, G. Gucciardo, A. Quintavalla, M. Pori and D. Giacomini, *Org. Biomol. Chem.*, 2010, **8**, 4117.
2. H. Li, J. Moncecchi and M. D. Truppo, *Org. Proc. Res. Dev.*, 2015, **19**, 695.
3. T. Saravan, S. Jana and A. Chadha, *Org. Biomol. Chem.*, 2014, **12**, 4682.
4. M. Toesch, M. Schober and K. Faber, *Appl. Microbiol. Biotechnol.*, 2014, **98**, 1485.
5. P. B. Brondani, H. Dudek, J. S. Reis, M. W. Fraaije and L. H. Andrade, *Tetrahedron: Asymmetry*, 2012, **23**, 703.
6. E. Henke, U. T. Bornscheuer, R. D. Schmid and J. Pleiss, *ChemBioChem.*, 2003, **4**, 485.
7. M. Breuer, K. Ditrich, T. Habicher, B. Hauer, M. Keßeler, R. Stürmer and T. Zelinski, *Angew. Chem., Int. Ed.*, 2004, **43**, 788.
8. M. Fuchs, D. Koszelewski, K. Tauber, J. Sattler, W. Banko, A. K. Holzer, M. Pickl, W. Kroutil and K. Faber, *Tetrahedron*, 2012, **68**, 7691.
9. M. J. Abrahamson, E. Vázquez-Figueroa, N. B. Woodall, J. C. Moore and A. S. Bommarius, *Angew. Chem., Int. Ed.*, 2012, **51**, 3969.
10. F. G. Mutti, T. Knaus, N. S. Scrutton, M. Breuer and N. J. Turner, *Science*, 2015, **349**, 1525–1529.
11. E. O'Reilly, V. Köhler, S. L. Flitsch and N. J. Turner, *Chem. Commun.*, 2011, **47**, 2490.
12. L. Yang, S. L. Koh, P. W. Sutton and Z.-X. Liang, *Catal. Sci. Technol.*, 2014, **4**, 2871.
13. J. M. Foulkes, K. J. Malone, V. S. Cooker, N. J. Turner and J. R. Lloyd, *ACS Catal.*, 2011, **1**, 1589.
14. C. Li, W. Wu, K. B. Cho and S. Shaik, *Chemistry*, 2009, **15**, 8492.
15. J. H. Schrittwieser, S. Velikogne and W. Kroutil, *Adv. Synth. Catal.*, 2015, **357**, 1655.
16. G. A. Aleku, S. P. France, J. Mangas-Sanchez, F. Leipold, S. Hussain, S. L. Montgomery, H. Man, G. Grogan and N. J. Turner, *Nature Chem.*, 2017, 9, DOI: 10.1038/NCHEM.2782.
17. H. Chen, J. Moore, S. J. Collier, D. Smith, J. Nazor, G. Hughes, J. Janey, G. Huisman, S. Novick, N. Agard, O. Alvizo, G. Cope, W. L. Yeo, J. Sukumaran, S. Ng, 2013, *Engineered imine reductases and methods for the reductive amination of ketone and amine compounds. US Pat.* Application 20130302859.

18. R. E. Deasy, M. Brossat, T. S. Moody and A. R. Maguire, *Tetrahedron: Asymmetry*, 2011, **1**, 47.
19. A. J. Carnell, S. McKenna, S. Leimkühler, S. Herter and N. J. Turner, *Green Chem.*, 2015, **17**, 3271.
20. P. Konst, H. Merkens, S. Kara, S. Kochius, A. Vogel, R. Zuhse, D. Holtmann, I. W. C. E. Arends and F. Hollmann, *Angew. Chem., Int. Ed.*, 2012, **51**, 9914.
21. K. McClean, C. Preston, D. Spence, P. W. Sutton and J. Whittall, *Tetrahedron Lett.*, 2011, **52**, 215.
22. T. Nuijens, E. Piva, J. A. W. Kruijtzer, D. T. S. Rijkers, R. M. J. Liskamp and P. J. L. M. Quaedflieg, *Adv. Synth. Catal.*, 2011, **353**, 1039.
23. J. Pietruszka and M. Scholzel, *Adv. Synth. Catal.*, 2012, **354**, 751.
24. S. van Pelt, M. Zhang, L. G. Otten, J. Holt, D. Y. Sorokin, F. van Rantwijk, G. W. Black, J. J. Perry and R. A. Sheldon, *Org. Biomol. Chem.*, 2011, **9**, 3011.
25. F. Escalettes and N. J. Turner, *ChemBioChem*, 2008, **9**, 857.
26. S. Gargiulo, D. J. Opperman, U. Hanefeld, I. W. C. E. Arends and F. Hollmann, *Chem. Commun.*, 2012, **48**, 6630.
27. T. Li, J. Liang, A. Ambrogelly, T. Brennan, G. Gloor, G. Huisman, J. Lalonde, A. Lekhal, B. Mijts, S. Muley, L. Newman, M. Tobin, G. Wong, A. Zaks and X. Zhang, *J. Am. Chem. Soc.*, 2012, **134**, 6467.
28. A. Rioz-Martinez, M. Kopacz, G. de Gonzalo, D. E. Torres Pazmiño, V. Gotor and M. W. Fraaije, *Org. Biomol. Chem.*, 2011, **9**, 1337.
29. S. P. Hanlon, A. Camattari, S. Abad, A. Glieder, M. Kittelmann, S. Lütz, B. Wirz and M. Winkler, *Chem Commun.*, 2012, **48**, 6001.
30. P. Ayhan and A. S. Demir, *Adv. Synth. Catal.*, 2011, **353**, 624.
31. H. Zhao and W. A. van der Donk, *Curr. Opin. Biotechnol.*, 2003, **14**, 583.
32. L. Babich, A. F. Hartog, M. A. van der Horst and R. Wever, *Chem. – Eur. J.*, 2012, **18**, 6604.
33. N. Li, H. Norgaard, D. M. Warui, S. J. Booker, C. Krebs and J. M. Bollinger Jr., *J. Am. Chem. Soc.*, 2011, **133**, 6158.
34. C. K. Winkler, D. Clay, N. G. Turrini, H. Lechner, W. Kroutil, S. Davies, S. Debarge, P. O'Neill, J. Steflik, M. Karmilowicz, J. W. Wong and K. Faber, *Adv. Synth. Catal.*, 2014, **356**, 1878.
35. Z. Rui, N. C. Harris, X. Zhu, W. Huang and W. Zhang, *ACS Catal.*, 2015, **5**, 7091.
36. R. Kourist, Y. Miyauchi, D. Uemura and K. Miyamoto, *Chem. – Eur. J.*, 2011, **17**, 557.

37. E. Zandvoot, E. M. Geertsema, B. Baas, W. J. Quax and G. J. Poelarends, *ChemBioChem.*, 2012, **13**, 1869.
38. A. L. Concia, L. Gómez, J. Bujons, T. Parella, C. Vilaplana, P. J. Cardona, J. Joglar and P. Clapés, *Org. Biomol. Chem.*, 2013, **11**, 2005.
39. F. Tao, Y. Zhang, C. Mar and P. Zhu, *Appl. Microbiol. Biotechnol.*, 2010, **87**, 1281, and have been used in the synthesis of the neuraminidase inhibitor zanamavir.
40. M. Cheriyan, M. J. Walters, B. D. Kang, L. L. Anzaldi, E. J. Toone and C. A. Fierke, *Bioorg. Med. Chem.*, 2011, **19**, 6447.
41. M. Muller, *Angew. Chem,. Int. Ed.*, 2005, **44**, 362.
42. G. Grogan, G. A. Roberts, D. Bougioukou, N. J. Turner and S. L. Flitsch, *J. Biol. Chem.*, 2001, **276**, 12565.
43. C. R. Müller, M. Pérez-Sánchez and P. D. de María, *Org. Biomol. Chem.*, 2013, **11**, 2000.
44. C. Dresen, M. Richter, M. Pohl, S. Lüdeke and M. Müller, *Angew. Chem., Int. Ed.*, 2010, **49**, 6600.
45. A. Glieder, R. Weis, W. Skranc, P. Poechlauer, I. Dreveny, S. Majer, M. Wubbolts, H. Schwab and K. Gruber, *Angew. Chem., Int. Ed.*, 2003, **42**, 4815.
46. T. Purkarthofer, K. Gruber, M. Gruber-Khadjawi, K. Waich, W. Skranc, D. Mink and H. Griengl, *Angew. Chem., Int. Ed.*, 2006, **45**, 3454.
47. Z. Guan, J.-P. Fu and Y.-H. He, *Tetrahedron Lett.*, 2012, **53**, 4959.
48. X. Chen, G. Chen, J. Wang, Q. Wu and X. Lin, *Adv. Synth. Catal.*, 2013, **355**, 864.
49. Y.-M. Seo, S. Mathew, H.-S. Bea, Y.-H. Khang, S.-H. Lee, B.-G. Kim and H. Yun, *Org. Biomol. Chem.*, 2012, **10**, 2482.
50. F. G. Mutti, C. S. Fuchs, D. Pressnitz, N. G. Turrini, J. H. Sattler, A. Lerchner, A. Skerra and W. Kroutil, *Eur. J. Org. Chem.*, 2012, 1003.
51. C. K. Savile, J. M. Janey, E. C. Mundorff, J. C. Moore, S. Tam, W. R. Jarvis, J. C. Colbeck, A. Krebber, F. J. Fleitz, J. Brands, P. N. Devine, G. W. Huismann and G. J. Hughes, *Science*, 2010, **329**, 305.
52. B. de Lange, D. J. Hyett, P. J. D. Maas and D. Mink, *ChemCatChem*, 2011, **3**, 289.
53. H. Raj, W. Szymanski, J. de Villiers, H. J. Rozeboom, V. P. Veetil, C. R. Reis, M. de Villiers, F. J. Dekker, S. de Wildeman, W. J. Quax, A.-M. W. H. Thunnissen, B. L. Feringa, D. B. Janssen and G. J. Poelarends, *Nat. Chem.*, 2012, **4**, 478.
54. N. J. Weise, F. Parmegiani, S. Ahmed and N. J. Turner, *J. Am. Chem. Soc.*, 2015, **137**, 12977.

55. L. N. Monsalve, F. Gillanders and A. Baldessan, *Eur. J. Org. Chem.*, 2012, 1164.

56. R. L. Hanson, S. L. Goldberg, D. B. Brzozowski, T. P. Tully, D. Cazzulino, W. L. Parker, O. K. Lynberg, T. C. Vu, M. K. Wong and R. N. Patel, *Adv. Synth. Catal.*, 2007, **349**, 1369.

57. M. M. Elenkov, I. Primoþiè, T. Hrenar, A. Smolko, I. Dokli, B. Salopek-Sondi and L. Tang, *Org. Biomol. Chem.*, 2012, **10**, 5063.

58. C. K. Prier, T. K. Hyster, C. C. Farwell, A. Huang and F. H. Arnold, *Angew. Chem., Int. Ed.*, 2016, **55**, 4711.

59. D. B. Janssen, *Adv. Appl. Microbiol.*, 2007, **61**, 233.

60. A. Westerbeek, W. Szymanski, H. J. Wijma, S. J. Marrink, B. L. Feringa and D. B. Janssen, *Adv. Synth. Catal.*, 2011, **353**, 931.

61. A. Westerbeek, J. G. E. van Leeuwen, W. Szymañski, B. L. Feringa and D. B. Janssen, *Tetrahedron*, 2012, **68**, 7645.

62. G. DeSantis, Z. Zhu, W. A. Greenberg, K. Wong, J. Chaplin, S. R. Hanson, B. Farwell, L. W. Nicholson, C. L. Rand, D. P. Weiner, D. E. Robertson and M. J. Burk, *J. Am. Chem. Soc.*, 2002, **124**, 9024.

63. S. Baum, F. van Rantwijk and A. Stolz, *Adv. Synth. Catal.*, 2012, **354**, 113.

64. J. S. Yadav, T. Srinivasa Rao, N. Nath Yadav, K. V. Raghavendra Rao, B. V. Subba Reddy and A. Al Khazim Al Ghamdi, *Synthesis*, 2012, **44**, 788.

65. D. Chang, Z. Wang, M. F. Heringa, R. Wirthner, B. Witholt and Z. Li, *Chem. Commun.*, 2003, 960.

66. A. Knezevic, G. Landek, I. Dokli and V. Vinokovic, *Tetrahedron: Asymmetry*, 2011, **22**, 936.

67. S. Alatorre-Santamaria, V. Gotor-Fernandez and V. Gotor, *Eur. J. Org. Chem.*, 2011, **6**, 1057.

68. P. D'Arrigo, L. Cerioli, A. Fiorati, S. Servi, F. Viani and D. Tessaro, *Tetrahedron: Asymmetry*, 2012, **23**, 938.

69. L.-B. Zhang, D.-X. Wang, L. Zhao and M.-X. Wang, *J. Org. Chem.*, 2012, 77, 5584.

70. K. Engstrom, M. Shakeri and J. Backvall, *Eur. J. Org. Chem.*, 2011, 1827.

71. L. Kiss, E. Forró and F. Fülöp, *Tetrahedron*, 2012, **68**, 4438.

72. L. T. Torres, A. Schließmann, M. Schmidt, N. Silva-Martin, J. A. Hermoso, J. Berenguer, U. T. Bornscheuer and A. Hidalgo, *Org. Biomol. Chem.*, 2012, **10**, 3388.

73. L. Wang, W. Wang, J. Cuui, W. Ren, N. Meng, J. Wang and X. Qian, *Tetrahedron Asymmetry*, 2010, **21**, 825.

74. N. D. Shen, Y. Ni, H. M. Ma, L. J. Wang, C. X. Li, G. W. Zheng, J. Zhang and J. H. Xu, *Org. Lett.*, 2012, **14**, 1982.
75. R. N. Patel, A. Banerjee, C. G. McNamee, D. B. Brzozowski and L. J. Szarka, *Tetrahedron Asymmetry*, 1997, **8**, 2547.
76. G. Xu, H. Yu, Z. Zhang and J. Xu, *Org. Lett.*, 2013, **15**, 5408.
77. X. Wu, J. Jiang and Y. Chen, *ACS Catal.*, 2011, **1**, 1661.
78. T. Fischer and J. Pietruszka, *Adv. Synth. Catal.*, 2012, **354**, 2521.
79. S. Luttenberg, T. D. Ta, J. vod der Heyden and J. Scherkenbeck, *Eur. J. Org. Chem.*, 2013, 1824.
80. H. S. Toogood and N. S. Scrutton, *Curr. Opin. Chem. Biol.*, 2014, **19**, 107.
81. R. Agudo, G.-D. Roiban and M. T. Reetz, *ChemBioChem*, 2012, **13**, 1465.
82. H. Lin, Y. Liu and Z. L. Wu, *Chem. Commun.*, 2011, **47**, 2610.
83. J. M. Neary, A. Powell, L. Gordon, C. Milne, F. Flett, B. Wilkinson and J. Micklefield, *Microbiology*, 2007, **153**, 768.
84. M. Sandy, J. N. Carter-Franklin, J. D. Martins and A. Butler, *Chem. Commun.*, 2011, **47**, 12086.
85. H. Lechner, D. Pressnitz and W. Kroutil, *Biotechnol. Adv.*, 2015, **33**, 457.
86. Z. J. Wang, H. Renata, N. E. Peck, C. C. Farwell, P. S. Coelho and F. H. Arnold, *Angew. Chem., Int Ed.*, 2014, **53**, 6810.
87. C. C. Farwell, R. K. Zhang, J. A. McIntosh, T. K. Hyster and F. H. Arnold, *ACS Cent. Sci.*, 2015, **1**, 89.
88. M. Korpaka and J. Pietruszkaa, *Adv. Synth. Catal.*, 2011, **353**, 1420.
89. F. Leipold, S. Hussain, D. Ghislieri and N. J. Turner, *ChemCatChem.*, 2013, **5**, 3505.
90. E. O'Reilly, C. Iglesias, D. Ghislieri, J. Hopwood, J. L. Galman, R. C. Lloyd and N. J. Turner, *Angew. Chem., Int. Ed.*, 2014, **53**, 2447.
91. J. A. McIntosh, P. S. Coelho, C. C. Farwell, Z. J. Wang, J. C. Lewis, T. R. Brown and F. H. Arnold, *Angew. Chem., Int. Ed.*, 2013, **52**, 9309.
92. C. N. Jensen, J. Cartwright, J. Ward, S. Hart, J. P. Turkenburg, S. T. Ali and G. Grogan, *ChemBioChem.*, 2012, **13**, 872.
93. B. Bechi, S. Herter, S. McKenna, C. Riley, S. Leimkühler, N. J. Turner and A. J. Carnell, *Green Chem.*, 2014, **16**, 4524.
94. S. C. Hammer, P.-O. Syrén, M. Seitz, B. M. Nestl and B. Hauer, *Curr. Opin. Chem. Biol.*, 2013, **17**, 293.
95. S. P. France, S. Hussain, A. M. Hill, L. J. Hepworth, R. M. Howard, K. R. Mulholland, S. L. Flitsch and N. J. Turner, *ACS Catal.*, 2016, **6**, 3753.

96. B. M. Ruff, S. Bräse and S. E. O'Connor, *Tetrahedron Lett.*, 2012, **53**, 1071.

97. J. Schrittwieser, B. Groenendaal, V. Resch, D. Ghisleri, S. Wallner, E.-M. Fischereder, E. Fuchs, B. Grischek, J. H. Sattler, P. Macheroux, N. J. Turner and W. Kroutil, *Angew. Chem., Int. Ed.*, 2014, **53**, 3731–3734.

98. Y. Lin, M. Liang, Y. Lin and C. Chen, *Chem. – Eur. J.*, 2011, **17**, 13059.

99. C. Wuensch, J. Gross, G. Steinkellner, A. Lyskowski, K. Gruber, S. M. Glueck and K. Faber, *RSC Adv.*, 2014, **4**, 9673.

100. D. Wischang and J. Hartung, *Tetrahedron*, 2011, **67**, 4048.

101. J. T. Payne, M. C. Andorfer and J. C. Lewis, *Angew. Chem., Int. Ed.*, 2013, **52**, 5271.

102. A.-W. Struck, M. L. Thompson, L. S. Wong and J. Micklefield, *ChemBioChem*, 2012, **13**, 2642.

103. V. de la Sovera, A. Bellomo and D. Gonzalez, *Tetrahedron Lett.*, 2011, **52**, 430.

104. M. A. Schätzle, S. Flemming, S. M. Husain, M. Richter, S. Günther and M. Müller, *Angew. Chem., Int. Ed.*, 2012, **51**, 2643.

105. A. N. Tsoligkas, M. Winn, J. Bowen, T. W. Overton, M. J. H. Simmons and R. J. M. Goss, *ChemBioChem*, 2011, **12**, 1391.

106. L. S. Mazzaferro, W. Hüttel, A. Fries and M. Müller, *J. Am. Chem. Soc.*, 2015, **137**, 12289.

107. M. Liebhold, X. Xie and S.-M. Li, *Org. Lett.*, 2012, **14**, 4882.

108. C. Wuensch, J. Gross, G. Steinkellner, K. Gruber, S. M. Glueck and K. Faber, *Angew. Chem., Int. Ed.*, 2013, **52**, 2293.

109. C. Wuensch, S. M. Glueck, J. Gross, D. Koszelewski, M. Schober and K. Faber, *Org. Lett.*, 2012, **14**, 1974.

110. H.-H. Nguyen-Tran, G.-W. Zheng, X.-H. Qian and J.-H. Xu, *Chem. Commun.*, 2014, **50**, 2861.

111. E. Busto, R. C. Simon and W. Kroutil, *Angew. Chem., Int. Ed.*, 2015, **54**, 10899.

112. J. B. Rannes, A. Ioannou, S. C. Willies, G. Grogan, C. Behrens, S. L. Flitsch and N. J. Turner, *J. Am. Chem. Soc.*, 2011, **133**, 8436.

113. M. Ferrero and V. Gotor, *Chem. Rev.*, 2000, **100**, 4319.

114. L. A. Condezo, J. Fernandez-Lucas, C. A. Garcia-Burgos, A. R. Alcantara and J. V. Sinisterra, *Enzymatic Synthesis of Modified Nucleosides. Biocatalysis in the Pharmaceutical & Biotechnology Industries*, CRC Press, Boca Raton, USA, Capitulo, 2007, vol. 14, pp. 401–426.

115. (a) A. Liese and L. Pesci, Enzyme Classification and Nomenclature and Biocatalytic Retrosynthesis, in *Science of Synthesis: Biocatalysis in Organic Synthesis*, ed. K. Faber, W.-D. Fessner and N. J. Turner, Georg Thieme Verlag, Stuttgart, 2015, vol. 1, pp. 41–74; (b) M. Bertau and G. E. Jeromin, Resolution of Alcohols, Acids and Esters by Hydrolysis, in *Science of Synthesis: Biocatalysis in Organic Synthesis*, ed. K. Faber, W.-D. Fessner and N. J. Turner, Georg Thieme Verlag, Stuttgart, 2015, vol. 1, pp. 129–188; (c) M. Rodríguez-Mata and V. Gotor-Fernández, Resolution of Alcohols, Amines, Acids and Esters by Nonhydrolytic Processes, in *Science of Synthesis: Biocatalysis in Organic Synthesis*, ed. K. Faber, W.-D. Fessner and N. J. Turner, Georg Thieme Verlag, Stuttgart, 2015, vol. 1, pp. 189–222; (d) R. Wever, L. Babich and A. F. Hartog, Transphosphorylation, in *Science of Synthesis: Biocatalysis in Organic Synthesis*, ed. K. Faber, W.-D. Fessner and N. J. Turner, Georg Thieme Verlag, Stuttgart, 2015, vol. 1, pp. 223–254; (e) Y. Asano, Hydrolysis of Nitriles to Amides, in *Science of Synthesis: Biocatalysis in Organic Synthesis*, ed. K. Faber, W.-D. Fessner and N. J. Turner, Georg Thieme Verlag, Stuttgart, 2015, vol. 1, pp. 255–276; (f) L. Martínková and A. B. Veselá, Hydrolysis of Nitriles to Carboxylic Acids, in *Science of Synthesis: Biocatalysis in Organic Synthesis*, ed. K. Faber, W.-D. Fessner and N. J. Turner, Georg Thieme Verlag, Stuttgart, 2015, vol. 1, pp. 277–302; (g) M. Hall, K. Faber and G. Tasnádi, Hydrolysis of Amides, in *Science of Synthesis: Biocatalysis in Organic Synthesis*, ed. K. Faber, W.-D. Fessner and N. J. Turner, Georg Thieme Verlag, Stuttgart, 2015, vol. 1, pp. 303–328; (h) J. W. Schmidberger, L. J. Hepworth, A. P. Green and S. L. Flitsch, Enzymatic Synthesis of Amides, in *Science of Synthesis: Biocatalysis in Organic Synthesis*, ed. K. Faber, W.-D. Fessner and N. J. Turner, Georg Thieme Verlag, Stuttgart, 2015, vol. 1, pp. 329–372; (i) C. Slomka, U. Engel, C. Syldatk and J. Rudat, Hydrolysis of Hydantoins, Dihydropyrimidines and Related Compounds, in *Science of Synthesis: Biocatalysis in Organic Synthesis*, ed. K. Faber, W.-D. Fessner and N. J. Turner, Georg Thieme Verlag, Stuttgart, 2015, vol. 1, pp. 373–414; (j) K. Faber and S. M. Glueck, Isomerizations: Racemization, Epimerization and *E/Z*-Isomerization, in *Science of Synthesis: Biocatalysis in Organic Synthesis*, ed. K. Faber, W.-D. Fessner and N. J. Turner, Georg Thieme Verlag, Stuttgart, 2015, vol. 1, pp. 415–482; (k) B. Cobuci-Ponzano and M. Moracci, Glycosidases and Glycosynthases, in *Science of Synthesis: Biocatalysis in Organic Synthesis*, ed. K. Faber,

W.-D. Fessner and N. J. Turner, Georg Thieme Verlag, Stuttgart, 2015, vol. 1, pp. 483–506; (l) J. Voglmeir and S. L. Flitsch, Glycosyltransferases, in *Science of Synthesis: Biocatalysis in Organic Synthesis*, ed. K. Faber, W.-D. Fessner and N. J. Turner, Georg Thieme Verlag, Stuttgart, 2015, vol. 1, pp. 507–542.

116. (a) K. Steiner, A. Glieder and M. Gruber-Khadjawi, Cyanohydrin Formation/Henry Reaction, in *Science of Synthesis: Biocatalysis in Organic Synthesis 2*, ed. K. Faber, W.-D. Fessner and N. J. Turner, Georg Thieme Verlag, Stuttgart, 2015, pp. 1–30; (b) P. Clapes, Aldol Reactions, in *Science of Synthesis: Biocatalysis in Organic Synthesis 2*, ed. K. Faber, W.-D. Fessner and N. J. Turner, Georg Thieme Verlag, Stuttgart, 2015, pp. 31–92; (c) M. Pohl, C. Wechsler and M. Muller, Acyloin, Benzoin and Related Reactions, in *Science of Synthesis: Biocatalysis in Organic Synthesis 2*, ed. K. Faber, W.-D. Fessner and N. J. Turner, Georg Thieme Verlag, Stuttgart, 2015, pp. 93–132; (d) R. Lewin, M. L. Thompson and J. Micklefield, Enzymatic Carboxylation and Decarboxylation, in *Science of Synthesis: Biocatalysis in Organic Synthesis 2*, ed. K. Faber, W.-D. Fessner and N. J. Turner, Georg Thieme Verlag, Stuttgart, 2015, pp. 133–158; (e) A. Ilari, A. Bonnamore and A. Boffi, Addition to $C=N$ Bonds, in *Science of Synthesis: Biocatalysis in Organic Synthesis 2*, ed. K. Faber, W.-D. Fessner and N. J. Turner, Georg Thieme Verlag, Stuttgart, 2015, pp. 159–176; (f) L. A. Wessjohann, H. F. Schreckenback and G. N. Kaluđerović, Enzymatic C-Alkylation of Aromatic Compounds, in *Science of Synthesis: Biocatalysis in Organic Synthesis 2*, ed. K. Faber, W.-D. Fessner and N. J. Turner, Georg Thieme Verlag, Stuttgart, 2015, pp. 177–212; (g) K. Faber and M. Hall, Addition of Hydrogen to $C=C$ Bonds; Alkene Reduction, in *Science of Synthesis: Biocatalysis in Organic Synthesis 2*, ed. K. Faber, W.-D. Fessner and N. J. Turner, Georg Thieme Verlag, Stuttgart, 2015, pp. 213–260; (h) V. Resch and U. Hanefeld, Addition of Water to $C=C$ Bonds, in *Science of Synthesis: Biocatalysis in Organic Synthesis 2*, ed. K. Faber, W.-D. Fessner and N. J. Turner, Georg Thieme Verlag, Stuttgart, 2015, pp. 261–290; (i) S. Bartsch and A. Vogel, Addition of Ammonia and Amines to $C=C$ Bonds, in *Science of Synthesis: Biocatalysis in Organic Synthesis 2*, ed. K. Faber, W.-D. Fessner and N. J. Turner, Georg Thieme Verlag, Stuttgart, 2015, pp. 291–312; (j) E. M. Geertsema and G. J. Poelarends, Enzymatic Carbon-Carbon Bond Forming Michael-Type Additions, in *Science of Synthesis: Biocatalysis in Organic Synthesis 2*, ed. K. Faber, W.-D. Fessner

and N. J. Turner, Georg Thieme Verlag, Stuttgart, 2015, pp. 313–334; (k) A. M. Bommarius and S. K. Au, Amino Acid and Amine Dehydrogenase, in *Science of Synthesis: Biocatalysis in Organic Synthesis 2*, ed. K. Faber, W.-D. Fessner and N. J. Turner, Georg Thieme Verlag, Stuttgart, 2015, pp. 335–358; (l) F. Leipold, S. Hussain, S. P. France and N. J. Turner, Imine Reductases, in *Science of Synthesis: Biocatalysis in Organic Synthesis 2*, ed. K. Faber, W.-D. Fessner and N. J. Turner, Georg Thieme Verlag, Stuttgart, 2015, pp. 359–382; (m) R. C. Simon, E. Busto, E.-M. Fischereder, C. S. Fuchs, D. Pressnitz, N. Richter and W. Kroutil, ω-Transaminases, in *Science of Synthesis: Biocatalysis in Organic Synthesis 2*, ed. K. Faber, W.-D. Fessner and N. J. Turner, Georg Thieme Verlag, Stuttgart, 2015, pp. 383–420; (n) T. S. Moody, S. Mix, G. Brown and D. Beecher, Ketone and Aldehyde Reduction, in *Science of Synthesis: Biocatalysis in Organic Synthesis 2*, ed. K. Faber, W.-D. Fessner and N. J. Turner, Georg Thieme Verlag, Stuttgart, 2015, pp. 421–458; (o) A. S. Lamm, P. Venkitasubramanian and J. P. N. Rosazza, Carboxylic Acid Reductase, in *Science of Synthesis: Biocatalysis in Organic Synthesis 2*, ed. K. Faber, W.-D. Fessner and N. J. Turner, Georg Thieme Verlag, Stuttgart, 2015, pp. 459–478; (p) A. T. Li and Z. Li, Asymmetric Synthesis of Enantiopure Epoxides Using Monooxygenases, in *Science of Synthesis: Biocatalysis in Organic Synthesis 2*, ed. K. Faber, W.-D. Fessner and N. J. Turner, Georg Thieme Verlag, Stuttgart, 2015, pp. 479–506; (q) M. Malerić-Elenkov, W. Szymański and D. B. Janssen, Reactions Catalyzed by Halohydrin Dehalogenases, in *Science of Synthesis: Biocatalysis in Organic Synthesis 2*, ed. K. Faber, W.-D. Fessner and N. J. Turner, Georg Thieme Verlag, Stuttgart, 2015, pp. 507–528; (r) R. Wohlgemuth, Epoxide Hydrolysis, in *Science of Synthesis: Biocatalysis in Organic Synthesis 2*, ed. K. Faber, W.-D. Fessner and N. J. Turner, Georg Thieme Verlag, Stuttgart, 2015, pp. 529–556.

117. (a) C. C. R. Allen, Dihydroxylation of Aromatics and Alkenes, in *Science of Synthesis: Biocatalysis in Organic Synthesis 3*, ed. K. Faber, W.-D. Fessner and N. J. Turner, Georg Thieme Verlag, Stuttgart, 2015, pp. 1–20; (b) J. C. Nolte and V. B. Urlacher, Cytochrome P450 in the Oxidation of Alkanes, in *Science of Synthesis: Biocatalysis in Organic Synthesis 3*, ed. K. Faber, W.-D. Fessner and N. J. Turner, Georg Thieme Verlag, Stuttgart, 2015, pp. 21-64; (c) S. Herter and N. J. Turner, Oxidation Other Than with P450s, in *Science of Synthesis: Biocatalysis in Organic*

Synthesis 3, ed. K. Faber, W.-D. Fessner and N. J. Turner, Georg Thieme Verlag, Stuttgart, 2015, pp. 65–114; (d) F. Hollmann, Oxidation Using Dehydrogenases, in *Science of Synthesis: Biocatalysis in Organic Synthesis 3*, ed. K. Faber, W.-D. Fessner and N. J. Turner, Georg Thieme Verlag, Stuttgart, 2015, pp. 115–138; (e) S. Herter and N. J. Turner, Oxidation Using Laccases, in *Science of Synthesis: Biocatalysis in Organic Synthesis 3*, ed. K. Faber, W.-D. Fessner and N. J. Turner, Georg Thieme Verlag, Stuttgart, 2015, pp. 139–156; (f) T. A. Ewing, M. W. Fraaije and W. J. H. van Berkel, Oxidation Using Alcohol Oxidases, in *Science of Synthesis: Biocatalysis in Organic Synthesis 3*, ed. K. Faber, W.-D. Fessner and N. J. Turner, Georg Thieme Verlag, Stuttgart, 2015, pp. 157–186; (g) G. de Gonzalo, W. J. H. van Berkel and M. W. Fraaije, Baeyer-Villiger Oxidation, in *Science of Synthesis: Biocatalysis in Organic Synthesis 3*, ed. K. Faber, W.-D. Fessner and N. J. Turner, Georg Thieme Verlag, Stuttgart, 2015, pp. 187–234; (h) Pollegioni and G. Molla, C-N Oxidation with Amine Oxidases and Amino Acid Oxidases, in *Science of Synthesis: Biocatalysis in Organic Synthesis 3*, ed. K. Faber, W.-D. Fessner and N. J. Turner, Georg Thieme Verlag, *Stuttgart*, 2015, pp. 235–284; (i) G. Grogan, Oxidation at Sulfur, in *Science of Synthesis: Biocatalysis in Organic Synthesis 3*, ed. K. Faber, W.-D. Fessner and N. J. Turner, Georg Thieme Verlag, Stuttgart, 2015, pp. 285–312; (j) S. Grüschow, D. R. M. Smith, D. S. Gkotski and R. J. M. Goss, Halogenases, in *Science of Synthesis: Biocatalysis in Organic Synthesis 3*, ed. K. Faber, W.-D. Fessner and N. J. Turner, Georg Thieme Verlag, Stuttgart, 2015, pp. 313–360; (k) M. B. Quinn, C. M. Flynn, J. J. Ellinger and C. Schmidt-Dannert, Isoprenoids, Polyketides and (Non)ribosomal Peptides, in *Science of Synthesis: Biocatalysis in Organic Synthesis 3*, ed. K. Faber, W.-D. Fessner and N. J. Turner, Georg Thieme Verlag, Stuttgart, 2015, pp. 361–402.

11 Comparison of Different Biocatalytic Routes to Target Molecules

11.1 Introduction

In the final part of this book, we will begin to put into practice the ideas and information presented in Chapters 1–9, particularly in terms of designing new synthetic routes to target molecules based on the use of biocatalysis. Before we start to practice applying biocatalytic retrosynthesis (Chapter 12), it is instructive and didactic to study how some important target molecules, particularly active pharmaceutical intermediates (APIs), have been previously synthesised using enzyme-catalysed transformations. In order to reduce the production costs of APIs, pharmaceutical companies often develop multiple routes to a given compound, thereby exploring all of the possible ways that a particular molecule can be synthesised. Academic groups also contribute to this exercise, often in order to exemplify a newly developed biocatalyst, thereby identifying previously unseen disconnections that also in turn serve to showcase particular biocatalytic platforms by demonstrating real applications. The case studies examined in this chapter represent the amalgamation of many years of published research in biocatalysis. It is interesting to note that initially many of the routes were based primarily on the use of hydrolases or oxidoreductases, simply because these were the dominant biocatalysts available. Increasingly there is an emphasis on using new, more recently developed biocatalysts. A case in point is the synthesis of chiral

Biocatalysis in Organic Synthesis: The Retrosynthesis Approach
By Nicholas J. Turner and Luke Humphreys
© Nicholas J. Turner and Luke Humphreys 2018
Published by the Royal Society of Chemistry, www.rsc.org

amines. 10 years ago the most popular approach for the synthesis of enantiomerically pure amines employing biocatalysis would have been to start with the ketone and carry out an asymmetric reduction to the chiral secondary alcohol, followed by chemical transformation to the amine using a sequence such as mesylate formation followed by azide displacement and then reduction. Such a sequence requires multiple unit operations and clearly suffers from low atom economy and the use of hazardous reagents. Increasingly, such conversions of ketone to amine are now attempted using transaminases, in which the amine functionality can be installed and the stereochemistry established in the same step. Emerging biocatalysts such as imine reductases (IREDs) permit the direct conversion of cyclic imines to amines and thereby offer an alternative approach to chiral amine synthesis. As you work through the examples presented, it is interesting to compare and contrast the various routes and to understand which of the different approaches is perhaps superior and why.

11.2   Duloxetine

Duloxetine, sold under the brand name Cymbalta among others, is mostly prescribed for major depressive disorder, generalized anxiety disorder, fibromyalgia and neuropathic pain. It is a thiophene deriva- tive, and a selective neurotransmitter re-uptake inhibitor for serotonin, norepinephrine, and to a lesser degree dopamine. It belongs to a class of heterocyclic antidepressants known as serotonin–norepinephrine re-uptake inhibitors and was first introduced by Eli Lilly.

Several routes to key enantiomerically pure intermediates have been developed, many of them based on the use of KREDs, or whole cell biocatalysts containing oxidoreductases, for enantioselective reduction of an appropriate ketone precursor (Figure 11.1). The monomethyl ketone precursor can be reduced with *Candida tropicalis* in high yield and e.e.[1] The same chiral secondary alcohol can also be prepared by asymmetric reduction of the β-ketoamide. Whole cells of *Rhodotorula glutinis* were used to reduce *N*-methyl-3-oxo-3-(thiophen-2-yl) propanamide to (*S*)-*N*-methyl-3-hydroxy-3-(2-thienyl) propionamide in 48 h. The reaction proceeded with excellent enantio-selectivity (>99.5% e.e.) and >95% conversion to give the corresponding β-hydroxyamide, which was then treated with borane to generate the amine.[2] The β-ketonitrile has also been used as a precursor and shown to undergo enantioselective reduction using *Rhodotorula rubra*.[3] The racemic β-hydroxynitrile has been stereoselectively acetylated in the presence of CALB and a ruthenium complex required for racemisation of the unreacted alcohol to effect a dynamic kinetic resolution. This

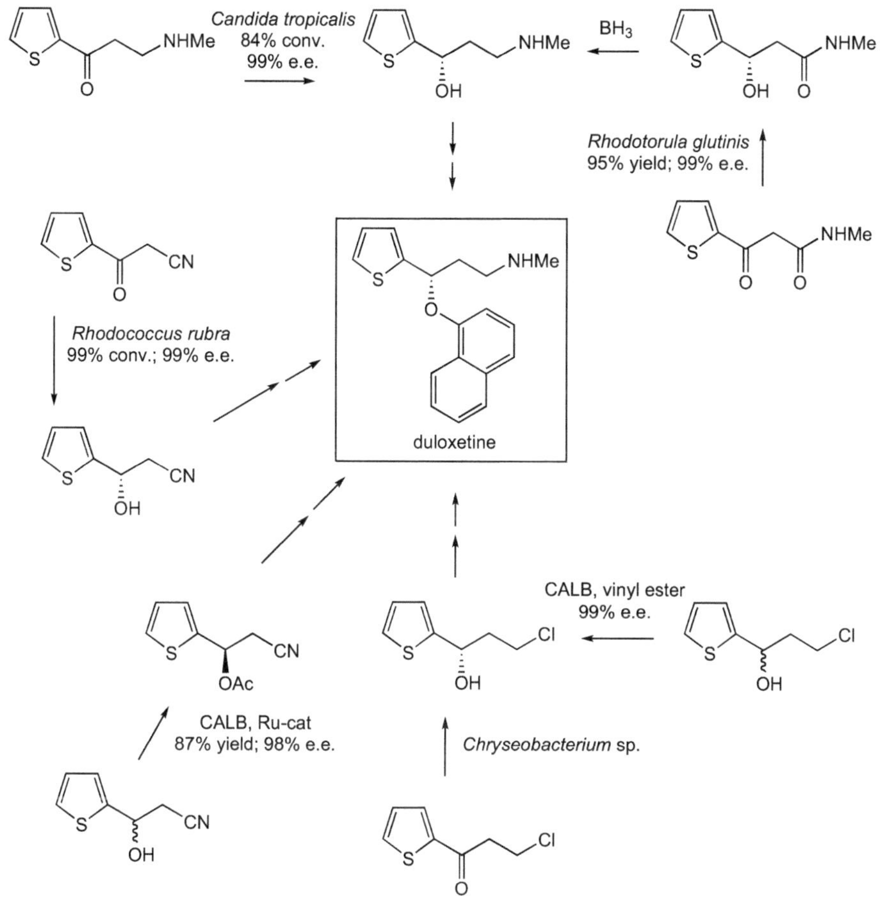

Figure 11.1 Synthesis of duloxetine.

approach generated the corresponding (R)-cyanoacetate in 87% yield and 98% e.e.[4] Finally, the thiophenyl-chloroketone has also been employed as a precursor to duloxetine using two complementary approaches. In the first, asymmetric reduction of the ketone to the chiral alcohol in high e.e. has been demonstrated using a KRED from *Chryseobacterium* sp.[5] The same secondary alcohol can also be prepared by CALB-catalysed kinetic resolution of the racemic alcohol using vinyl butanoate as the acylating agent.[6]

11.3 Rivastigmine

Rivastigmine (sold under the trade name Exelon) is a para-sympathomimetic or cholinergic agent for the treatment of mild to moderate dementia of the Alzheimer's type and dementia due to Parkinson's disease.

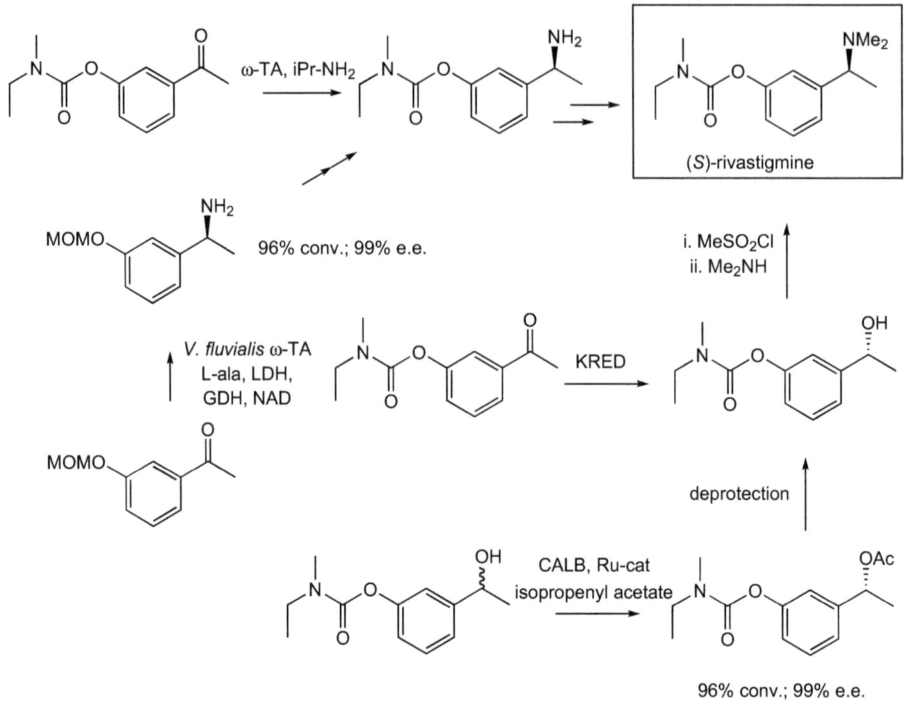

Figure 11.2 Synthesis of rivastigmine.

Rivastigmine is an excellent example of where the availability of recently developed transaminase technology has presented additional opportunities for biocatalytic retrosynthesis (Figure 11.2). Thus the ω-transaminase from *Paracoccus denitrificans* has been used to convert the immediate ketone precursor to the primary amine in 78% yield and >99% e.e.[7] In this case, L-alanine was used as the amine source together with NADH, glucose, LDH and GDH. The MOM-protected ketone has also been converted to the corresponding amine using the ω-TA from *Vibrio fluvialis*.[8] Subsequent di-methylation using standard procedures yielded rivastigmine. Another approach is based on the use of a lipase (CALB) dynamic kinetic resolution (DKR) process. The precursor racemic alcohol was treated with the lipase, isopropenyl acetate and a ruthenium catalyst to yield the product acetate in 96% conversion and 99% e.e. Subsequent mesylation with methane sulfonyl chloride and displacement with dimethylamine yielded (*S*)-rivastigmine.[9] This same alcohol has also been obtained by asymmetric reduction of the precursor ketone using a KRED.[10]

11.4   D-*para*-Hydroxyphenylglycine

D-*para*-Hydroxyphenylglycine is a large volume chiral D-amino acid used principally as a building block for amoxicillin. Amoxicillin is an antibiotic useful for the treatment of a number of bacterial infections. It is the first line treatment for middle ear infections. It may also be used for strep throat, pneumonia, skin infections, and urinary tract infections, among others. It is taken by mouth, or less commonly by injection.

The major process used for the production of 4-hydroxyphenylglycine involves the combined use of an enantioselective D-hydantoinase and D-carbamoylase (Figure 11.3). D-hydantoinases have been isolated from a number of organisms including *Agrobacterium* sp. and give a high e.e.

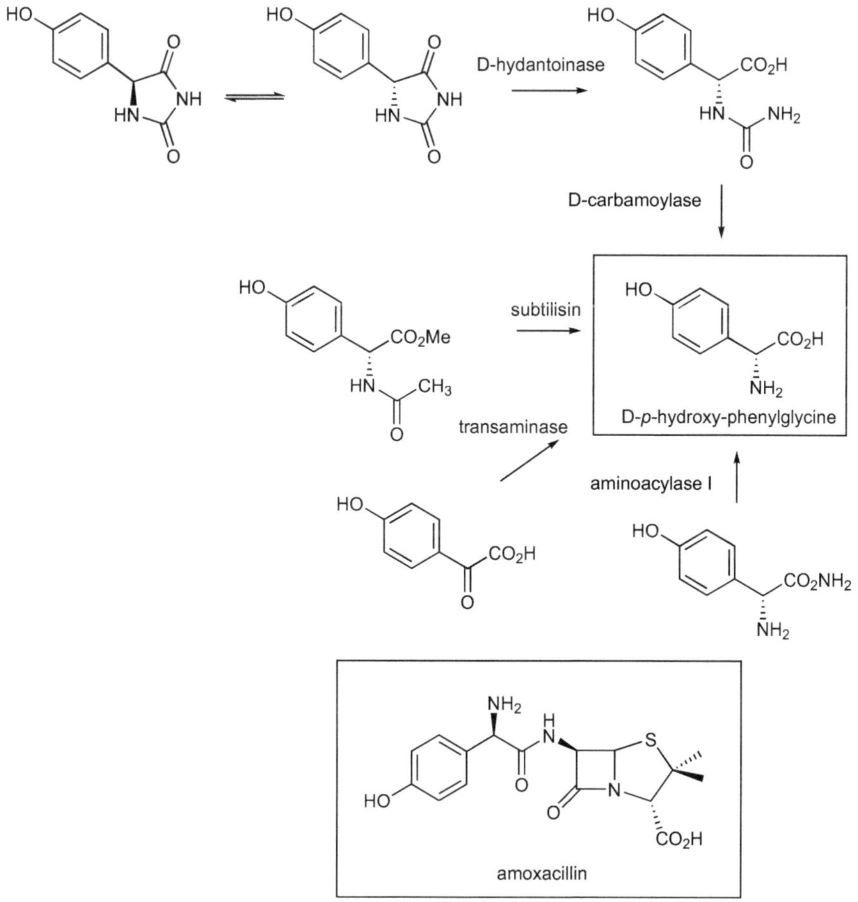

Figure 11.3   Synthesis of D-*para*-hydroxyphenylglycine.

of the *N*-carbamoyl derivative, which is then converted to the target using the D-carbamoylase.[11] The unreacted L-hydantoin spontaneously racemises under the reaction conditions to regenerate the racemate. Other approaches that have been investigated include (i) the use of an α-transaminase with glutamate as the amine donor on the keto acid,[12] (ii) kinetic resolution of *N*-acyl and amide derivatives using amino-acylase I,[13] and (iii) kinetic resolution of the *N*-acetyl methyl ester using subtilisin (Bayer).[14] However, it seems that these alternative routes are not competitive with the hydantoinase approach.

11.5 Salsolidine

Salsolidine is a tetrahydroisoquinolone alkaloid found in *Salsola* and other plants. Salsolidine inhibits the enzymes monoamine oxidase (MAO), acetylcholinesterase (AChE), and buytlcholinesterase (BChE), and may also inhibit catechol-*O*-methyltransferase (COMT). Derivatives of salsolidine are neurotoxic and cytotoxic and have been the subject of several synthetic strategies.

Scientists at NPIL Pharmaceuticals developed an elegant process based on dynamic kinetic resolution of the corresponding racemic starting material using a lipase from *Candida rugosa* together with an iridium catalyst (SCRAM catalyst) for *in situ* racemisation of the amine (Figure 11.4).[15] They optimised the choice of acylating agent and eventually chose to use the carbonate shown, which generated the *n*-propylcarbamate derivative of (*R*)-salsolidine, which was

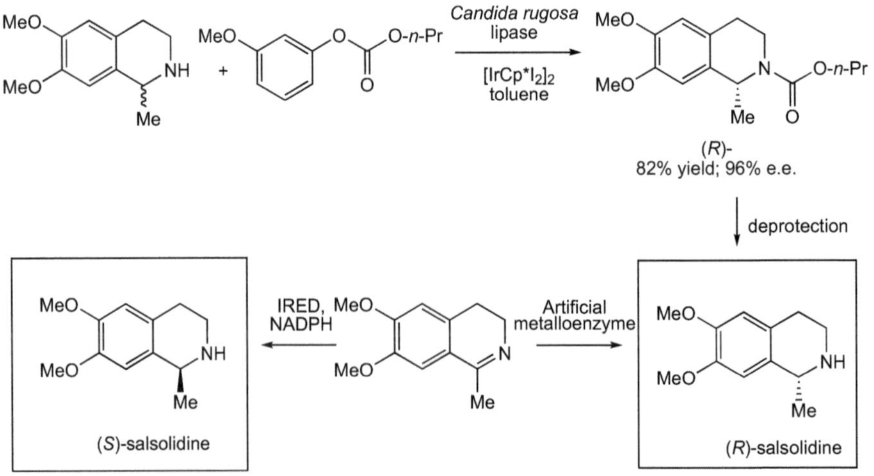

Figure 11.4 Synthesis of salsolidine.

subsequently deprotected. The (R)-enantiomer has also been obtained by reduction of the precursor imine using an artificial metalloenzyme based on Ru and Rh metal centres.[16] Finally the same imine can be reduced directly to (S)-salsolidine using an NADPH-dependent (S)-selective imine reductase (IRED) from *Streptomyces* sp. with co-factor recycling.[17]

11.6 (R)-Fluazifop

Fluazifop-*p*-butyl, marketed under the trade name Fusilade, is a selective phenoxy herbicide used for post-emergence control of annual and perennial grass weeds. It is used on soybeans and other broad-leaved crops such as carrots, spinach, potatoes, and ornamentals. Initially Fluazifop was introduced as a racemate, but this has now been replaced by the single enantiomer.

Scientists at ICI developed an efficient and large-scale method for production of enantiomerically pure 2-chloro-lactic acid by treatment of the corresponding racemate with an enantioselective dehalogenase (Figure 11.5). The (R)-specific dehalogenase was isolated from *Pseudomonas putida*. Interestingly, the contaminating (S)-selective dehalogenase was eliminated by mutation. This biocatalytic kinetic

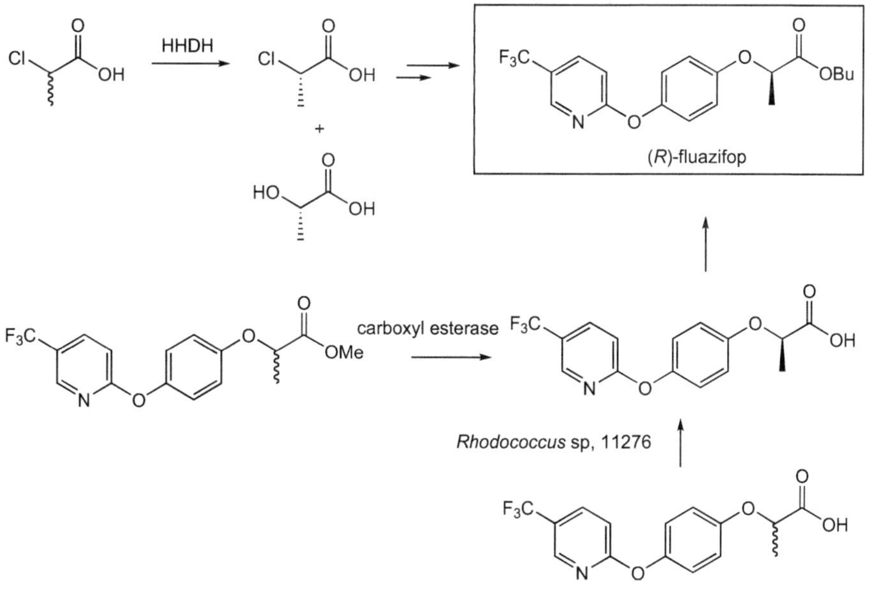

Figure 11.5 Synthesis of (R)-fluazifop.

resolution produced (*S*)-2-chloropropionic acid together with (*S*)-lactic acid. The former was then incorporated into the target molecule. One attraction of this approach is that the (*S*)-2-chloro-lactic acid produced can also be used as a building block for related commercially important herbicides. ICI built a full-scale production plant at Huddersfield in the UK with a capacity of 2000 tonnes per annum.

An alternative approach involves the use of an esterase on a later stage intermediate. A novel esterase (FpbH) was cloned from *Aquamicrobium* sp. FPB-1 and shown to be able to resolve a range of aryloxy propionic acid esters. FpbH was expressed in *E. coli*, purified and used to resolve a range of different aryloxyphenoxypropionate herbicides.[18] An intriguing approach reported is the use of *Rhodococcus* sp. 11276 to effect the whole cell deracemisation of the racemic acid to the enantiopure acid.

11.7 Vince Lactam

Vince lactam is the commercial name given to the bicyclic molecule γ-lactam 2-azabicyclo[2.2.1]hept-5-en-3-one. This lactam is a versatile chemical intermediate used in organic and medicinal chemistry. It is used as a synthetic precursor for three drugs (approved or in clinical trials) and is named after Robert Vince, who used the structural features of this molecule for the preparation of carbocyclic nucleosides. Vince's work with this lactam eventually led to his synthesis of abacavir. The synthesis of peramivir is also dependent on Vince lactam as a starting material.

Because of the commercial importance of abacavir as an antiviral drug, a number of important biocatalytic routes have been developed for the manufacture of the Vince lactam (Figure 11.6). The original biocatalytic processes were based on the use of stereocomplementary γ-lactamases from either *Rhodococcus equi* (ENZA-1)[19] or *Pseudomonas solanacearum* (ENZA-20). Subsequently it was shown that the racemic *N*-acetyl amino acid methyl ester could be kinetically resolved using pig liver esterase (PLE). Other routes were developed based on the use of savinase from *Bacillus lentus*, using either the *N*-Boc or *N*-acetyl racemic lactam as a substrate. Finally, approaches were also developed using either the *N*-hydroxymethyl acetate or a variation using the racemic hydroxymethyl derivative and then the use of a lipase (MML) in the presence of vinyl acetate to prepare the enantiomerically pure acetate, which can be subsequently converted back to the parent lactam.[20]

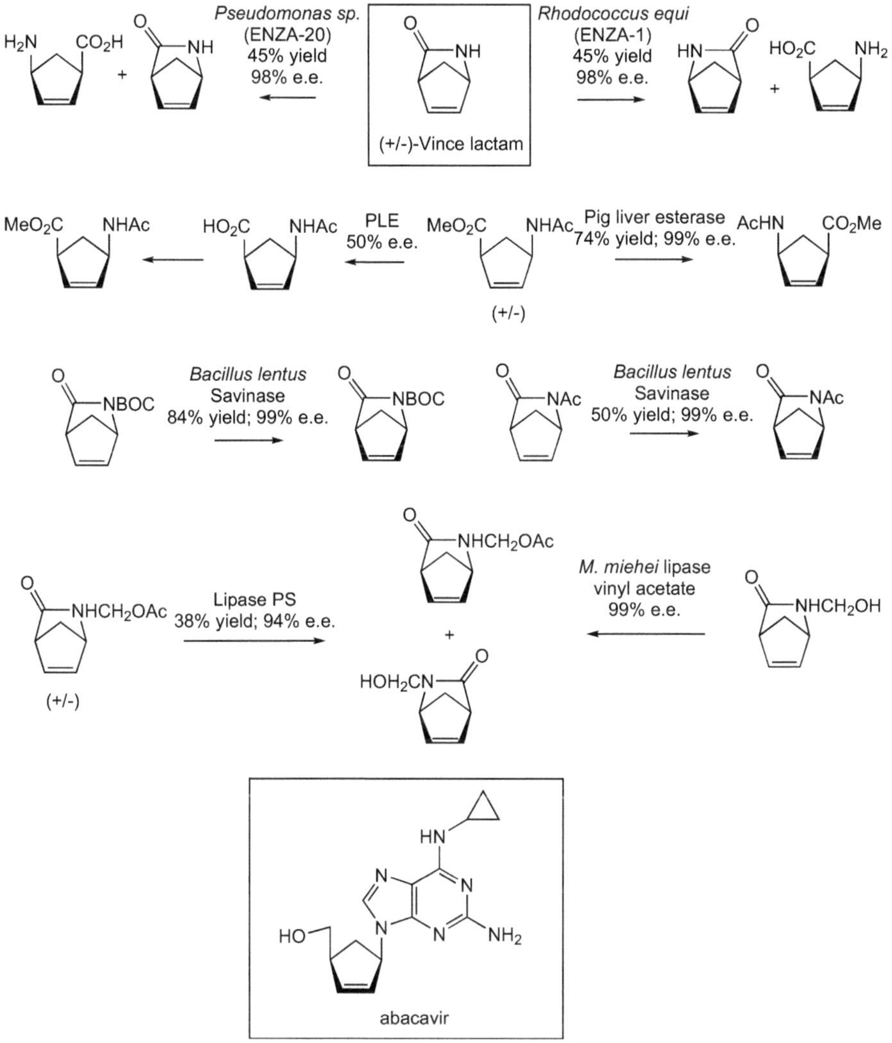

Figure 11.6 Synthesis of Vince lactam and some derivatives.

11.8 Clopidogrel

Clopidogrel, sold under the brand name Plavix among others, is a medication that is used to reduce the risk of heart disease and stroke in those at high risk. It is also used together with aspirin in heart attacks and following the placement of a coronary artery stent (dual antiplatelet therapy). Clopidogrel was initially developed by Sanofi-Aventis but has recently come off patent, and hence there has

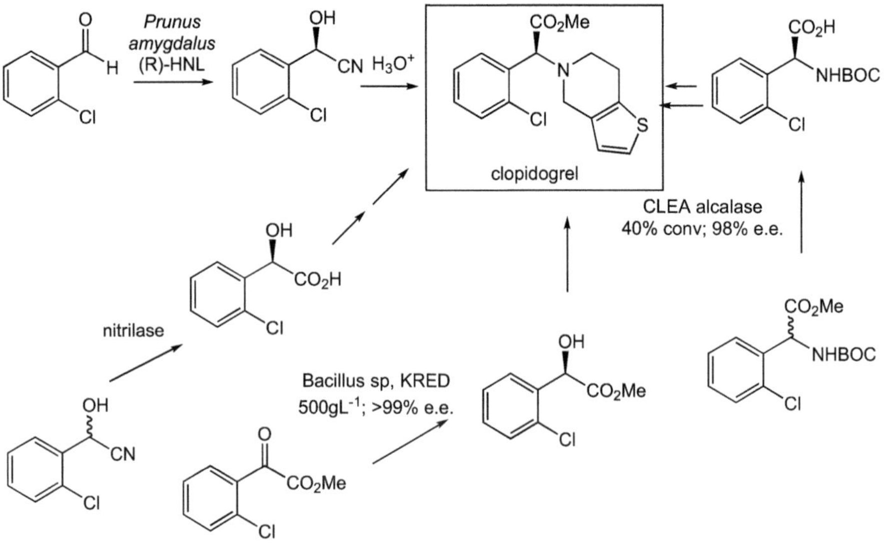

Figure 11.7   Synthesis of clopidogrel.

been intense competition to develop low-cost routes for its manufacture as a generic pharmaceutical. It is now on the World Health
Organisation's list of Essential Medicines.

Two complementary strategies have been developed that rely
on biocatalysis (Figure 11.7). In one approach, the *ortho*-Cl-mandelic
acid fragment is generated followed by construction of the six-
membered N-containing ring. In another approach, the *N*-atom is
included in the chiral precursor. One over-riding challenge is the
presence of the *ortho*-Cl, which creates issues with steric demand
during the biotransformation. One elegant approach is based on
the use of an (*R*)-selective hydroxynitrile lyase (HNL) from *Prunus
amygdalus* to catalyse enantioselective addition of cyanide from HCN
to *ortho*-chloro-benzaldehyde.[21] The activity of the wild-type HNL
biocatalyst was enhanced by site-directed mutagenesis at the active-
site, replacing an alanine residue with a glycine in order to create
more space to accommodate the chlorine atom in the substrate.
The same group also used a nitrilase to convert the racemic *ortho*-
chloro-mandelonitrile to *ortho*-chloro-mandelic acid. Several groups
have reported the reduction of methyl o-chloro-benzoylformate using
either a carbonyl reductase[22] or keto reductase (KRED). Particularly
impressive is the use of an aldo-keto reductase from *Bacillus* sp., with
co-expressed glucose dehydrogenase (GDH), which was carried out at
$[S] = 500 \text{ g L}^{-1}$ and yielded the desired alcohol in >99% e.e.[23] Finally,

racemic Boc-protected phenylglycine methyl ester was resolved using a CLEA-alcalase to give the corresponding (S)-acid in high e.e.[24]

11.9   L-(S)-*tert*-leucine

L-(S)-*tert*-leucine (or L-*t*-butyl-glycine) is a non-proteinogenic acid that is frequently used as a chiral auxiliary and also as a key building block in the development of biologically active peptidomimetics. The bulky and lipophilic nature of the *t*-butyl group presents particular challenges in developing efficient biocatalytic routes for its synthesis.

A number of years ago, scientists at Degussa developed an elegant and scalable process for the production of L-(S)-*tert*-leucine using trimethylpyruvic acid as the precursor and L-leucine dehydrogenase (LeuDH) as the biocatalyst (Figure 11.8).[25] The reaction was carried out in a membrane reactor to localise both the biocatalyst and also the NADH co-factor, which was PEGylated. NADH was regenerated

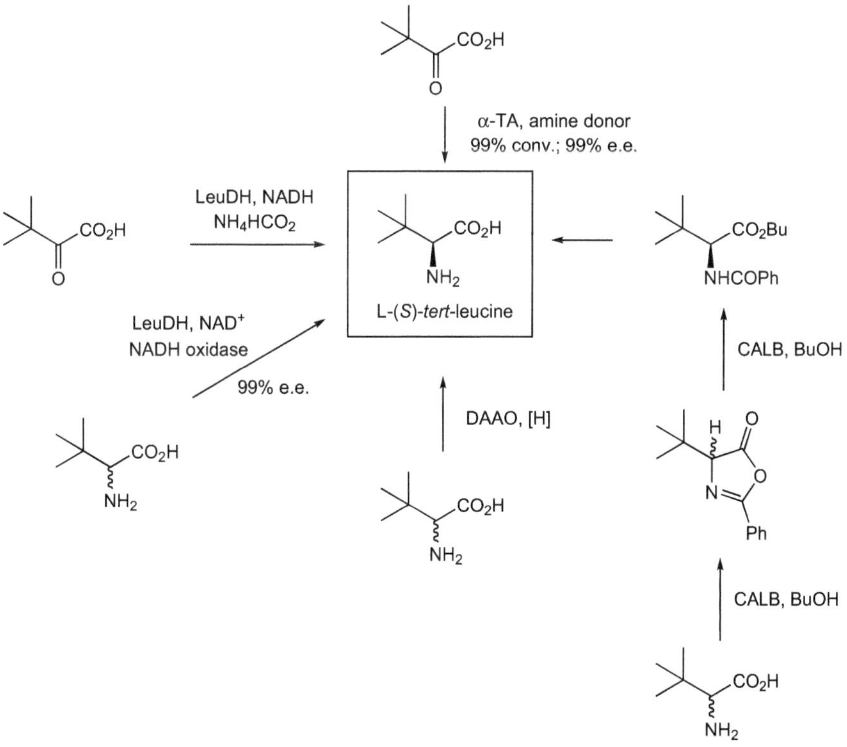

Figure 11.8   Synthesis of L-(S)-*tert*-leucine.

using formate dehydrogenase (FDH) and ammonium formate was used as the source of the amine and also formate. Ingenza developed a scaled process for the production of L-(S)-*tert*-leucine *via* deracemisation of the racemate using an engineered D-amino acid oxidase in combination with a chemical reducing agent.[26] In order to obtain high enough biocatalyst activity it was necessary to subject the D-AAO to several rounds of directed evolution. Another approach starting from the racemic amino acid relies on the use of an enantioselective amino acid dehydrogenase (leucine DH) for a kinetic resolution reaction in which the D-enantiomer is converted to the keto acid leaving the L-enantiomer in *ca.* 50% yield and 99% e.e. NAD$^+$ is required as the co-factor, which was recycled using NADH oxidase.[27] The same α-keto acid can be converted to L-(S)-*tert*-leucine by the use of an α-transaminase (α-TA) from *Ochrobactrum anthropi* to yield the product in 99% conversion and 99% e.e.

Finally, a conceptually different approach was developed based on the use of a lipase-catalysed dynamic kinetic resolution.[28] D/L-*tert*-leucine was first converted to the racemic oxazolone by sequential treatment with benzoyl chloride and acetic anhydride. Treatment of this oxazolone with Lipozyme (lipase from *Mucor miehei*) in butanol and toluene as a solvent led to a high conversion (94%) of the *N*-benzoyl-L-*tert*-leucine butyl ester in 99.5% e.e. The C–H proton at C-2 on the oxazolone has a low pKa due to the stability of the conjugate base. This biotransformation was scaled by Chirotech to 200 g L^{-1}. In order to complete the synthesis, it was necessary to remove the protecting groups and here problems were encountered. Treatment of the protected *tert*-leucine with 6 M HCl led to partial racemisation of the product (73% e.e.) presumably due to partial reformation of the oxazolone, which rapidly racemises. This problem could be overcome by a 2-stage deprotection strategy in which alcalase was first added to hydrolyse the butyl ester followed by 6 M HCl.

11.10 Rasagiline

Rasagiline (Azilect, TVP-1012, *N*-propargyl-1(*R*)-aminoindan) is an irreversible inhibitor of monoamine oxidase-B and is used as a monotherapy to treat symptoms in early Parkinson's disease or as an adjunct therapy in more advanced cases.

Rasagiline is a relatively simple target molecule and indeed several biocatalytic routes have been developed, especially towards the alcohol and amine precursors (Figure 11.9). Thus, racemic indanol

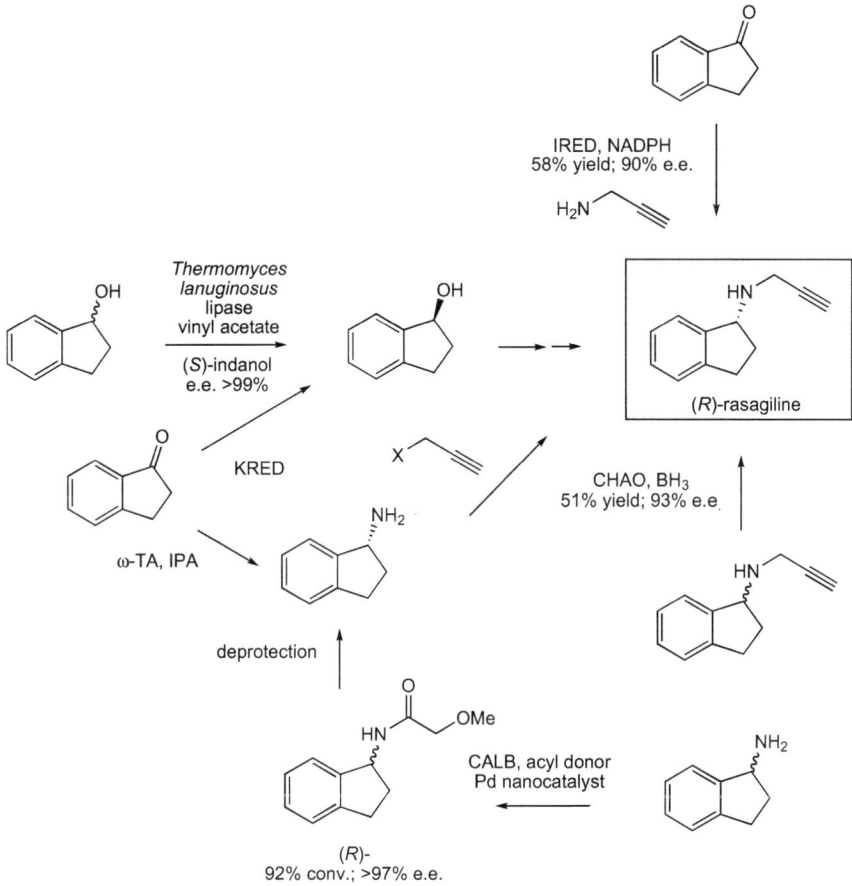

Figure 11.9 Synthesis of rasagiline.

can be converted to (S)-indanol using *Thermomyces lanuginosus* lipase-catalysed resolution in the presence of vinyl acetate.[29] The TL lipase acylates (R)-indanol leaving the (S)-enantiomer in high e.e., which can then be converted to (R)-rasagiline. The same (S)-configured alcohol can also be obtained by a ketoreductase (KRED)-mediated reduction of indanone. Conversion of indanone directly to the (R)-amine has been reported using an ω-transaminase (ω-TA), although indanone is known to generally be a poor substrate for transaminases.[30] Another route to the same enantiomerically pure amine is *via* dynamic kinetic resolution of the racemic amine using CALB in the presence of a palladium nanocatalyst and vinyl acetate as the acylating agent.[31] Recently, two biocatalytic routes that generate (R)-rasagiline directly have been reported. The first involves deracemisation of racemic rasagiline, to give the (R)-enantiomer, by the use of an (S)-selective cyclohexylamine oxidase (CHAO) variant in combination with ammonia borane giving the product in 51% yield and 93% e.e.[32] This approach is especially interesting since (R)-rasagiline is itself an inhibitor of amine oxidases. Finally, the enantioselective reductive amination of indanone and propargylamine has recently been reported using imine reductases (IREDs).[33] IREDs were identified for this conversion with the optimal IRED producing 58% yield and 90% e.e. of the product. Although the use of an IRED represents a very attractive option, the reaction requires further optimisation since currently it uses a large excess of propargylamine (50 equivalents) and a high biocatalyst loading.

11.11   Chiral Biaryl Atropisomers

Enantiomerically pure atropisomers, which possess axial chirality, are widely used as chiral auxiliaries in asymmetric synthesis. For example, BINAP, which was first introduced by Noyori, is regarded as a privileged structure and is widely used as a ligand in transition metal catalysis. Biaryl ethers are more recently developed atropisomers and methods for their synthesis are not so well developed.

In this comparative study, three different approaches for the synthesis of the chiral aldehyde were explored (Figure 11.10).[34] The first

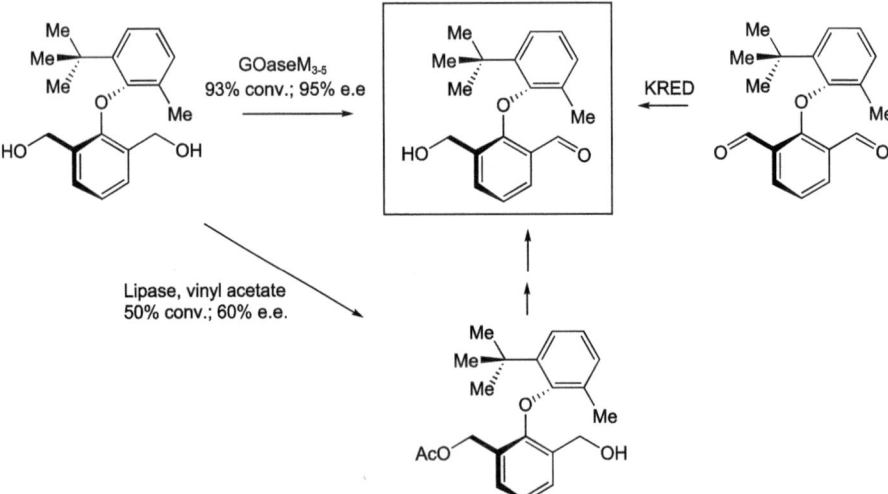

Figure 11.10   Synthesis of chiral biaryl atropisomers.

approach relied upon the use of CALB to catalyse a desymmetrisation of the corresponding diol using vinyl acetate as the acylating agent in an organic solvent. This transformation was found to proceed slowly (>24 h) and gave only moderate e.e.'s (up to 60%) with conversions up to 50%. An alternative approach relied upon starting with the symmetric dialdehyde and screening a panel of KREDs from Codexis to identify biocatalysts with high enantioselectivity. Interestingly, KREDs were discovered that gave both the (P)- (91% conversion; 77% e.e.) and (M)- (39% conversion; 78% e.e.) enantiomers of the product aldehyde in good, but not perfect, enantiomeric excess. The final, and ultimately most successful approach, relied on the use of an engineered variant of galactose oxidase (GOase M_{3-5}). This variant had previously been shown to oxidise racemic phenethanol derivatives with high enantioselectivity. With the diol shown here, the GOase M_{3-5} variant showed high enantioselectivity for the (P)-enantiomer (e.e. >95%). Interestingly, if the reaction was allowed to proceed for longer time periods, some of the di-aldehyde could be detected. Since this second transformation, *i.e.* oxidation of the mono-aldehyde to di-aldehyde, is a kinetic resolution reaction, the yield and the e.e. of the mono-aldehyde were found to change as a function of time. Optimally, a yield of 93% and an e.e. of 95% of the aldehyde were obtained.

11.12 Pregabalin

Pregabalin, marketed under the brand name Lyrica by Pfizer, is a medication used to treat epilepsy, neuropathic pain, fibromyalgia, and generalized anxiety disorder. Its use for epilepsy is as an add-on therapy for partial seizures with or without secondary generalization in adults. Pregabalin is structurally related to g-amino butyric acid (GABA) and works by increasing the density of GABA transporter proteins and thereby increasing the rate of functional GABA transport. It also increases extracellular GABA concentrations in the brain by producing a dose-dependent increase in l-glutamic acid decarboxylase.

The early routes to pregabalin were based on the use of hydrolytic enzymes (Figure 11.11). Thus the γ-nitro ester was shown to undergo kinetic resolution with *Candida antarctica* lipase B (CALB) to yield the optically pure ester, which was then reduced to give (S)-pregabalin.[35] A second approach, which was ultimately commercialised, was also based on a kinetic resolution reaction using

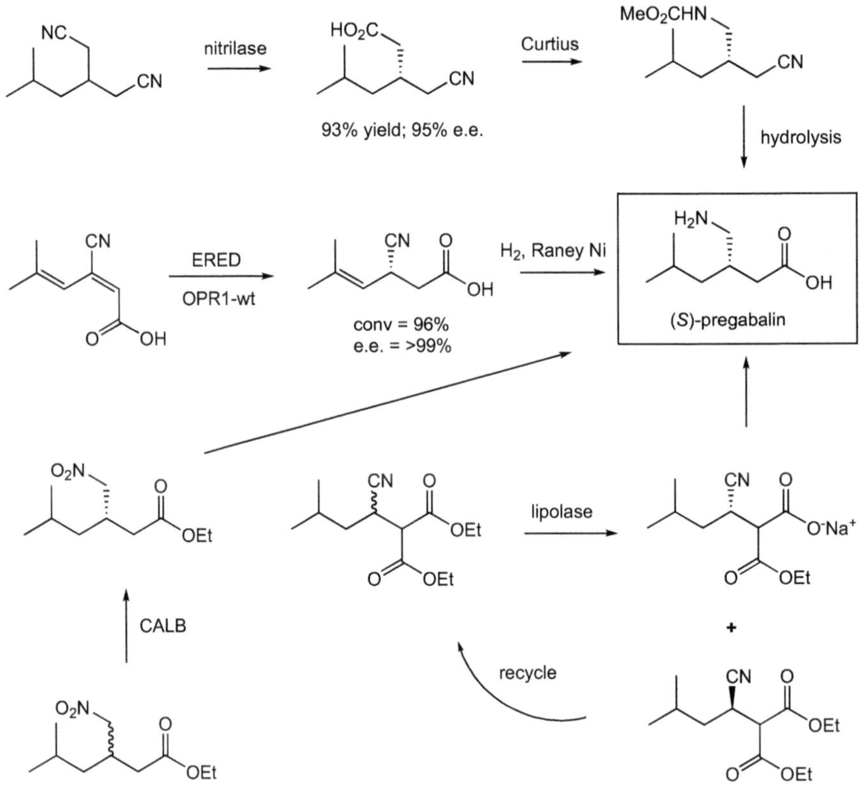

Figure 11.11 Synthesis of pregabalin.

Lipolase, which is a 4% liquid formulation of the lipase from *Thermomyces lanuginosus*. This process employed an interesting strategy of resolving the key stereogenic centre by hydrolysis of the corresponding diester to the mono-carboxylic acid, which then underwent decarboxylation.[36] The nitrile was subsequently reduced to the amine to furnish pregabalin. The unreacted (*R*)-enantiomer was recycled back to the racemic substrate. Finally, two asymmetric approaches to (*S*)-pregabalin have recently been developed. The first involves the use of an ene reductase (ERED) for the asymmetric reduction of the diene carboxylic acid. This reaction is selective for the C=C containing the CN substituent and gives a high conversion and e.e. (>99%) of the nitrile, which is then reduced with Raney nickel. The second approach involves desymmetrisation of the corresponding dinitrile using a nitrilase. Subsequent Curtius rearrangement following chemical hydrolysis of the remaining nitrile group yielded pregabalin.[37]

11.13 Ticagrelor

Ticagrelor (trade names Brilinta, Brilique, and Possia) is a platelet aggregation inhibitor produced by AstraZeneca and is one of the most potent drugs on the market for the treatment of acute coronary syndrome. Ticagrelor is used for the prevention of thrombotic events (for example stroke or heart attack) in people with acute coronary syndrome or myocardial infarction with ST elevation.

The enantiomerically pure *trans*-cyclopropyl amine represents a challenging target, in particular to explore and ultimately compare alternative biocatalytic routes for its synthesis (Figure 11.12). One of the

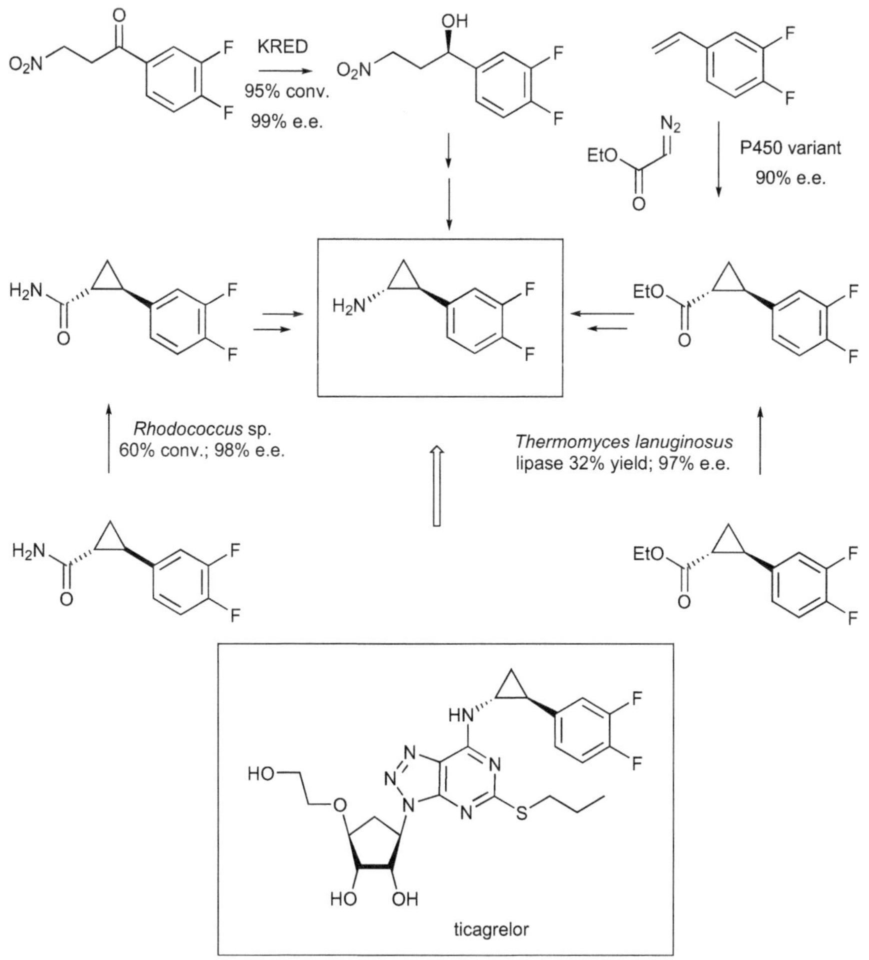

Figure 11.12 Synthesis of ticagrelor.

earliest chemical routes was based on the chiral nitro alcohol, which could be subsequently activated (Ph$_3$P/CBr$_4$) and then treated with a base (DBU) to effect formation of the cyclopropane ring followed by reduction of the nitro group to an amine. The required (S)-alcohol could be obtained in high conversion and e.e. (99%) by the use of KRED-130 from Codexis.[38] Cofactor recycling was achieved by the use of glucose/GDH. A second approach involved the use of an amidase to effect kinetic resolution of the corresponding racemic amide. This transformation was carried out using whole cell *Rhodococcus rhodochronus*, which is known to catalyse the enantioselective hydrolysis of cyclopropyl-containing amides. The unreacted amide of correct absolute configuration was obtained in up to 98% e.e. at a conversion of 60%. A third approach involved the use of a lipase from *Thermomyces lanuginosus* for kinetic resolution of the racemic *trans*-cyclopropyl ester. In this case, it was shown that the product carboxylic acid had the required absolute configuration. The TL lipase gave high e.e. and was eventually scaled up to provide the required acid in good yield (32%) and high e.e. (97%).[39] This carboxylic acid has been shown to undergo Curtius rearrangement *via* the acyl azide with high stereoselectivity to yield the required amine. A conceptually distinct approach to the same ester has recently been reported using a P450 variant which is able to catalyse enantioselective cyclopropanation of the difluorostyrene using ethyl diazoacetate as the carbene precursor. This reaction proceeds in good yield and high e.e. and from the point of view of biocatalytic retrosynthesis presents a completely new way of assembling the target molecule.[40]

11.14   Lipitor/Crestor

Atorvastatin, marketed under the trade name Lipitor among others, is a member of the drug class known as statins, which are used primarily as a lipid-lowering agent and for prevention of events associated with cardiovascular disease. Like all statins, atorvastatin works by inhibiting HMG-CoA reductase, an enzyme found in liver tissue that plays a key role in production of cholesterol in the body. Rosuvastatin, marketed under the tradename Crestor, is another member of the statins, which is used in combination with exercise, diet, and weight-loss to treat high cholesterol and related conditions, and to prevent cardiovascular disease.

Lipitor and Crestor represent excellent examples of the way in which different biocatalysts from the tool-box can be combined to generate new synthetic routes to complex molecules. Both drugs, in common with all statins, contain a *syn*-3,5-dihydroxy acid side-chain pharmacophore, which can be assembled by the use of biocatalysts for both asymmetric FGIs and C–C bond formation starting from simple building blocks (Figure 11.13).[41] Codexis have used gene

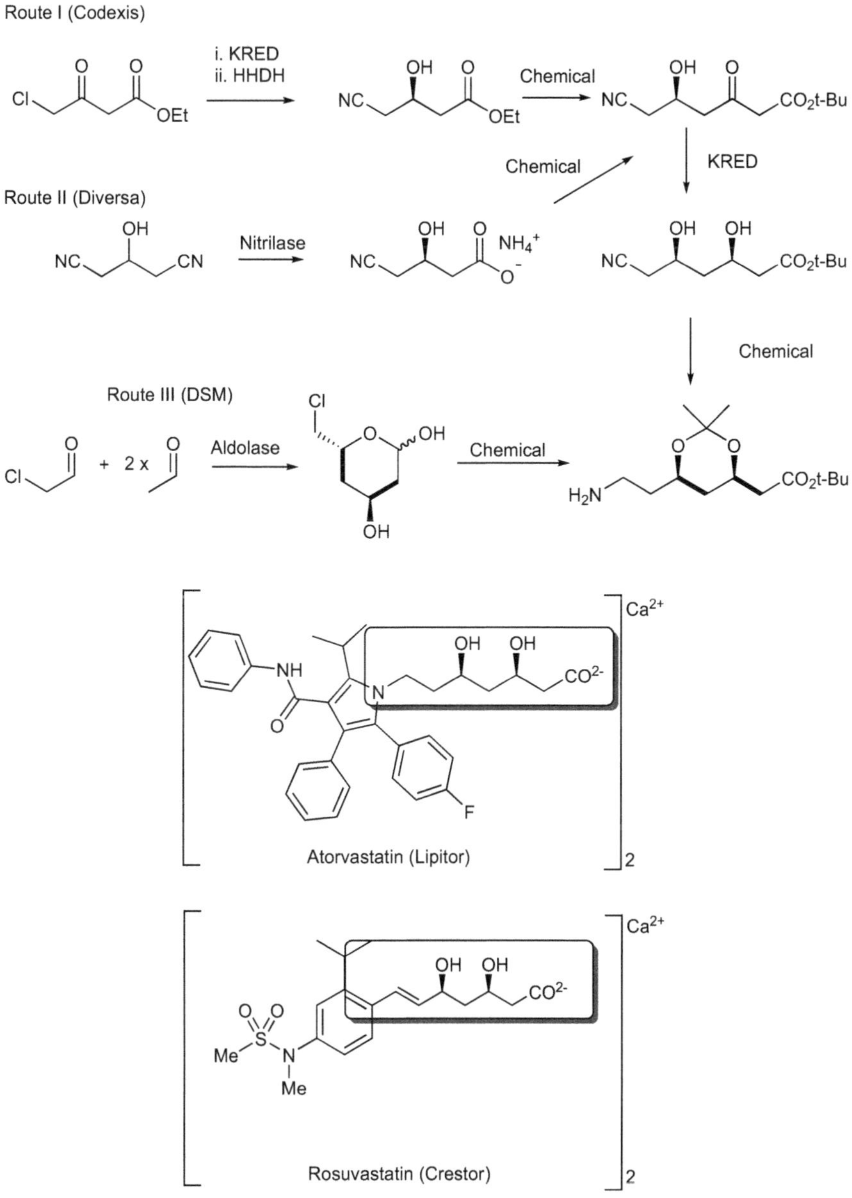

Figure 11.13 Synthesis of atorvastatin (Lipitor) and rosuvastatin (Crestor).

shuffling technologies to enhance the stability and activity of both a ketoreductase and a halohydrin dehalogenase (HHDH), which catalyse an asymmetric reduction and subsequent cyano-substitution of ethyl 4-chloro-3-hydroxybutyrate, providing the chiral intermediate (Route I). Diversa have employed a nitrilase for the desymmetrisation of pro-chiral 3-hydroxyglutaronitrile, which can easily be accessed from commercially available epichlorohydrin (Route II). In common with route I, this strategy also employs a KRED for the selective reduction of the β-ketoester and preparation of the advanced pharmaceutical intermediate. Finally, the DSM approach (Route III) relies on the asymmetric C–C bond forming ability of 2-deoxyribose-5-phosphate aldolase for the synthesis of the pyran, starting from the cheap, non-chiral bulk chemicals acetaldehyde and chloroacetaldehyde. A number of alternative routes employing lipases and esterases have also been reported for the preparation of this chiral side-chain.

References

1. P. Soni and U. C. Banerjee, *Appl. Microbiol. Biotechnol.*, 2005, **67**, 771.
2. C. G. Tang, H. Lin, C. Zhang, Z. Q. Liu, T. Yang and Z. L. Wu, *Biotechnol. Lett.*, 2011, **33**, 1435.
3. I. Rimoldi, G. Facchetti, D. Nava, M. L. Contente and R. Gandolfi, *Tetrahedron Asymm.*, 2016, **27**, 389.
4. A. Träff, R. Lihammar and J. E. Bäckvall, *J. Org. Chem.*, 2011, **76**, 3917.
5. Z. Q. Ren, Y. Liu, X. Q. Pei, H. B. Wang and Z. L. Wu, *J. Mol. Cat B: Enzym.*, 2015, **113**, 76.
6. H. Liu, B. H. Hoff and T. Anthonsen, *Chirality*, 2000, **12**, 26.
7. M. Fuchs, D. Koszelewski, K. Tauber, J. Sattler, W. Banko, A. K. Holzer, M. Pickl, W. Kroutil and K. Faber, *Tetrahedron*, 2012, **68**, 7691.
8. M. Fuchs, D. Koszelewski, K. Tauber, W. Kroutil and K. Faber, *Chem. Commun.*, 2010, **46**, 5500.
9. K. Han, C. Kim, J. Park and M.-J. Kim, *J. Org. Chem.*, 2010, **75**, 3105.
10. M. K. Sethi, R. Bhandya, N. Maddur, R. Shukla, A. Kumar and V. S. N. Jayalakshmi Mittapalli, *Tetrahedron Asymm.*, 2013, **24**, 374.
11. C. Hartley, F. Manford, S. Burton and R. Dorrington, *Appl. Microbiol. Biotechnol.*, 2001, **57**, 43.
12. U. Müller, F. van Assema, M. Gunsior, S. Orf, S. Kremer, D. Schipper, A. Wagemans, C. A. Townsend, T. Sonke, R. Bovenberg and M. Wubbolts, *Metabol. Eng.*, 2006, **8**, 196.

13. M. I. Youshko, L. M. van Langen, R. A. Sheldon and V. K. Švedas, *Tetrahedron Asymm.*, 2004, **15**, 1933.
14. H. Schutt, US Patent 4260684 A.
15. A. J. Blacker, M. J. Stirling and M. I. Page, *Org. Proc. Res. Dev.*, 2007, **11**, 642.
16. M. Dürrenbergcr, T. Ilcinisch, Y. M. Wilson, T. Rossel, E. Nogueira, L. Knörr, A. Mutschler, K. Kersten, M. J. Zimbron, J. Pierron, T. Schirmer and T. R. Ward, *Angew. Chem. Int. Ed. Engl.*, 2011, **50**, 3026.
17. F. Leipold, S. Hussain, D. Ghislieri and N. J. Turner, *ChemCatChem*, 2013, **5**, 3505.
18. C. Wang, J. Qiu, Y. Yang, J. Zheng, J. He and S. Li, *Biotechnol. Lett.*, 2017, **39**, 553.
19. S. J. C. Taylor, A. G. Sutherland, C. Lee, R. Wisdom, S. Thomas, S. M. Roberts and C. Evans, *J. Chem. Soc. Chem. Commun.*, 1990, 1120.
20. E. Xun, J. Wang, H. Zhang, G. Chen, H. Yue, J. Zhao, L. Wang and Z. Wang, *J. Chem. Technol. Biotechnol.*, 2013, **88**, 904.
21. A. Glieder, R. Weis, W. Skranc, P. Poechlauer, I. Dreveny, S. Majer, M. Wubbolts, H. Schwab and K. Gruber, *Angew. Chem. Int. Ed.*, 2003, **42**, 4815.
22. T. Ema, N. Okita, S. Ide and T. Sakai, *Org. Biomol. Chem.*, 2007, **5**, 1175.
23. Y. Ni, J. Pan, H.-M. Ma, C.-X. Li, J. Zhang, G.-W. Zheng and J.-H. Xu, *Tetrahedron Lett.*, 2012, **53**, 4715.
24. P. Ferraboschi, M. De Mieri and F. Galimberti, *Tetrahedron Asymmetry*, 2010, **21**, 2136.
25. A. S. Bommarius, M. Schwarm, K. Stingl, M. Kottenhahn, K. Huthmacher and K. Drauz, *Tetrahedron Asymmetry*, 1995, **6**, 2851.
26. P. P. Taylor and R. E. Speight, US20110059503 A1.
27. W. Hummel, M. Kuzu and B. Geueke, *Org. Lett.*, 2003, **5**, 3649.
28. N. J. Turner, J. R. Winterman, R. McCague, J. S. Parratt and S. J. C. Taylor, *Tetrahedron Lett.*, 1995, **36**, 1113.
29. T. de Sousa Fonseca, M. R. da Silva, M. da Conceição Ferreira de Oliveira, T. L. G. de Lemos, R. de Araújo Marques and M. C. de Mattos, *Appl. Catal. A: Gen.*, 2015, **492**, 76.
30. M. S. Malik, E. S. Park and J. S. Shin, *Green Chem.*, 2012, **14**, 2137.
31. G. Ma, Z. Xu, P. Zhang, J. Liu, X. Hao, J. Ouyang, P. Liang, S. You and X. Jia, *Org. Proc. Res. Dev.*, 2014, **18**, 1169.
32. G. Li, P. Yao, P. Cong, J. Ren, L. Wang, J. Feng, P. C. K. Lau, Q. Wu and D. Zhu, *Scientific Rep.*, 2016, DOI: 10.1038/srep24973.
33. P. Matzel, M. Gand and M. Höhne, *Green Chem.*, 2017, **19**, 385.

34. B. Yuan, A. Page, C. P. Worrall, F. Escalettes, S. C. Willies, J. J. W. McDouall, N. J. Turner and J. Clayden, *Angew. Chem. Int. Ed.*, 2010, **122**, 7164.
35. F. Felluga, G. Pitacco, E. Valentin and C. D. Venneri, *Tetrahedron Asymmetry*, 2008, **19**, 945.
36. C. A. Martinez, S. Hu, Y. Dumond, J. Tao, P. Kelleher and L. Tully, *Org. Proc. Res. Dev.*, 2008, **12**, 392.
37. Y. Duan, P. Y. Yao, J. Ren, C. Han, Q. Li, J. Yuan, J. H. Feng, Q. Q. Wu and D. Zhu, *Sci. China Chem.*, 2014, **57**, 1164.
38. K. G. Hugentobler, H. Sharif, M. Rasparini, R. S. Heath and N. J. Turner, *Org. Biomol. Chem.*, 2016, **14**, 8064.
39. K. G. Hugentobler, M. Rasparini, L. A. Thompson, K. E. Jolley, A. J. Blacker and N. J. Turner, *Org. Proc. Res. Dev.*, 2017, **21**, 195.
40. K. E. Hernandez, H. Renata, R. D. Lewis, S. B. Jennifer Kan, C. Zhang, J. Forte, D. Rozzell, J. A. McIntosh and F. H. Arnold, *ACS Catal.*, 2016, **6**, 7810.
41. For an overview of different approaches see: M. Müller, *Angew. Chem. Int. Ed.*, 2005, **44**, 362

12 Applications of Biocatalytic Retrosynthesis

12.1 Summary

With the increasing availability of a broader range of different biocatalysts, more opportunities for the application of enzymes in organic synthesis present themselves. However, for molecules of even moderate complexity there will be 'hidden' disconnections that need to be applied to the molecule in a retrosynthetic manner before it is apparent which biocatalyst could potentially be used. For target molecules containing two or more stereogenic centres, the level of complexity in terms of retrosynthesis will rapidly increase. In addition, in some cases it may be that more than one type of biocatalyst could in principle be used to achieve the desired conversion. This situation often arises when thinking about the synthesis of chiral alcohols or amines, since there may be several options available. For example, in earlier chapters in this book, we met a number of different methods for the synthesis of 2° alcohols including ketone reduction, resolution of esters *etc.* Similarly, for the synthesis of chiral amines, again there are an increasing number of options including the use of transaminases, amine dehydrogenases, imine reductases and amine oxidases. Increasingly, biocatalysts for functional group interconversions (FGI) are being considered alongside those for making C–C and C–X bonds in order to develop multi-enzymatic approaches to target molecules.

Chapter 10 presented a different approach to the use of biocatalysis, in which different enzyme-catalysed reactions that generated the same functional group or motif within a molecule were collected

Biocatalysis in Organic Synthesis: The Retrosynthesis Approach
By Nicholas J. Turner and Luke Humphreys
© Nicholas J. Turner and Luke Humphreys 2018
Published by the Royal Society of Chemistry, www.rsc.org

under one heading, thereby allowing the reader to instantly gain an appreciation of how many different options present themselves. This paper exercise, in which biocatalytic retrosynthesis is applied to suggest various potential biocatalysts, and hence starting materials, must then of course be followed up by some type of assessment regarding (i) the availability of the starting materials, (ii) the availability of the biocatalyst, and (iii) the likelihood or precedent that the available biocatalyst(s) will be able to catalyse the desired transformation. Assuming that some kind of proof of principle experiment can be undertaken to satisfy and answer these questions, then the researcher can move quickly to optimisation studies with a view to initiating preparative-scale reactions and ultimately process intensification for scale-up. This kind of approach is now very prevalent in the field of biocatalysis, particularly within the pharmaceutical industry, where speed is increasingly of the essence in order to rapidly progress the development of compounds through clinical trials. Interestingly, with the development of biocatalyst panels that are able to address a broader substrate scope, as detailed in Chapters 3–9, there is growing interest and awareness within the discovery part of the pharmaceutical industry that enzymes may have a role to play in medicinal chemistry. For example, the development of ketoreductases (KREDs) has reached the point where panels of >100 biocatalysts are now available from commercial suppliers and these panels offer very broad substrate coverage giving access to both (*R*)- and (*S*)-enantiomers. Under this scenario, it perhaps makes better sense to use KREDs for hit identification in the knowledge that the biocatalyst that is used to generate the lead compound may then be used to prepare larger amounts as the compound progresses through clinical trials. The application of protein engineering and directed evolution, coupled with genome mining, will undoubtedly result in biocatalyst panels with broader substrate coverage, increasingly able to transform substrates that are of interest to the medicinal chemist.

Chapter 11 set out to reinforce the ideas of biocatalytic retrosynthesis by presenting a series of case studies in which commercially important target pharmaceuticals have stimulated the development of multiple solutions based on the use of biocatalysis. Biocatalytic retrosynthesis is most powerfully applied when it leads to the redesign of the manufacturing route to the API. A new synthetic route based on biocatalysis can present many benefits, including patent protection, reduced cost of goods, improved safety and reduced waste. The use of biocatalysis is inherently complementary to chemical synthesis and chemo-catalysis and hence the introduction of enzymes into synthesis can in some cases lead to a step change in the manufacturing route to a target molecule.

The aim of Chapter 12 is to provide the reader of this book with an opportunity to test their understanding and knowledge of the biocatalytic reactions presented in Chapters 3–9 as a prelude to gaining experience in disconnecting target molecules based on the principles of biocatalytic retrosynthesis provided. Chapters 3–9 should be studied alongside Chapter 10, since in essence they represent two sides of the same coin; the material present in Chapters 3–9 is simply presented in a different format to that in Chapter 10. Chapter 12 deliberately takes a gentle step forward into the application of biocatalytic retrosynthesis in order to build confidence for those readers for whom biocatalysis is still a black art. In order to provide a quick revision of some of the material already covered in this book, the first set of problems simply present biocatalytic reactions in the context of the conversion of substrates to products, exactly as covered earlier the book. For each of these enzyme-catalysed transformations, you should think about choosing a suitable biocatalyst (*e.g.* transaminase, lipase, amine oxidase) together with the accompanying reagents (*e.g.* amine donor for a transamination reaction; acyl donor for a lipase-catalysed acylation reaction) and any additional co-factors (*e.g.* NADH, ATP *etc.*) required to achieve the transformation with high conversion and enantioselectivity. In some cases, the reactions involve the conversion of racemic substrates to products of high enantiomeric purity in >50% yield and hence here it is important to think about the application of dynamic kinetic resolution or deracemisation processes, which have already been discussed in various sections of the book. Other questions are based on asymmetric synthesis or desymmetrisation of prochiral or *meso*-substrates. Where co-factors are required, try thinking about the reagents and additional enzymes required for cost-effective and economic co-factor recycling, which would be important to introduce on a preparative scale. Once you have attempted this first set of questions, you can immediately check your answers on the following page and then perhaps go back and check the relevant section in Chapters 3–9. Thereafter, you should then be in a good position to start applying retrosynthetic disconnections to target molecules in order to begin to design synthetic routes to target molecules in which a biocatalyst is employed in one or more key bond forming steps. If you feel confident enough, then simply jump straight to the worked examples at the end of the chapter where a set of 25 problems, of increasing difficulty and complexity, are presented, together with answers, to help develop skills in applying biocatalytic retrosynthesis.

12.2   Questions (a)–(r)

(a)

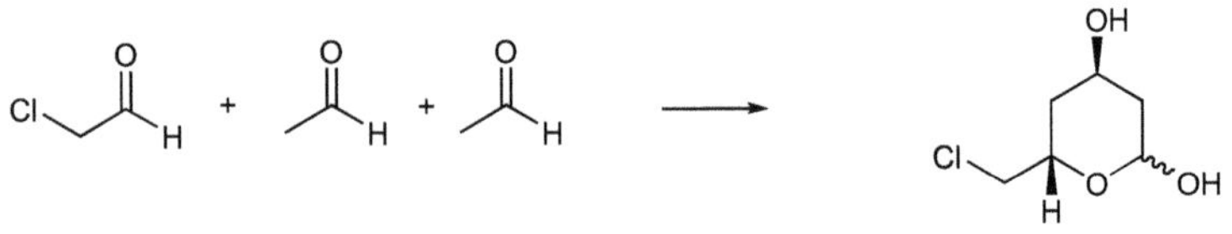

(b)

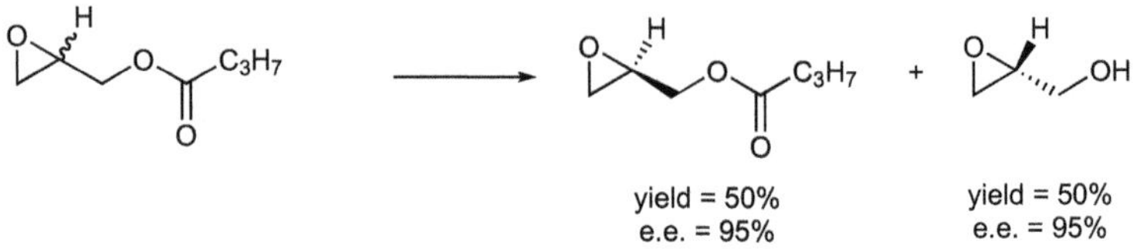

yield = 50%
e.e. = 95%

yield = 50%
e.e. = 95%

(c)

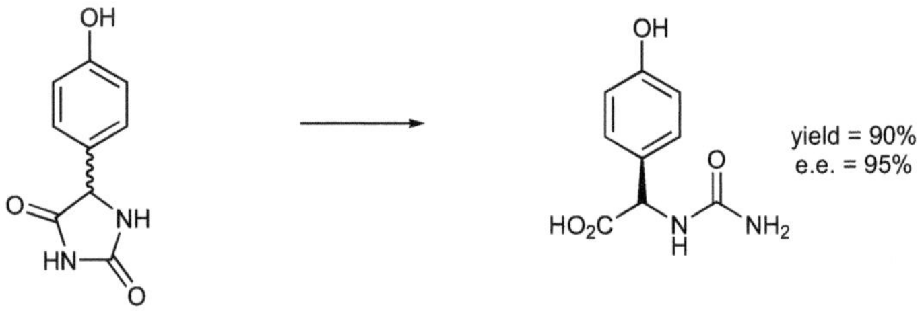

yield = 90%
e.e. = 95%

(d)

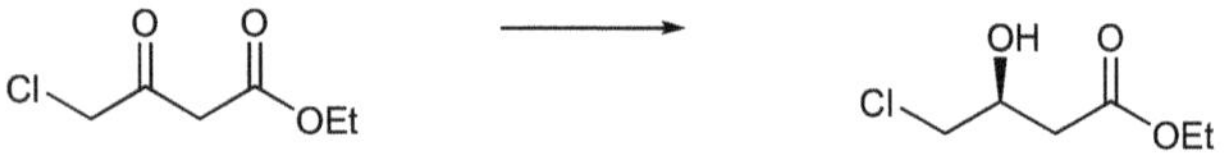

(e)

(f)

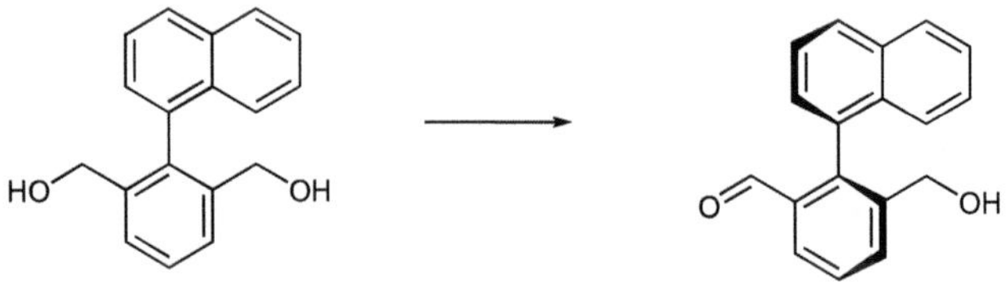

(g)

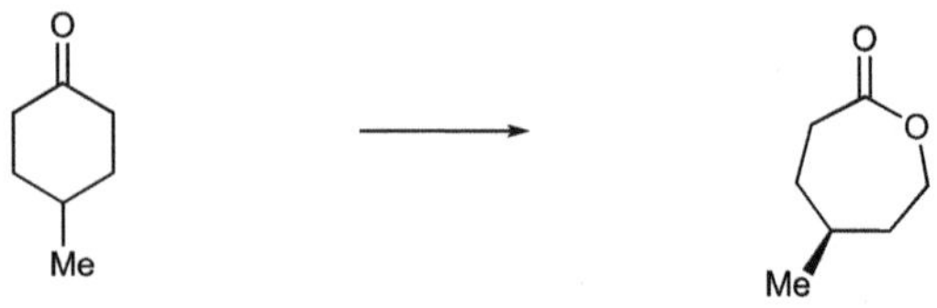

(h)

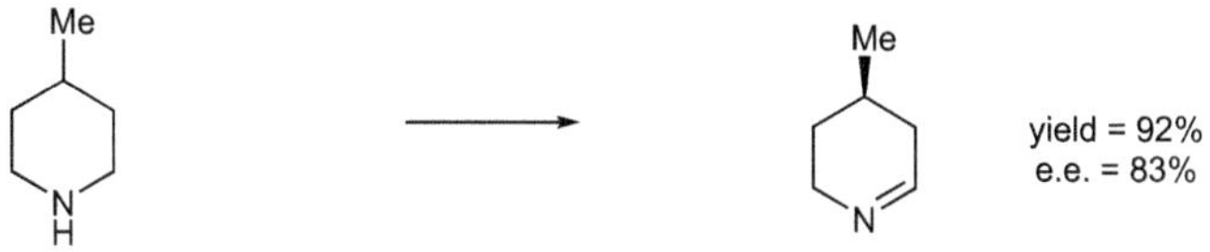
Me
N
H
Me
N
yield = 92%
e.e. = 83%

(i)

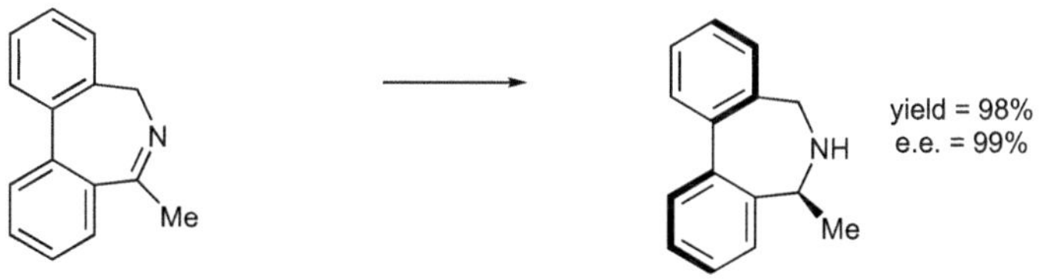
N
Me
NH
Me
yield = 98%
e.e. = 99%

(j)

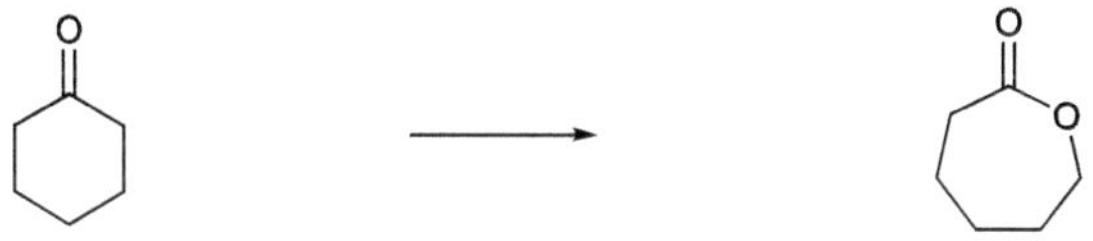
O
O
O

(k)

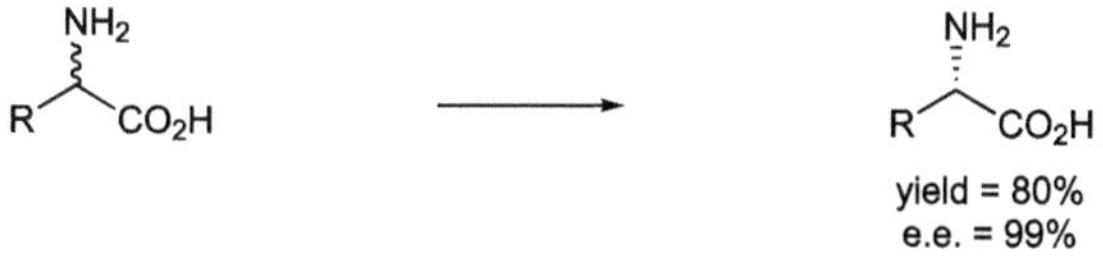
NH2
R
CO2H
NH2
R
CO2H
yield = 80%
e.e. = 99%

(l)

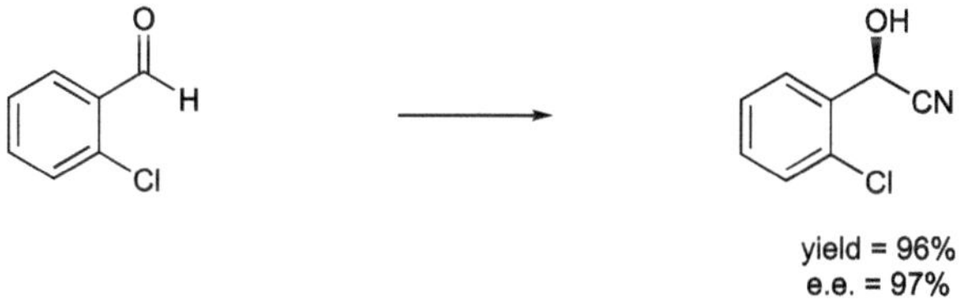
O
H
Cl
OH
CN
Cl
yield = 96%
e.e. = 97%

(m)

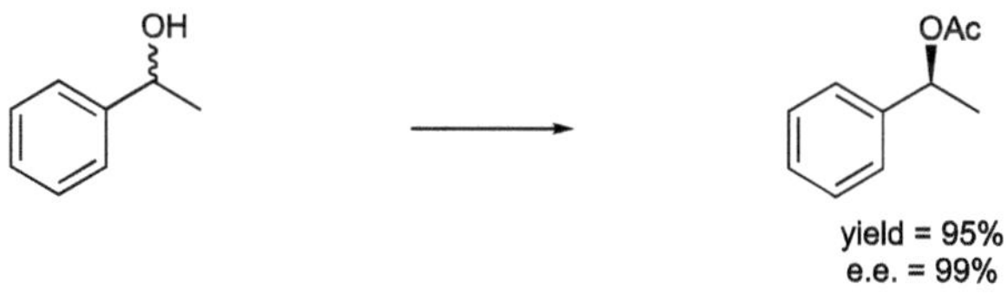
OH
OAc
yield = 95%
e.e. = 99%

(n)

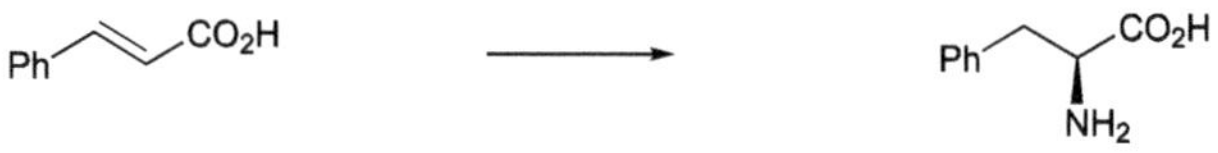
Ph
CO2H
Ph
CO2H
NH2

(o)

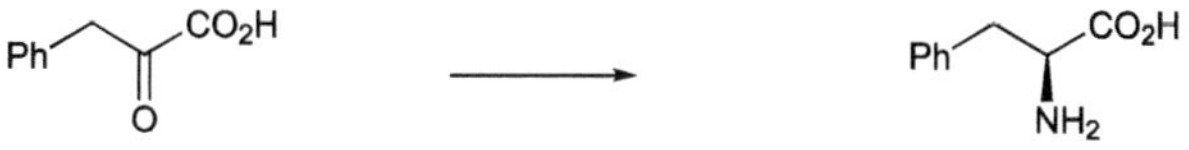
Ph
CO2H
O
Ph
CO2H
NH2

(p)

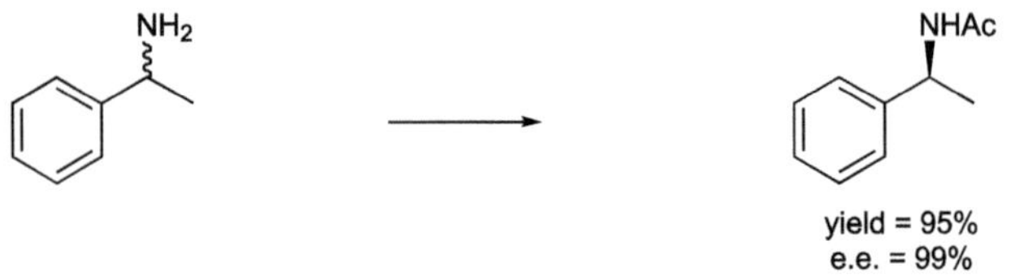
NH2
NHAc
yield = 95%
e.e. = 99%

(q)

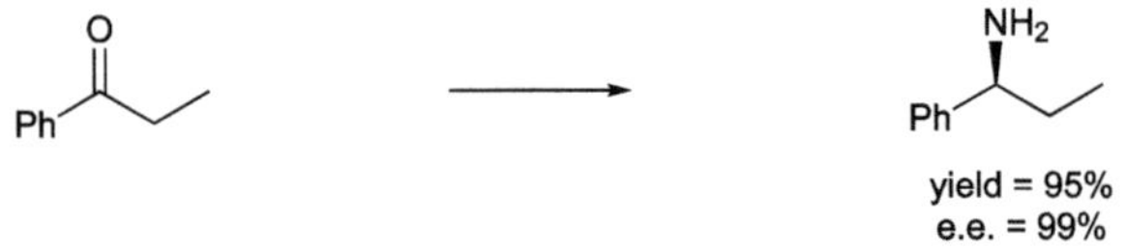
O
Ph
NH2
Ph
yield = 95%
e.e. = 99%

(r)

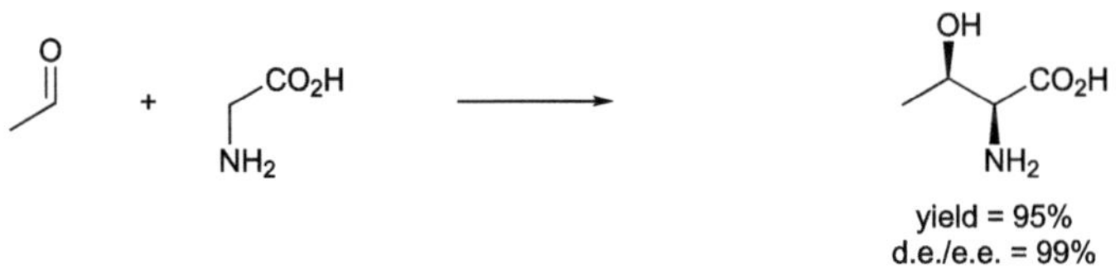
O
+
CO2H
NH2
OH
CO2H
NH2
yield = 95%
d.e./e.e. = 99%

12.3 Answers (a)–(r)

(a)

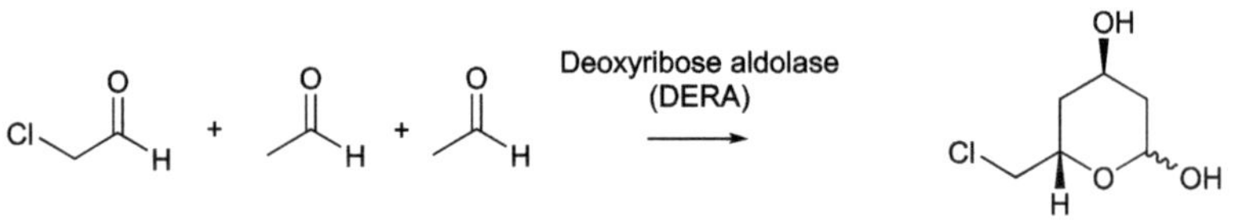
Cl
O
H
+
O
H
+
O
H
Deoxyribose aldolase
(DERA)
OH
Cl
H
O
OH

(b)

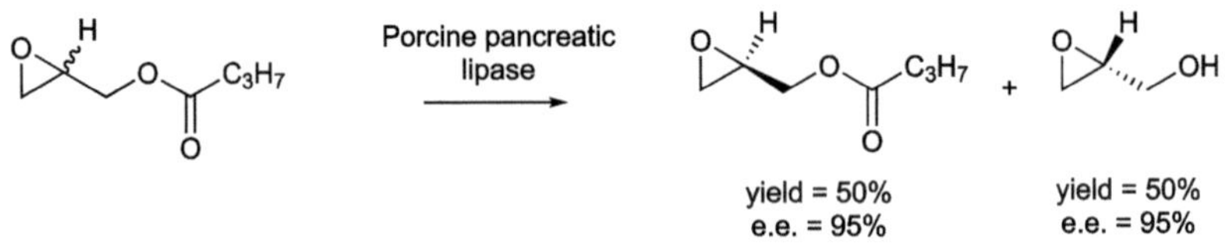
H
O
O
C3H7
O
Porcine pancreatic
lipase
O
H
O
C3H7
O
+
O
H
OH
yield = 50%
e.e. = 95%
yield = 50%
e.e. = 95%

(c)

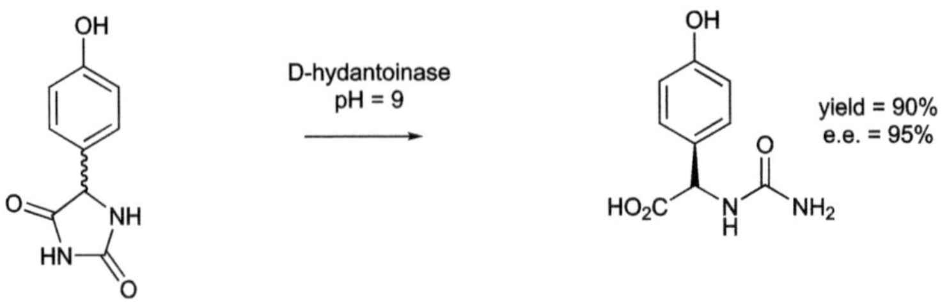
OH
O
NH
HN
O
D-hydantoinase
pH = 9
OH
O
HO2C
N
H
NH2
yield = 90%
e.e. = 95%

(d)

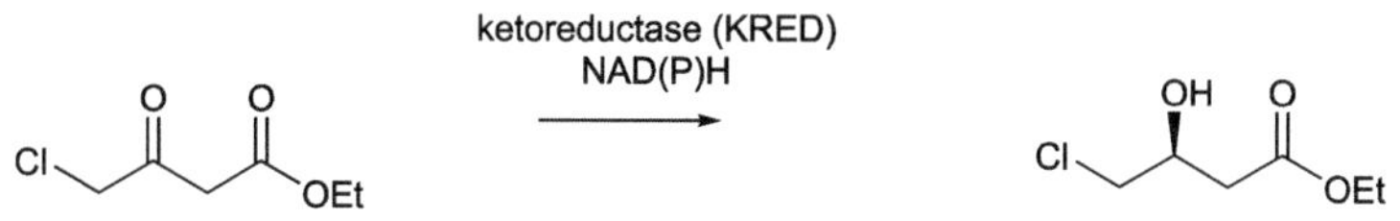

ketoreductase (KRED)
NAD(P)H

(e)

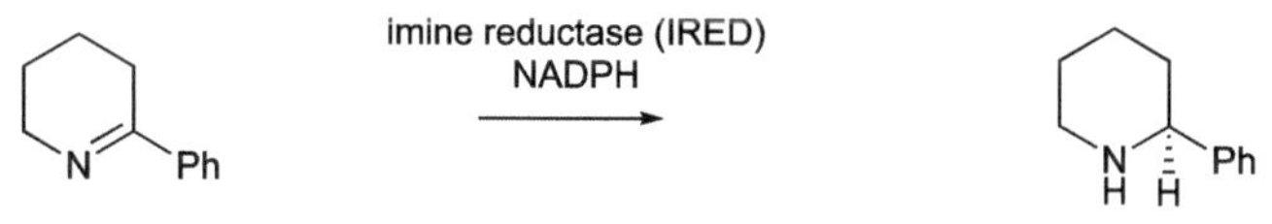

imine reductase (IRED)
NADPH

(f)

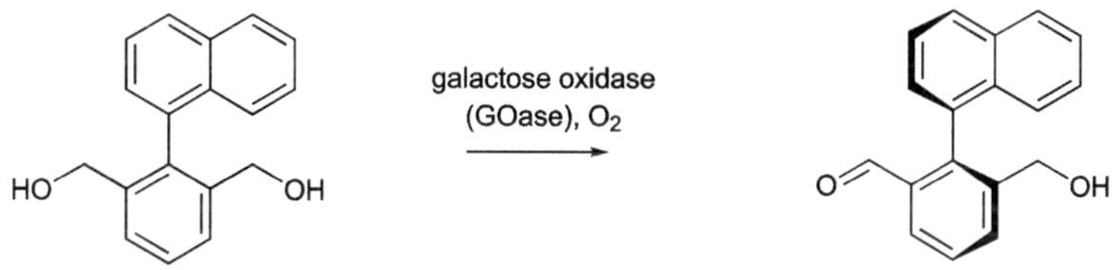

galactose oxidase
(GOase), O_2

(g)

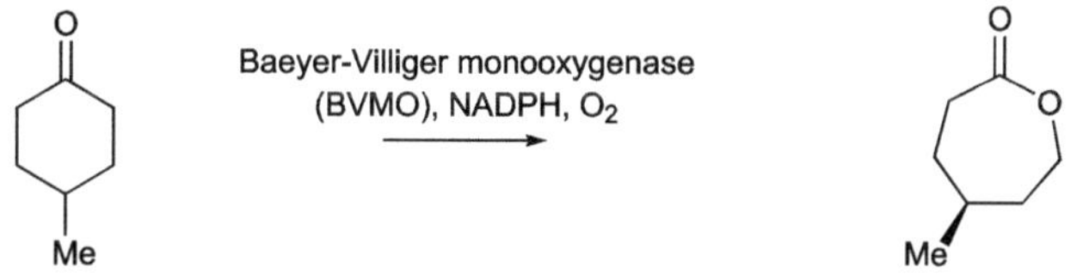

Baeyer-Villiger monooxygenase
(BVMO), NADPH, O_2

(h)

monoamine oxidase
(MAO-N), O_2

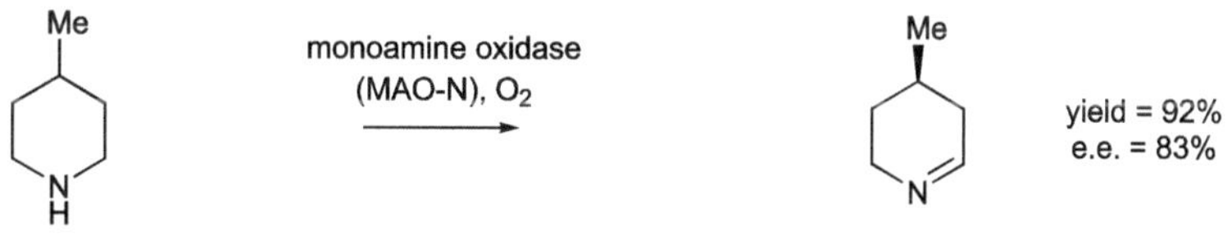

yield = 92%
e.e. = 83%

(i)

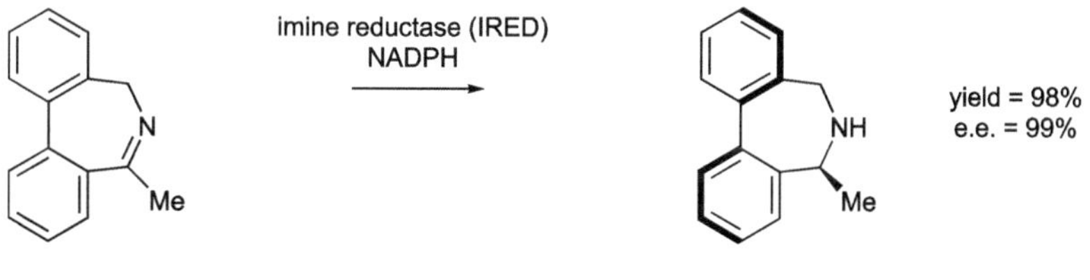

imine reductase (IRED)
NADPH

yield = 98%
e.e. = 99%

(j)

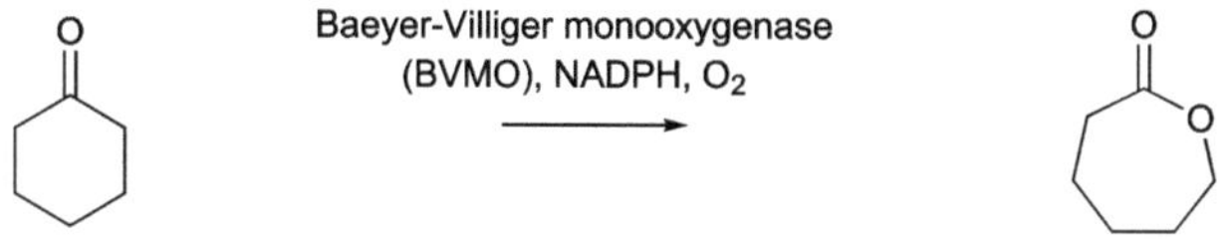

Baeyer-Villiger monooxygenase
(BVMO), NADPH, O_2

(k)

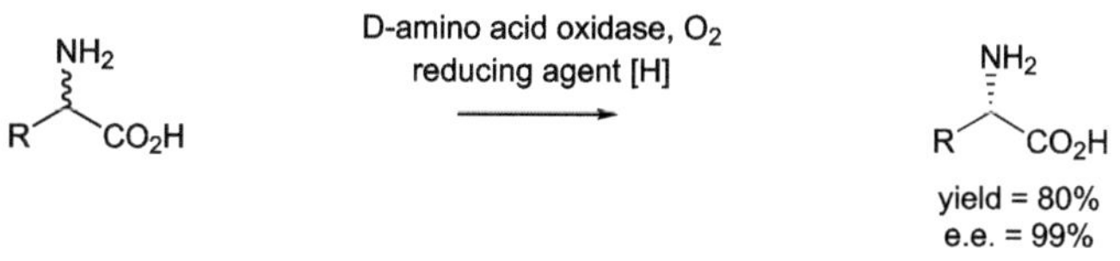

D-amino acid oxidase, O_2
reducing agent [H]

yield = 80%
e.e. = 99%

(l)

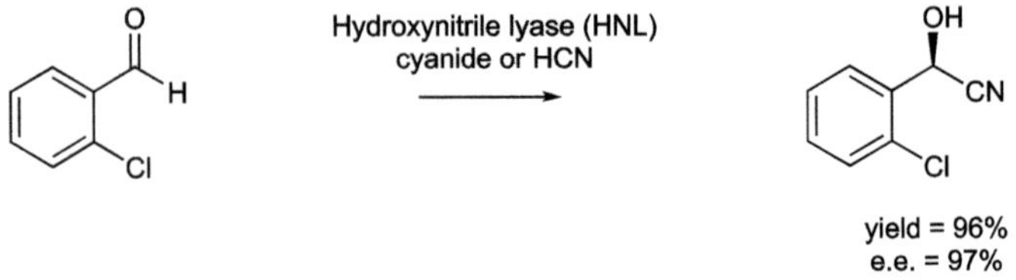

Hydroxynitrile lyase (HNL)
cyanide or HCN

yield = 96%
e.e. = 97%

(m)

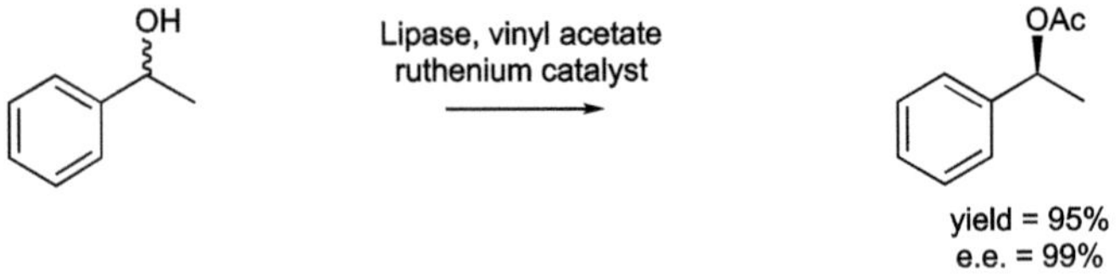

Lipase, vinyl acetate
ruthenium catalyst

yield = 95%
e.e. = 99%

(n)

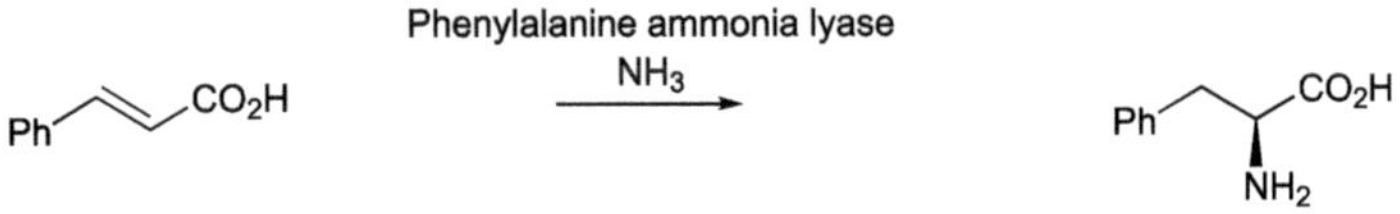

Phenylalanine ammonia lyase
NH$_3$

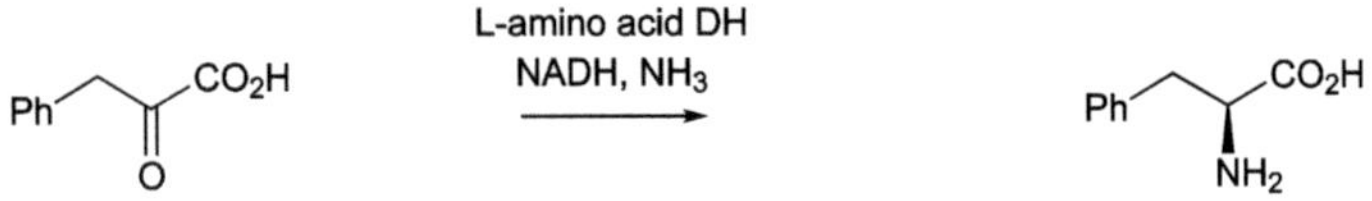

(o)

L-amino acid DH
NADH, NH$_3$

(p)

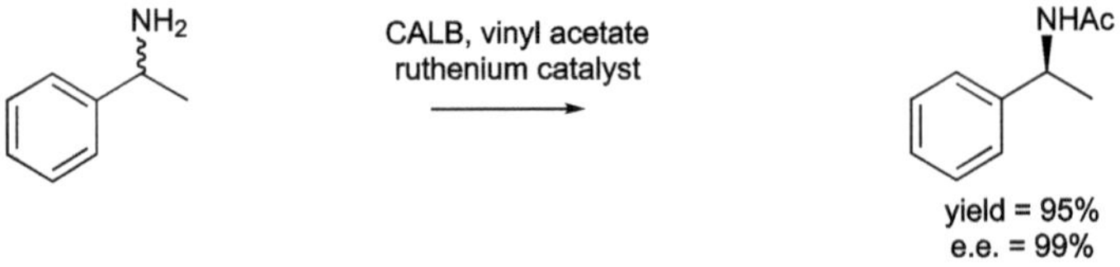

CALB, vinyl acetate
ruthenium catalyst

yield = 95%
e.e. = 99%

(q)

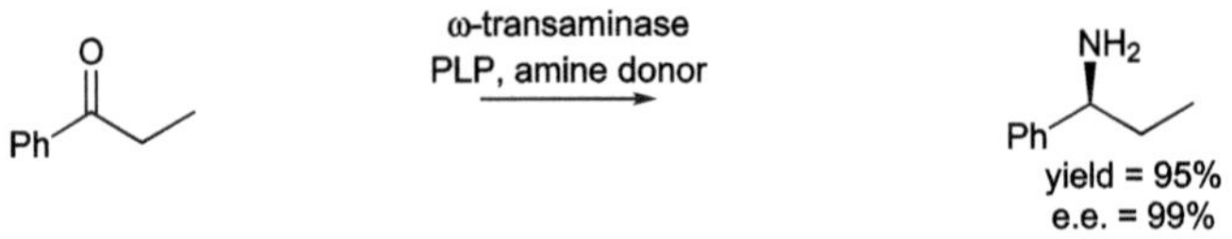

ω-transaminase
PLP, amine donor

NH$_2$

Ph
yield = 95%
e.e. = 99%

(r)

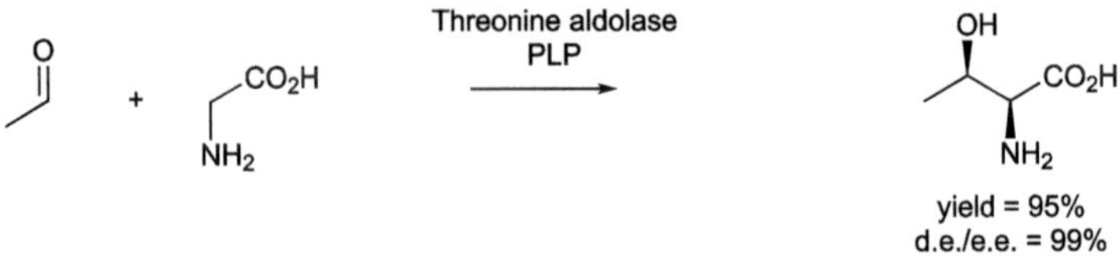

Threonine aldolase
PLP

yield = 95%
d.e./e.e. = 99%

12.4 Designing Synthetic Routes to Target Molecules

We now come to the final part of the book, where the application of biocatalytic retrosynthesis will be used to design potential synthetic routes to target molecules, in particular intermediates or building blocks for pharmaceuticals. This section is organised in a way that the target molecules become progressively more challenging in nature and eventually require more than one biocatalyst, together with the ability to recognise 'hidden' disconnections.

12.4.1 Easy Retrosynthetic Targets

The first set of examples is designed to introduce the reader gently to biocatalytic retrosynthesis and to revise some of the material that has previously been presented in the book. These questions require only a single retrosynthetic disconnection involving the use of a biocatalyst, *i.e.* no additional synthetic manipulations are required other than to recognise which biocatalyst needs to be employed to achieve the key reaction. In each case, the bond that needs to me made, or the functional group that needs to be introduced, is clearly indicated, to help focus the mind. As indicated above, in several cases there will be more than one option available, so you should determine the possibilities and consider the advantages or disadvantages of the different approaches, based on considerations such as availability of the starting material, reaction conditions *etc.*

Questions

Answers

For each of these chiral building blocks, there are clearly multiple solutions as already highlighted earlier in the book. This situation is especially true when the target is a secondary alcohol, chiral amine or amino acid. In these cases, it is a good exercise to use biocatalytic retrosynthesis to generate all possible solutions, even if it is just a paper exercise to ensure that all potential starting materials have been considered.

Amino acid oxidase, [H]   L-AADH, NH₃

OR

Ammonia lyase, NH3

IRED, NADPH   MAO-N, [H]

Dehalogenase   ERED, NAD(P)H   Oxidase

12.4.2   More Advanced Biocatalytic Retrosyntheses

In the final section of this chapter, and indeed the book, the reader will be presented with more challenging target molecules, which increasingly will require the disconnection of more than one bond in order to reveal where a biocatalytic step may be introduced. The format for this set of 25 problems is that initially the target molecule will be introduced together with the ideal or desired starting material. Thereafter, the published route to the target molecule will be presented together with a brief analysis of why this particular approach was developed. The primary reference is also given if the reader wishes to obtain more detail regarding the route, particularly in terms of practical details, including scale-up. The questions start fairly easy, we hope, and then become decidedly more tricky towards the end. Remember the solutions given are those that have been published and are not necessarily the only solutions, or the best. As we have emphasised throughout the book, the increased availability of more biocatalysts, with broader substrate scope, will result in more retrosynthetic disconnections, as is the case for classical organic synthesis, where for example the formation of a carbon–carbon double bond can be achieved in many different ways.

Question 1

(*R*)-Flurbiprofen is a non-steroidal anti-inflammatory drug which contains a stereogenic methyl group alpha- to a carboxylic acid. Consider how you might prepare the target molecule from the corresponding unsaturated ester.

(*R*)-Flurbiprofen

Answer 1[1]

In this example, an ene reductase (ERED) from *Bacillus subtilis* (YqjM) was employed as part of a chemoenzymatic synthesis of (*R*)-flurbiprofen. The catalytic system comprised a glucose dehydrogenase (GDH) co-factor recycling system and granted access to enantiopure propionic ester in good yield.

Question 2

This is an example of a non-asymmetric transformation in which the target molecule is a lactam and the starting material is either an amino alcohol or a cyclic amine. This transformation requires the use of two biocatalysts for sequential oxidation reactions. Hint: In both examples, the intermediate is the corresponding cyclic imine.

Answer 2[2]

Starting from the acyclic amino alcohol, a galactose oxidase variant (GOase M3-5) was used to selectively oxidise the alcohol group to the aldehyde without oxidation of the amine. The amino aldehyde product spontaneously cyclises to give the cyclic imine, which is then subjected to oxidation to the lactam using xanthine dehydrogenase (XDH). Alternatively, the same cyclic imine can be generated by oxidation of the cyclic amine using monoamine oxidase N (MAO-N). Interestingly, all three biocatalysts use molecular oxygen as the oxidant, providing a green approach to lactam synthesis.

Question 3

Consider how the 1,5-diketone could be converted to the 2,6-disubstituted piperidine. The synthesis should be ideally able to generate all four possible stereoisomers of the product, *i.e.* both enantiomers of the *syn-* and *anti-*diastereoisomers. One important issue to consider is the need to carry out some type of regioselective amination on the starting diketone.

Answer 3[3]

Treatment of the 1,5-diketone with an ω-transaminase (ω-TA) results in highly regioselective amination of the less sterically demanding methyl ketone in the presence of the phenyl ketone. This reaction is also highly enantioselective and can generate either the (S)- or (R)-amine by appropriate choice of the ω-TA. The intermediate amino ketone undergoes spontaneous cyclisation to the imine, which is then reduced to the amine using an IRED. Since both (*S*)- and (*R*)-IREDs are available, all four possible stereoisomers can be generated *via* this approach.

1,5-diketone

2,6-disubstituted piperidine

Question 4

2-Substituted piperidines are valuable building blocks in the synthesis of pharmaceuticals. In this case, the specific target molecule is 2-phenyl piperidine and the requisite starting material is the corresponding δ-keto acid. Things to consider here include (i) what biocatalyst would be best for establishing the chiral amine stereochemistry and (ii) what is the best source of the amine group?

Answer 4[4]

Treatment of the 1,5-keto acid with carboxylic acid reductase (CAR) gives a clean and high yielding conversion to the corresponding 1,5-ketoaldehyde. This reaction is carried out with lyophilised whole cells in the presence of glucose to facilitate recycling of the ATP and NADPH co-factors. Subsequent addition of an ω-transaminase results in regioselective amination of the aldehyde in the presence of the ketone. ω-TAs generally have low reactivity towards phenyl ketones and hence you would not expect to generate the isomeric amine. The amino ketone intermediate spontaneously cyclises to the cyclic imine, which then undergoes asymmetric reduction to the piperidine in the presence of an imine reductase (IRED) and NADPH.

1,5-ketoacid

1,5-ketoaldehyde

Question 5

How might a slight variation on the approach outlined above be employed to achieve the synthesis of 2-phenyl-3-methyl piperidine as both a single diastereoisomer and single enantiomer? Here there is the added challenge that the starting keto acid is racemic and hence the synthesis needs to include a step in which both enantiomers can be converted to a single enantiomer of the product.

Answer 5[5]

The principal retrosynthetic disconnections are the same as above, *i.e.* the target piperidine can be derived from reduction of the imine, which can in turn be derived from transamination of the ketone followed by spontaneous cyclisation. As before, the starting material is the corresponding keto acid. However, the imine in this particular case is in equilibrium with the corresponding enamine, which is stabilised by the 2-phenyl group. Generation of the enamine leads to equilibration of the stereochemistry of the 2-methyl group and since the imine reductase (IRED) step is both enantioselective as well as highly diastereoselective, the product is obtained in high d.e. (95%) and good e.e. (81%).

Question 6

An interesting variation on the use of ω-TAs to generate chiral 2,5-disubstituted piperidines can be used to prepare target molecules containing a ketone functionality in the side-chain. The requisite starting material for this synthesis is now a 1,7-diketone. Again, the key to success here is the regioselectivity of the transaminase.

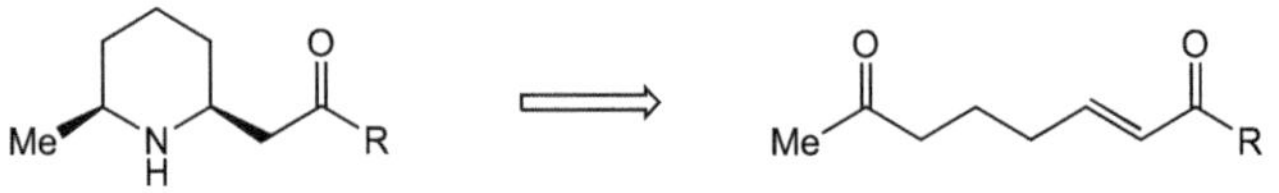

Answer 6[6]

This elegant synthesis exploits the fact that ω-TAs display low reactivity towards α,β-unsaturated ketones and hence it is possible to carry out a regio- and enantioselective amination of the substrate to give an enantiomerically pure amine, which undergoes spontaneous conjugate addition of the amine to the enone to generate the piperidine. This cyclisation reaction is under thermodynamic control and results in the more stable 1,3-*syn*-diastereomeric products.

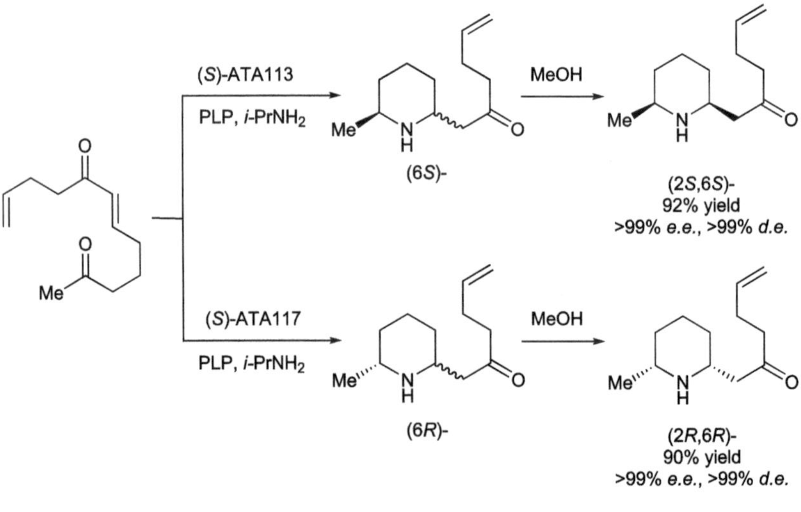

Question 7

5- and 6-membered ring lactones and lactams are extremely versatile building blocks in organic synthesis, and moreover feature very strongly in biologically active compounds and also as flavour and fragrance components. Suggest a general synthesis of both lactones and lactams starting from the corresponding keto esters.

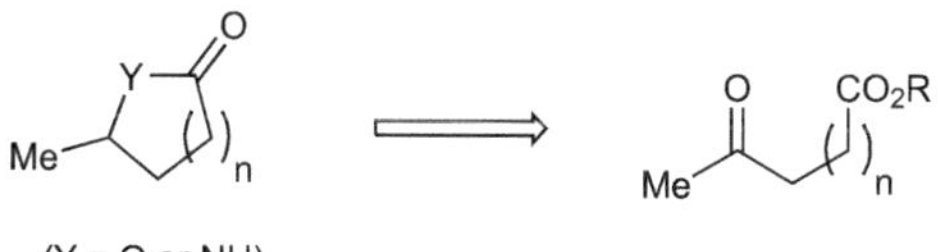

(Y = O or NH)

Answer 7[7]

For the lactone synthesis, treatment of the keto ester with either a ketoreductase (KRED) or alcohol dehydrogenase (ADH) will result in highly enantio- and chemo-selective reduction of the ketone group to give the hydroxy ester, which spontaneously cyclises. Use of an ω-TA will result in ketone amination, followed again by spontaneous cyclisation to the lactam.

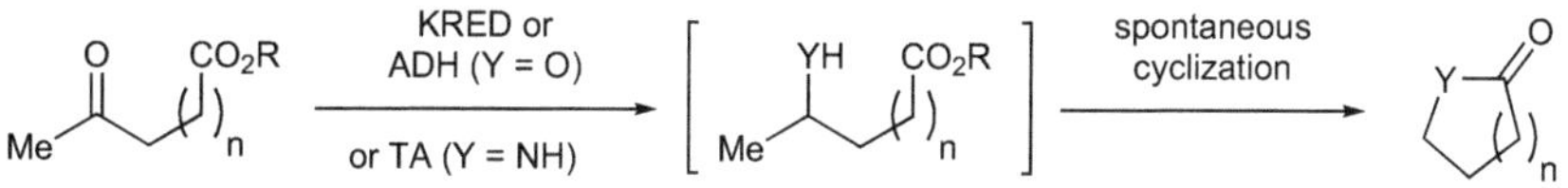

Question 8

How could you modify the synthesis outlined above to convert the enone shown below to the 5-membered ring lactone in which both methyl groups in the product are *syn*?

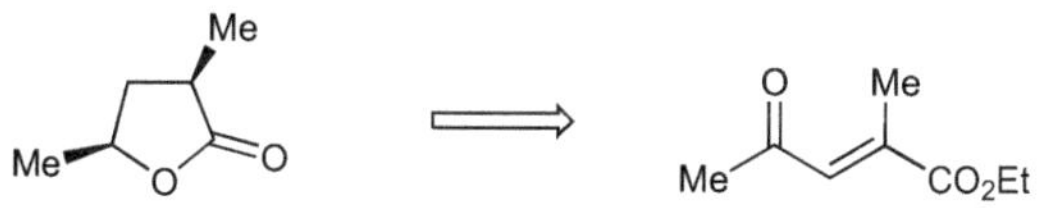

Answer 8[8]

This is an elegant example of the use of a combination of ADHs and EREDs. A one-pot method has been developed for the formation of two different stereoisomers of α,γ-dimethyl γ-lactones from the readily available γ-keto butyric ester with excellent enantio- and dia-stereocontrol. This one-pot method involved initial reduction of the ketone by OYE1 (Old Yellow Enzyme 1, *Saccharomyces carlsbergensis*) before adding an alcohol dehydrogenase to give the hydroxyester, which underwent lactonisation. The use of **ADH-T** (*Thermo-anaerobacter* sp.) or **ADH-LK** (*Lactobacillus kefir*) gave the *syn-* and *anti-* products, respectively.

Question 9

How would you convert a β-keto ester into an α-hydroxy ester with the added requirement of starting with a racemic substrate and trying to convert this racemate into an enantiomerically pure product?

Answer 9[9]

Baeyer–Villiger monooxygenases (BVMOs) are FAD-dependent enzymes that catalyse the oxidative conversion of a wide range of cyclic and acyclic ketones to the corresponding lactones and esters, respectively. While mono- and bicyclic ketones represent a popular substrate class for enantioselective biocatalytic Baeyer–Villiger reactions, the oxidation of acyclic ketones, in particular aliphatic acyclic ketones, is less well explored. One important contribution in this area is the discovery that the BVMO phenylacetone monooxygenase (PAMO) from *Thermobifida fusca* is able to catalyse the resolution of a variety of racemic α-substituted β-keto esters to the corresponding α-acyloxy esters with good conversion and in high enantioselectivity. The authors also demonstrated that the latter could be hydrolysed to the corresponding α-hydroxy esters without loss of enantiomeric purity.

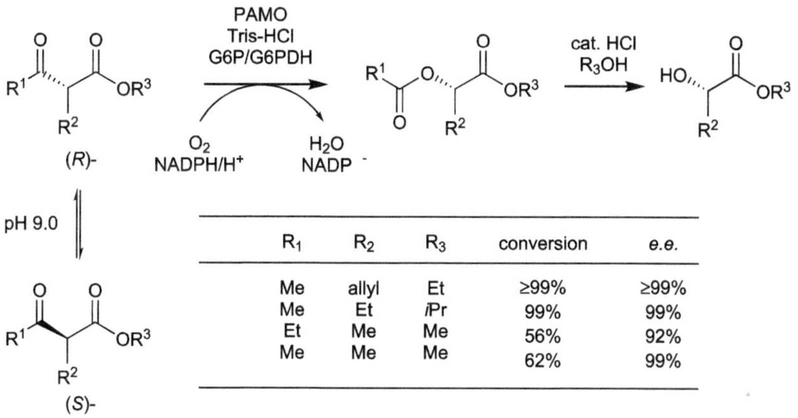

	R_1	R_2	R_3	conversion	e.e.
	Me	allyl	Et	≥99%	≥99%
	Me	Et	iPr	99%	99%
	Et	Me	Me	56%	92%
	Me	Me	Me	62%	99%

Question 10

How would you convert the sulfonic acid into the corresponding sultam?

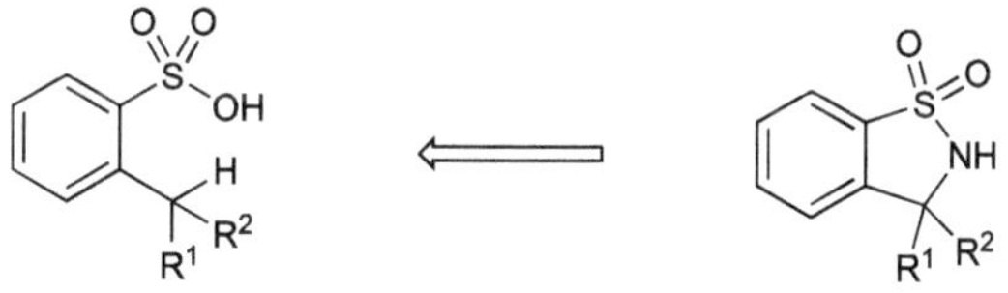

Answer 10[10]

P450 variants, in which the axial heme ligand Ser is exchanged for His, have recently been developed and used for the generation of carbenes and nitrenes from appropriate precursors. For example, a P450 variant was used for the synthesis of enantiomerically enriched allylic amines *via* conversion of an allylic sulfide to a sulfinimine followed by [2,3]-sigmatropic rearrangement. In this example, a complementary approach is illustrated for the synthesis of γ-sultam by insertion of a sulfonyl nitrene into a pendant benzylic C–H bond.

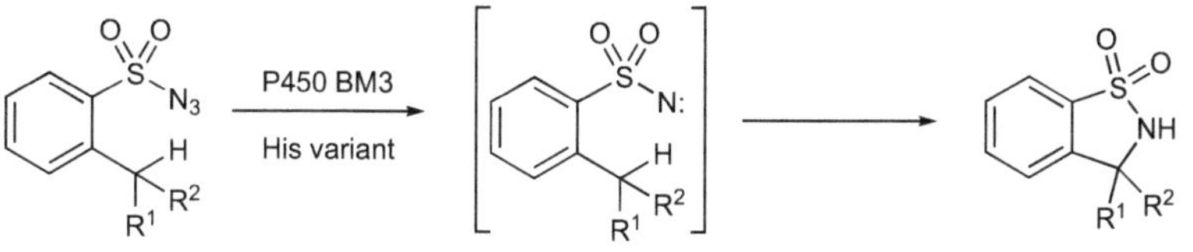

Question 11

Aza- or imino-sugars are biologically active molecules that often show the ability to inhibit carbohydrate processing enzymes such as glycosidases. A particular challenge is establishing the stereo-chemistry of the hydroxyl groups, which are required for biological activity. How would you synthesise the pyrrolidine shown below from the corresponding precursors dihydroxyacetone and the requisite aldehyde?

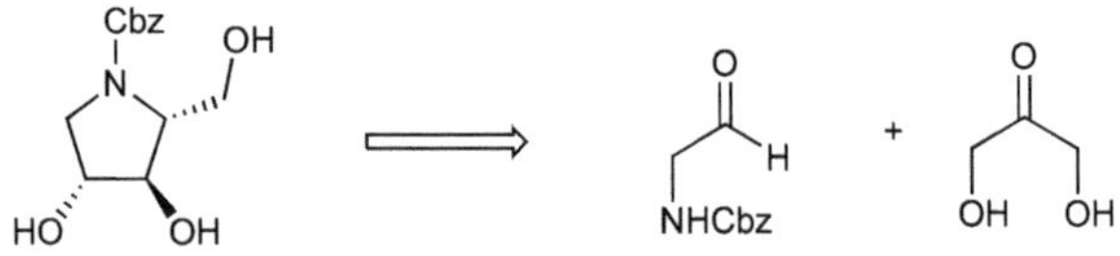

Answer 11[11]

The synthesis of this target molecule involves the use of a stereo-selective aldolase (FSA) to create the key C–C bond with complete control of both stereogenic centres. The product triol was isolated and then treated with hydrogen and palladium, followed by protein to yield the final product.

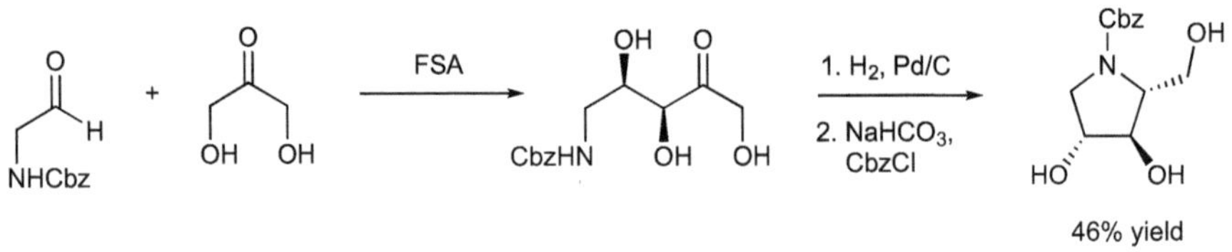

Question 12

Zanamivir (Relenza) is the API of a drug developed by GSK for the treatment of influenza. The drug works by blocking a key enzyme (neuraminidase), which is required for trafficking by the flu virus. A key building block in the synthesis of this target molecule is N-acetyl neuraminic acid. How would you convert N-acetylmannosamine and pyruvate into NANA? For an extra point, can you think of how the synthesis could be modified and improved by using N-acetyl-glucosamine as the precursor?

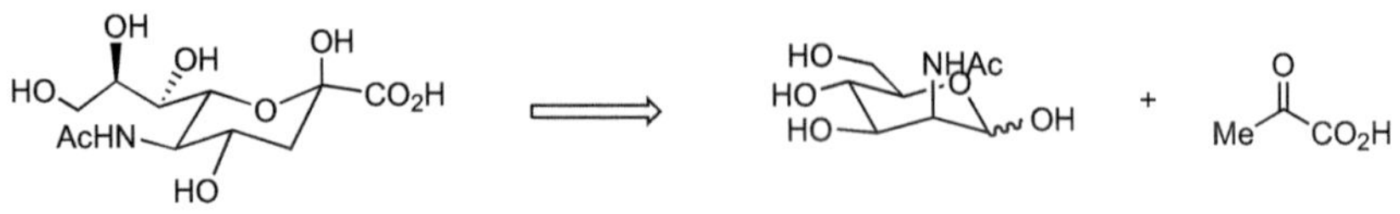

Answer 12[12]

The key biocatalyst here is N-acetyl-neuraminic acid (NANA) aldolase for the stereoselective condensation of pyruvic acid with N-acetyl-mannosamine. N-acetylmannosamine is a hexose sugar that exists as an equilibrium mixture of the ring-opened and ring-closed forms, the former generating the aldehyde required for the aldolase reaction. *N*-acetylmannosamine is relatively expensive and hence was generated *in situ* from *N*-acetylglucosamine by base-catalysed epimerisation at C-2.

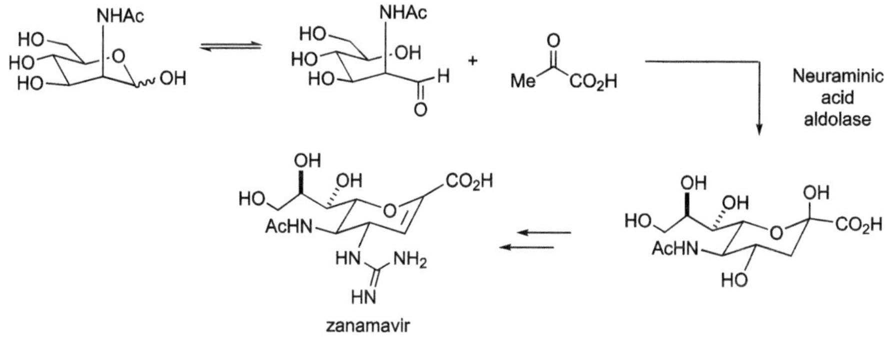

zanamavir

Question 13

The SMO inhibitor shown below is currently in development by Pfizer and is a particularly challenging target molecule. The intermediate chiral amine possesses a β-stereogenic centre, which must also be established during the synthesis. How might you convert the ketone to the amine using a biocatalyst and also incorporating dynamic kinetic resolution (DKR) into the process?

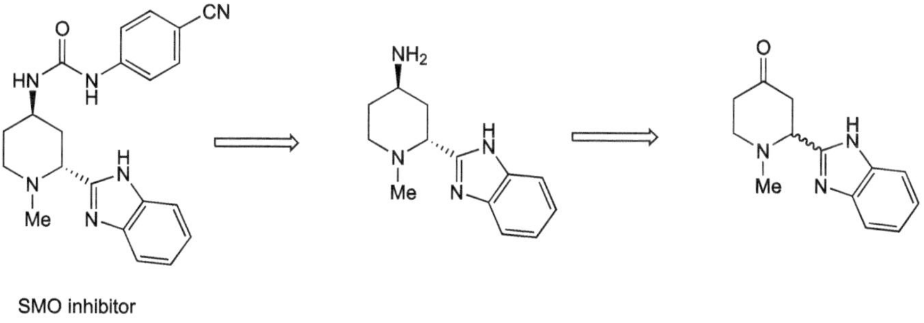

SMO inhibitor

Answer 13[13]

As is increasingly the case with chiral amines, the solution is to use an ω-TA to convert the precursor ketone to the corresponding chiral amine. In this case, if the reaction is operated a high pH, then it is possible to racemise the β-stereogenic centre *via* a retro-elimination process. The Pfizer scientists established that the ring-opened product was indeed the source of racemisation by trapping with another nucleophilic amine.

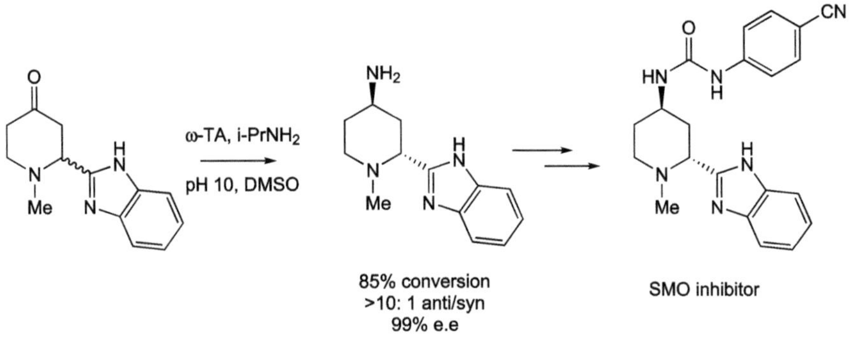

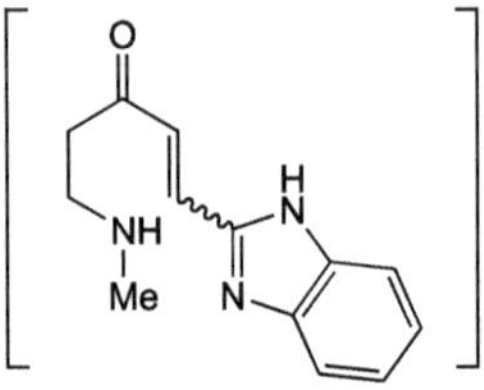

85% conversion
>10: 1 anti/syn
99% e.e

SMO inhibitor

Question 14

Niraparib (originally MK-4827) is an orally active small molecule PARP inhibitor being developed by Tesaro and Merck to treat ovarian cancer. A key intermediate is the chiral 4-bromophenyl-substituted lactam, which is a challenging target in view of the remote nature of the stereogenic centre in relation to the carbonyl moiety. How would you prepare this lactam from the corresponding acyclic racemic aldehyde with the added challenge of effecting dynamic kinetic resolution?

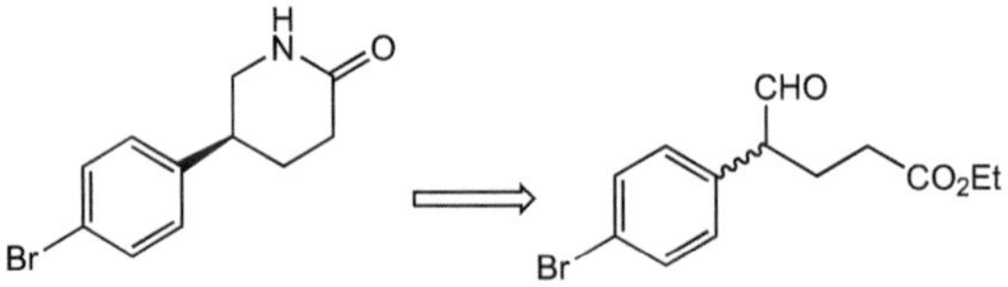

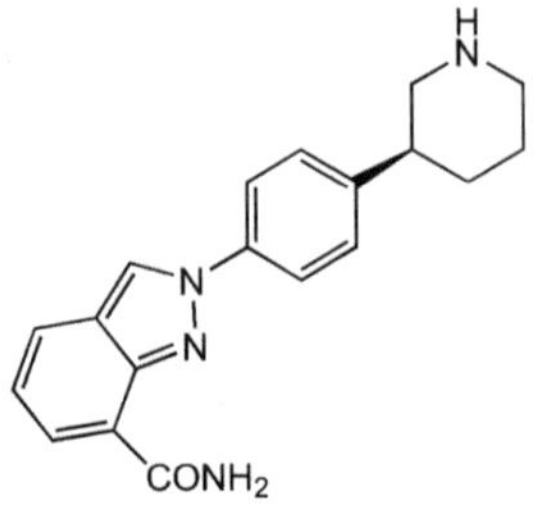

niraparib

Answer 14[14]

The racemic aldehyde with an α-aryl group looks like it should be a good candidate for transamination with DKR. Indeed, prior to this report, there was a precedence in the literature for such transformations with ω-TAs. The product amine in this case was obtained in high e.e. and yield and could subsequently be converted to the corresponding lactam.

Question 15

Vernakalant is an investigational drug currently being developed by Merck for the treatment of atrial fibrillation. The molecule possesses three stereogenic centres and requires the construction of a chiral tertiary amine, which represents a synthetic challenge. How might vernakalant be prepared from the racemic α-alkoxy cyclohexanone?

vernakalant

Answer 15[15]

Interestingly the Merck scientists have reported two different biocatalytic routes to vernakalant, both of which utilise a racemic α-alkoxy cyclohexanone building block. In the first synthesis, they use an ω-TA to carry out amination of the ketone under conditions of dynamic kinetic resolution, which leads to a high yield (>50%) and e.e. of the desired primary amine. The hydroxypyrrolidine ring is then constructed chemically by double alkylation of the amine. The second route is more convergent and involves the use of an engineered opine dehydrogenase for a diastereoselective reductive amination of 2-hydroxypyrrolidine with the requisite ketone. This reaction proceeds with high diastereoselectivity (75% d.e.).

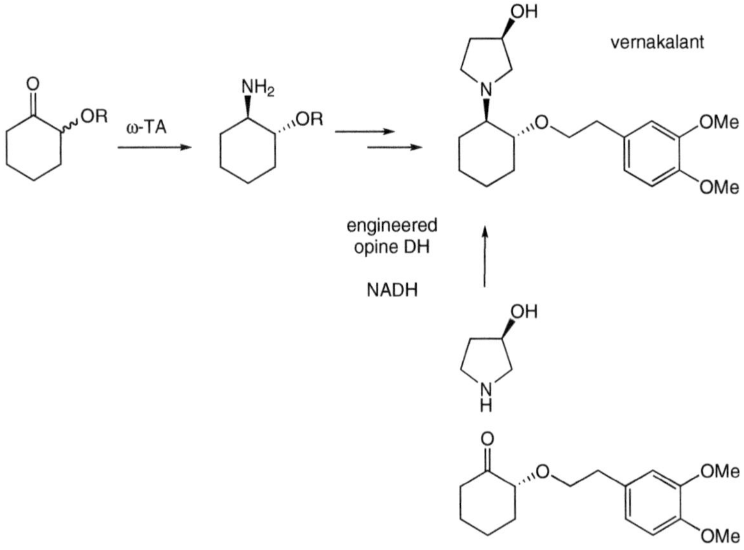

Question 16

How would you synthesise the tetracyclic tertiary amine harmicine from the corresponding 2-substituted indole? On the face of it, this is equivalent to an enantioselective Pictet–Spengler reaction involving C–C bond formation at C-1 of the indole together with the generation of a new stereogenic centre.

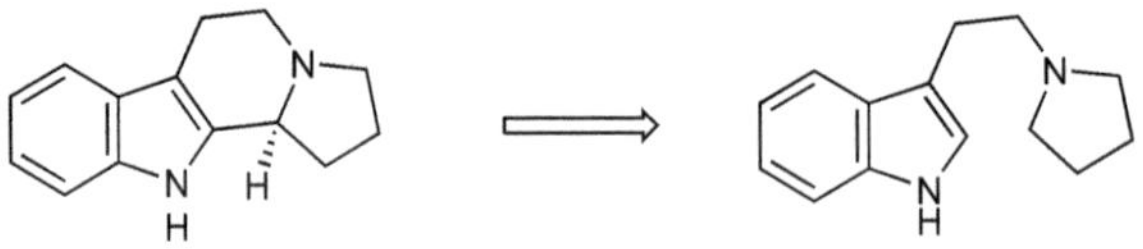

Answer 16[16]

The first step of this synthesis involves the use of an engineered monoamine oxidase (MAO-N) for the oxidation of the substrate to the corresponding iminium ion, which then undergoes a spontaneous Pictet–Spengler reaction to generate racemic harmicine. However, harmicine is also a substrate for MAO-N and hence a second oxidation takes place to give the corresponding iminium ion. This second oxidation is highly enantioselective and hence the addition of the reductant ammonia borane results in deracemisation *via* cycles of oxidation and reduction to give enantiomerically pure harmicine.

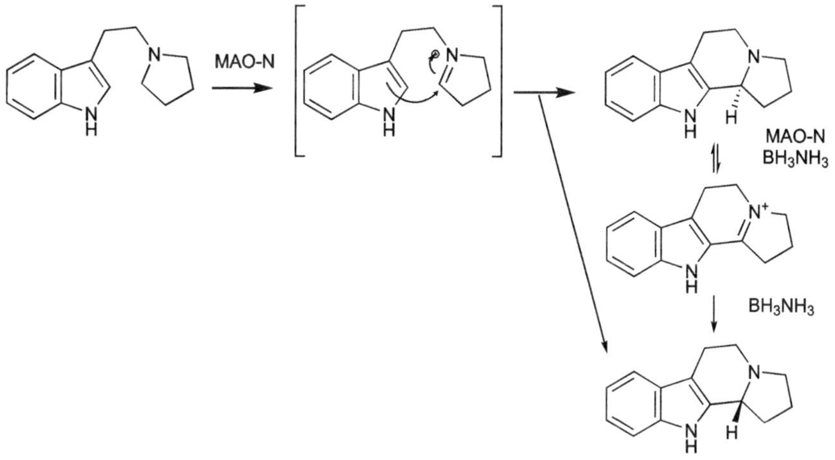

Question 17

The bicyclic amino acid shown below is a key intermediate in the synthesis of the anti-hypertensive drug perindopril. How would you convert *ortho*-bromo-cinnamic acid to the enantiomerically pure bicyclic amino acid? The challenge here is to construct two C–N bonds with control of enantioselectivity.

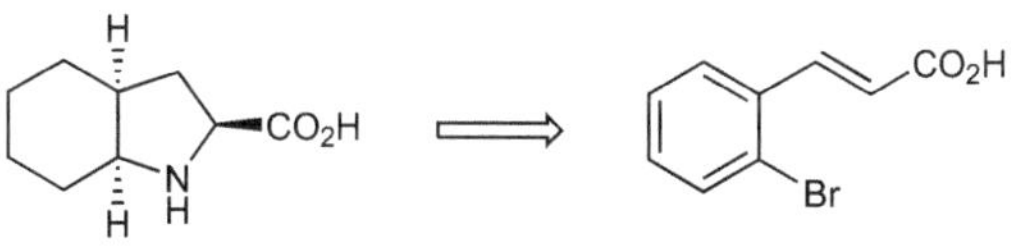

Answer 17[17]

This example of a practical application of a phenylalanine ammonia lyase (PAL) was developed by scientists at DSM and scaled-up for the commercial manufacture of perindopril. PAL was initially used to catalyse the regio- and enantioselective amination of *ortho*-bromocinnamic acid to yield the corresponding *ortho*-bromo-phenylalanine with perfect e.e. After intramolecular Ullman coupling and subsequent reduction, octahydro-1H-indole-2-carboxylic acid was obtained as a key advanced intermediate.

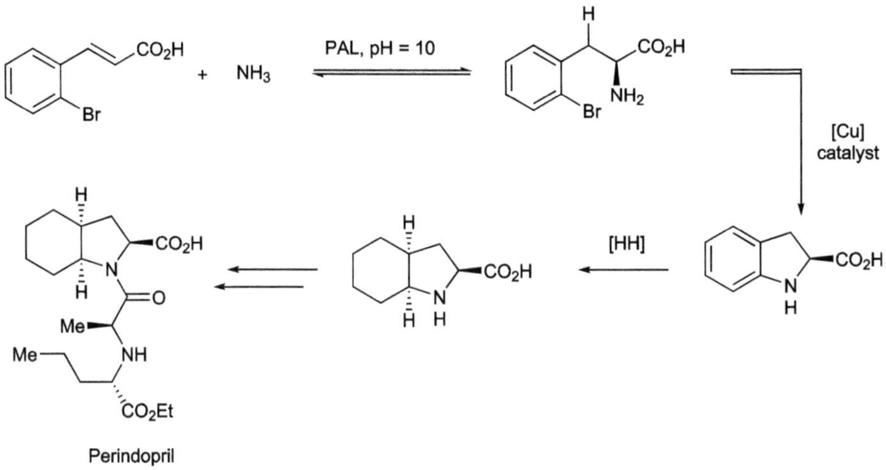

Perindopril

Question 18

How would you prepare the decongestants norephedrine and nor-pseudo-ephedrine using benzaldehyde and pyruvic acid as the common precursors? This synthesis requires both C–C bond formation and also enantioselective amination.

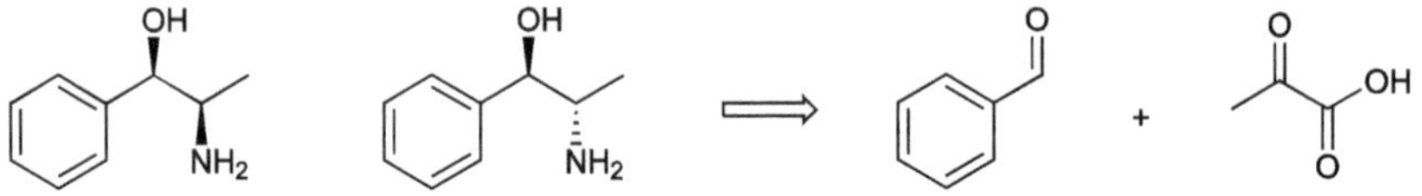

Answer 18[18]

The thiamine diphosphate (ThDP)-dependent acetohydroxyacid synthase I (AHAS-I) was employed in the first step to decarboxylate pyruvate and perform a ligation to benzaldehyde, affording the intermediate (R)-phenylacetylcarbinol with high stereoselectivity (>98%). Subsequently an (R)- or (S)-selective transaminase converted (R)-phenylacetylcarbinol into the final (1R,2R) and (1R,2S) products, respectively, using alanine as a co-substrate. Interestingly, the cascade could be operated in a 'recycling' mode in which the pyruvate by-product generated by the transaminase step could then re-enter the cascade, either directly as a substrate in the first step or *via* the reversible formation of an acetolactate intermediate.

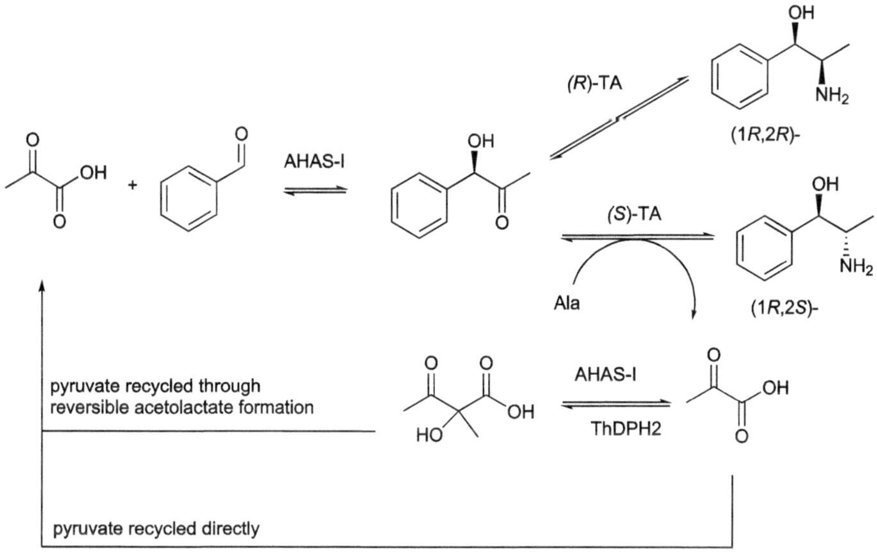

Question 19

Target molecules containing β-amino alcohols are very prevalent in pharmaceuticals and invariably need to be prepared in enantiomerically pure form. Suggest a method for preparing chiral β-amino alcohols as shown below using glycine and substituted benzaldehydes as precursors. Hint: try adding a carboxylic acid group back into the target molecule.

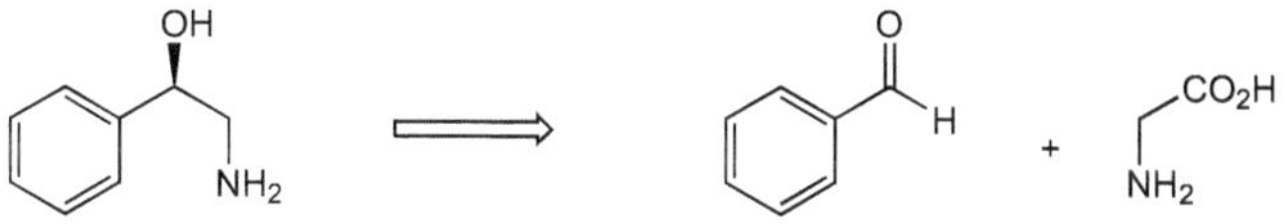

Answer 19[19]

The two key biocatalysts used here are L-threonine aldolase and L-tyrosine decarboxylase. The aldolase uses glycine as the 'enolate' equivalent and is able to condense a range of different benzaldehyde derivatives to give the corresponding β-hydroxy-α-amino acids. Subsequent treatment with L-tyrosine decarboxylase yields the desired β-amino alcohols.

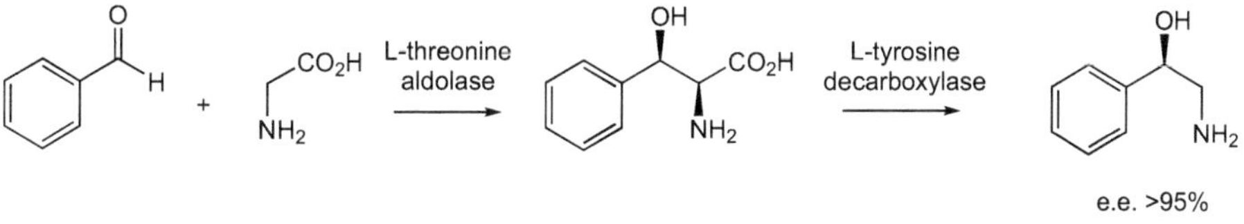

e.e. >95%

Question 20

How might the enantiomerically pure aryl-substituted pyrrolidine be prepared from the corresponding γ-chloroketone?

Answer 20[20]

For the synthesis of the target pyrrolidine the reaction involves initial enantioselective transamination to the amine using a transaminase in the presence of isopropylamine as the amino donor. This amine then undergoes spontaneous cyclisation to the product. In this example, the use of an engineered transaminase allows transamination of an aryl ketone.

Question 21

Suggest a two-step method, involving two biocatalysts, for the conversion of the 1,4-diketone to the various diastereoisomers of the corresponding 2,5-disubstituted pyrrolidines.

Answer 21[21]

The conversion of 1,n-diketones to cyclic imines can be achieved *via* regioselective amination with an ω-transaminase, followed by spontaneous cyclisation. Subsequent treatment with MAO-N/ammonia borane yields 2,5-disubstituted pyrrolidines in high enantiomeric and diastereomeric purity. Alternatively, the intermediate imines can be reduced directly to the amine using an NADPH-dependent imine reductase (IRED).

Question 22

This problem involves the synthesis of (*S*)-scoulerine from (*R/S*)-reticuline and requires two key transformations, namely conversion of a racemate to a single enantiomer and also construction of a key C–C bond between the *N*-methyl group of reticuline and the aromatic ring. (*S*)-scoulerine is a key intermediate in the biosynthesis of the berberine alkaloids.

(*S*)-scoulerine   (*R/S*)-reticuline

Answer 22[22]

The solution to this problem involves the successful application of a monoamine oxidase (MAO-N)-based biocatalytic deracemisation strategy in order to achieve a one-pot synthesis of (*S*)-scoulerine from racemic reticuline. Sequential addition of MAO-N and ammonia borane, followed by berberine bridge enzyme (BBE), allowed deracemisation to take place before the *in situ* formed (*S*)-reticuline was converted into the desired product in excellent yield and enantioselectivity. Crucially, the two flavin-dependent, oxygen-requiring oxidases employed act on different parts of the benzylisoquinoline substrate.

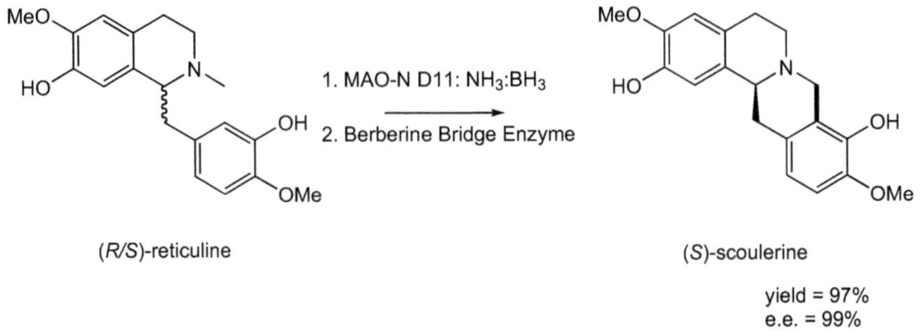

(R/S)-reticuline   (S)-scoulerine

                                                          yield = 97%
                                                          e.e. = 99%

Question 23

Suggest a method for achieving *para*-vinylation of a substituted phenol using a combination of three different biocatalysts. This is a particularly intriguing and difficult to envisage disconnection since the key functional group required for the chemistry to be successful is not present in the final product. Hint: try adding a carboxylic acid group back into the final product at the terminus of the double bond.

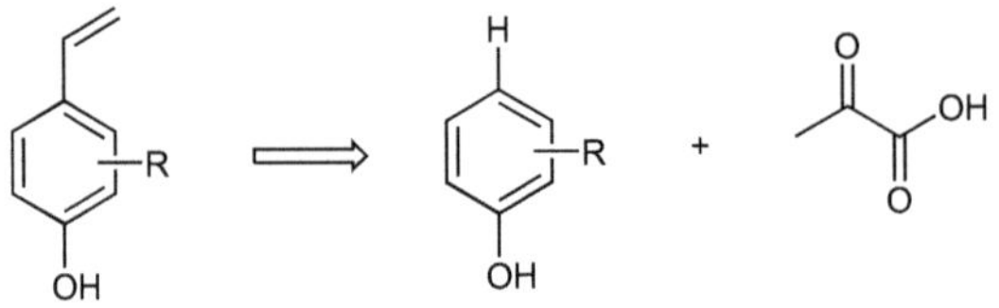

Answer 23[23]

Kroutil and co-workers developed a highly selective biocatalytic route to perform the formal *para*-vinylation of phenols *via* a three-enzyme *in vitro* cascade. The important initial C–C bond forming step was catalysed by a tyrosine phenol lyase (TPL) that coupled the phenol substrate with pyruvate and ammonia to generate an L-tyrosine intermediate. This compound was then deaminated by a tyrosine ammonia lyase (TAL) to afford a coumaric acid derivative that was finally decarboxylated by a ferulic acid decarboxylase (FAD).

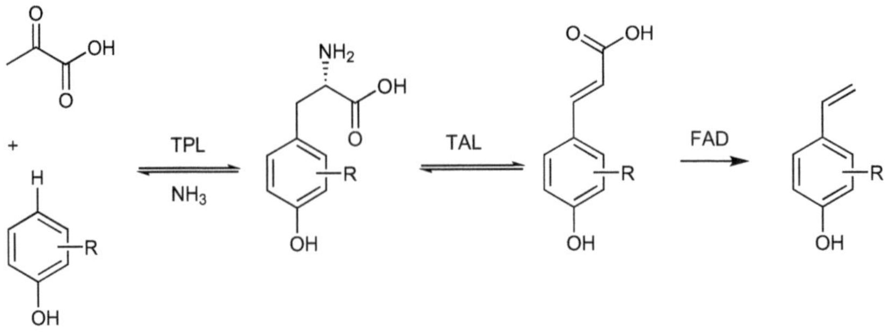

R = 2-F, 2-Cl, 2-Br, 2-Me, 3-F, 3-Cl, 2,3-F₂, 2-F-3-Cl

Question 24

Try to devise a multi-enzyme process for the synthesis of carvolactone from limonene. In this transformation, which involves four distinct biocatalysts, there are a number of challenges including the formation of the lactone and also control of the stereochemistry of the methyl group.

carvolactone   limonene

Answer 24[24]

Initially cumulene dioxygenase (CumDO) was employed for re-gioselective allylic oxidation of limonene. Thereafter, a nonselective alcohol dehydrogenase (ADH) continued the cascade process by converting the allylic alcohol into the α,β-unsaturated ketone. An ene-reductase (ERED) selectively reduced the double bond functionality of the α,β-unsaturated ketone to give a saturated ketone, which was fi-nally oxidized further by a Baeyer–Villiger monooxygenase (CHMO) to yield the lactone product. A potential bottleneck of the cascade, the ADH-catalysed oxidation of the substrate to the unsaturated ketone intermediate, was avoided because the ERED successfully shifted the reaction equilibrium towards product formation.

Question 25

Telaprevir is a drug used for treating hepatitis C. It is a somewhat complex molecule containing several stereogenic centres. Using the disconnections suggested below, develop a synthesis of telaprevir in which the achiral bicyclic pyrrolidine is a key building block. In order to attempt this question, you will need to be familiar with the Ugi reaction, which is a way of combining an imine, an isonitrile and a carboxylic acid in one step. Once you have identified the chiral imine that you require, try to think how you might prepare it from the symmetrical pyrrolidine using a biocatalyst.

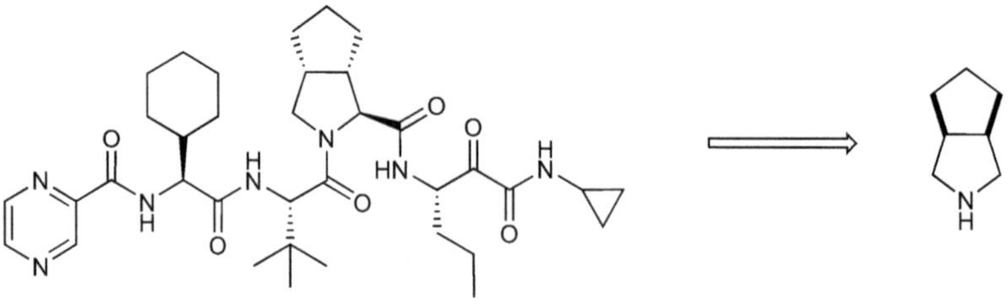

Telaprevir

Answer 25[25]

The key chiral building block for the Ugi reaction is the enantiomerically pure bicyclic imine. This imine can be obtained by monoamine oxidase (MAO-N)-catalysed desymmetrisation of the bicyclic pyrrolidine. Reaction of the imine with the appropriate isonitrile and carboxylic acid leads to an advanced synthetic intermediate for the preparation of telaprevir. Interestingly in this example the other two building blocks generated, namely the isonitrile and the protected amino acid, also present further opportunities for biocatalysis.

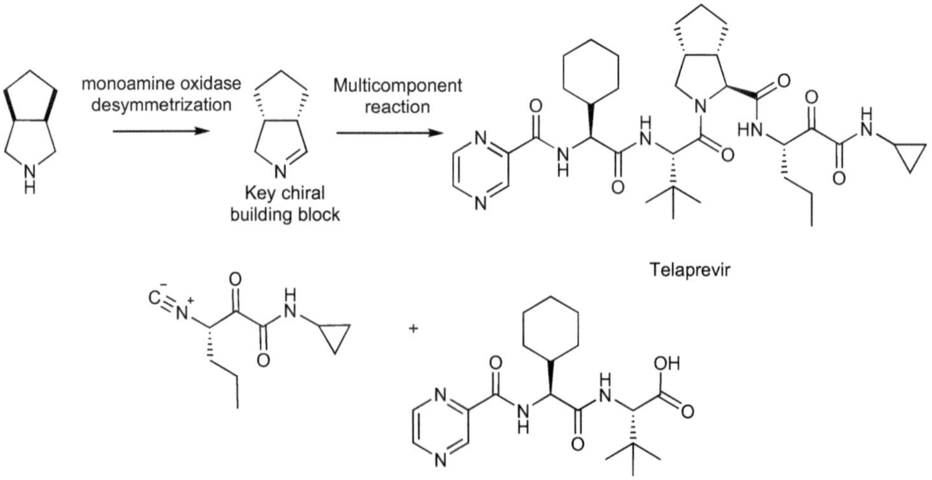

monoamine oxidase
desymmetrization

Multicomponent
reaction

Key chiral
building block

Telaprevir

References

1. J. Pietruszka and M. Schölzel, *Adv. Synth. Catal.*, 2012, **354**, 751.
2. B. Bechi, S. Herter, S. McKenna, C. Riley, S. Leimkühler, N. J. Turner and A. J. Carnell, *Green Chem.*, 2014, **16**, 4524.

3. S. P. France, S. Hussain, A. M. Hill, L. J. Hepworth, R. M. Howard, K. R. Mulholland, S. L. Flitsch and N. J. Turner, *ACS Catal.*, 2016, **6**, 3753.
4. S. P. France, S. Hussain, A. M. Hill, L. J. Hepworth, R. M. Howard, K. R. Mulholland, S. L. Flitsch and N. J. Turner, *ACS Catal.*, 2016, **6**, 3753.
5. S. P. France, S. Hussain, A. M. Hill, L. J. Hepworth, R. M. Howard, K. R. Mulholland, S. L. Flitsch and N. J. Turner, *ACS Catal.*, 2016, **6**, 3753.
6. J. Ryan, M. Šiaučiulis, A. Gomm, B. Maciá, E. O'Reilly and V. Caprio, *J. Am. Chem. Soc.*, 2016, **138**, 15798.
7. A. Díaz-Rodríguez, W. Borzęcka, I. Lavandera and V. Gotor, *ACS Catalysis*, 2014, **4**, 386.
8. M. Korpak and J. Pietruszka, *Adv. Synth. Catal.*, 2011, **353**, 1420.
9. A. Rioz-Martínez, A. Cuetos, C. Rodríguez, G. de Gonzalo, I. Lavandera, M. W. Fraaije and V. Gotor, *Angew. Chem. Int. Ed.*, 2011, **50**, 8387.
10. J. A. McIntosh, P. S. Coelho, C. C. Farwell, Z. J. Wang, J. C. Lewis, T. R. Brown and F. H. Arnold, *Angew. Chem. Int. Ed.*, 2013, **52**, 9309–9312.
11. A. L. Concia, L. Gomez, J. Bujons, T. Parella, C. Vilaplana, P. J. Cardona, J. Joglar and P. Clapes, *Org. Biomol. Chem.*, 2013, **11**, 2005.
12. F. Tao, Y. A. Zhang, C. Q. Ma and P. Xu, *Appl Microbiol Biot.*, 2010, **87**, 1281.
13. Z. Peng, J. W. Wong, E. C. Hansen, A. L. A. Puchlopek-Dermenci and H. J. Clarke, *Org. Lett.*, 2014, **16**, 860.
14. C. K. Chung, P. G. Bulger, B. Kosjek, K. M. Belyk, N. Rivera, M. E. Scott, G. R. Humphrey, J. Limanto, D. C. Bachert and K. M. Emerson, *Org. Proc. Res. Dev.*, 2014, **18**, 215.
15. (i) J. Limanto, E. R. Ashley, J. Yin, G. L. Beutner, B. T. Grau, A. M. Kassim, M. M. Kim, A. Klapars, Z. Liu, H. R. Strotman and M. D. Truppo, *Org. Lett.*, 2014, **16**, 2716; (ii) H. Chen, S. J. Collier, J. Nazor, J. Sukumaran, D. Smith, J. C. Moore, G. Hughes, J. Janey, G. W. Huisman, S. J. Novick, N. J. Agard, O. Alvizo, G. A. Cope, W. L. Yeo and S. N. Minor, Engineered imine reductases and methods for the reductive amination of ketone and amine compounds US 9487760 B2.
16. D. Ghislieri, A. P. Green, M. Pontini, S. C. Willies, I. Rowles, A. Frank, G. Grogan and N. J. Turner, *J. Am. Chem. Soc.*, 2013, **135**, 10863.
17. B. de Lange, D. J. Hyett, P. J. D. Maas, D. Mink, F. B. J. van Assema, N. Sereinig, A. H. M. de Vries and J. G. de Vries, *ChemCatChem*, 2011, **3**, 289.

18. T. Sehl, H. C. Hailes, J. M. Ward, R. Wardenga, E. Von Lieres, H. Offermann, R. Westphal, M. Pohl and D. Rother, *Angew. Chem., Int. Ed.*, 2013, **52**, 6772.
19. J. Steinreiber, M. Schürmann, M. Wolberg, F. van Assema, C. Reisinger, K. Fesko, D. Mink and H. Griengl, *Angew. Chem. Int. Ed.*, 2007, **46**, 1624.
20. C. K. Savile, J. M. Jane, E. C. Mundorff, J. C. Moore, S. Tam, W. R. Jarvis, J. C. Colbeck, A. Krebber, F. J. Fleitz, J. Brands, P. N. Devine, G. W. Huisman and G. J. Hughes, *Science*, 2010, **329**, 305.
21. E. O'Reilly, C. Iglesias, D. Ghislieri, J. Hopwood, J. L. Galman, R. C. Lloyd and N. J. Turner, *Angew. Chem. Int. Ed.*, 2014, **53**, 2447.
22. J. H. Schrittwieser, B. Groenendaal, V. Resch, D. Ghislieri, S. Wallner, E.-M. Fischereder, E. Fuchs, B. Grischek, J. H. Sattler, P. Macheroux, N. J. Turner and W. Kroutil, *Angew. Chem. Int. Ed.*, 2014, **53**, 3731.
23. E. Busto, R. C. Simon and W. Kroutil, *Angew. Chem., Int. Ed.*, 2015, **54**, 10899.
24. N. Oberleitner, C. Peters, J. Muschiol, M. Kadow, S. Saß, T. Bayer, P. Schaaf, N. Iqbal, F. Rudroff, M. D. Mihovilovic and U. T. Bornscheuer, *ChemCatChem*, 2013, **5**, 3524.
25. A. Znabet, M. M. Polak, E. Janssen, F. J. J. de Kanter, N. J. Turner, R. V. A. Orru and E. Ruijter, *Chem.Commun.*, 2010, **46**, 7918.

Subject Index